RE-THINKING WATER AND FOOD SECURITY

Re-thinking Water and Food Security

Fourth Botín Foundation Water Workshop

Editors

Luis Martínez-Cortina
Spanish Geological Survey (IGME), Spain

Alberto Garrido
Research Centre for the Management of Agricultural and Environmental Risks (CEIGRAM), Spain; Technical University of Madrid (UPM), Spain; Marcelino Botin Foundation – Water Observatory (FMB-WO), Spain

Elena López-Gunn
Marcelino Botin Foundation – Water Observatory (FMB-WO), Spain; London School of Economics (LSE), UK

CRC Press
Taylor & Francis Group
Boca Raton London New York Leiden

CRC Press is an imprint of the
Taylor & Francis Group, an **informa** business

A BALKEMA BOOK

Cover photo:
Courtesy of M. Campos and M. Rica Izquierdo.

CRC Press/Balkema is an imprint of the Taylor & Francis Group, an informa business

Typeset by MPS Ltd. (A Macmillan Company), Chennai, India

Printed and bound in Poland by Poligrafia Janusz Nowak, Poznán

Library of Congress Cataloging-in-Publication Data

Marcelino Botín Water Forum (4th : 2009 : Santander, Spain)
Re-thinking water and food security : fourth Marcelino Botín Foundation water workshop / editors, Luis Martínez-Cortina, Alberto Garrido, Elena López-Gunn.
p. cm.
Includes index.
ISBN 978-0-415-58790-7 (hard cover : alk. paper) 1. Water-supply–Security measures–Congresses. 2. Water resources development–Congresses. 3. Food security–Congresses. I. Martínez-Cortina, Luis. II. Garrido, Alberto, 1964– III. López-Gunn, Elena. IV. Title.

TD201.M375 2009
363.6′1—dc22

2010031757

Published by: CRC Press/Balkema
P.O. Box 447, 2300 AK Leiden, The Netherlands
e-mail: Pub.NL@taylorandfrancis.com
www.crcpress.com – www.taylorandfrancis.co.uk – www.balkema.nl

ISBN: 978-0-415-58790-7 (Hbk)
ISBN: 978-0-203-82982-0 (eBook)

Table of Contents

Foreword

CHAPTER LVIII – PART TWO "DON QUIXOTE"

Which tells how adventures came crowding thick and fast on Don Quixote

"Freedom", he said, turning to Sancho, "is one of the most precious gifts that the heavens have bestowed on men; with it the treasures locked in the earth or hidden in the depths of the sea are not to be compared; for the sake of freedom, as for the sake of honor, one may and should risk one's life, and captivity, on the other hand, is the greatest evil that can befall a human being. You have seen, Sancho, the abundance and luxury in that castle we have just left; yet I assure you that in the midst of those delicious banquets and snow-cooled beverages it seemed to me as though I were in the straits of hunger, since I did not enjoy them with the same freedom as if they had been my own. The obligation to return benefits and favors received is a shackle on the liberty-loving spirit of man. Happy he to whom Heaven gives a slice of bread without his being obliged to thank any other person but only Heaven itself!"

Brief memorandum on the three previous Workshops and their approach

This book contains most of the papers presented in the Fourth Marcelino Botin Foundation Water Workshop, after the authors reviewed their initial presentations for debate in the forum.

There are a number of threads that sum up the approach that has always inspired Botin Workshops. A first aspect is the search for innovation; "thinking out of the water box" together with a clear spirit of "freedom of thought and speech" (see Don Quixote quotation at the beginning). A second aspect is ensuring, as far as possible, that the participants come from a range of disciplinary backgrounds with persons coming from the natural and social sciences, not only from the Academia but also from the corporate world, NGOs and the political arena.

The main goal of the Workshops is not to produce one more declaration or statement to be added to the hundreds that already exist, but to provide true added value by facilitating brain storming among a group of experts, often on purpose selected with clear different perspectives to incentivize high level constructive debate and productive discussion. Presentations of Spanish case studies are relatively frequent in this book. This is due, not only to the Spanish roots of the Botin Foundation but also to the fact that Spain is the driest country in Europe and it has experienced very significant political, social and economic changes during the last half century, becoming in a relatively short space of time an industrialized and democratic country.

I was co-editor of the three books with the Proceedings of the corresponding Botin Foundation Water Workshops. This time I decided not to be co-editor because: a) the current co-editors are significantly younger than me and more skilful for this job; and b) this task is time-consuming and the current research projects of the Water Observatory demand all my time.

The *hot issues* and glossary suggested to the participants in the fourth water workshop

Several months before the Workshop I sent to all the participants a rather long list of what I considered *hot issues*, which could provide points to be discussed during the Workshop. Participants had the complete freedom to comment on them either in favor or against, or simply to forget them, since the main aim was to stimulate debate, as a warm up for the Workshop. Here I have not

Covers corresponding to the three previous books of the Marcelino Botin Foundation Water Workshops, published by Taylor & Francis

reproduced the complete list; some of these hot issues are briefly commented below. Others will be probably dealt with in a future article.

Additionally, and in order to facilitate the use of a common language, a tentative Glossary was sent to all the participants. It was prepared by Dr. M. Aldaya with the collaboration of some of the Workshop participants and it has been included at the end of this book.

The relevance of green water and some questions not solved yet

The relevance of green water in agricultural production has been emphasized in many of the chapters. This means that water and food security is influenced by land use and rain-fed agriculture. This emphasis however, was not equally shared by all the participants and this is an important finding in this Fourth Workshop. One main conclusion perhaps is the need for a better assessment of green water needs for the good health of the ecosystems. Another relevant pending question is how to cope with the great variability of rain-fed agriculture due to the normal climate variability, independently of the relevance of climate change.

The pros and cons of the extended water footprint and the analyses of food (virtual water) trade

A good number of presentations used the *Water Footprint* tool to analyze different cases. Some of them were mainly related to hydrologic values but others also included economic values together with the hydrological values. In one case – the Doñana National Park case – a preliminary attempt was done to economically evaluate the needs of water to be bought to conserve this important wetland.

It seems clear that significant uncertainties still exist in the methods to calculate the *extended water footprint*. Nevertheless, it is a useful and transparent method to facilitate an *Integrated Water Resources Management*. It seems that the main obstacle to solve the water and food security problems in arid and semiarid countries is the persistence in the minds of most policy makers of the idea of the strategic need of food and water self-sufficiency. It seems that one main way to change the pervasive paradigm of water and food self-sufficiency to the paradigm of water and food security is to show to key political actors that the regulations for international food trade must not allow the existence of great international food corporations oligopolies or the political pressure of certain powerful countries against developing or emerging countries. Therefore, it is clear that achieving water and food security for the poor countries is mainly a global ethical issue.

Conflicts between food and fiber production and environmental degradation. Is there a solution?

Several authors call attention to the fact that food (virtual water) trade usually does not consider the potential environmental impact that the export of food may create in the exporting country. This is an important aspect that deserves more attention in the near future. Nevertheless, this demands more research because the water demands for the good health of the ecosystems are still poorly known.

The nexus water-food and energy is also tackled in a few chapters. It deserves also more analyses. A preliminary indication is that the production of biofuels with edible crops is not a good solution in arid and semiarid countries. On the other hand, it seems that generally blue water consumption for production of renewable energy (hydropower, wind, solar) and non-renewable energy is irrelevant in comparison with the consumptive use of water for agriculture. Nevertheless, in specific cases conflicts may appear. The energy needs for irrigated agriculture (i.e. energy for water) however may be significant in some countries and may constitute an obstacle in poor countries to achieve food security.

Nevertheless, although in the Workshop the issue was barely tackled, it seems clear that one of the most important issues in current and future water policy is how to cope with the problem of diffuse pollution due to the use of agrochemicals in intensive agriculture (rain-fed and irrigated). This is a serious topic both in humid and dry countries not yet solved.

Is the new proposed motto *more cash and care of nature per drop* realistic?

In a side-event during the Fifth World Water Forum (Istanbul, March 2009) the Water Observatory proposed that the usual motto *More Crops and Jobs per Drop* should be changed to the new motto *More Cash and Care of Nature per Drop*. This change was suggested mainly for industrialized and emergent economies countries. In the poor or developing countries the former motto should be maintained.

The experience shows that farmer lobbies are very strong almost everywhere in the world (both developed, emergent and developing countries) and it is necessary to find *win–win* solutions. Otherwise the mentioned main problem in water policy – i.e. intensive agricultural diffuse pollution – will continue.

Why the silent revolution of groundwater intensive development is not frequently considered?

The First Botin Foundation Water Workshop on *Intensive Use of Groundwater: Challenges and Opportunities*, already showed the great relevance of groundwater irrigation, mainly in arid and semiarid countries. This has been described as a *silent revolution* because it has been done by millions of modest farmers with scarce or no planning and control by conventional governmental water agencies. Still today, because of a blend of mental inertia, ignorance, professional bias or corruption, most water policy makers consider almost exclusively solutions based on surface water infrastructures. This situation begins to be better known in some countries like India and Spain, as described in several chapters in this book.

Ethical issues

Many of the problems or conflicts described in the book are linked to and conditioned by the systems of values and these, in every country, are related to its cultural and normative value frameworks, and their religious background. The ignorance of this fact may lead to clamorous failures. There is not blue print or standard solution for water and food security. Solutions should be tailor-made taking into account the physical, social, economic and cultural environment.

It is worth mentioning some specific aspects:

a) It is appropriate distinguish between Social Ethics (relations among humans) and Environmental Ethics (relations between humans and Nature).
b) For instance, the right to water and sanitation is a typical social ethics issue. It is very important from a humanitarian perspective, mainly for its incidence in public health. However, the amount of global water necessary for solving this problem is a very small proportion (less than 5%) of all the consumptive water uses for humans. These are mainly for agriculture.
c) The increasing demands for water and food are due to the growth in population and the standard of life (a better food diet). The second – changes in diet – is being much more significant and obviously it cannot be stopped. The great difference on the standard of life between rich and poor countries is the main ecological problem of this planet.
d) The reduction of the huge gap between rich and poor countries may require, among other things, a more just and equitable regulation of international food trade.

Potential future advances in *Water and Food Security*

The advances in Science and Technology during the last half century give us today new means for solving many difficult water and food problems. A few decades ago these new means were unthinkable. Among them can be quoted desalination, the food (virtual water) trade, and the silent revolution of intensive groundwater use. This book with the Proceedings of the Fourth Botin Foundation Water Workshop tries to be a modest contribution to these advances.

Almost for sure new advances in different areas will appear in the near future. *Natural resources are limited but human ingenuity is boundless.*

May 2010
M. Ramón Llamas
Director, Water Observatory of the M. Botin Foundation

Participants in the workshop[1]

JOSÉ ALBIAC

José Albiac is Research Fellow at the *Agrifood Research and Technology Centre* (CITA-Government of Aragon, Spain). He has a BSc in Economics (*University of Zaragoza*), and an MSc and PhD in Agricultural Economics (*University of Illinois*). His fields of interest are Environmental and natural resource economics, Environmental policies, and Bioeconomic modelling. His applied research work deals with water resources management, nonpoint pollution, and forest economics. He has been involved in studies on water planning in Spain, irrigation investments, aquifer management, and nonpoint pollution. Major contributions include: i) collaboration with the *University of California* (Riverside) in a project to compare water institutions, policies and technologies between countries; ii) presentation on the *Water Framework Directive* at the *Yellow River Forum*; iii) study for the *OECD Water Workshop* in Australia; and iv) study on the Ebro water transfer for the Spanish government.

MAITE M. ALDAYA

Dr. Maite M. Aldaya is currently a researcher at the *University of Twente* (the Netherlands), *Water Footprint Network* and a Research Associate at the *Water Observatory* of the *Marcelino Botin Foundation*.

Maite is a biologist and has completed a PhD in Ecology from the *University of Navarre* and a Master in Environmental Policy and Regulation at the *London School of Economics*. During and after her studies she has engaged in different jobs and internships. She was a researcher at the *General Ecology Laboratory* of the *National Museum of Natural History* in Paris – which is attached to the *Center National de Recherche Scientifique* (CNRS) and at the *Biological Oceanographic Laboratory*, *University of Bordeaux* (France). She has also been teaching in a variety of subjects, including water resources management and ecology. She has worked in several international organizations, such as the *Agriculture and Soil Unit* at the European Commission in the adoption of the *Soil Thematic Strategy* and implementation of cross compliance, and within the *Land and Water Development Division* at the *Food and Agriculture Organization* of the United Nations. She also worked in the *NeWater European Project* at the *Complutense University of Madrid*, where she undertook the first water footprint of a catchment which also included its economic aspects: the water footprint analysis (hydrologic and economic) of the Guadiana river basin.

She is currently involved in different projects in relation to the water footprint and virtual water trade, a research field addressing the relations between water management, consumption and trade. Her research activities are focused both on government and business water accounting and management. Maite's background is very much based on an interdisciplinary research in the field of water, ecology, and its application to integrated water resource management through the use of the water footprint.

J.A. [TONY] ALLAN

Tony Allan [BA Durham 1958, PhD London 1971] heads the *Water Research Group* at *King's College London* and *SOAS*. He specialises in the analysis of water resources in semi-arid regions

[1]The *Fourth Marcelino Botin Foundation Workshop on Water and Food Security* was held in the Marcelino Botin Foundation headquarters (Santander, Spain) from September 22 to 24, 2009. The draft manuscripts of the chapters of this book were discussed during this Workshop.

and on the role of global systems in ameliorating local and regional water deficits. In his early career he was concerned with hydrological and environmental issues but gradually turned his attention to the social and political, when it became evident that environmental science could not explain why people manage water as they do. He pointed out that the water short economies achieve water and food security mainly importing water intensive food commodities. He established the concept of *virtual water*. He provides advice to governments and agencies especially in the Middle East on water policy and water policy reform. His most recent book is entitled "The Middle East water question: hydropolitics and the global economy". In 2008 he was awarded the *Stockholm Water Prize* in recognition of his contribution to water science and water policy.

MARGARET CATLEY-CARLSON

Margaret Catley-Carlson was Chair and is now a Patron of the *Global Water Partnership*, a working partnership among all those involved in water management formed in 1996 by the *World Bank*, the *United Nations Development Program* and the *Swedish International Development Cooperation Agency*.

In 1978, she was appointed Vice President (Multilateral) of the *Canadian International Development Agency* (CIDA), and was Senior Vice President/Acting President from 1979 to 1980. In 1981, she was assistant secretary general in the *United Nations*, serving as deputy executive director of operations for the *United Nations Children's Fund* (UNICEF). From 1983 to 1989, she was President of the CIDA. From 1989 to 1992, she was the Deputy Minister of Health and Welfare in Ottawa, Canada. From 1992 to 1996, she was the Chair of the *Water Supply & Sanitation Collaborative Council* at the WHO. From 1993 to 1998, she was president of the *Population Council*.

In 1984, she was appointed to the Board of Governor of the *International Development Research Centre* (IDRC), a Canadian crown corporation that supports researchers from the developing world in their search for the means to build healthier, more equitable, and more prosperous societies. She has served for four non-consecutive terms. She is a member of the Board of Trustees of the *International Institute for Environment and Development.*

In 2002, she was made an Officer of the Order of Canada in recognition of "her distinguished public service career". She has received honorary degrees from the *University of Regina* (1985), *Saint Mary's University*, *Ryerson Polytechnical Institute* (1986), *Concordia University* (1989), *Mount Saint Vincent University* (1990), *University of British Columbia* (1993), *University of Calgary* (1994), *Carleton University* (1994) and *University of Dundee*, as a Doctor of Law, Honoris Causa (2007).

JAMEL CHAHED

Jamel Chahed is now Professor at the National Engineering School of Tunis (Civil Engineering Department) and researcher at the Laboratory of Modelling in Hydraulics and Environment (LMHE). He holds a "Third Cycle Doctorate" in 1989 and a "Doctorate ès Sciences" in Hydraulics in 1999. With his engineering background, he has embraced a scientific career in a variety of fields. His work ranges from theoretical and applied fluid mechanics to many environmental questions from hydraulics, hydrology, water resource planning and management, sustainable energy, environmental economics. His research has appeared as articles, book chapters, technical reports, proceedings of conferences and symposia in diverse specialized literature such as hydrodynamics, turbulence and transfers, chemical engineering, water resources and environmental issues.

Over the past 25 years he has been involved as a teacher, educator, researcher, and expert in number of national and international research, academic and technical programs. As visiting professor, associate researcher, guest speaker he joined number of research teams at academic institutes, research laboratories, R&D centres in Canada, France, Portugal, Spain, Sweden, USA. During his latest career, he takes part to a strong-minded task group which made more than a decade of technical and policy contributions to water sustainable management and governance with a thematic focus on global water balance assessment. The major contribution in this field concerns the enlargement of fresh water concept to all water resources at the national scale including the "*Virtual Water*" content of rainfed agriculture and of foodstuffs trade balance. This global water vision opened new avenues in facing the great challenges of water resources in arid countries like Tunisia.

ASHOK CHAPAGAIN

Dr Ashok Chapagain is a water footprint researcher working on issues related to sustainable consumption and fresh water at *WWF-UK*. His main expertise lies in analysing the global linkages of production and consumption and the related impacts on environment. With a background in civil engineering and work experience of more than a decade in irrigation, Chapagain has an MSc degree in Water Resources and Environmental Management and a PhD degree in the field of Water Systems and Policy Analysis from *UNESCO-IHE*, the Netherlands. His research is mainly focused on developing and analysing the concept of virtual water and water footprint, and its applicability in policy making at different levels, published as a book titled "Globalisation of water: Opportunities and threats of virtual water trade". During his PhD at *UNESCO-IHE*, he refined and developed the concept and methods of Water Footprint together with Prof. Arjen Hoekstra, and they co-authored a book titled "Globalization of water: Sharing the planet's fresh water resources". Before joining *WWF-UK* in 2007, he was working as a post doc researcher at the *University of Twente*.

JOPPE CRAMWINCKEL

Joppe Cramwinckel is Sustainable Development Lead for *Shell International Exploration and Production B.V.* In 1985 he joined *Shell* and after initial induction assignments, he served as environmental lead in two very different operating environments and cultures (Brunei and Oman). From 2001 to 2003 he was part of *Shell's* Corporate Sustainable Development team, with a specific

Headquarters of the Marcelino Botin Foundation in Santander (Spain)

focus on integrating Sustainable Development in *Shell's* internal decision-making process. In 2003 he became the *Shell* Issue Manager for Water, which he continued in his current job.

As a member of the *World Business Council for Sustainable Development* (WBCSD) *Water Working Group*, he was the driving force behind the development of the WBCSD "Business in the World of Water" scenarios, and the associated WBCSD work activities.

EMILIO CUSTODIO

Dr. Industrial Engineer, full professor of Geo-engineering (Groundwater Hydrology) in the *Civil Engineering School* of the *Technical University of Catalonia*, Barcelona, Spain. Corresponding Member of the *Spanish Academy of Sciences*, head of the Advisory Council to the *Foundation International Centre for Groundwater Hydrology*.

Former Director General of the *Geological and Mining Institute of Spain*, and former president of the *International Association of Hydrogeologists*. Doctor *Honoris Causa* by the *University of Tucuman* (Argentina) and Honorary Professor of the *Universities of Santa Fe and la Pampa* (Argentina).

Main teaching, study and research subjects: groundwater hydrology and water resources, with emphasis on general evaluations, hydrogeochemistry, environmental isotope techniques, aquifer recharge, coastal aquifers, volcanic islands hydrogeology and groundwater-dependent wetlands.

Author or co-author of 25 books, close to 500 papers and published communications, and about 50 extensive technical reports on groundwater resources and hydrogeology.

MALIN FALKENMARK

Malin Falkenmark is Professor of Applied and International Hydrology. In 1991 she joined the *Department of Systems Ecology*, *Stockholm University*, and the *Stockholm Water Company* (later *Stockholm International Water Institute*, SIWI), chairing for 13 years the *Scientific Programme Committee* for the annual *Stockholm Water Symposia* (later *World Water Week* in Stockholm). In March 2007 she joined *Stockholm Resilience Center*. She is a future-oriented water scientist, hydrologist by training, with broad interdisciplinary interests and a large number of publications. She introduced the *water crowding* indicator, the concepts *hydrosolidarity* and *green and blue water*. She has received a number of international prizes.

ALBERTO GARRIDO

Professor of Agricultural and Natural Resource Economics. Director, *Research Centre for the Management of Agricultural and Environmental Risks* (CEIGRAM), *Technical University of Madrid*.

His work focuses on natural resource and water economics and policy. He has been consultant for OECD, IADB, European Parliament, European Commission, FAO, and various Spanish Ministries and Autonomous Communities. He is the author of 110 scientific papers. He serves in the Advisory Committee of the *Rosenberg International Water Forum* since its foundation in 1996. His most recent books are: "Water Policy in Spain" co-edited with Prof. M.R. Llamas (Taylor & Francis, Leiden, 2009), and "Water Footprint and Virtual Water Trade in Spain: Policy Implications" co-authored with Llamas, Varela-Ortega, Novo, Rodríguez-Casado and Aldaya (Springer, New York, 2010).

MAGDY HEFNY

Director of the *Regional Center for Studies and Research of Water Uses Ethics*, Cairo, since 2003. Vice Chairman of the *Egyptian Council on Ethics of Scientific Research* since 2004. Advisor to H.E. the Minister of State for Environmental Affairs, 2002/2003.

Diplomat by Profession: Ambassador to the Royal Kingdom of Norway (1997–2000), and Ambassador of Egypt to Ethiopia (1991–1995). Vice Chairman of the *Egyptian Association of Strategic Thinking* (EAST), since 2002.

Visiting Professor at the *Mediterranean Academy of Diplomatic Studies*, University of Malta, since 2004. Visiting Professor at Cairo University, *African Studies Institute*, 2001/02. Professor of *Negotiation Skills Cross Culture*, for MIBA program at the *Arab Academy for Science and Technology* (*Advanced Management Institute*), Cairo (2001 and 2002). He is lecturing in the *Egyptian Diplomatic Institute* on various issues on *Diplomacy and International Affaires* as well as on *International Water of the Nile*. Guest Researcher at the University of Bergen, Norway, (fall semester of 2000). Chancellor for Africa, and Member of the *International Water Academy*, Oslo, Norway, 2001 and 2002, and member of the Advisory Committee of the Board of the *International Water Academy*, Oslo (Norway).

Member of *Global Society of Organizational Learning Network*, Boston, USA. Hefny is the coordinator of SOL Egypt within the *Global SOL Network*. Facilitator, Consultant and Trainer in areas of: International Trade, International Relations, International Waters of Africa and the Nile, Water Conflict Management and Negotiation Skills, Human Resources Management, Knowledge Management, Organizational Learning.

Member of the Egyptian Permanent Delegation to the United Nations-Geneva, as Counselor of Economic Affairs (1981–1985) and Part of Egyptian Delegations to the United Nations Economic Fora in Geneva and New York.

Hefny participated in the Summit Conferences of the *Organization of African Unity*, and Ministerial Conferences of the *Economic Commission for Africa*. He was the First Chairman of the *Central Organ of Conflict Prevention, Management and Resolution* within the *Organization of African Unity* (1993/1994).

PhD in Economics – *Hochschule fur Okonomie*, Berlin, Germany (1973).

M. RAMÓN LLAMAS

Prof. M. Ramón Llamas is Emeritus Professor at the *Complutense University of Madrid*. Fellow of *Spain's Royal Academy of Sciences* (1986). Fellow of the *European Academy of Sciences and Arts* (2005). He received the *Cannes International Great Prize for Water* (2006). Member of the *French Academie de l'Eau* (2006). President of the *International Association of Hydrogeologists* (IAH) (1984–1989). Honorary Fellow of the *Geological Society of the United Kingdom* (1992). Vice-President of the *International Association of Water Resources* (IWRA) (2001–2003).

He is author or co-author of nearly hundred books or monographs and about three hundred scientific papers. His last book, as co-editor: "Water Policy in Spain" (Taylor and Francis) has been published in 2009.

Director of the Marcelino Botin Foundation Water Observatory. This activity begun with a large research program launched (1999–2003) on the role of groundwater resources on water policy. A total of nine books and thirteen monographs were published within this project. They included the Proceedings of the First Botin Foundation Water Workshop: "Intensive Use of Groundwater". In 2004, organized the Second Botin Foundation Water Workshop: "Water Crisis: Myth or Reality?" In 2006 organized the Third Botin Foundation Water Workshop: "Water Ethics".

ELENA LÓPEZ-GUNN

Dr. Elena López-Gunn is a Senior Research Fellow at the *Botin Foundation Water Observatory* and Visiting Senior Fellow at the *London School of Economics and Political Science*. She was an Alcoa Research Fellow from 2001 until 2009. Previously she worked as lecturer in Environmental Policy at the *University of Hertfordshire*, and also as Research Assistant at the *University of Cambridge* for the EU's MEDALUs project on land degradation in the Mediterranean. Elena has a PhD from the *King's College of London*, a Masters in Environment and Development from the *University of Cambridge* and a Bachelor of Sciences (Economics and Politics) from the *University of Wales*. Elena specializes in environmental policy and in institutional and governance aspects of water management. Much of her current research focuses on the groundwater management and she has recently authored a number of papers on how to rethink global water scarcity (co-authored with Professor Ramon Llamas), transboundary groundwater management and the use of alternative instruments in the implementation of the *EU Water Framework Directive*. She is currently undertaking research on regulatory reform in Bolivia, the role of the EU as a normative global leader in EU water policy, revisiting the problem of collective action in groundwater management from a comparative perspective, and on Spanish water policy and politics.

JAN LUNDQVIST

Jan Lundqvist, born May 18 1942, was trained as a social scientist, with a PhD in Human Geography (*University of Göteborg*, 1975). From 1980 to 2009, he served as a professor at the *Department of Water and Environmental Studies*, *Linköping University*. He is currently senior scientific advisor at *Stockholm International Water Institute*, where he also chairs the Scientific Programme committee for the *World Water Week* in Stockholm. He has supervised a large number of PhD candidates and Masters students. He has been employed and been involved in research projects and consultancies in many countries in Africa, South Asia, Middle East and in Norway (seven years as a senior lecturer and *dosent* before joining *University of Linköping*). His research profile includes aspects of water resources management and policy and their fundamental role in human socio-economic development and environmental considerations. Food and nutrition security in a context of increasing resource competition and in a food chain perspective is a current theme in his research. Losses and waste of water and food are important elements in this regard. Transboundary water management is another topic in which he is engaged.

He has been given assignments from various development and policy agencies (Sida, Min. of Foreign Affairs, UN organisations). He has published widely in scientific and other Journals and has been interviewed in leading media: BBC, CNBC, Canadian TV, Xiamen TV and Scandinavian media.

LUIS MARTÍNEZ-CORTINA

Dr. Luis Martínez-Cortina has a PhD in Civil Engineering at the *University of Cantabria*, Spain. He has worked as a researcher of several European Union Research Projects, and also for the *Groundwater Project*, launched in 1999 by the Spanish *Marcelino Botín Foundation* (four years long), focused on the role of groundwater resources on water policy. He has carried out hydrogeological studies, groundwater flow numerical models, other empirical tools and its application to the practical management of water resources. He is co-author of four books and monographs, and is the author or co-author of about forty scientific articles. He was co-editor of the previous books corresponding to the *Marcelino Botin Foundation Water Workshops*, related to Water and Ethics, and Water Crisis. He was the Coordinator of the *Spanish Association of Groundwater Users*. Currently, he is a full researcher of the *Spanish Geological Survey* (IGME), at the *Department of Groundwater Resources Research*.

ADITI MUKHERJI

Dr. Aditi Mukherji is a Researcher (Social Scientist) at the *International Water Management Institute* (IWMI), Colombo. IWMI is a CGIAR center dedicated to research on water, food and environment. Aditi received her PhD degree from the *University of Cambridge*, United Kingdom, in 2007. Her doctoral study at Cambridge was funded by the prestigious *Gates Cambridge* scholarship. In her doctoral dissertation she has looked into the functioning of informal groundwater based irrigation services markets in West Bengal, India. She has written more than 30 academic papers on social and economic aspects of groundwater irrigation in South Asia. She has also participated in a number of national and international conferences on groundwater issues. In 2006, she co-edited a special issue of the *Hydrogeology Journal*. This theme issue of the *Hydrogeology Journal* was dedicated to social and economic aspects of groundwater governance. In 2008, she was awarded the coveted *Global Development Network Award* for best paper under the category of Natural Resources Management. In 2009, she was nominated for CGIAR's Young Promising Scientist of the Year award by IWMI.

IGNASI NIETO

Ignasi Nieto is the Chairman of INECO (*Economy and Engineering of Transport*), a company owned by the Ministry of Public Works of Spain. He is a Senior Industrial Engineer (*Technical University of Catalonia*), PhD in Economics (*University of Barcelona*) and Master in Special Studies in

Participants in the Fourth Marcelino Botin Foundation Water Workshop (Santander, September 2009)

Administration and Management (*Harvard University*, Cambridge). From 2008 to 2010 he was Executive Vice-President for Isdefe (*System Engineering for the Defence of Spain*). From 2006 to 2008 he was Secretary General for Energy, President of IDAE (*Instituto para la Diversificación y Ahorro Energético*) and President of IRMC (*Instituto para la Reestructuración de la Minería del Carbón*). Before becoming Secretary General, he was Director for Regulation and Competition of *Comisión Nacional de Energía* (CNE) from 2005 to 2006, Deputy Director of *Instituto Catalán de la Energía* (ICAEN) from 2004 to 2005, and Director for Trade to *FECSA-ENDESA* from 1998 to 2002. In 2002 he published the book: "The Spanish electricity liberalization from the industrial perspective". He received the prize of the *Catalan Economic Society* 2004. Ignasi Nieto has participated also in several masters and University courses on issues related to energy.

STUART ORR

Stuart has been with WWF since 2006 and engages with the private sector on a range of water related activities, from water footprint measures to public policy engagement. Stuart has published numerous papers on water measurement, agricultural policy and water-related risk. He is a member of the *World Economic Forums*' agenda council for water security and is currently co-drafting water policy guidelines for the private sector as part of the *UN Global Compact*. Previous to life with WWF, he researched agricultural rice systems in West Africa and worked for many years in the private sector in Asia and the USA. Stuart holds an MSc in Environment and Development from the *School of International Development* at the *University of East Anglia* and is currently based in Switzerland where he works at *WWF-International* as a Manager in the Freshwater Team.

JUAN A. SAGARDOY

Juan Antonio Sagardoy is a water management specialist with more than 40 years of international professional experience and worked in more than 45 countries of Latin America, Asia and the Mediterranean region. He worked 30 years with FAO as leader of the *Water Management Section*. He worked 2 years with the EC (DG I and DG VIII) as Senior Technical Administrator responsible for water resources development programmes in several countries of Asia, and worked 9 years as independent consultant working for the WB, FAO, WMO, UNDP, IDB, CIHEAM and other national organizations. At present, he is Project Manager of a European Research Project financed by the EC coordinated by the *Mediterranean Agronomic Institute of Bari* (CIHEAM) and concerned with Gender Mainstreaming and Water resources in the Mediterranean Region.

The main areas of expertise are: management, operation and maintenance of irrigation systems, participatory irrigation management and irrigation management transfer, water resource management, gender issues in relation to water management, water pricing, capacity building and he has ample teaching experience in several institutions.

CONSUELO VARELA-ORTEGA

Consuelo Varela-Ortega, Prof. of Agricultural Economics at the *Universidad Politécnica de Madrid* (UPM), Spain, is currently involved in research in Spanish, European (EU projects) and International networks in the fields of agricultural policy and the environment, natural resources (water and land), and institutions. Also collaborating with international organizations (FAO, IDB, WB, CIHEAM, EU) she has been directly implicated in the analysis and workshops of irrigation water policies in a number of less developed countries (Syria, Jamaica, Mexico, Georgia, Lebanon, Bolivia). Actively involved in national and international associations and congresses in Agricultural Economics and Environmental and Resource Economics, she has published extensively in scientific journals and books. She has been a member of the editorial board of the *European Review of Agricultural Economics* and of several Spanish journals. Prof. Varela-Ortega is the country representative for Spain in the *International Association of Agricultural Economists* (IAAE), and she belongs to the scientific committee of several international institutions and to the advisory panel of the *Global and National Food and Water System* at IFPRI for the CGIAR *Challenge Program on Water and Food*. She has been an invited member in the international FAO panel for water scarcity, of the Netherlands research program on Climate Change and in other international organizations. In the EU research networks she has been and is directly involved as scientific coordinator for the Spanish research team in a number of EU projects related to water, agricultural policies and climate change, and she is presently a member of the executive board of the EU integrated project SCENES where she coordinates the Mediterranean region. She has collaborated extensively with the EU Commission as expert evaluator of research projects with the *General Directorate of Research* of the EU Commission, as a member of the Advisory group of the *7th FP for Environment and Climate Change* and with the *Directorate of Social Sciences and Humanities* in the joint research panel EU-China, all of which are comprised by high-level representatives of science in Europe.

KAREN G. VILLHOLTH

Karen Villholth is associated with the *Geological Survey of Denmark and Greenland*, Denmark, as a senior researcher with primary responsibilities related to groundwater research and management in a developmental context. She has a strong academic background combining chemical engineering, environmental studies and hydrological, hydrodynamic and numerical disciplines. Her technical scientific interests and experiences emphasises themes like soil and groundwater hydraulics and contamination. Karen Villholth has collaborated broadly, with a host of international partners, resulting in several scientific papers.

Since 1999, Karen Villholth has been involved in broader issues of water resources management. She was responsible for introducing the first course on *Integrated Water Resources Management* at the *Technical University of Denmark*. She has been assigned to several international projects concerning water resources management, focusing on river basin and groundwater management, institutional revisions, hydrological and water resources decision support modelling, capacity building, data collection and *Geographical Information Systems*.

Karen has ample experience in teaching and training, from the *Technical University of Denmark* and the *Asian Institute of Technology* for M.Sc. and Ph.D. students, and short term courses for professionals within water, water quality, and environmental disciplines.

From 2004 to 2007, Karen joined IWMI, Sri Lanka, as a senior researcher within groundwater modelling and management. In this position, she developed work related to the physical impacts of the tsunami on groundwater and the rehabilitation and coping strategies in coastal areas of Sri Lanka. She also directed the development of a large training and research capacity building program on groundwater governance in Asia, encompassing five South and South East Asian nations. She was involved in the *Comprehensive Assessment of Water Management in Agriculture* and was co-editor of the book *The Agricultural Groundwater Revolution: Opportunities and Threats to Development.* Lately, Karen Villholth is devoted to integrated and interdisciplinary climate change and water resources research and management.

HONG YANG

Dr. Hong Yang is currently a senior scientist at *Swiss Federal Institute for Aquatic Science and Technology*, Switzerland. She is the leader of the *Water, Food and Environment Studies Group* in the *Department of Systems Analysis, Integrated Assessment and Modelling* at the Institute. She also holds a visiting professorship at the *Institute of Geographical Science and Natural Resources Research*, in the *Chinese Academy of Sciences*. Dr. Hong Yang's main research interests include integrated land and water resources management, water scarcity and its impacts on food security and implications for the global food trade, regional development, poverty alleviation, and water and agricultural policies.

I

Are global water resources a limitation for food production and security?

CHAPTER 1

Food security in water-short countries – Coping with carrying capacity overshoot

Malin Falkenmark
Stockholm International Water Institute, Sweden; Stockholm Resilience Center, Sweden

Mats Lannerstad
Stockholm Environment Institute, Sweden

ABSTRACT: Given that water is the core resource for plant production, the study analyses the carrying capacity overshoot threatening a large part of the planet by 2050 AD, in view of current dietary tendencies towards 3,000 kcal/day per person with 20% animal-based food. It shows the enormous unbalance caused by population growth and the ambitions of the Millennium Development Goals (MDGs) of hunger alleviation if environmental sustainability should be met. The world is moving towards a future where indeed 2/3 of the world population because of water constraints will be depending on the remaining 1/3 for supplying them with part of their food needs, so that they can feed their population at the assumed dietary level. This would necessitate more than doubling of today's trade-based virtual water transfer to compensate for foreseeable food water deficits. But food security in some poor countries would still remain problematic due to lack of purchasing power to import. Half of these countries might solve their future food supply by aiming at lower animal based food. Others will have to seek support from other nations, in terms of emergency aid to avoid horizontal expansion, or continued undernourishment. It will thus be essential for the world to steer away from current dietary tendencies in terms of water-consuming animal-based components. Much can be gained already by moving from today's 20% to the 5% considered satisfactory for human health.

Keywords: water-based carrying capacity, virtual water transfer, food water deficits, purchasing power, dietary tendencies, water productivity

1 FOOD SECURITY CHALLENGE

Humanity faces an unprecedented challenge of meeting rapidly growing food demands on a planet with shrinking per capita availability of water and land resources that can be sustainably used, and where water constraints will in many parts of the world rise due to anthropogenic climate change. Molden (2007) estimated that water for food production will have to increase from current 7,000 km^3/yr to some 9,000–11,000 by 2050. This increase involves a challenge of a scale nothing less than a new green revolution.

1.1 *Sequence of approaches*

The issue of water and food supply for a growing humanity was raised by Swedish scientists already at the UN Population Conference 1974 in Bucharest, based on an informal document (Falkenmark & Lindh, 1974). While the UN Food Conference in Rome also in 1974 addressed the foreseeable need to expand irrigation, the UN Water Conference in Mar del Plata in 1977 broadened the perspective of blue water, i.e. liquid water in rivers and aquifers; the UN information material included the publication "Water for a starving world" (Falkenmark & Lindh 1976).

Attention to the link between food and water shortages continued through the following decades. In the 1980s, the severe droughts in the 1970s and 1980s drew attention to the links between water shortage and hunger (Figure 1). It was observed that around 20 African countries hit by the hunger catastrophe caused by the 1984–85 drought (Figure 1, Map 1) in fact shared a number of underlying water shortage characteristics (Falkenmark & Rockström, 1993):

- Water-related soil problems: short growing season (Figure 1, Map 2), infiltration/land degradation problems (Figure 1, Map 4).
- Water availability problems: low runoff generation (Figure 1, Map 5), large rain variability (Figure 1, Map 3).

In the 1990s, international attention was directed to the possibility to import food to countries too water-short to base food supply on irrigation; the word *virtual water* was introduced to denote the hidden water transferred by export (Allan, 1993; 1994). Numerous studies have later quantified the amounts of virtual water transferred between regions in this way (e.g. Chapagain *et al.*, 2006; Yang *et al.*, 2006; Hoekstra & Hung, 2005; Hoekstra, 2003). Attention was also drawn to the fact that most water involved in food production was in fact not blue (liquid), but green, i.e. infiltrated rain, accessible as soil moisture in the root zone (Falkenmark, 1995; Rockström, 1997). Later, climate change added a new perspective, raising an interest in the *agricultural green water potential*, and the *carrying capacity of the planet* if adhering to the Millennium Development Goals (MDGs) of hunger alleviation while securing environmental sustainability (Rockström *et al.*, 2005). Attention was also drawn to the *consumptive use* component of irrigation. It was observed that with increased irrigation development total withdrawals in many basins exceeded renewable supply, and with intensified reuse the depleted fraction of the utilizable flow would approach 100% in many basins. The concept *basin closure* was introduced to describe the path and stage of river depletion (e.g., Keller *et al.*, 1998; Molden *et al.*, 2005).

To eradicate hunger in line with the MDGs is an increasing challenge. In the first decade of the 2000s, a central question that caught increasing attention was which parts of the world will have enough water to be food self-sufficient, and which parts will not. More and more attention went into projections of future water requirements for feeding the world by 2050 when population would be expected to stabilise (Rockström *et al.*, 2007; Molden, 2007; Rockström *et al.*, 2009; Falkenmark *et al.*, 2009). Already today total consumptive water use for food production has been estimated at around 7,000 km^3/yr (Molden, 2007). From present world population of about 6,900 million in year 2010, projections point at an expected increase of about 32% to 9,100 million by 2050 (UN medium projection), with almost the entire increase located to the less and least developed countries (UN, 2010, 2008 revision). The absolute number of undernourished have oscillated around 850 million since the beginning of the 1970s. After the food and economic crisis 2007–2009 the estimates for year 2008 showed for the first time a total above 900 million, and estimates for 2009 indicated a continued rapid increase to more than 1,000 million (1,020 million) (FAO, 2009a).

In short, these 40 years of global food security analysis basically lead up to the conclusion that particular attention has to be paid to the implications of continued population growth in water short countries when combined with the MDGs, in particular in terms of hunger alleviation.

1.2 *This study*

Large differences in terms of hydroclimatic preconditions for an expanded crop production between different geographic regions have raised further questions (Rockström *et al.*, 2010). What will be the relation between the water-rich regions with potential to produce a surplus of food to be exported to water-short regions with continuing population increase, unable to produce the amount needed to feed their populations on an acceptable nutritional level? What parts of the food-deficient regions will be able to cover food deficits by imports? What parts will have to depend on food aid? How large virtual water flows will this transfer of food from well-supplied to deficit regions imply? What options are available to countries unable to pay for imports?

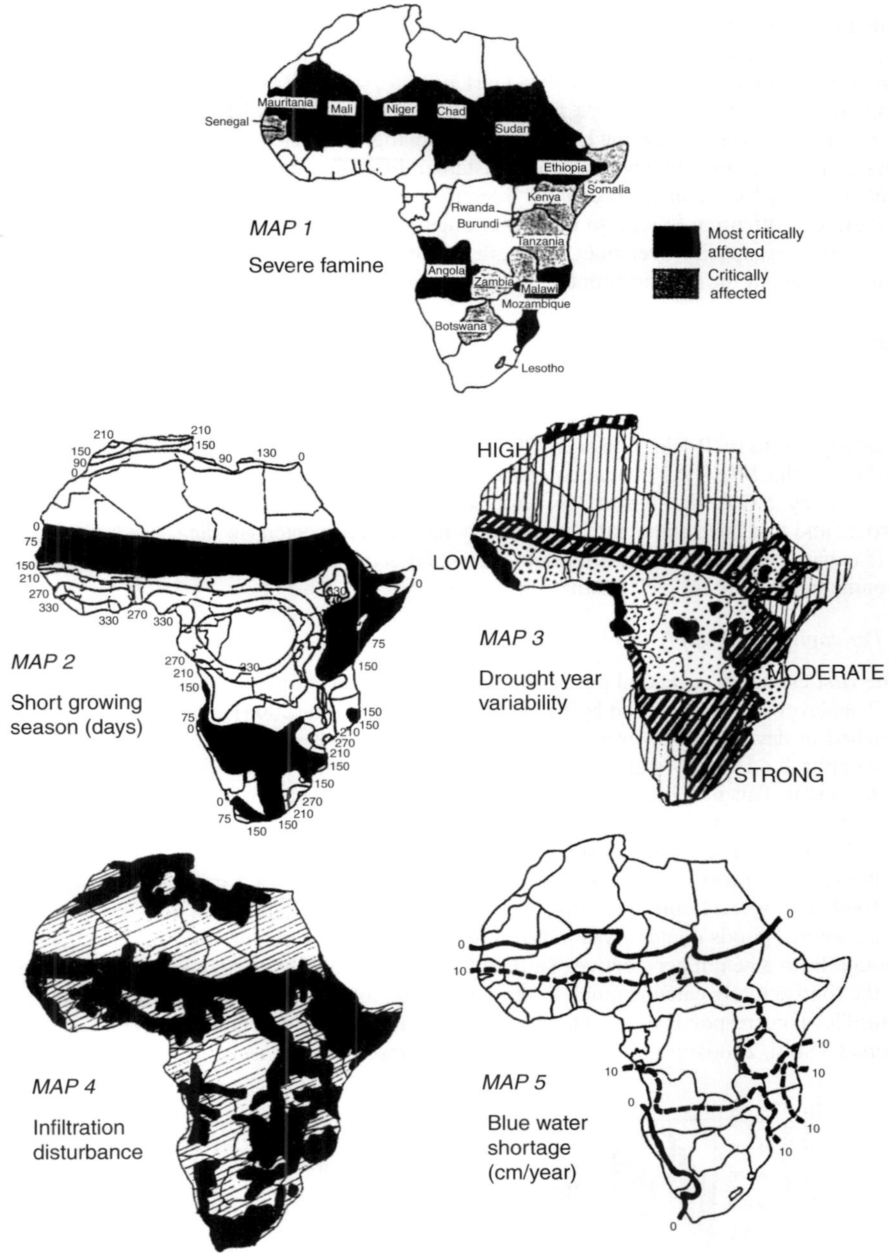

Figure 1. Links between sub-Saharan proneness to famine and complex water scarcity.
Source: Falkenmark & Rockström (2004).
Legend: Map 1: Severe famine (1984–85 drought); Map 2 and 4: Green water shortage (2: growing season below 150 days; 4: infiltration disturbance/land degradation); Map 3: High green and blue water variability (striped: two severe drought years in a row 5–7 times in 50-years period); Map 5: Blue water shortage (rainfall below 100 mm/yr).

Our study is a further advancement of a recent model-based studies by Rockström *et al.* (2009, 2010) which was a country-based analysis based on the LPJmL dynamic global vegetation and water balance model (Sitch *et al.*, 2003; Gerten *et al.*, 2004; Rost *et al.*, 2008). It was a back-casting study analysing the options for resilient and sustainable water resource supply for food production by 2050, assuming the UN medium population projection and no cropland expansion in line with the MDG goal on environmental sustainability. The present study goes on from there by further analysing the water short country dilemmas and realistic options to close the water deficit gap. Particular attention is paid to the importance of respectively economic development, water related carrying capacity overshoot, emerging water deficits, and the importance of the animal protein component for the consumptive use water requirements.

2 METHOD

2.1 *Scenario*

The underlying study (Rockström *et al.*, 2010) assumed that by 2050 climate change has proceeded according to the SRES A2 scenario, and the population grown to 9,100 million [UN Medium Population] by 2050); that the MDGs have been reached, i.e. hunger alleviation achieved in all countries; and that environmental sustainability has been respected (in the sense both of avoiding further expansion of croplands to protect terrestrial ecosystems and of respecting the needs for environmental flow to protect aquatic ecosystems).

2.2 *Per capita food supply level*

For the first estimates a general per capita food supply demand of 3,000 kcal/day per person was used. This level was earlier used by Falkenmark & Rockström (2004) and is projected by FAO to be reached in developing countries by 2030 (Alexandratos *et al.*, 2006). An average of 20% of the per capita food energy supply is assumed to originate from animal foods to ensure sufficient protein content. This is in line with the present rising global average of 17% and the fact that many less developed countries with rapid economic development have already reached above 20% (e.g. China 22%) (Lannerstad, 2009; FAOSTAT, 2009) (see Figure 2).

In the following analysis of countries with water-related carrying capacity overshoot we use three food supply level combinations. First is the general level specified above. For the second level the animal foods content was lowered to a minimum acceptable level of 5% considered to be enough from a health perspective (Smil, 2000), but the 3,000 kcal/day per person level retained since the food actually eaten is much lower due to food losses from the market to the dinner table. The third level corresponds to what is considered the necessary food intake level of 2,200 kcal/day per person (Smil, 2000), assuming a *loss free* food consumption system.

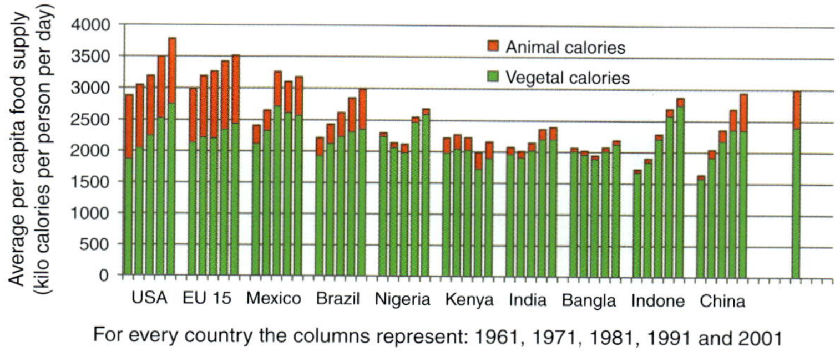

Figure 2. Average per capita food supply per day for 1961–2001 separated into vegetal and animal calories. Last pillar to the right shows assumed food supply of 3,000 kcal/day per person, 20% animal calories. *Source*: Adapted from Lannerstad (2009); data source: FAOSTAT.

2.3 *Agricultural water requirements*

Based on previous global assessments under current water productivity levels (Falkenmark & Rockström, 2004; Rockström *et al.*, 1999) we assume that the evapotranspiration on average required to produce the equivalent of 1,000 kcal of vegetal foods is 0.5 m^3, while 4 m^3 per 1,000 kcal is required for animal products. More water is required to produce animal foods, because only part of the vegetal energy consumed by animals is transformed into e.g. meat, milk, or eggs. The values used here are equal to or lower compared with earlier estimates (Pimentel *et al.*, 1997; Cai & Rosegrant, 2003; Chapagain & Hoekstra, 2003). The result is annual per capita agricultural water requirements of 1,300 m^3 for our food supply level of 3,000 kcal/day per person with 20% originating from animal foods.

2.4 *Agricultural water availability*

Blue and green water availability on current managed agricultural lands (i.e. croplands and permanent pasture) was estimated on a country-by-country basis to distinguish countries with surplus of water, able to export food to countries with a water deficit, unable to support their populations (Rockström *et al.*, 2010).

2.5 *Water surplus/deficit analysis*

When cropland green water availability was compared with food water requirements on a country by country basis, a gross water deficit of some 4,500 km^3/yr was identified. Table 1 shows the implications of plausible water productivity improvement, and irrigation expansion and their effects in bringing down the gross water deficit to a *net agricultural deficit* of 2,150 km^3/yr in water short countries (23% of the water requirements). In water rich countries, on the other hand, the surplus above what will be needed for food self-sufficient production amounts to almost the double.

A conclusion from these results is evidently that there will be enough water in the world to support a population of 9,100 million by 2050. The future problem will therefore not be to produce enough food, but how to allocate it to the people in dry and overpopulated countries, heading for carrying capacity overshoot.

2.6 *Food trade analysis*

In our study, the implications of food trade have been analysed for two different projections in terms of economic trends in South and East Asia and Africa, respectively. Case A assumes no economic development in low income countries, case B assumes an economic development in line with a recent World Bank categorisation (Income group: Economies are divided according to 2008 GNI per capita, calculated using the World Bank Atlas method. The groups are: low income, US$ 975 or less; lower middle income, US$ 976–3,855; upper middle income, US$ 3,856–11,905; and high income, US$ 11,906 or more) (WB, 2009), and allowed a new differentiation of developing countries as compared to the earlier one in Rockström *et al.* (2010).

Table 1. Country-based estimates of agricultural water deficits and surpluses for food production in 2050 (UN Medium Population scenario).

	Deficit (km^3/yr)	Surplus (km^3/yr)
Water deficit/surplus	4,471	2,052
Water productivity improvements	−1,973	532
Irrigation expansion	−348	1,379
Net deficit/surplus	**2,150**	**3,963**

Source: Rockström *et al.* (2010).

The study has furthermore analysed the water/food security pathway by looking closer at the country based food security gap, and the relative size of the main options by which the food water deficit may be met.

3 ECONOMICALLY BASED COUNTRY COMPARISONS

Table 1 clarifies that water productivity improvement and irrigation expansion more than halved the original water deficit based on current water productivity. What remains is a net water deficit gap of 2,150 km^3/yr in vulnerable water short countries. Possible options to meet this food water gap include virtual water transfers through trade from water surplus to water deficit nations. Countries without purchasing power will be left to national solutions like horizontal expansions, risking unsustainable land use, or to aim for less water intensive diets by lowering the per capita food supply calorie level and/or the animal food ratio. In some cases food aid, another virtual water transfer solution, might be the only solution.

3.1 *Importance of economic development*

Since food trade expects people to pay for the transferred food, the 2009 study (case A) suggested that only populations in medium and high income countries would be able to do so (see Table 2). Case A is based on the assumption that the relative economic situation in the water short countries would not change very much in the next 40 years from the situation around 2000 AD. This would imply that only some 750 km^3/yr of the deficit could be covered by virtual water transfer through trade, leaving the food supply of the 3,800 million poor and water short to so-called *other solutions*, in terms of conversion of pasture to croplands, cropland expansion, reduced diet expectations, or aid, case A in Figures 3 and 4.

However, considering past economic trends in many developing countries with rapid economic growth during the last decades, e.g. India, it may be reasonable to assume that the relative purchasing power of some of the low income countries will in fact increase till 2050 (case B). Assessments of economic growth in different regions (Sulser *et al.*, 2009) clarifies the need to look separately at South Asia with an assessed economic development 2000–2050 of 5.1% per year as opposed to sub-Saharan Africa with only 2.8% per year. Within regions there may be individual countries with a stronger development as opposed to countries with a lower growth. The World Bank categorisation (WB 2009) reflects this economic development perspective by sorting all countries into four income categories: *high*, *medium high*, *medium low*, and *low*, thus moving some of the

Table 2. Comparison of country level net annual water deficits and surpluses and population for Case A and B, summarized for different country income categories.

Case A				Case B			
Country income groups (WB, 2005)	Net deficit 2,150	Net surplus 3,963	km^3/yr	Country income groups (WB, 2009)	Net deficit 2,150	Net surplus 3,963	km^3/yr
Low	1,404	407	km^3/yr	Low	602	381	km^3/yr
	3,785	477	million		1,492	464	million
Medium	487	2,680	km^3/yr	Medium Low	1,242	646	km^3/yr
	2,115	1,614	million		4,246	659	million
				Medium High	46	2,044	km^3/yr
					160	950	million
High	259	876	km^3/yr	High	260	892	km^3/yr
	522	631	million		524	649	million

countries from *low* in case A to *medium low*. This would radically change the purchasing power assumptions in some water short countries with large populations, turning the global food security outlook to the classification as shown in Figure 4, case B. In case B the 3,800 million people *deficit other solutions* sector from case A have been rearranged so that 2,300 million have been moved up

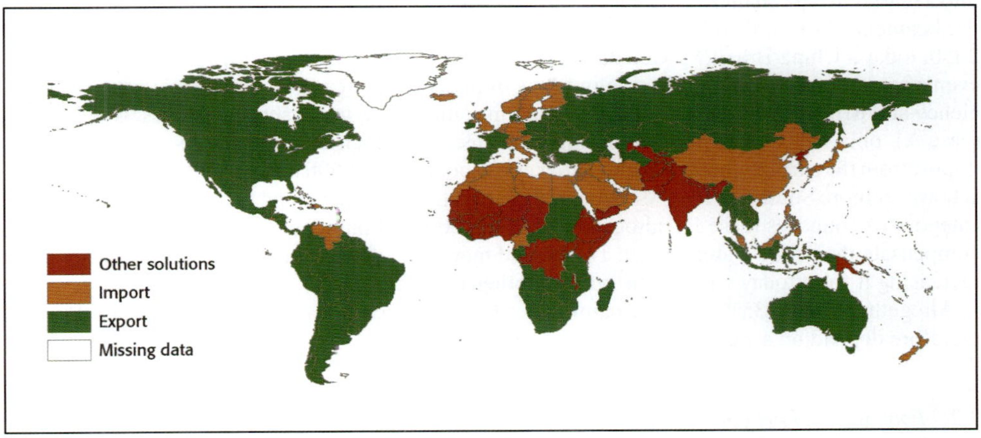

Figure 3. Food importing and exporting countries from a water availability perspective in 2050, including low-income, water short countries which may have to rely on other solutions.
Source: Falkenmark *et al.* (2009).

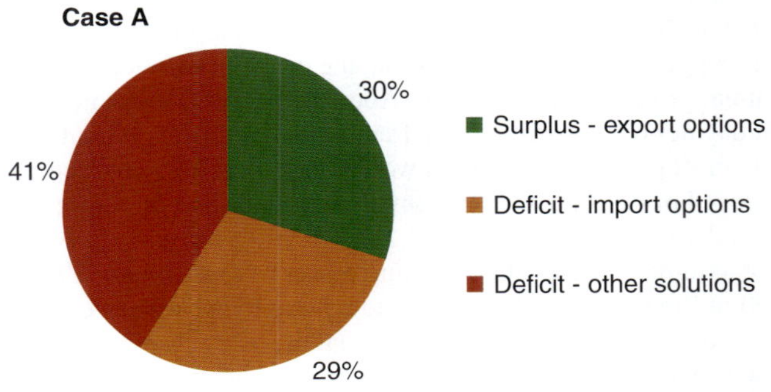

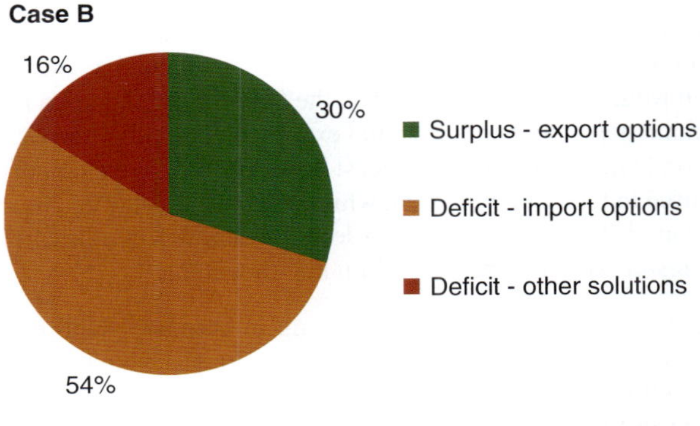

Figure 4. Comparison of population vulnerability by 2050 in water short countries for Cases A and B. The circles show percent of world population living in countries with export options, import options, and having to rely on *other solutions* to meet food demands under different assumptions on rate of economic development and purchasing strength on the global food market. Case A: no advancement of Low Income countries; Case B: WB 2009 land categorisation.

to the category *medium low*, expected to manage a certain food import. This leaves only a future population of about 1,500 million that because of limited purchasing power on the international market will have to find other solutions but trade to secure a sufficient food supply.

Comparing now the two cases A and B, we may conclude that global water security is growing into a massive trade challenge. Water constraints will increasingly be shaping the basic food supply predicament. Two thirds of the world population will be living in countries where tomorrow, i.e. 2050, today's China/Brazil/Mexico diet (cf. Figure 2) cannot be provided from current croplands, even after agricultural modernisation in terms of plausible increase of agricultural water use efficiency and irrigation increase. If today's economic situation in water short countries would remain (case A), only half of that population might be able to compensate their food security deficit by import from the surplus countries (altogether corresponding to 2,600 million and some 750 km^3/yr). If however more South and East Asian countries are categorised to belong to the *medium low* income category of countries (case B), altogether almost 5,000 million people might be able to import to compensate their food water deficit. This would more than double the virtual water flow by trade, increasing it from today's 1,300 km^3/yr by another 1,550 km^3/yr.

Altogether around 35% of the remaining food water deficit in case A and 70% in case B would therefore depend on a well functioning system of food trade.

3.2 *Poor water short country options*

For case A, Rockström *et al.* (2010) analysed how much food the 3,800 million in the 39 poor water short countries by 2050 might in fact be able to produce within their agricultural water availability constraints. The first option is to intensify agricultural production by expanding cropland onto present pasture. By doing this, a water deficit of about 400 km^3/yr in countries with 390 million people can be met. For the remaining 20 countries the second option was to reduce the dietary expectations. It was suggested that for 600 million, the problem was not really serious but an issue of reducing the animal protein part of the diet. About 1,900 million would be able to produce on average 2,500 kcal/day per person with a 10% fraction derived from animal foods. The remaining 1,300 million in the most precarious situation would only be able to produce some 2,000 kcal/day per person with reduced or no contribution from animal foods and therefore remain dependent on aid to avoid broad undernutrition.

As a result of the assumed increased purchasing power in case B, only 29 countries with a total population of 1,500 million would still continue to belong to the water deficit *low income* group. After expansion of croplands onto existing pasture lands, as in case A, about 230 km^3/yr can be gained. For the case B only 17 countries with a total population of 1,200 million would still belong to the water deficit *low income* country group. Nine of these countries would be able to manage their national food supply just by heading for an Indonesian average food supply of about 3,000 kcal/day per person and 5% animal foods (see Table 3 and Figure 2). Altogether 8 countries with a population of almost 500 million would remain with less than the 560 m^3/yr per person needed to sustain such a diet, and would remain dependent on horizontal expansion into other terrestrial biomes wherever possible, or indeed food aid. Bangladesh, the least water short of these countries, lacks only about 2% of the water required to feed its population, while Togo, the most water short in this group, lacks 55% of the water required. The remaining water deficit for these 8 countries amounts to 49 km^3/yr.

Summarising the results so far, our comparison has shown that:

- Only 30% of the 2050-population would be living in countries with enough water for food self-sufficiency on the assumed dietary level (current Brazil-China diet level).
- 2,600 (case A) or 4,900 million (case B) depending on to the economic development in Asia would be able to cover their food water deficit by import from water rich regions of the world.
- 3,800 (case A) or 1,500 million (case B) would be left as poor and water short. In case B 12 countries in this group can meet their water deficit by converting pasture to croplands, 9 countries would be able to manage on a 5% animal protein level, leaving the remaining 8 countries dependent on aid or horizontal expansion to cover their deficit. Four of these countries might

be able to produce the food needed provided that there are no food losses (the *loss free* option) which is however mainly to be seen as a theoretical possibility only, i.e. an untapped option to be explored in the future.

4 FOOD SECURITY PATHWAY

The above analysis has revealed three key obstacles relevant for global food security by 2050:

a) Assumed preferences for animal-based food (20%) will not be achievable and are exaggerated as compared to health motivated amounts (5%) (countries 1–9 in Table 3).
b) Food losses, demanding an over-production of food items to eliminate undernourishment of parts of the population in a country (illustrated by countries 10–13).
c) Overpopulation in comparison with the water-controlled carrying capacity of the countries (countries 14–17).

In this section we will carry the analysis further and compare, on the one hand, the different components by which the revealed global food deficit gap may be closed. On the other hand, we will look at the main factors that contribute to accelerating that gap.

Table 3. Food supply options to meet national net water deficits for the 17 different countries with net supply below 1,000 m^3/yr per person under Case B. Left part of the table shows country wise data; in right part striped areas indicate country dietary options.

	Total Food Supply Level		kcal/day per person		3,000	3,000	2,200		
	Proportion Animal Foods		%		20	5	5		
		Total population 2050	Per capita Net Deficit 2050	Per capita Net Supply 2050	Full Food Supply	Full Energy Supply Minimum Animal	Supply only intake "Loss Free"	Other Solutions	Diet and Other Solutions Water Volumes
Country		million	m^3/yr per person		1,000	560	410	<410	km^3/yr
1	Congo, Dem. Rep.	187	20	980					
2	Afghanistan	79	90	910					
3	Ethiopia	183	210	790					
4	Nepal	52	240	760					
5	Uganda	93	250	750					Less animal foods
6	Eritrea	11	250	750					
7	Korea, Dem. Rep.	25	270	730					321
8	Gambia, The	4	290	710					
9	Burkina Faso	38	370	630					
10	Bangladesh	254	450	550					
11	Rwanda	23	580	420					
12	Niger	53	590	410					
13	Yemen, Rep.	58	590	410					
14	Malawi	32	630	370					Horizontal expansion or Aid
15	Benin	23	690	310					
16	Burundi	28	730	270					49
17	Togo	14	740	260					
Total		1,156							370

4.1 *Closing the water deficit gap in water short countries*

Figure 5 shows the relative role of different ways to close the country-based water deficit gap in the country-level food security pathway curve. The gap may principally be closed *from above* with increased efficiency or lowered food demand thus reducing the consumptive water use requirements through water productivity improvement, diet expectation adaptation, and if possible food loss reductions, or *from below* by additional water through irrigation, virtual water from surplus regions (trade and aid), making use of non-productive evaporation on current pasture lands, or horizontal cropland expansion.

4.1.1 *Water productivity increase*

The importance of water productivity increase was demonstrated in the study by Rockström *et al.* (2010) on which this study builds. As already stressed, the gross water deficit of 4,500 km^3/yr with present water productivity can with improvements be brought down by almost half to 2,500 km^3/yr (Table 1). This implies that the current water requirement of 1,300 m^3/yr per capita can be reduced to 1,000 m^3/yr per capita by 2050 by implementing practices that improve soil, crop and water management (conservation tillage, soil fertility management, soil and water conservation, water harvesting, integrated pest management and crop improvements) (Rockström *et al.*, 2007). There is a particularly great opportunity to improve water productivity at the low-yield range (Rockström, 2003; Rockström *et al.*, 2007; Molden *et al.*, 2010). This highlights a very important point –that the largest untapped potential to save water in food production is in the lowest yielding savannah regions of the world and these are also the regions where escalation in food requirements is fastest due to population growth and where a green revolution is therefore most needed.

4.1.2 *Irrigation*

In the past much attention has been paid to irrigation as the primary way to find relief from agricultural water shortage. In recent years, it has however become evident that the irrigation expansion during the green revolution had very serious environmental impacts. One serious effect is the flow depletion in numerous rivers in a broad land ribbon from North-East China in the east, to South-West USA in the west (Smakhtin *et al.*, 2004). Some well known and alarming examples

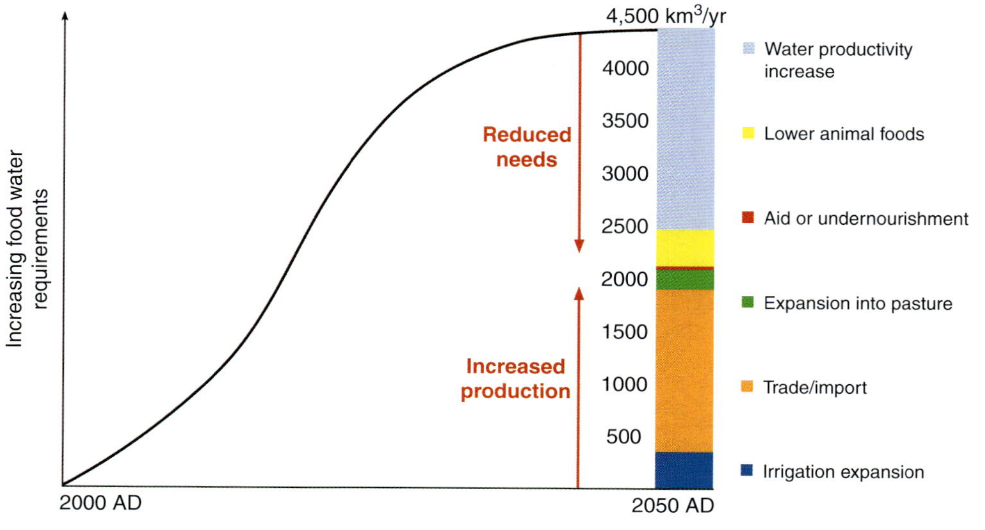

Figure 5. Global food security pathway outlook for agricultural water deficit countries, and options for closing the food water deficit gap.

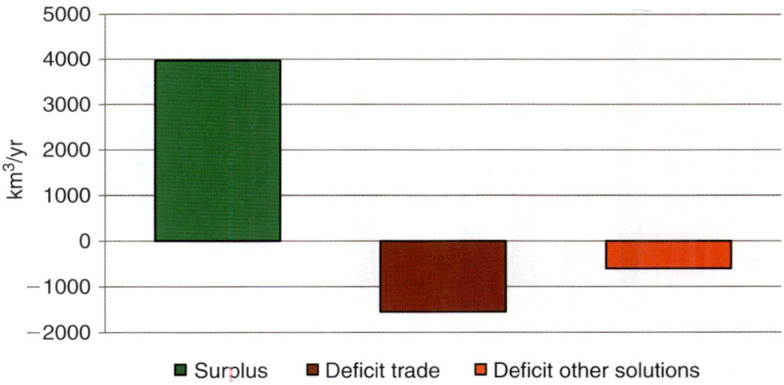

Figure 6. Case B country group categories, showing agricultural water surplus and deficits by 2050. Depending on level of economic development and expected purchasing power the deficit can be met by either trade or other solutions.

of rivers that are running more or less dry during parts of the year are the Nile, the Colorado River, the Yellow River, and tributaries Amu Darya and Syr Darya to the depleted Aral Sea, but there are numerous others (Lannerstad, 2002; Falkenmark & Lannerstad, 2005). A river basin with no usable outflow has been described as evolving from an *open* to a *closed* state (Seckler, 1992; 1996). The term *basin closure* is now widely used in the sense of very little additional water being left in the river for further human uses and in terms of enough water being left for the aquatic ecosystems (generally 30% of the total available blue water resources) (Falkenmark & Molden, 2008; Molle *et al.*, 2010).

Groundwater overexploitation is a parallel and increasing dilemma. Rosegrant *et al.* (2002) estimated total groundwater use to more than 800 km^3 in 1995, out of which as much as 200 km^3, was assessed as overdraft of non-recharged groundwater (Postel, 1999). Well documented areas with long-term groundwater depletion are the North China Plain (Kendy *et al.*, 2003), the USA Great Plains with the Ogallala aquifer (Postel, 1999) and several states in India, e.g. Gujarat (Moench *et al.*, 2003).

As a consequence of these effects of consumptive water use and overexploitation, the contribution that irrigation may offer to close the water deficit gap is surprisingly limited in relation to improvements in water productivity, although the assessments vary between different authors. Given the assumptions in this study, Rockström *et al.* (2010) assessed that irrigation might in the deficit areas contribute in the order of 350 km^3/yr, i.e. of the order of only 8% of the total deficit, an assessment in agreement with what de Fraiture & Wichelns (2010) estimated as an optimistic, but plausible scenario.

4.1.3 *Virtual water transfers through trade, land grabbing and aid*

The relation between water surplus and water deficits for case B is shown in Figure 6. Virtual water flows through trade would meet a deficit of 1,550 km^3/yr, leaving 600 km^3/yr to be met by other solutions. Our estimated trade increase will thus stand for 35% of the total deficit. Already today some water scarce regions are heavily dependant on imports. One example is Northern Africa where about 50% of the cereal food consumption is based on imports (FAOSTAT, 2009). Compared to present virtual water flows through trade of agricultural products (some 1,300–1,600 km^3/yr) (Chapagain *et al.*, 2006; UNESCO, 2009), our figure indicates more than a doubling of the global needs.

Besides the necessary purchasing power of the importing countries, three requirements will have to be met. The basic prerequisite is that surplus countries manage to produce enough food for export. A large production increase poses a true challenge even if agricultural water availability does not

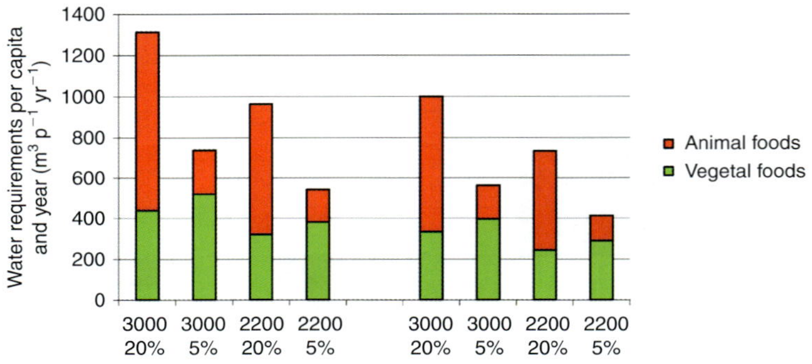

Figure 7. Different ways of bringing down food water demand from a calorie and animal *vs.* vegetal foods water perspective. [See text].
Four columns to the left: present water productivity of 1,300 m^3/yr per person.
Four columns to the right: expected water productivity by 2050 of 1,000 m^3/yr per person.

involve any constraint. The second necessity is an upscaling of the global food transport infrastructure to handle doubled trade volumes. This will mean investments in land and sea transports as well as other infrastructure like harbours and food storage. The third and most difficult factor is a well functioning global food market. During the food crisis 2007–2008 many countries implemented export restrictions, e.g. 40–44% in East Asia and South Asia (FAO, 2009b). As pointed out by von Braun & Meinzen-Dick (2009), the food crisis and the volatility in food markets have undermined the trust in trade and is a main cause behind foreign investment in agricultural land, often called *land grabbing*. This trend of state controlled virtual water flow started already some years ago. China began to lease land for food production in Cuba and Mexico 10 years ago and continues to search for new opportunities to feed its large population (*ibid.*).

4.1.4 *Cropping pasture*

Some of the water deficits can be met by expanding croplands on to permanent pasture lands. In the underlying green blue water flow analysis (Rockström *et al.*, 2010) it was assumed that livestock production only utilises 50% of the vapour flow from permanent pasture lands. When converting these areas to croplands, it is assumed that the remaining 50%, presently sustaining other terrestrial ecosystems can be appropriated by crops and meet a deficit of about 230 km^3/yr. Out of the total remaining 600 km^3/yr water deficits, only 370 km^3/yr would be left to be met by other measures. Although cropping pasture will have negative effects on non-agricultural ecosystems co-existing on pasture lands, it maximises production on already domesticated lands and minimises the expansion and environmental degradation into areas that remain largely untransformed for human agricultural needs.

4.1.5 *Less water intensive diets*

As described above, three average per capita food supply combinations are used in this chapter. For those countries that because of water deficits cannot produce a full diet our study point at the possibility to reduce the animal foods content. The Indonesian case in Figure 2 offers an alternative of almost 3,000 kcal/day per person and only a minor fraction of animal foods.

Figure 7 visualises the water requirement gains for the different food supply levels (vegetal *vs.* animal foods fractions), and food loss reduction for the two alternative water productivity levels. By reducing the animal content from 20 to 5% for the 3,000 kcal level a reduction of almost 45% in agricultural water requirements can be achieved. This shows the crucial multiplying impact of increased animal foods contents in our diets.

The importance of the animal foods factor calls for further research. In line with Rockström *et al.* (1999) it has been assumed that 0.5 m^3 of evapotranspiration is required to produce 1,000 kcal

Table 4. Relative scale of potential water deficit alleviation components (km^3/yr).

Food security pathway components	km^3/yr
Water productivity increase	1,973
Trade across national borders	1,548
Irrigation expansion	348
Lower animal foods proportion	321
Conversion of pasture to croplands	232
Aid or undernourishment	49
Total water deficit alleviated	**4,471**

of vegetal foods and 4 m^3 per 1,000 kcal of animal foods, i.e. that animal foods production needs eight times as much water per produced calorie. This relation however depends on the kind of and combination of animal production systems (grazing, cultivated fodder, and cultivated feed), and the feed conversion efficiency for different species and the breeds of these species (e.g. traditional or modern) (Lannerstad, 2009).

4.1.6 *Horizontal expansion of croplands and pasture into other terrestrial biomes*

Even if non desirable from the perspective of environmental sustainability, horizontal expansion into forests and other terrestrial ecosystems is nevertheless an option to enlarge agricultural lands and increase food production. In a number of countries, hosting altogether some 600 million people (Bangladesh, Burundi, Nigeria, Rwanda, Togo), the populations are however so large that more terrestrial land (forestry and other land uses) than what is even totally available would be required to meet dietary water needs. That is to say that even if the whole terrestrial area of the country could be converted to croplands, this would still not meet the total water requirement to feed the whole population. Horizontal expansion is therefore not considered in our analysis.

4.1.7 *Scale overview*

The above overview of the relative contributions to closing the food deficit gap in water short countries identified in Section 3 gives the relations shown in Table 4. The single most effective contribution is water productivity increase, which reduces the food water deficit gap of the water short countries by 44%. Food trade would be able to contribute to meeting 35% of the water deficit. Irrigation expansion is much less effective as seen on the global scale. Lowering the animal food component of the current dietary tendencies and cropping of pasture land can also only add minor volumes. In our analysis, these options are considered only for the residual low income countries without trade opportunities. The aid needed to support the eight countries not able to meet their food needs even after water productivity increase, crop cultivation on pasture lands, and lowered animal-based food expectations amounts to no more than 50 km^3/yr, or only about 1% of the total global food water deficit.

4.2 *Major factors driving the water deficit gap*

Turning now to the main factors that tend to accelerate the global food deficit gap of water short countries by 2050 AD, we will highlight three main aspects:

- Urbanisation, economic development and diet composition.
- Losses along the food chain.
- Population growth.

4.2.1 *Economic development, urbanisation and diet composition*

By 2050 about 70% of the estimated world population of 9,100 million will live in urban areas (UN, 2010). Higher disposable incomes will drive a trend towards higher per capita calorie levels and a global convergence of diets, in terms of e.g. the increasing poultry consumption and similar eating habits, with fast and convenience food. The urban dwellers lose the traditional connection to the local agricultural production, and gradually become consumers on the urban, and thus also on the international, market (Lannerstad, 2009). Relatively monotonous diets based on indigenous staple grains or starchy roots, locally grown vegetables, other vegetables and fruits, and limited foods of animal origin will be replaced by more varied diets that include more pre-processed food, more foods of animal origin, more added sugar and fat, and often more alcohol (Steinfeld *et al.*, 2006). Consumption of meat and milk and dairy products in all developing countries (not considering China and Brazil) may double till 2050.

The importance of economic development for changed average supply patterns is clearly visible in Figure 2. USA and the EU15 represent economically strong regions with a high urban majority. Mexico, Brazil and China show a rapid increase in both average calorie levels and fast rise in the proportion of animal foods. India, also a BRIC country like Brazil and China, has a more recent economic boom and probably has only begun the climb up the food and animal foods ladder.

4.2.2 *Food losses –an unnecessary agricultural consumptive water use*

Producing more food than actually eaten is equivalent to a consumptive use of large amounts of a scarce water resource. Today, food losses are in fact considerable (Lundqvist *et al.*, 2008). According to Smil (2000), in 1990 vegetarian produce equivalent to some 4,600 kcal/day per person were being harvested while the amount food consumed was estimated to only 2,000 kcal/day per person. Out of the difference about 1,400 kcal/day per person, or 30%, represents post harvest and food losses. If all losses along the food chain, from farm level to the actual intake could be eliminated, future average per capita food consumption could be optimally brought down to the average food energy requirement of 2,200 kcal/day per person (Smil, 2000), the estimated minimum need for a healthy life. Since food losses are practically very difficult to avoid, a future food supply level of only 2,200 kcal/day per person is not really a very realistic option (Lannerstad, 2009). Figure 7 however clearly visualises the importance of food loss reduction. When the average food supply level at 3,000 kcal/day per person is reduced to only 2,200 kcal/day per person, but with the same animal food proportion the water requirements can be reduced by about 27%. Although loss reduction is desirable, Figure 7 visualises that reduction of animal foods is much more important. With both loss and animal foods reduction the per capita water requirements can be reduced by about 60% down to 400 m^3/yr per person.

The considerable benefits that could be reaped if food losses could be minimised however suggests that efforts to reduce losses must be immediately started in both developed and developing countries. The character of losses differs considerably between developing and developed countries. With modern harvest, transport and storage techniques, it would be a realistic option for many developing countries to dramatically reduce losses on the way from field to the market. In the USA and Europe the major losses and waste take place in the latter half of the food supply chain (Lundqvist *et al.*, 2008). However, if the countries rapidly becoming middle-class societies like China, Brazil and Mexico follow the same market and consumption system path as already is established in the developed world they risk replacing one wasteful food system with another (Lannerstad, 2009).

4.2.3 *Population growth*

As shown by Falkenmark *et al.* (2009), the difference in food water requirement by 2050 between the two population projections SRES A2 (10,900 million) and the UN medium population (9,100 million) contributes an amount to the water deficit gap of water short countries equal to low water productivity, i.e. of the order of 2,000 km^3/yr. Finding acceptable ways to keep population growth under control –especially in water short countries– would therefore be an effective way of limiting the food water deficit gap that will have to be met.

Basically, population growth is however not an issue of fertility decline only but also of momentum and rising live length. The age structure of a population is important. Even after the total fertility rate reaches replacement level in a country (two children per woman), the population continues to grow as the decline in mortality increases the number of old people in the population.

Although population growth rate has in fact slowed from some 2% per year in the 1960s to only 1.2% per year at present, the absolute growth and therefore the pressures on the carrying capacity of the Planet continues to be very high (Sachs, 2008). The continued population increase in countries, scarce in water and other natural resource, like India, is therefore a crucial factor when analysing the future global food challenge. The importance of population growth control is illustrated when comparing population development projections for China and India. With its one-child policy since 1979 China will reach its peak population of around 1,460 million about 2030 which will decrease to 1,420 million by 2050. India will on the other hand be the most populous country by 2050 with a population of about 1,610 million after an increase of more than 400 million (UN, 2010).

5 SUMMARY AND DISCUSSION

The above analysis has thrown interesting light on global food security by mid 21st century, given climate change, population growth and current dietary tendencies. It is however essential to stress that the study has been based on climatic averages only, with no attention to increasing variability. This variability will probably increase food supply challenges even further. Neither is any attention paid to soil fertility or to biomass production activities competing for production on agricultural land, such as fibre and biofuel production.

5.1 *Carrying capacity overshoot*

Although the international debate on global food security during the last three decades of the 20th century developed only slowly, it however revealed the links between water shortage, hunger and population growth. This leads to a growing interest in the global hotspot regions and the potential for these areas to live up to the Millennium Development Goals (MDGs), which is the focus of this study.

We have shown that global food security in line with the diet composition of China, Brazil and Mexico, involving 3,000 kcal/day per person and 20% animal-based components, would imply some 60% larger food water requirements than at present. A world-wide *modernisation of agriculture* (including irrigation expansion), in the developing world, would bring down the additional water requirement and leave a summarized country-based estimate of net water deficit in water short regions of altogether 2,150 km^3/yr which represents some 45% of the total additional global water requirements.

The study has put a finger on a foreseen *carrying capacity overshoot* in countries hosting 70% of world population by 2050, if looked at from the perspective of current dietary tendencies shown in Figure 2, which include around 20% animal based food. We have shown the enormous unbalance caused by population growth and the problems regarding realism in the MDGs of hunger alleviation if environmental sustainability should be met. The world is moving towards a future where indeed 2/3 of the world population will be depending on the remaining 1/3 for supplying them with part of their food needs, so that they can feed their population at the assumed dietary level. This would necessitate more than doubling of today's trade-based virtual water transfer only 40 years from now, to compensate for foreseeable food water deficits.

With the present dietary tendencies (cf. Figure 2), global food security by 2050 is in other words developing into an issue of *massive food trade*. This highlights the fundamental importance of rising income in water short countries to make them able to import the food deficit for feeding their populations (Liu *et al.*, 2009). Future research is however needed to clarify the realism of the necessary food surplus production in the well-watered regions for export to the deficit regions. This issue includes besides surplus agricultural production also the necessary infrastructure, transport and trade involved, but also proper global governance of the food trade system. But food security in some poor countries may still remain unresolved since they might not afford to import what is missing, unless they can secure an economic development that opens up a window of purchasing power. Half of these may in fact manage by steering away from the current international food demand tendency and solve by aiming at an Indonesian type diet, reducing the animal-based food component of the assumed diet. Others will –unless they can seek support to lift their economic capacity– depend on aid, or choose horizontal expansion wherever possible, or indeed endure continued undernourishment.

The scale of the challenges suggests that it would be wise to steer away from current dietary tendencies in terms of *water-consuming animal-based components*, especially from the assumed 20% towards the 5% considered satisfactory for human health. Animal-based food preferences have enormous implications in terms of water needs difficult to meet, of carrying capacity overshoot in water-short countries, and of expanding trade dependency of around 5,000 million people, and of exposing the most overpopulated countries to the risk of mass emigration.

The *food water deficit gap* behind the carrying capacity overshoot can be closed in two principal ways: from above in terms of bringing down the food needs by water productivity improvement, by low animal-based diet, and by loss reduction, should that be possible. Closure from below will involve different measures to bring additional water to the deficit countries by irrigation, trade/virtual water transfer, and cropping pasture lands. One may state that expected animal-based food preferences tend to threaten global food security, in the sense that they multiply the pressure in terms of consumptive water use on the water resources system. In terms of scales of the different measures to close the gap, trade and economic development to achieve purchasing power would be the most efficient activities next after agricultural modernisation, corresponding to 72% of the net water deficit gap. Cropping pasture and diet adaptation (bringing down the animal-based food component from 20% to 5%) would close another 26% of the gap. Remaining will be 2% of the gap representing food aid or undernutrition.

One key obstacle complicating global food security is the large scale of current food losses from field to fork. However, as shown in Table 3, even a loss-free production would not help the 6 most precarious countries (altogether some 200 million people).

5.2 *Importance of clear concepts and fresh courage to tackle core components*

There has in the past been a *poor conceptualisation* of water use. Looking back to the alarms generated already in the 1970s, it is evident that approaches to global food security have remained conceptually truncated. For some 30 years, the approach remained supply-oriented and for a long time technically oriented with focus on irrigation requirements. Only in the late 1990s did it become clear that most food was in fact produced by infiltrated rain (green water) rather than by irrigation water (blue water) except for certain regions with large scale irrigation This chapter has demonstrated the implications of adding to the analysis of future food security the new concepts of green water and virtual water. This made possible a fuller understanding of the water challenges and regional dimensions involved in feeding humanity in the future. We now need to make use of the many possible ways to combine blue and green water management, today referred to by the concept green/blue continuum (Vidal *et al.*, 2010).

When analysing future global food supply, there is every reason to remember the wisdom formulated by the Nobel Laureate in Physics 1937, Sir Georg Paget Thomson: "All science depends on its concepts. These are ideas which receive names. They determine the questions one asks, and the answers one gets. They are more fundamental than the theories which are stated in terms of them".

Besides conceptual development it will be essential to also bring the *unspeakable* issue of *population growth* to the surface. Unless the poor and most water short countries can find ways to secure economic development enough for food import, emigration may be a non-neglectable threat. The world seems not yet to have realised the overloading of their natural resources in an unsustainable way, and that water availability is now a limiting factor in large parts of the world. The fact that the poorest countries with the most rapid population growth are located in the dry climate low latitude regions is an extremely serious dilemma. The longer it will take to reach the replacement level, the more difficult will it be to alleviate hunger. After all, more calories are equivalent to more water consumption, and the hungry and undernourished are primarily living in the water short regions of the world, moving towards a carrying capacity overshoot. Restraining the ongoing population increase is a fundamental key to limit the constantly increasing exploitation of natural resources and environmental degradation. Many developing countries are currently caught in a demographic trap by having developed far enough economically and socially to reduce mortality, but not far enough to quickly reduce fertility (Brown, 2009). Others have during the last decades rapidly falling birth rates (Rosling, 2010) and this is a positive sign that might possibly make the expected population increase come to halt earlier than the projected 9,000 million.

Already 40 years ago the father of the green revolution, Norman Borlaug, who passed away in September 2009 at the age of 95, made exactly this point in 1970 when he accepted the Nobel Peace Prize: "There can be no permanent progress in the battle against hunger until the agencies that fight for increased food production and those that fight for population control unite in a common effort" (Sachs, 2009).

6 CONCLUSIONS

In terms of country level food security, we may conclude:

- That global food security with present dietary tendencies develops into a 1/3–2/3 world of massive food trade from water surplus countries to almost 5,000 million people in countries with agricultural water deficit
- The basic option for the economically more developed group of developing countries will either be to pay for food import allowing a 20% animal food level diet, or alternatively heading for a lower level to free economic resources for other more preferable purposes.
- That 17 poor countries without purchasing power to pay for import might however be left with food security problems and face a different set of options: half of them can be food self-sufficient by heading for only 5% animal food rather than the global tendency of 20%. To the other half, the only remaining choice stands between horizontal cropland expansion and aid, in order to avoid undernutrition and outmigration.

Other main conclusions are the following:

- That most future food production can take place on current croplands provided that agriculture in all countries be modernised to reduce avoidable water losses.
- That water productivity increases are implemented globally is extremely important. If today's water productivity would prevail till 2050, the global food water deficit would be twice the surplus, compared to our positive projections where the global surplus is twice as big as the deficit.
- That food transfers across national borders will have to grow considerably and be divided in two parts: a major trade flow to countries on an acceptable economic level, and some aid to avoid undernutrition of the poorest parts of the populations in the remaining water deficit countries.
- That it will be absolutely essential to generate economic development in poor countries to provide purchasing power to import the food required.
- That the ongoing trend with more and more water intensive diets in terms of animal food is a global environmental threat and must be restrained, and in developed countries reversed.
- That loss reduction will be essential; both water related losses by continued water productivity improvements, and the so far untapped possibility to reduce food losses in the chain from field to fork.

REFERENCES

Alexandratos, N.; Bruinsma, J.; Bödeker, G.; Schmidhuber, J.; Broca, S.; Shetty, P. & Ottaviani, M.G. (2006). World agriculture: towards 2030/2050. Interim report, prospects for food, nutrition, agriculture and major commodity groups. Global Perspective Studies Unit, FAO. Rome, Italy.

Allan, J.A. (1993). Fortunately there are substitutes for water otherwise our hydro-political futures would be impossible. In: *Priorities for water resources allocation and management*. ODA, London, UK: 13–26.

Allan, J.A. (1994). Overall perspectives on countries and regions. In: P. Rogers & P. Lydon (eds.), *Water in the Arab World: perspectives and prognoses*. Harvard University Press, Cambridge, Massachusetts, USA: 65–100.

Brown, L.R. (2009). *Plan B 4.0: mobilizing to save civilization*. Norton, New York, USA.

Cai, X. & Rosegrant, M. (2003). Limits and Opportunities for Improvements. In: J.W. Kijne, R. Barker & D. Molden (eds.), *Water Productivity in Agriculture*. CABI International, Cambridge, Massachusetts, USA: 163–178.

Chapagain, A.K. & Hoekstra, A.Y. (2003). Virtual Water Flows between Nations in relation to Trade in Livestock and Livestock Products. Value of Water Resources Report Series. United Nations Educational, Scientific and Cultural Organization (UNESCO); Intangible Heritage, Delft, the Netherlands).

Chapagain, A.K.; Hoekstra, A.Y. & Savenije, H.H.G. (2006). Water saving through international trade of agricultural products. *Hydrology and Earth System Sciences*, 10: 455–468.

de Fraiture, C. & Wichelns, D. (2010). Satisfying future water demands for agriculture. *Agricultural Water Management*, 97: 502–511.

Falkenmark, M. (1995). Land-water linkages: a synopsis. Land and Water Integration and River Basin Management. *FAO Land and Water Bulletin*, 1: 15–16. Rome, Italy.

Falkenmark, M. & Lannerstad, M. (2005). Consumptive water use to feed humanity – curing a blind spot. *Hydrology and Earth System Sciences*, 9: 15–28.

Falkenmark, M. & Lindh, G. (1974). *Impact of water resources on population*. Swedish contribution to the UN World Population Conference, Bucharest, Romania.

Falkenmark, M. & Lindh, G. (1976). *Water for a starving world*. [In cooperation with the United Nations/UNESCO as part of the information component of the United Nations Water Conference]. Westview Press, Boulder, Colorado, USA.

Falkenmark, M. & Molden, D. (2008). Wake up to realities of river basin closure. International Journal of *Water Resources Development*, 24: 201–215.

Falkenmark, M. & Rockström, J. (1993). Curbing rural exodus from tropical drylands. *Ambio*, 22: 427–443.

Falkenmark, M. & Rockström, J. (2004). *Balancing Water for Humans and Nature*. Earthscan, London, UK. 247 pp.

Falkenmark, M.; Rockström, J. & Karlberg, L. (2009). Present and future water requirements for feeding humanity. *Food Security*, 1: 59–69.

FAO (2009a). The state of food insecurity in the world (SOFI), Economic crises – impacts and lessons learned. FAO, Rome, Italy. Available at: [http://www.fao.org/publications/sofi/en/].

FAO (2009b). The State of Agricultural Commodity Markets 2009. FAO, Rome, Italy. Available at: [http://www.fao.org/docrep/012/i0854e/i0854e00.htm].

FAOSTAT (2009). FBS (Food Balance Sheet) data for year 2005. [http://faostat.fao.org/default.aspx].

Gerten, D.; Schaphoff, S.; Haberlandt, U.; Lucht, W. & Sitch, S. (2004). Terrestrial vegetation and water balance: Hydrological evaluation of a dynamic global vegetation model. *Journal of Hydrology*, 286: 249–270. doi:10.1016/j.jhydrol.2003.09.029.

Hoekstra, A.Y. (2003). Virtual water trade. Proceedings of the International Expert Meeting on Virtual Water Trade. Value of Water Research Report Series No. 12 (Ed.: A.Y. Hoekstra). Delft, the Netherlands.

Hoekstra, A.Y. & Hung, P.Q. (2005). Globalisation of water resources: international virtual water flows in relation to crop trade. *Global Environmental Change-Human and Policy Dimensions*, 15: 45–56.

Keller, J.; Keller, A. & Davids, G. (1998). River basin development phases and implications of closure. *Journal of Applied Irrigation Science*, 33(2): 145–163.

Kendy, E.; Molden, D.; Steenhuis, T.S. & Liu, C. (2003). Policies drain the North China Plain – Agricultural policy and groundwater depletion in Luancheng County, 1949–2000. Research Report 71, International Water Management Institute, Colombo, Sri Lanka.

Lannerstad, M. (2002). *Consumptive water use feeds the world and makes rivers run dry*. MSc Thesis. Royal Institute of Technology (KTH), Stockholm, Sweden.

Lannerstad, M. (2009). *Water realities and development trajectories: global and local agricultural production dynamics*. Doctoral Thesis, Department of Water and Environmental Studies, Linköping University, Sweden.

Liu, J.; Zehnder, A.J.B. & Yang, H. (2009). Global consumptive water use for crop production: The importance of green water and virtual water. *Water Resources Research*, 45.

Lundqvist, J.; de Fraiture, C.; Molden, D.; Berndes, G.; Falkenmark, M.; Holmen, H.; Karlberg, L. & Lannerstad, M. (2008). *Saving water: From field to fork – Curbing losses and wastage in the food chain*. SIWI Policy Brief, Stockholm, Sweden.

Moench, M.; Burke, J. & Moench, Y. (2003). *Rethinking the approach to groundwater and food security*. FAO Water Report 24. FAO, Rome, Italy.

Molden, D. (ed.) (2007). *Water for food, water for life: a comprehensive assessment of water management in agriculture*. Earthscan, London, UK.

Molden, D.; Sakthivadivel, R.; Samad, M. & Burton, M. (2005). Phases of river basin development: the need for adaptive institutions. In: Svendsen, M. (ed.), *Irrigation and River Basin Management: Options for Governance and Institutions*. IWMI, Colombo, Sri Lanka; CABI, Wallingford, UK: 19–29.

Molden, D.; Oweis, T.; Steduto, P.; Bindraban, P.; Hanjra, M.A. & Kijne, J. (2010). Improving agricultural water productivity: Between optimism and caution. *Agricultural Water Management*, 97: 528–535.

Molle, F.; Wester, P. & Hirsch, P. (2010). River basin closure: Processes, implications and responses. *Agricultural Water Management*, 97: 528–535.

Pimentel, D.; Houser, J.; Preiss, E.; White, O.; Fang. H.; Mesnik, L.; Barsky, T.; Tariche, S.; Schreck, J. & Alpert, S. (1997). Water resources: Agriculture, the environment, and society. An assessment of the status of water resources. *Bio Science*, 47: 97–106.

Postel, S. (1999). *Pillar of sand: Can the irrigation miracle last?* W.W. Norton & Co., New York, USA.

Rockström, J. (1997). *On farm agrohydrological analysis of the Sahelian yield crisis: Partitioning, soil nutrients, and water use efficiency of pearl millet*. PhD Thesis, Department of Systems Ecology, Stockholm University, Sweden.

Rockström, J. (2003). Water for Food and Nature in the Tropics: Vapour Shift in Rainfed Agriculture. *Philosophical Transactions of the Royal Society Biological Sciences*, 358: 1997–2009. DOI: 10.1098/rstb.2003.1400. Available at: [http://rstb.royalsocietypublishing.org/content/358/1440/1997.full.pdf+html].

Rockström, J.; Gordon, L.; Folke, C. Falkenmark, M. & Engwall, M. (1999). *Conservation Ecology*, 3(2): 5. Available at: [http://www.consecol.org/vol3/iss2/art5/].

Rockström, J.; Axberg, G.A.; Falkenmark, M.; Lannerstad, M.; Rosemarin, A.; Caldwell, I.; Arvidson, A. & Nordström, M. (2005). *Sustainable pathways to attain the Millennium Development Goals – Assessing the role of water, energy and sanitation*. SEI, Stockholm, Sweden: 114 pp.

Rockström, J.; Lannerstad, M. & Falkenmark, M. (2007). Assessing the water challenge of a new green revolution in developing countries. *Proceedings of the National Academy of Sciences of the United States of America*, 104: 6253–6260.

Rockström, J.; Falkenmark, M.; Karlberg, L.; Hoff, H.; Rost, S. & Gerten, D. (2009). Future water availability for global food production: The potential of green water for increasing resilience to global change. *Water Resources Research*, 45.

Rockström, J.; Karlberg, L. & Falkenmark, M. (2010). *Global food production in a water-constrained world: Exploring green and blue challenges and solutions*. Cambridge University Press, UK.

Rosegrant, M.W.; Cai, X. & Cline, S.A. (2002). *World Water and Food to 2025*. International Food Policy Research Institute (IFPRI), Washington, D.C., USA.

Rosling, H. (2010). UN statistics over time visualized at: [http://www.gapminder.org/].

Rost, S.; Gerten, D.; Bondeau, A.; Lucht, W.; Rohwer, J. & Schaphoff, S. (2008). Agricultural green and blue water consumption and its influence on the global water system.*Water Resources Research*, 44, W09405. doi:10.1029/2007WR006331.

Sachs, J.D. (2008). *Common wealth. Economics for a crowded planet*. Penguin Books, London, UK.

Sachs, J.D. (2009). Transgressing Planetary Boundaries. *Scientific American*, 301: 36.

Seckler, D. (1992). The Sardar Sarovar Project in India: A commentary on the report of the independent review. Water Resources and Irrigation Division, Discussion Paper 8. Winrock International, Arlington, Virginia, USA.

Seckler, D. (1996). *The new era of water resources management: from* dry *to* wet *water savings*. IWMI Research Report 1. IWMI, Colombo, Sri Lanka.

Sitch, S.; Smith, B.; Prentice, I. C.; Arneth, A.; Bondeau, A.; Cramer, W.; Kaplan, J.O.; Levis, S.; Lucht, W.; Sykes, M.T.; Thonicke, K. & Venevsky, S. (2003). Evaluation of ecosystem dynamics, plant geography and terrestrial carbon cycling in the LPJ dynamic global vegetation model. *Global Change Biology*, 9(2): 161–185. doi:10.1046/j.1365-2486.2003.00569.x.

Smakhtin, V.; Revenga, C.M. & Döll, P. (2004). A pilot global assessment of environmental water requirements and scarcity. *Water International*, 29: 307–317.

Smil, V. (2000). *Feeding the world: A challenge for the twenty-first century*. MIT Press, Cambridge, Massachusetts, USA.

Steinfeld, H.; Gerber, P.; Wassenaar, T.; Castel, V.; Rosales, M. & de Haan, C. (2006). *Livestock's long shadow: environmental issues and options*. FAO, Rome, Italy.

Sulser, T.B.; Ringler, C.; Zhu, T.; Msangi, S.; Bryan, E. & Rosegrant, M.W. (2009). Green and blue water accounting in the Limpopo and Nile basins: Implications for food and agricultural policy. IFPRI Discussion Paper 00907. IFPRI, Washington, D.C., USA. 36 pp.

UN (United Nations) (2010). *World Population Prospects: The 2008 Revision*. Population Division of the Department of Economic and Social Affairs of the United Nations Secretariat. Available at: [http://esa.un.org/unpp].

UNESCO (2009). *Water in changing world: The United Nations world water development report 3*. UNESCO, Paris, France; Earthscan, London, UK.

Vidal, A.; van Koppen, B. & Blake, D. (2010). The Green-to-Blue Water Continuum: An approach to improve agricultural systems' resilience to water scarcity. Manuscript.

von Braun, J. & Meinzen-Dick, R. (2009). Land Grabbing by Foreign Investors in Developing Countries: Risks and Opportunities. IFPRI Policy Brief, 13. IFPRI, Washington, D.C., USA.

WB (World Bank) (2005). World Bank, Data and Statistics. Data from 2005. Accessed in October, 2008.

WB (World Bank) (2009) Data and statistics: country classification. Available at: [http://go.worldbank.org/K2CKM78CC0].

Yang, H.; Wang, L.; Abbaspour, K.C. & Zehnder, A.J.B. (2006). Virtual water trade: an assessment of water use efficiency in the international food trade. *Hydrology and Earth System Sciences*, 10: 443–454.

CHAPTER 2

Global agricultural green and blue water consumptive uses and virtual water trade

Junguo Liu
School of Nature Conservation, Beijing Forestry University, Beijing, China

Hong Yang
Swiss Federal Institute of Aquatic Science and Technology (Eawag), Duebendorf, Switzerland

ABSTRACT: An accurate estimate of global water uses with high spatial resolution is crucial for assessing global water scarcity and for understanding human's interference with the ecosystems. As agricultural production is the single largest water user in most of the areas of the world, it is especially important to have a spatially explicit assessment of both green and blue water uses in agriculture and virtual water flows associated with international food trade. In this chapter, we estimated consumptive water use (CWU) in cropland on a global scale with a spatial resolution of 30 arc-minutes. A GIS-based version of the EPIC model, GEPIC, is used for the estimation. The results show that in crop growing periods, global CWU was 5,938 km^3/yr in cropland around the year 2000, of which green water contributed to 84%. On an annual basis, global CWU was 7,323 km^3/yr in cropland, of which green water contributed to 87%. Almost 95% of the world crop related virtual water trade has its origin in green water. High levels of net virtual water import (NVWI) generally occur in countries with low CWU on a per capita basis. A virtual water trade strategy could be an attractive water management option to compensate for domestic water shortage for food production for these countries. However, NVWI is constrained by income. Low-income countries generally have a low level of NVWI. Strengthening low-income countries economically will allow them to develop a virtual water trade strategy to mitigate malnutrition of their people.

Keywords: water scarcity, consumptive water use, virtual water trade, GEPIC

1 INTRODUCTION

Agriculture is the single largest water user amongst all the economic sectors in most of the countries in the world. There is an intrinsic linkage between water availability, food production and food trade. The virtual water concept, emerged in the mid 1990s, specifically addresses this linkage from the perspective of food trade as a means of redistributing global water resources and as a possible policy option in managing local water resources, particularly in water scarce countries and regions (Allan, 1994). This study takes a green and blue water perspective in the investigation of water-food-trade relations across different geographical regions and on a high spatial resolution with a global coverage.

The concept of green water was first introduced by Falkenmark (1995), referring to the total crop evaporation during crop growth. Later, green water resource has been generally used to refer to the water that comes from precipitation, is stored in the soil, and subsequently released to the atmosphere through crop evaporation. In contrast, blue water refers to the water in rivers, lakes, reservoirs, ponds and aquifers. Both green and blue water resources are important for food production. Rainfed agriculture uses green water only, while irrigated agriculture uses both green and blue water.

In many water scarce countries, an increasing amount of food is imported to meet domestic food demand (Yang *et al*., 2006). For these countries, importing food is equivalent to importing virtual water to mitigate the physical lack of water for domestic food production. With population growth and economic development, water resources are under pressure in an increasing number of countries. Unravelling the relationship between a country's consumptive water use (CWU) in crop production and virtual water trade can improve the understanding of water-food-trade relationship, and help formulate appropriate policies to deal with water scarcity.

In this study, we quantify CWU in crop production and investigate CWU–virtual water trade relations. CWU at the global level is assessed with a spatial resolution of 30 arc-minutes (about 50 km × 50 km in each grid near the equator). Special attention is given to the green water component of CWU for the production of 22 major crops. Virtual water trade is quantified for each crop and is summed up as a common yardstick in investigating CWU–virtual water trade relations. The green water proportion in both domestic crop production and virtual water trade is calculated, and the virtual water trade and CWU for low-, middle- and high-income countries are examined.

2 CALCULATION OF CROP YIELD, EVAPORATION AND CROP WATER PRODUCTIVITY

2.1 *Methodology*

A GEPIC model is used to simulate crop yield, CWU (defined as evapotransporation, ET, in this study), and crop water productivity (CWP) for individual crops in each grid cell at the spatial resolution of 30 arc-minutes covering the entire world. The GEPIC model is a GIS-based EPIC model designed to simulate the spatial and temporal dynamics of the major processes of the soil-crop-atmosphere-management system (Liu *et al*., 2007a; Liu *et al*., 2007b; Liu *et al*., 2008). CWU in a country is calculated as the sum of the CWU of all grid cells within this country. CWP is calculated as the ratio of crop yield to ET.

In this study, CWU refers to the total amount of water consumed by crops in terms of evapotranspiration. In each grid cell, CWU is calculated as:

$$CWU = CWU_r + CWU_i \tag{1}$$

$$CWU_r = \sum_c CWU_r^c = 10 \times \sum_c (ET_r^c \times A_r^c) \tag{2}$$

$$CWU_i = \sum_c CWU_i^c = 10 \times \sum_c (ET_i^c \times A_i^c) \tag{3}$$

Where CWU is consumptive water use in m^3/yr in one grid cell, subscript r and i refer to rainfed and irrigated agricultural systems, respectively. The subscript c represents the crop code. ET is evapotranspiration of crop c under rainfed conditions (r) or irrigated conditions (i) in mm/yr, while A is area of crop c under rainfed or irrigated conditions in ha. The constant 10 converts mm into m^3/ha.

For rainfed crops, CWU_r is all from green water. For irrigated crops, CWU_i is partly from green water and partly from blue water. In order to estimate the proportion of green and blue water uses in irrigated agriculture, two different soil water balances are performed for irrigated crops according to FAO (2005).

1) Soil water balance I is carried out by assuming that the soil does not receive any irrigation water. Seasonal evapotranspiration computed using this soil water balance is referred to as *SET1*.
2) Soil water balance II is carried out by assuming the soil receives irrigation water. Seasonal evapotranspiration computed using this soil water balance is referred to as *SET2*.

For a specific crop under irrigated conditions, according to FAO (2005), green water use is equal to *SET1*, while blue water use is equal to the difference between *SET2* and *SET1*, or *SET2-SET1* in crop growing periods. Hence, for a specific crop under irrigated conditions, the proportion of blue water in crop growing periods is calculated as (Liu *et al.*, 2009; Liu & Yang, 2010):

$$b_i^c = \frac{SET2^c - SET1^c}{SET2^c} \tag{4}$$

Where b is the blue water proportion of crop c under irrigated conditions i.

In a grid cell, consumptive blue water use (*CBWU*) for all crops can be estimated as:

$$CBWU = \sum_c (b_i^c \times CWU_i^c) \tag{5}$$

The blue water proportion (B) and green water proportion (G) in each grid cell are calculated as follows:

$$B = \frac{CBWU}{CWU} \tag{6}$$

$$G = 1 - B \tag{7}$$

With *Eq. 6*, blue water proportion in both crop growing periods and in the entire year is calculated. It is assumed that irrigation is not applied in non-growing periods. Hence, *CBWU* remains the same for the crop growing periods and the entire year. *CWU* during the growing periods differs from that during the entire year, leading to different blue water proportion in the two calculations.

2.2 *High-resolution data of harvested area*

In this study, 22 crop categories in rainfed and irrigated areas are considered (Table 1). The four crop types ("citrus", "date palm", "grapes/vine", and "others perennial") are combined into one category "fruits". Grape is the most planted fruit in terms of harvested area (FAO, 2006); hence, it is used as a representative crop for the simulation of fruits by the GEPIC model. Similarly, tomato is the most planted vegetable in terms of harvested area (FAO, 2006), and it is selected as a representative crop type for the simulation of vegetables. Fruits and vegetables only account for 3.7% and 3.4% of the total cropland (Ramankutty *et al.*, 2008). They account for 6.0% and 6.4% of the total irrigated cropland. Hence, the use of representative crops will not significantly affect the simulation results of CWU. For the assessment of virtual water trade, data for trade quantities of some minor crops are not always available for the period the study covers. Hence, only 17 most important crops (marked with * in Table 1) are considered in the calculation of virtual water trade.

Two data sources are used in this study for the harvest area of crops. One source is the Center for Sustainability and the Global Environment (SAGE) of the University of Wisconsin at Madison, USA (Ramankutty *et al.*, 2008). The SAGE dataset provides harvested area of 175 primary crops in the year 2000 with spatial resolutions of 30 arc-minutes. In this dataset, the harvested area is the sum of the rainfed and irrigated crop area. Another source is the Institute of Physical Geography of the University of Frankfurt (Main), Germany (hereafter referred to as "MIRCA2000 dataset"). The MIRCA2000 dataset provides harvested area of 26 irrigated crops around 2000 with a spatial resolution of 30 arc-minutes (Portmann *et al.*, 2008). The harvested area of these irrigated crops are calculated mainly based on the SAGE harvested area data and the global map of irrigated areas (Siebert *et al.*, 2007). The harvested area (rainfed plus irrigated) of the 26 crops is also integrated from the SAGE dataset by Portmann *et al.* (2008). For these crops, the harvested area of a rainfed crop is assumed to be the difference between the harvested area and the irrigated harvested area of the corresponding crop in each grid cell. In case that the irrigated harvested area is higher than the total harvested area, we assume there is no rainfed harvested area for the corresponding crop.

Table 1. The 22 crop categories used in this study.

Crop category in this study	Corresponding crop category in MIRCA 2000	Representative crop for simulation	Potential heat unit (°C)
Wheat*	Wheat	Wheat	1,750
Maize*	Maize	Maize	1,000
Rice*	Rice	Rice	1,500
Barley*	Barley	Barley	1,000
Rye*	Rye	Rye	1,750
Millet*	Millet	Millet	1,500
Sorghum*	Sorghum	Sorghum	1,500
Soybeans*	Soybeans	Soybean	1,800
Sunflower*	Sunflower	Sunflower	1,500
Potatoes*	Potatoes	Potato	1,500
Cassava*	Cassava	Cassava	1,500
Sugar cane*	Sugar cane	Sugar cane	1,500
Sugar beets*	Sugar beets	Sugar beet	1,500
Oil palm	Oil palm	–	–
Rapeseed/canola*	Rapeseed/canola	Rapeseed	1,500
Groundnuts/peanuts*	Groundnuts/peanuts	Groundnut	1,500
Cotton*	Cotton	Cotton	1,500
Pulses*	Pulses	Peas	1,600
Coffee and cocoa	Coffee, cocoa	Coffee	1,700
Fruits	Citrus, date palm, grapes/vine, others perennial	Grape	2,223
Vegetables	Others annual	Tomato	1,700
Managed grassland/pasture	Managed grassland/pasture	Pasture	2,000

* The 17 major crops

3 CWU AND GREEN WATER PROPORTION

3.1 *CWU of crops*

Spatial patterns of CWU during the growing period are demonstrated in Figure 1a. The global CWU in crop growing periods was 5,938 km^3/yr in cropland around the year 2000 (the average of the years 1998–2002). The highest CWU per grid cell (e.g. > 400 Mm^3/yr) was found in most part of India, Eastern part of China, some countries in Southeast Asia such as Indonesia, Mid Central of the USA, part of Argentina and Brazil, and very few countries in Africa (e.g. Nigeria, Ghana, and Ivory Coast). These regions represent the most intensive agricultural production area in the world. In Europe, CWU in most grid cells is between 300 and 400 Mm^3/yr. In other parts of the world, CWU was generally lower than 100 Mm^3/yr.

Spatial pattern of annual CWU in the entire year is similar to that of CWU in the crop growing periods (Figure 1b). The highest CWU per grid cell (e.g. >300 Mm^3/yr) was found in most part of India, in the river basins of the Yellow River, the Huai River, the Hai River, and the Yangtze in China, in the Mississippi river basin in North America, and some part of the Parana and Sao Francisco river basins in South America. These regions mainly contained grid cells with high fraction of arable lands and permanent crops. At the global level, CWU was 7,323 km^3/yr in cropland around the year 2000. This means that 81% of the annual CWU was used in the crop growing periods, while the remaining 19% occurred in the non-growing periods. At the river basin level, Mississippi, Yangtze, Ganges and Nile are the four river basins with the highest CWU both during the growing periods and for the entire year (Table 2). These four river basins account for around 20% of the global CWU.

Of the global CWU of 5,938 km^3/yr, over two thirds can be attributed to cereal crops. Wheat and rice account for two thirds of the CWU of cereal crops.

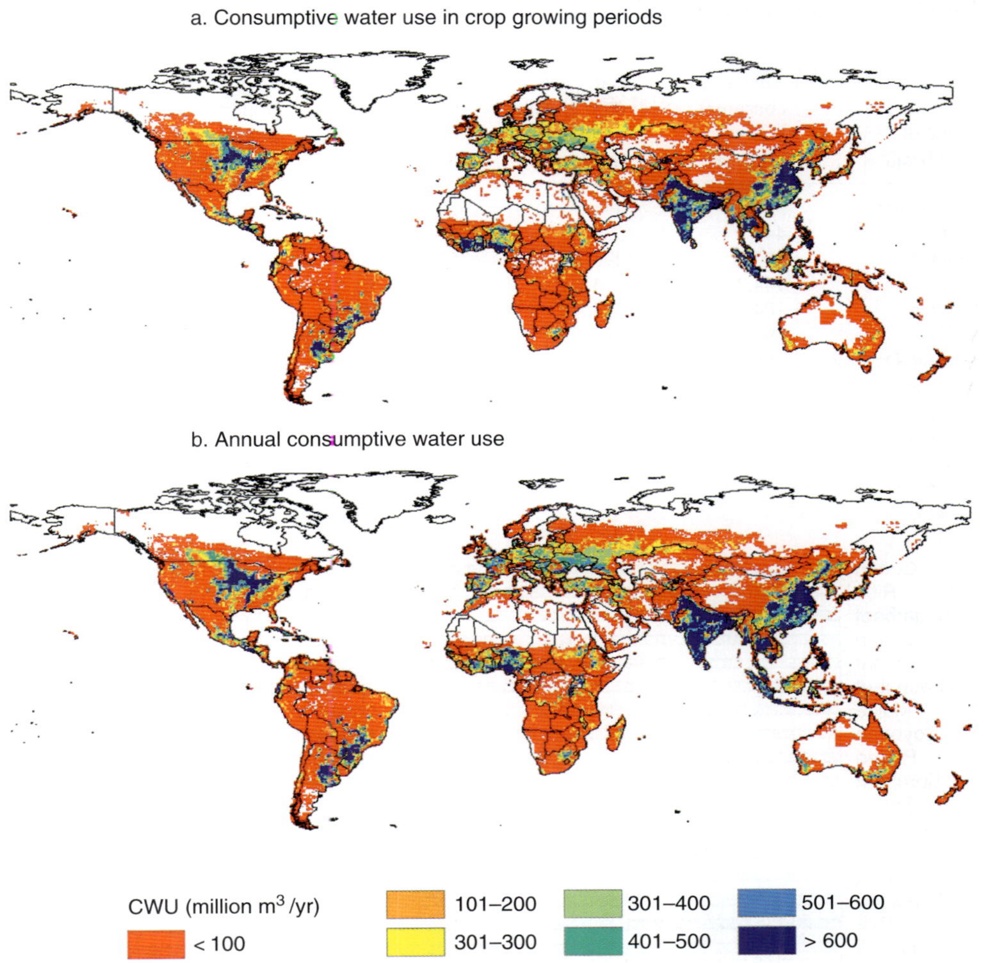

Figure 1. Spatial distribution of consumptive water use (CWU) for crop production per grid cell of 30 arc-minutes (average over 1998–2002).

There have been several estimates of global CWU in the literature. These estimates range from around 3,500 km^3/yr (Zehnder, 1997) to 7,400 km^3/yr (Postel *et al.*, 1996), depending on the land types and the methods used for the estimation. Postel *et al.* (1996) provided a CWU value of 7,370 km^3/yr in cultivated land in 1990. Cultivated land area refers to arable land and land under permanent crops. Cultivated land area is almost equal to cropland area; hence, the above estimate can be regarded as CWU for cropland. The estimation by Postel *et al.* (1996) is very rough with several strong assumptions. Rockström *et al.* (1999) calculated global CWU at 6,800 km^3/yr for the period 1992–1996 by using crop production and crop water productivity of 18 crop groups, with differentiation of tropical and temperate climate zones. Crop water productivity of various crop groups was based on extensive literature review. Chapagain & Hoekstra (2004) calculated the global CWU as 6,390 km^3/yr for 164 crops based on national average crop production and national average crop water productivity (CWP) for the period 1997–2001. This estimate considered crop production and crop water productivity in individual countries, but it did not take into account the variations within a country.

Table 2. Consumptive water use and blue water proportion in major river basins (average over 1998–2002).

Name of river basin	Annual consumptive water use (km^3/yr)	Consumptive water use in crop growing periods (km^3/yr)	Consumptive blue water use (km^3/yr)	Blue water proportion in the entire year (%)	Blue water proportion in crop growing periods (%)
Mississippi	538.3	445.6	58.4	10.8	13.1
Yangtze	441.7	338.5	65.0	14.7	19.2
Ganges	407.0	296.7	57.0	14.0	19.2
Nile	144.2	114.1	19.1	13.2	16.7
Danube	104.2	77.8	2.0	1.9	2.6
Yellow	94.7	73.3	24.5	25.9	33.4
Murray–Darling	56.1	37.0	9.3	16.6	25.1
Amazon	55.6	48.8	1.4	2.5	2.8
Orange	25.4	19.5	1.4	5.3	6.9
Mackenzie	7.2	5.5	0.002	0.0	0.0
Lena	0.24	0.19	0.043	17.9	22.6

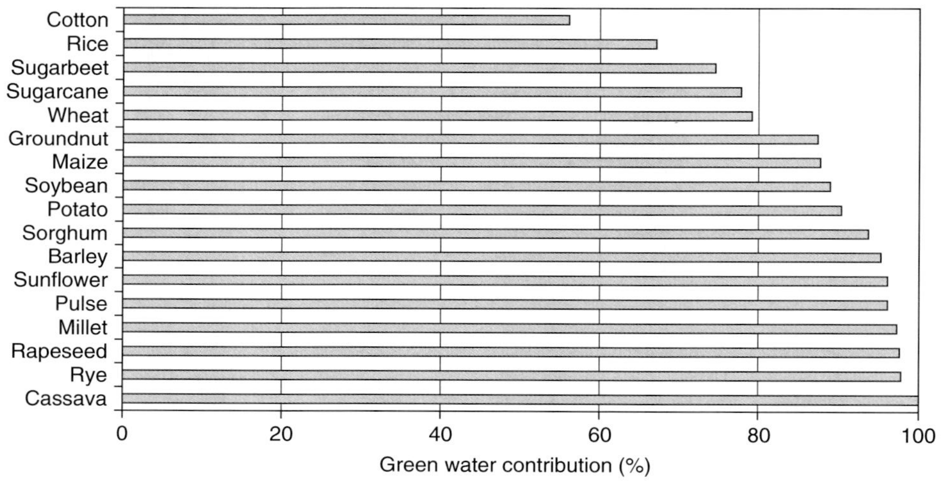

Figure 2. Global average green water portion for individual crops (average over 1998–2002).

3.2 *Green water proportion in CWU*

On global average, green water accounts for 84% of CWU during the crop growing period. For most crops, green water accounts for more than two thirds of the consumptive water use except for cotton (Figure 2). Cotton has the green water proportion of 56%, the lowest among all crops. This portion closes to the value of 48% reported by Chapagain *et al.* (2006). Globally, about 73% of the cotton production is from irrigated fields. The main cotton producers are arid regions such as Egypt, Uzbekistan, Pakistan, and Northwest China. Rice has a green water proportion of 67%, which represents the lowest green water proportion next to cotton. Cassava is a highly drought tolerant crop, and thus is less dependent on irrigation. With almost 100%, it has the highest green water proportion of all crops.

At the national levels, agricultural production greatly depends on green water (Figure 3). In Canada, Brazil, Argentina, many African and European countries, and Australia (90%), no less

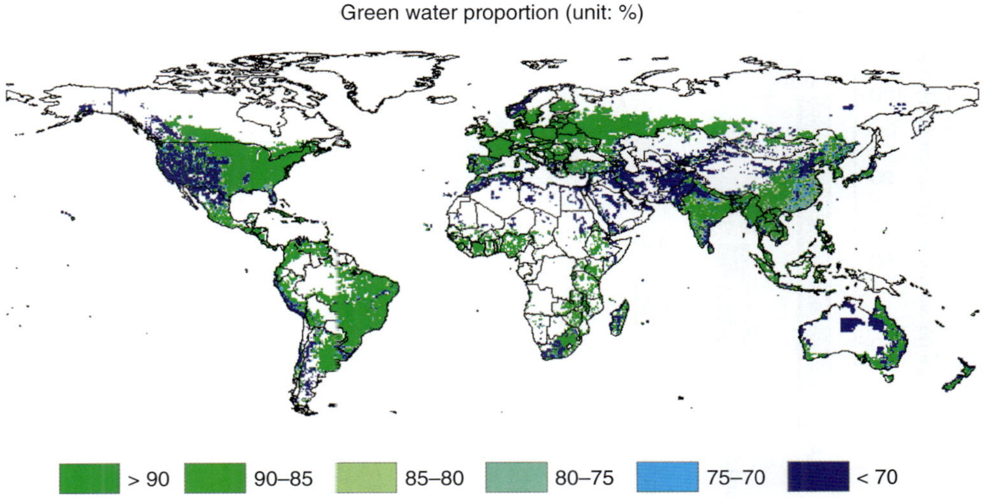

Figure 3. Green water proportion at the grid cell level (average over 1998–2002).

than 90% of CWU has its origin in green water. Arid and warm zones such as many countries in the MENA region show a low green water proportion. About 80% of the MENA region has annual rainfalls of less than 100 mm/yr. The lack of rainfall coupled with high evaporation makes irrigation crucially important for agriculture production.

The consumptive blue water use (CBWU) was 927 km^3/yr in cropland on a global scale around the year 2000 for the 22 crops considered. During crop growing period, blue water accounts for 16% of the global CWU, while green water accounts for 84%. On an annual basis, the figures are 13% and 87% for blue water and green water, respectively. High CBWU (or low CGWU) occurs in Northern and Southern India, Eastern part of China, and the Mid Central of the USA (Figure 3). These regions are the major agricultural production regions in the world, and they also have very high CWU. When irrigation infrastructure exists, these regions often use a large volume of blue water, mainly due to the large agricultural area there. As for the blue water proportion, regions with high values are located in the Northern part of China, several West Asian countries, Middle East and North Africa, the Western part of the USA, and Chile. These regions mostly have arid or semi-arid climate with low precipitation. Precipitation can only meet part of the water required by crops. In order to achieve high crop yields, irrigation water has to be supplied in addition to precipitation. Largely due to the low precipitation, irrigation depth is generally very high, resulting in high blue water proportion.

At the river basin level, the Yellow River, Lena and Murray-Darling river basins had the highest blue water proportion (Table 2). These river basins are located in arid or semi-arid climates with low precipitation. For example, the mean annual rainfall in the Yellow river basin is 452 mm/yr. Meanwhile, this river basin is an important food producing region in China (Yang & Jia, 2008), and almost three fourths of the population lived in rural areas of the basin in 2000. Irrigation is vital to maintain high agricultural productivity, leading to relatively higher blue water proportion compared to other river basins. In contrast, the Mackenzie, Danube and Amazon river basins had the lowest blue water proportions (Table 2).

4 RELATIONS BETWEEN CWU AND VIRTUAL WATER TRADE

The global virtual water export is dominated by few countries. Ten major virtual water exporting countries shown in Figure 4 account for 94% of global total virtual water export. On global average,

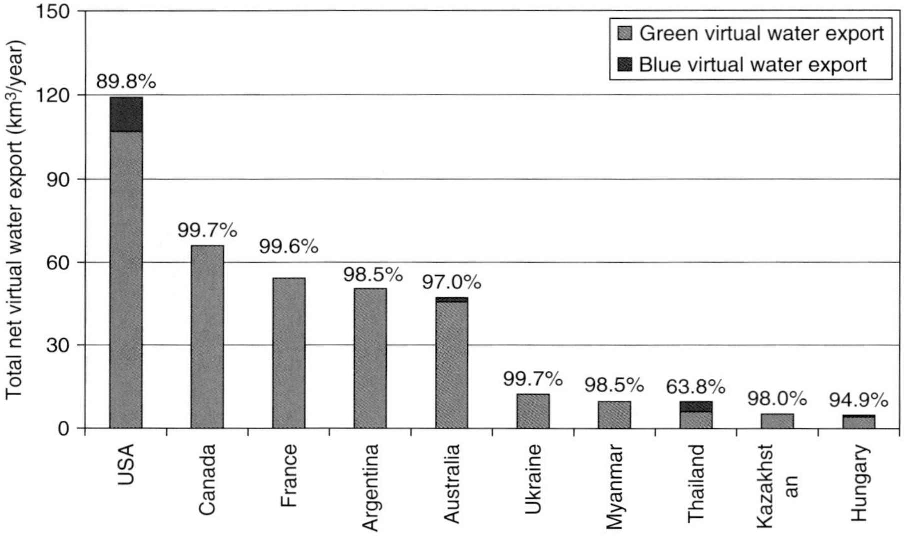

Figure 4. Total net blue and green virtual water export in major exporting countries, and green water proportion in total virtual water export (17 crops, average over 1998–2002). Green water proportions are marked as percentage above individual countries.

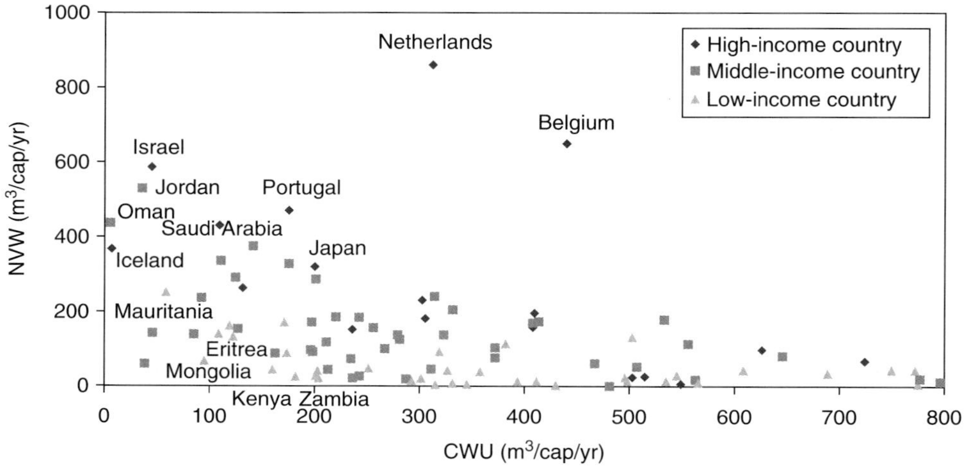

Figure 5. Relationship between net virtual water import (NVWI) and consumptive water use (CWU) in importing countries (17 crops, average over 1998–2002).

green water accounts for almost 95% of the global virtual water export. Hence, the international virtual water trade is dominantly *green*. With its almost 2,500 m^3/yr per capita, Australia is the most important net virtual water exporting country on a per capita basis. It is followed by Canada (2,137) and Argentina (1,372), and France, Paraguay, Hungary, the USA, and Denmark exporting between 350 and 900 m^3/yr per capita.

On the importing side, the Netherlands and Belgium are the two top net virtual water importing countries on a per capita basis. Net virtual water import (NVWI) into the two countries is about

860 and 650 m^3/yr per capita, respectively (Figure 5). Both countries are big meat exporters with over 120 kg/yr per capita, of meat exports. A large amount of imported crop products is used as feed for livestock (FAO, 2006).

Israel and Jordan are the third and fourth biggest importing countries with a NVWI of over 500 m^3/yr per capita. Both countries show a very low CWU of less than 50 m^3/yr per capita. Their NVWI is to compensate for the lack of water, a fact which is seen in all the MENA countries. Besides the 22 crops considered, Israel is importing also substantial amounts of meat and dairy products, which almost doubles the NVWI calculated here (Yang *et al.*, 2007).

Countries respond to CWU differently when CWU is below 250 m^3/yr per capita. NVWI is affected by the levels of incomes. High- and middle-income countries generally have larger NVWI than low-income countries with a similar level of CWU (Figure 5). For example, CWU in Japan and Zambia is about 200 m^3/yr per capita. The NVWI in Japan is over 300 m^3/yr per capita, while it is negligible in Zambia. Apparently, the economic situation of a country is decisive to satisfy internal nutritional needs.

For some low-income countries, NVWI remains at a low level even with a low CWU. This means that part of the population is undernourished or obtains their calories from other sources than the 17 major crops considered. For instance, in Eritrea, the sum of CWU and NVWI is 261 m^3/yr per capita. The calorie intakes from animal products and other vegetal foods are also low. In fact, this country is being confronted with serious food security problem and 73% of its population is undernourished.

The countries with large per capita NVWI are mainly located in the regions where poor climatic conditions do not allow large area of agricultural production (as a result, CWU is also low in these countries), e.g. the arid MENA region, and the low-temperature countries, e.g. Northern Europe and Mongolia. Particularly in the MENA countries, the current NVWI already reaches or even exceeds combined green and blue water uses in domestic agriculture. Virtual water imports play a vital role in mitigating the regional water scarcity and in guaranteeing the regional food security. Given the strong agreement among climate models for less precipitation in the future in the MENA region, virtual water trade will become more important for the regional water and food security.

5 CONCLUDING REMARKS

We quantified CWU in cropland in a spatially explicit way by taking into account both green and blue water components. The results show that the global CWU was 5,938 km^3/yr in the crop growing periods and 7,323 km^3/yr in the entire year in cropland around the year 2000. Green water contributed to 84% of the global CWU in the crop growing periods, and 87% of the global CWU on an annual basis. The high proportion of green water was in part due to the dominance of rainfed agriculture, which consumed 4,068 km^3/yr of water in the crop growing periods and 5,105 km^3/yr of water in the entire year. In addition, in irrigated cropland, green water contributed to 50% of the total CWU in the crop growing periods, and over 60% of the annual total CWU.

The important role of green water in crop production gives rise to the need for a better management of this water resource. However, in the past, water engineers and managers have mainly focused on expansion of irrigation infrastructure, particularly in many Asian countries. There is a general lack of green water management. Nowadays, further developing irrigation infrastructure becomes more and more difficult. There is not much potential to build large dams in most countries because water projects have been developed in the most suitable locations. Against this background, improving green water management should be an important option to guarantee world food security in the future.

Around 94% of the world crop-related virtual water trade has its origin in green water, which generally constitutes a low-opportunity cost of green water as opposed to blue water. High levels of net virtual water import (NVWI) generally occur in countries with low CWU on a per capita basis, where a virtual water strategy is an attractive water management option to compensate for domestic water shortage for food production. NVWI is constrained by income; low-income countries generally have a low level of NVWI. Strengthening low-income countries economically will allow them to develop a virtual water strategy to mitigate malnutrition of their people.

REFERENCES

Allan, J.A. (1994). Overall perspectives on countries and regions. In: P. Rogers & P. Lydon (eds.), *Water in the Arab World: perspectives and prognoses*. Harvard University Press, Cambridge, Massachusetts, USA: 65–100.

Chapagain, A.K. & Hoekstra, A.Y. (2004). *Water footprints of nations*. UNESCO-IHE, Delft, the Netherlands.

Chapagain, A.K.; Hoekstra, A.Y.; Savenije, H.H.G. & Gautam, R. (2006). The water footprint of cotton consumption: An assessment of the impact of worldwide consumption of cotton products on the water resources in the cotton producing countries. *Ecological Economics*, 60(1): 186–203.

Falkenmark, M. (1995). Coping with water scarcity under rapid population growth. Paper presented at Conference of SADC Ministers, Pretoria, South Africa, 23–24 November 1995.

FAO (2005). *Review of agricultural water use per country*. Food and Agriculture Organization of the United Nations, Rome, Italy.

FAO (2006). FAOSTAT: FAO statistical databases. Food and Agriculture Organization of the United Nations, Rome, Italy. [Available at http://faostat.fao.org/default.aspx].

Liu, J. & Yang, H. (2010). Spatial explicit assessment of global consumptive water uses in cropland: green and blue water. *Journal of Hydrology*, 384: 187–197.

Liu, J.; Wiberg, D.; Zehnder, A.J.B. & Yang, H. (2007a). Modelling the role of irrigation in winter wheat yield, crop water productivity, and production in China, *Irrigation Science*, 26(1): 21–33.

Liu, J.; Williams, J.R.; Zehnder, A.J.B. & Yang, H. (2007b). GEPIC – modelling wheat yield and crop water productivity with high resolution on a global scale. *Agricultural Systems*, 94(2): 478–493.

Liu, J.; Fritz, S.; van Wesenbeeck, C.F.A.; Fuchs, M.; You, L.; Obersteiner, M. & Yang, H. (2008). A spatially explicit assessment of current and future hotspots of hunger in Sub-Saharan Africa in the context of global change, *Global Planetary Change*, 64(3–4): 222–235.

Liu, J.; Zehnder, A.J.B. & Yang, H. (2009). Global consumptive water use for crop production: the importance of green water and virtual water. *Water Resources Research*, 45: W05428

Portmann, F.; Siebert, S.; Bauer, C. & Döll, P. (2008). Global data set of monthly growing areas of 26 irrigated crops, version 1.0. Institute of Physical Geography, University of Frankfurt (Main), Frankfurt, Germany.

Postel, S.L.; Daily, G.C. & Ehrlich, P.R. (1996). Human appropriation of renewable fresh water. Science, 271(5250): 785–788.

Ramankutty, N.; Evan, A.T.; Monfreda, C. & Foley, J.A. (2008). Farming the planet: 1. Geographic distribution of global agricultural lands in the year 2000. *Global Biogeochemical Cycles*, 22(1).

Rockström, J.; Gordon, L.; Folke, C.; Falkenmark, M. & Engwall, M. (1999). Linkages among water vapor flows, food production, and terrestrial ecosystem services. *Conservation Ecology*, 3(2): 5.

Siebert, S.; Döll, P.; Feick, S.; Frenken, K. & Hoogeveen, J. (2007). Global map of irrigation areas, version 4.0.1. University of Frankfurt, Main, Germany; and FAO, Rome, Italy.

Yang, H. & Jia, S. (2008). Meeting the basin closure of the Yellow river in China. *International Journal of Water Resources Development*, 24(2): 265.

Yang, H.; Wang, L.; Abbaspour, K.C. & Zehnder, A.J.B. (2006). Virtual water trade: an assessment of water use efficiency in the international food trade. *Hydrol Earth Syst Sc*, 10: 443–454.

Yang, H.; Wang, L. & Zehnder, A.J.B. (2007). Water scarcity and food trade in the Southern and Eastern Mediterranean countries. *Food Policy*, 32(5–6): 585–605.

Zehnder, A.J.B. (1997). Water – a commodity in short supply. *Eawag News*, 43: 3–4.

CHAPTER 3

Water scarcity and food security: A global assessment of water potentiality in Tunisia

Jamel Chahed & Mustapha Besbes
National School of Engineers of Tunis, Tunisia

Abdelkader Hamdane
National Agronomic Institute of Tunis, Tunisia

ABSTRACT: In arid countries, where water resources are scarce, the various aspects related to water management and water uses are interlinked at the national scale. Tunisia is fairly advanced in water resource planning and management and its scarce hydraulic resources are almost entirely mobilized. The country is therefore obliged to apply new concepts, new paradigms, to optimize the use of different types of water resources. A comprehensive model for water balance of Tunisia has been developed and adjusted from data at the national scale. This model takes into account all of water resources: the withdrawal water *Blue Water*, the Equivalent-Water of the rainfed agriculture *Green Water*, and the net contribution in Equivalent-Water of the import-export food balance *Virtual Water*. The model is used in order to simulate three scenarios for the prospective horizons 2025 and 2050. The first two scenarios are based on a traditional vision of the water resource which considers only the withdrawal water management *Blue Water*. The third scenario considers the total water resource potential involved in food production. The simulations indicate that the improvement of food safety will depend, in the future, on the capacity to manage all the available water resources, in particular by improving the potential of rainfed agriculture.

Keywords: water resources, water management, water balance, green water, blue water, virtual water, Tunisia

1 INTRODUCTION

The recent increases in prices of basic food products (cereals, rice, vegetable oil, etc.) have revived the debate on the sensitive issue of food security in importing countries. These countries suffer the effects of these increases on their trade balances that generate, in the poorest countries, real food crises. This situation challenges traditional patterns of agricultural development and food security. In fact, the spectacular results of the green revolution and its impact on agricultural production have led to a widespread acceptance, which is to admit that the use of intensive irrigated agriculture is essential to meet the food challenge.

In arid and semi-arid countries the scarcity of water resources is a highly limiting factor for increased food production. When water resources are limiting agricultural production, food importation seems to be a way which is used consciously or unconsciously to fill the water deficit. One speaks about *Virtual Water*. This concept means that the importation of foodstuffs is similar to importing an amount of water equivalent to the volume required to produce it locally (Allan, 1998). In general, the significant contribution of *Virtual Water* is not directly taken into account in water resources planning and is not accounted for characterizing water stress situations.

On the other hand, the Equivalent-Water of the agricultural production related to rainfed crops corresponds to the volume extracted by plants from water contained in the soil (*Green Water*). The amount of *Green Water*, often much more important than water used in irrigated agriculture, takes an important part in *Virtual Water* trade (cereals, oil, etc.) and its contribution is often decisive. Liu (2007) estimated that over 80% of the water involved in agricultural production comes from rainfed agriculture, which accounts for 90% to foodstuffs trade. The quantification of the trade in *Virtual Water*, taking care to distinguish between the water that comes from irrigated agriculture and trade from rainfed agriculture, is very instructive. It helps to clarify the relationship between agricultural and water policies and their impacts on the management and use of water resources.

Before progressing in the analysis, we must first note that the direct needs (urban, industrial and tourism) are generally moderate. Some authors like Gleick (1996) and Warner (1995), have tried to quantify the *Basic Water Requirements* (BWRs). BWRs define the minimum quantities of water needed to cover the basic water uses of a person: drinking, food preparation and hygiene. All these uses account for only relatively small water quantities estimated at 18 m^3/hab/yr (Gleick, 1996). The needs for basic water represent only a small part (few percent) of the overall water demand. In contrast, the part of water demand involved in foodstuffs production is relatively high.

Many recent works have been devoted to the evaluation of the quantity of water required in foodstuffs production (Hoekstra, 2003; Oki *et al.*, 2003; Renault & Wallender, 2000). In spite of their disparities, data resulting from research provide edifying information on the relation between water and foodstuffs production. It appears that the amount of water needed for food production depends greatly on the nutritional mode. Disregarding the origin of water used in agriculture, the production of 1 kg of cereals needs approximately 1 m^3 of water, while more than 20 m^3 are necessary to produce 1 kg of beef (Oki *et al.*, 2003). As a result, the water needed for food production is about 2.5 m^3/day per capita for a diet with low animal product intake, e.g. in North Africa; it exceeds 5 m^3/day per capita for a diet with high animal product intake such as in Europe or in the USA (Renault & Wallender, 2000).

The important quantities of water involved in food production indicate that a full understanding of water issues should consider the structure of agricultural production. Adequacy or shortage of water resources depends mainly on the role that society assigns to irrigation and the place it is supposed to play in development policies in general and in agricultural policies in particular. The concept of the *Water Footprint* introduced in the early 2000s suggested a comprehensive review to assess the potential of all water resources. The annual freshwater availability, which corresponds to the sum of *Green Water* availability and *Blue Water* availability, is equal to the total precipitation above land. This includes all resources used by the ecosystem (Besbes *et al.*, 2002; Hoekstra & Hung, 2002; Chapagain & Hoekstra, 2004; Hoekstra & Chapagain, 2007).

In an analysis of water resources through the globe, De Marsily (2006) specified the significant contribution of international food trade in water supply-demand adequacy, and drew attention to the important role of rainfed agriculture in food safety. This new presentation of water resources should be used today for planning water resources development and allocation in arid countries by explicitly taking into account all kinds of water (*Green, Blue* and *Virtual Water*) and all the current needs.

In previous papers we proposed a comprehensive water balance for Tunisia (Chahed *et al.*, 2007; 2008). In this balance one considers all kinds of water resources, e.g. *Blue Water*, *Green Water* as well as the contribution of the agro-alimentary trade balance *Virtual Water*. The adjustment of this balance has led to develop a model that compares the total of water demand in Tunisia with the potential of all water and soil resources.

Based on a holistic water vision, we developed a first prospective investigation for the horizon 2025 using a global water resource model (Besbes *et al.*, 2007; Besbes *et al.*, 2010). We propose, in the present work, to use this model in order to simulate scenarios for two prospective horizons (2025 and 2050). In these scenarios we attempt to assess the impacts of changing water needs, including food demand. These scenarios are then tested with different modes of management and development of all water resources including water resources involved in rainfed agriculture *Green Water*. Tunisia is an interesting example that lends itself well to this kind of exercise: it is a semi-arid country where water resources, which are structurally limited, have been fully mobilized and completely regulated. Tunisia is therefore required to find solutions to the problem of limited water resources by exploring the full development potential and enhancement of all water resources.

Figure 1. Tunisia.
Source: UNESCO (2009).

But first let's give an overview on the state of water resources and how Tunisia has addressed the question of future water use. Tunisia is situated in North Africa, bounded on the West by Algeria and on the Southeast by Libya; and bordered on the North and the East by the Mediterranean Sea along 1,200 km of coastline and on South by the Sahara (Figure 1).

Tunisia, which occupies an area of 164,420 km^2, has an uneven relief ranging from the mountainous regions in the Northwest to the desert regions of the Southeast through the plains of central Tunisia, the mountainous regions in the Northwest, the plains of central region and the desert in Southeastern regions. The population, estimated at 10.2 million in 2007, has nearly doubled in the space of 35 years, and should reach and stabilize at 13 million people by 2040 (INS, 2005). The rural population accounts for 35% and there has been rapid urbanization (expected at 75% in 2025) due to preferential migration to coastal cities.

In Tunisia, as in many other water scarce countries, the main water-resource management objective has been to provide sufficient quantities to municipalities, industry, tourism and agriculture by developing surface water and groundwater. Tunisia has positive results in terms of economic growth and productivity which have led to a significant improvement of living conditions. The country has reached respectable levels of GDP per capita and good social welfare, as evidenced by large coverage of water supply and sanitation in urban and rural zones. But at the same time, these performances are the origin of new contradictions between the imperatives of economic productivity which has been constantly increased, especially in agriculture, and the necessity of conserving natural resources, water and soil, scarce and fragile. Industrial development and urbanization also have important impacts on the development of water resources.

2 THE WATER RESOURCES OF TUNISIA

Spanning seven degrees of latitude, Tunisia offers a wide variety of hydro climatic regimes: a) sub humid in the North; b) semi-arid in the Northwest and in the Cap-Bon; c) arid in the central region; and d) hyper-arid in the desert region covering the entire South of the country. In the North, the Atlas mountains oriented SW-NE (between 500 and 1,500 m of altitude) delimitate the most fertile plains of the country. The main river, the Oued Medjerda, is now entirely controlled by a series of large dams. It crosses the North region from West to East and discharges in the Northern Gulf of Tunis.

The rainfall distribution is related to climate, to the disposition of relief and to the direction of prevailing NW winter winds. If the average rainfall reaches 1,500 mm/yr in the far North part of the country, it is about 500 mm/yr in the North, 250 mm/yr in the central regions, and it rarely exceeds 50 mm/yr in the extreme Southern regions located in the Sahara (Figure 2). The estimated rainfall in Tunisia amounted to 36,000 Mm^3/yr, corresponding to an average rainfall height of 220 mm/yr.

2.1 *Surface water resources*

For the three major natural regions, the contributions from surface water are as follows: a) the North, which covers 28% of the land area, provides regular and significant inputs into surface water evaluated at 2,200 Mm^3/yr, or 81% of the total potential of the country; b) the Center covers an area of 28% and has irregular resources estimated at 320 Mm^3/yr or 11% of the total surface water; c) the hyper-arid South, covering an area of 44% with only 180 Mm^3/yr or 8% of the total surface water. The national potential of surface water is estimated at 2,700 Mm^3/yr on average, of which 2,100 Mm^3/yr can be easily mobilized by dams and small lakes; the remaining part (600 Mm^3/yr) is more difficult to mobilize and requires small hydraulic structures like spates, reservoirs, cisterns . . .

Surface water resources have a very high inter-annual variability, with a minimum of 780 Mm^3/yr (observed in 1993–94) and a maximum of 11,000 Mm^3/yr (in 1969–70). The ratio of max/min of annual flows varies between 9 in the North and 180 in the South (Frigui, 2005). Moreover, the quality of surface water varies greatly by region: while in the North, 82% of surface water has salinity of less than 1.5 g/L, the proportion with this salinity is only 48% in the central region of the country and 3% in the South (Kallel, 1994).

In 2008, Tunisia had 27 large dams (height above 15 m), 200 small dams and 800 small lakes. All these water infrastructures can mobilize 1,800 Mm^3/yr.

2.2 *Groundwater resources*

Groundwater is related to the structure and geometry of geological formations. In Northern Tunisia, where strata are strongly pleated, there are many small layers, which might grow in the plains. In return, the central region of Tunisia is characterized by an important development of basins containing large aquifers which can be up to hundreds of meters thick.

The South of Tunisia is dominated by the Saharan platform where large aquifers extend over hundreds of thousands of km^2. Tunisia shares with Algeria and Libya these gigantic reserves, which are only partially usable and weakly renewable.

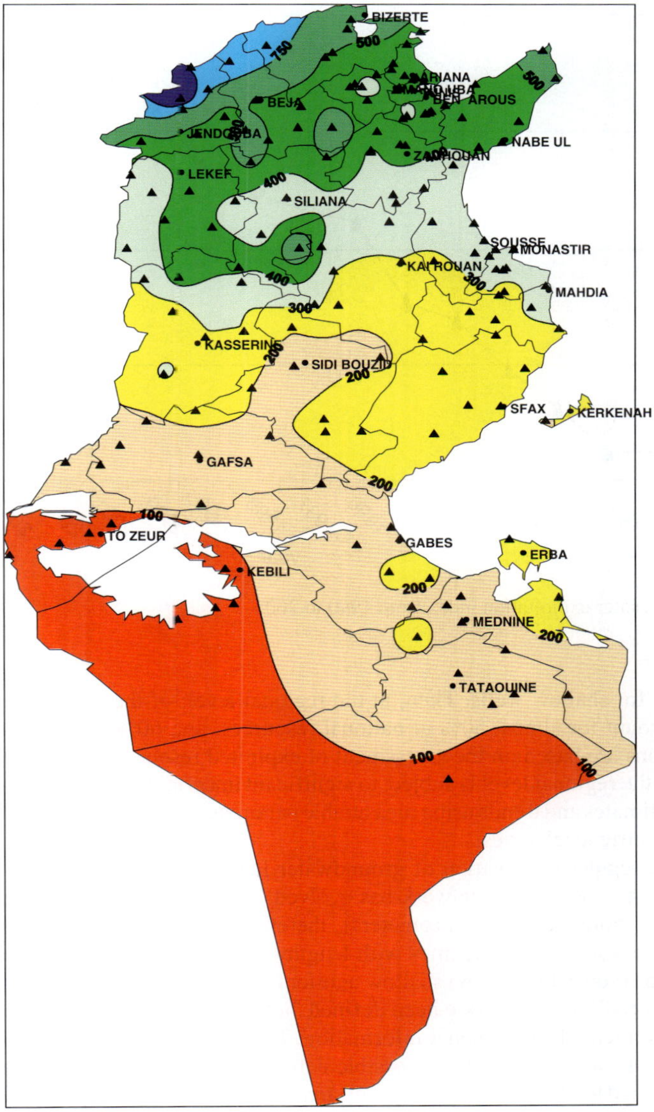

Figure 2. Mean Isohyets (mm/yr) and national rainfall Network.
Source: Frigui (2005).

Tunisian hydrogeologists conventionally distinguish between shallow aquifers exploited by large diameter dug wells and deep aquifers exploited by several hundred meters drillings. The exploitable groundwater resources of Tunisia are estimated at 2,150 Mm^3/yr (Hamza, 2006) [750 Mm^3/yr for shallow aquifers, 1,400 Mm^3/yr for deep aquifers]. Abstracted volumes in 2005 are estimated at 1,950 Mm^3/yr [800 in shallow aquifers, 1,150 in deep aquifers], which represent an exploitation index [ratio Exploitation/Resources] of 90%. Figure 3 shows how this index has changed over the past 15 years, with a steady increase from 1990 to 2001, then a stabilization between 2001 and 2005, although this is still too short a period to be considered significant.

However, the value of the exploitation index at the national scale hides huge discrepancies at two levels. First, between shallow and deep aquifers: while the index by deep wells is about 80%,

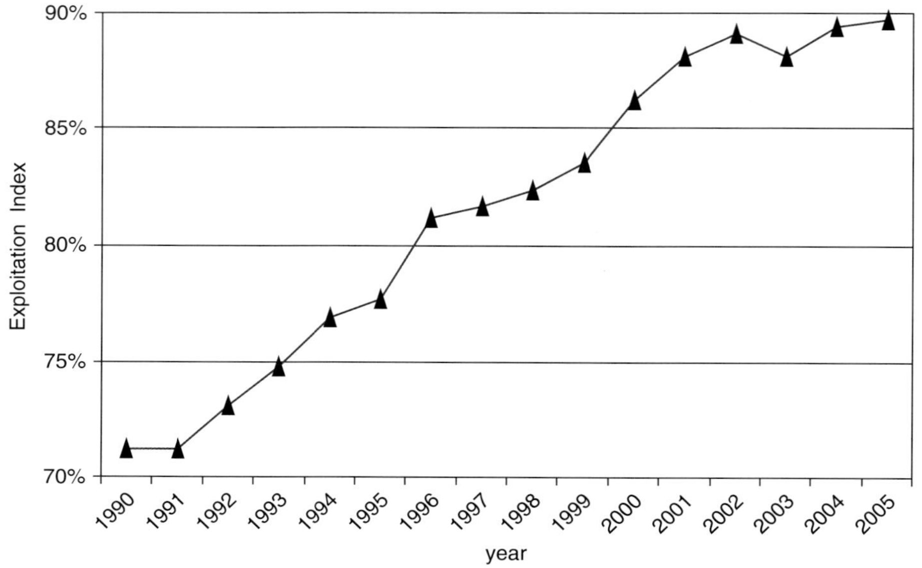

Figure 3. Groundwater exploitation index from 1990 to 2005.

it reaches 108% for shallow wells. Then, at the regional level: of the 24 regions (governorates), five of them account for almost 50% of national resources, and 60% of water mobilization. The above observations are based on the assessment of exploited resources, but the assessment of the water balance at the regional level is subject to significant uncertainties. It is therefore necessary to confirm these estimates and trends using measurable parameters, the most accessible and significant being the piezometric level series.

In this regard, regular observation of groundwater level began in Tunisia more than 60 years ago. Since then, the observation network has evolved. Now, it is composed of 3,800 monitoring points (dug wells, boreholes and piezometers), therefore the national piezometric network now includes a series of measures sometimes with lengthy 50 year series. It allows the observation of 150 water systems (consisting of 166 shallow aquifers and 124 deep aquifers). To characterize this network, a number of indicators have been defined, both for the quality of existing local networks and on the requirements for additional information. The classification of aquifers based on these indicators highlights aspects related to possible network improvements. The overall synthesis of various indicators led to identify the priorities in streamlining the piezometric networks (Horriche & Besbes, 2006).

For hydrogeological systems where tertiary formations predominate, are sometimes rich in clay and gypsum, leading to significant salts concentrations: only 10% of groundwater has a salinity less than 1.5 g/L, 60% between 1.5 and 5 g/L and 30% greater than 5 g/L, this last category being classified as brackish water.

2.3 *Soil water resources*

Water resources in the soil are defined as the proportion of infiltrated rainfall, temporarily stored in the soil, which is available: either for direct evaporation in the case of bare soil, or for crop consumption if the ground is covered. It corresponds to the available water in the soil, a notion used in the hydrological conceptual models. In theory, the total stock of soil water could be used by plants as in the case of very dense coverage such as forest or grass.

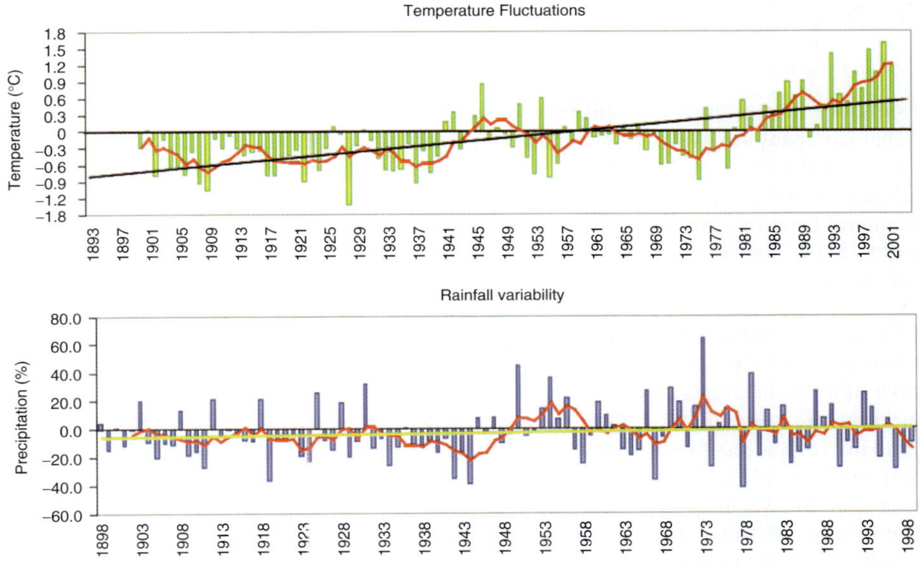

Figure 4. Anomalies in average temperature and precipitation over the past century reconstructed for Tunisia. From MARH–GTZ–Gopa–EXAConsult (2007).

Soil water resources in Tunisia includes arable land (4.5 million ha), rangeland (4.5 million ha) and forests (80,000 ha) on the surface of which the rainfall resource is produced. The evaluation of these resources across the country is a complex undertaking. Although some estimates have been proposed (Ennabli, 1993; Besbes, 1998), which estimate these resources at around 12,000 Mm^3/yr, which include the sum of the equivalent-water related to the rainfed crops, forests and rangelands, and the evaporation on bare soil for an average hydrological year.

2.4 *Climatic variations and impact on water resources*

The analysis of changes in temperature and rainfall averages over Tunisia during the last century (Figure 4) indicates a significant temperature increase estimated at +1.2°C (King *et al.*, 2007). This increase is greater than the increase in the global average (+0.7°C) indicated by the 2001 report of the Intergovernmental Panel on Climate Change (IPCC). For precipitation no significant trend is detected. However, it should be observed that the reference period 1961–1990 is characterized by a higher variability (higher standard deviation) in comparison with previous periods (1931–1960 and 1901–1930).

In the context of climate change, the future evolution of rainfall and temperature has been modeled on the global scale but with relatively large uncertainties. The extent and the accuracy nature of future changes remain imprecise (Lahache Gafrej, 2007a). Therefore, caution should be taken in order to implement predictions at the local or regional scales. The medium scenario used by the study confirms these difficulties: if the general increase in temperature is admitted, the evolution of rainfall is sometimes positive and sometimes negative, it varies by regions and season and, globally, the decrease is very low. This will probably not affect the runoff significantly and the inputs to dams. However the decrease in summer rainfall and the increase of temperature and potential evapotranspiration will augment the water deficit of the soil as well as agricultural needs. This could lead to more exploitation of groundwater and a worsening of their salinity. In addition, rising sea levels could exacerbate the salinization of coastal aquifers and might contribute to reduce the potential of groundwater in the long term.

3 MAIN WATER USES

3.1 *Water supply and sanitation*

The only operator in the sector of drinking water for urban, industrial and touristic uses is the National Company for Water Exploitation and Distribution (SONEDE). The rural operators are jointly the SONEDE for rural agglomerated centres of population, and the Administration of Rural Engineering (DGGR) in the rural dispersed areas through water users associations organized into Agricultural Development Groups (GDA): in 2005 there were 1,800 GDA for supplying drinking water to 45% of the rural population (1.6 million inhabitants).

The coverage rate of drinking water in urban areas has evolved from 98% in 1985 to 100% in 1993. These performances were obtained by using large transfers of water from the North, where surface water resources are relatively abundant, to the South, characterized by an arid climate; and through brackish water desalination for the benefit of the tourist areas of the Southeast. The number of urban localities currently supplied with drinking water is about 500 and the individual connection rate is around 99%. Over the past two decades, significant efforts have been made in order to improve access to drinking water in rural areas: the coverage rate has been increased from 30% in 1985 to 92% in 2006. The overall coverage rate for the whole country has evolved from 66% in 1985 to 97% in 2006 (Figure 5).

The volumes distributed by SONEDE networks have attained 405,106 m^3 in 2006. Moreover, part of the water demand is ensured by direct abstractions from groundwater. Applied to the total population, all direct water needs have increased from 30 m^3/yr per capita in 1990 to 40 m^3/yr in 2006.

Concerning urban sanitation, the National Office of Sanitation (ONAS) is the unique operator responsible for wastewater collection and treatment, sludge and solid waste. The rate of wastewater treatment increased from 32% in 1974 (creation of ONAS) to 97% in 2006. The connection rate in urban population reached 87% in 2006. In rural areas, sanitation is still traditional (like septic tanks, a discharge into the natural environment), but the National Office of Sanitation implementing new development programs. All together, many efforts are being made to minimize health risks by promoting behavioral changes in hygiene and by developing health education.

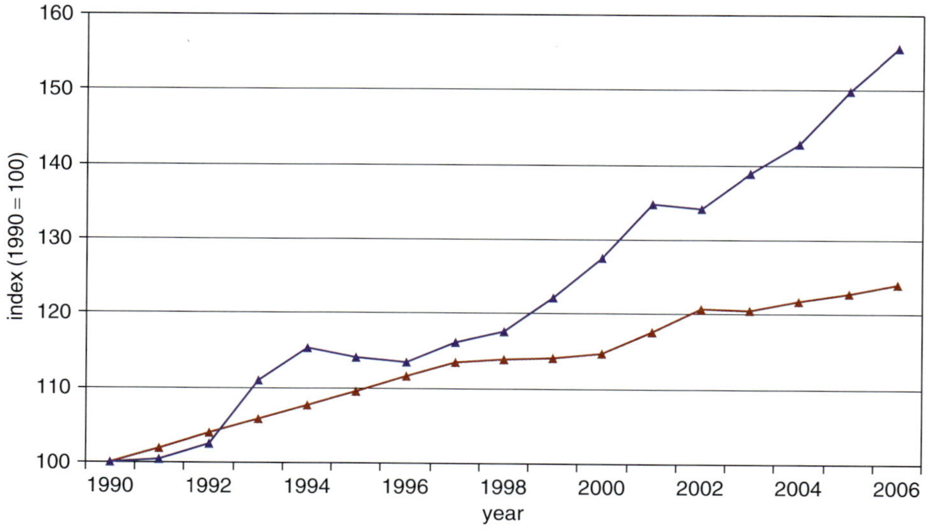

Figure 5. Evolution of the Tunisian population and potable water volumes distributed by public networks.

3.2 *The tourism sector*

Certainly the individual consumption of a tourist (550 L/day per bed occupied, and 900 L/day per bed for 5-star hotels is very high (see Lahache Gafrej, 2007b): this represents five to eight times the daily consumption of the average Tunisian user. However, the sum of consumption for the entire tourism sector reached 25 Mm^3/yr which represents only 1% of all water uses, including agriculture, and 6% of resources allocated to drinking water. This rate is slightly higher in the South where the tourist sector is subjected to strong growth; but this development could be controlled: first by applying new regulations on the audit of large consumers of drinking water (the target is saving 50% in hotels consumption), and secondly by using desalinated seawater.

3.3 *Water and Tunisian agriculture*

The role of agriculture in national economic growth remains important, although proportionally, the agricultural GDP tends to decline slowly. In 2006, agriculture accounted for 11% of GDP, employs 25% of the workforce and involves 475,000 farms with an average area of 11 ha. Industry, mining, and the service sector contribute respectively to 29% and 60% of the GDP. Exports of agricultural products (mainly olive oil and fruits) account for more than 20% of total exports, but agriculture remains dependent on uncertain climatic conditions. The total water abstraction reached 2,640 Mm^3/yr in 2006, including 2,140 Mm^3/yr allocated to irrigation, or 81% (15% for urban uses and 4% for industry). Irrigation water allocation is guaranteed by 75% of groundwater, 24% of surface water, and 1% of the treated wastewater.

Given the water allocated to agriculture, the irrigation potential is estimated at 560,000 ha, comprising 410,000 ha in full or partial control and 150,000 ha of additional irrigation and flood spreading. The total area was about 400,000 ha in 2005, comprising: a) 175,000 ha of informal irrigation by small private exploitations from surface wells and shallow drillings, or from pumping in watercourses; b) 225,000 ha equipped for collective irrigation from deep wells, dams and hillside dams; these equipments are made using public investments. The collective networks are generally modern and waterproof. The efficiency of these networks is estimated at 85%.

A comprehensive program from saving water irrigation began in 1995. Significant financial incentives are offered to promote efficient irrigation equipments. The water subsidies can represent, for small size farms, up to 60% of the facilities costs. Thus, the rate of irrigated area equipment using water saving systems was 80% in 2007 (25% with surface irrigation improvements, 27% with sprinklers, and 28% with drip irrigation). This strategy within ten years allowed the stabilization of the water demand from irrigation, despite the expansion of the irrigated area at the national scale (Figure 6).

On the other hand, irrigation has an important aspect related to *energy*. Indeed, 95% of irrigated areas are equipped with pumping systems and the total pumping capacity has been estimated at

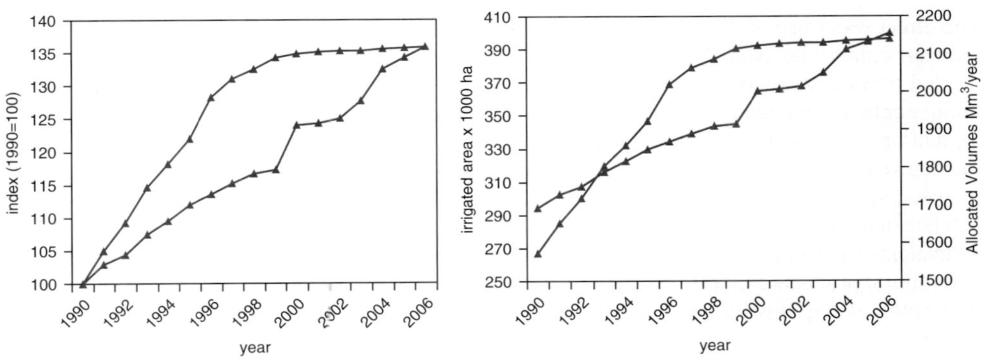

Figure 6. Evolution of irrigated area and water demand for irrigation.

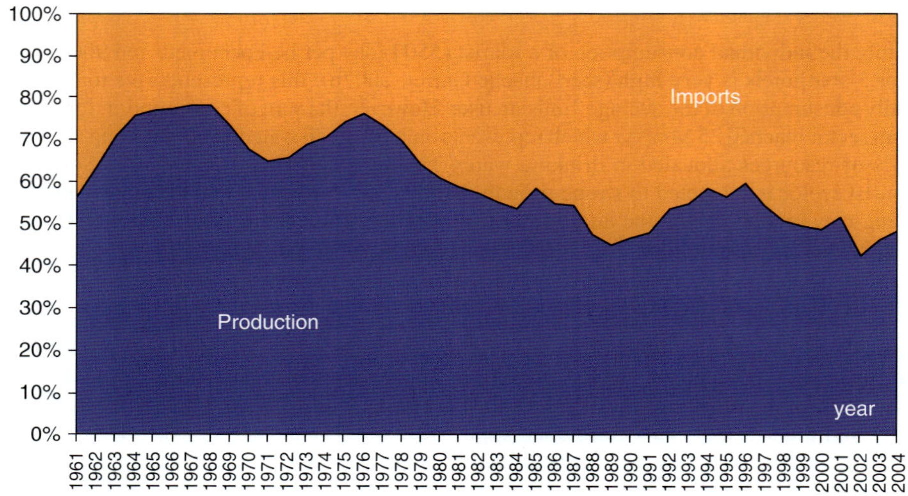

Figure 7. Structure of consumption of wheat in Tunisia.
Source: Besbes *et al.* (2007).

about 160,000 kW in 2005. The irrigation sector represents 27% of agricultural employment. The production of irrigated areas is estimated at 35% of the total agricultural production (in value) and its participation in the export of agriculture products is around 20%.

3.4 *Role of agriculture in food security*

In Tunisia, the concept of food security is understood as an objective to cover a number of commodities involved in food through national production: cereals, oils, meat, milk, potatoes, and sugar. But Tunisia is in deficit for several of these products. On the other hand, the climatic fluctuations induce strong fluctuations in rainfed agriculture production. The agri-food trade balance of Tunisia has been negative during for last two decades, except in wet years like 1999 and 2004. This balance is chronically dependent on cereals imports which represent nearly 45% in value of foodstuffs imports (Figures 7 and 8). However, the overall coverage is evolving positively; the reason for this positive development is that milk and meat are beginning to cover local demand.

3.5 *Foodstuffs trade and Virtual Water exchanges*

The direct water use of different sectors (urban, industry, tourism . . .) is relatively low compared with agricultural demand. Agricultural demand corresponds to all agricultural water requirements needed to produce food. The concept of water-equivalent (amount of water needed to produce a good) applied to the total food demand can highlight some key aspects of the relationship between agriculture and water resources. It appears that, in an average hydrologic year, nearly the half of the equivalent-water of Tunisian food demand is provided by rainfed crops *Green Water*, and that irrigated agriculture accounts for about one sixth, with almost a third filled by the agri-food trade balance in the form of *Virtual Water*. To supplement food needs, Tunisia imports foodstuffs with an equivalent-water estimated at 5,200 Mm^3/yr, mainly in the form of cereals, and at the same time, the agricultural products export (citrus, dates, vegetables, olive oil, etc.) represent an equivalent-water of 1,500 Mm^3/yr, leaving a balance deficit of 3,700 Mm^3/yr in an average year (Chahed *et al.*, 2007).

As in most arid countries, the contribution of *Virtual Water* to food security is essential, and the volume of food product trade depends closely on the production of rainfed agriculture, whose

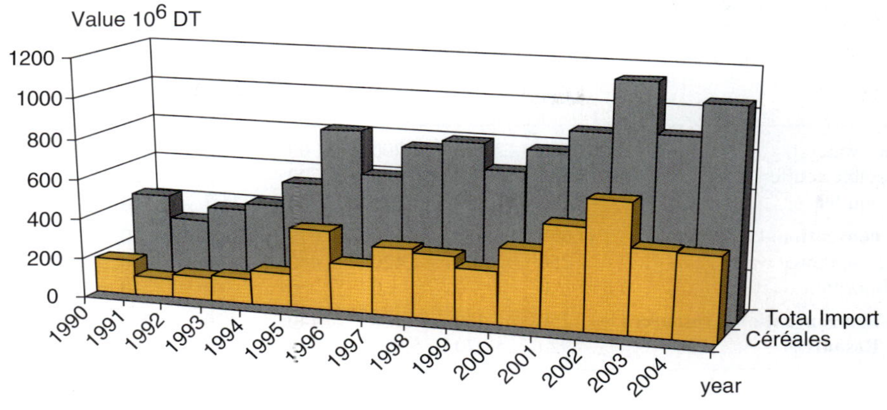

Figure 8. Part of cereals in food imports, in value, 10^6 DT [1 DT = 0.8 US$].
Source: MARH (2005); Besbes *et al.* (2007).

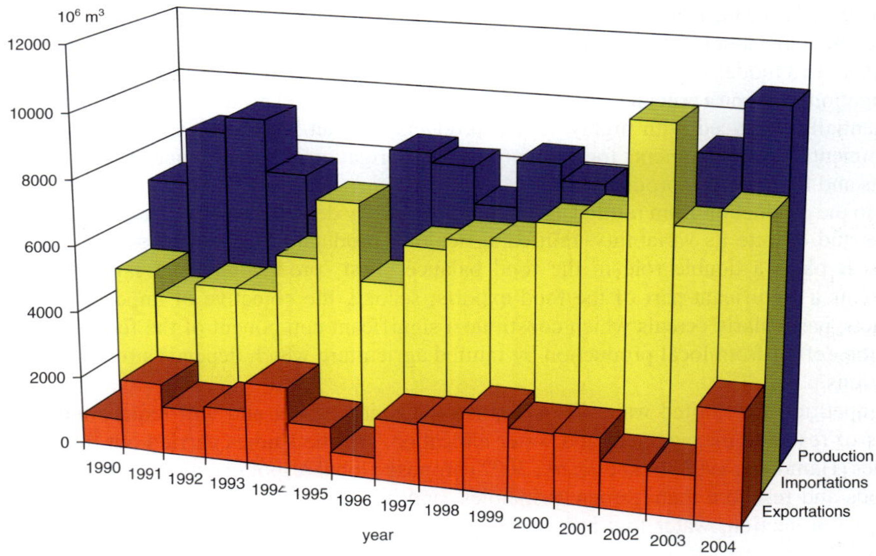

Figure 9. Production, imports, exports of food, expressed in equivalent-water from 1990 to 2004 (Mm^3).
Source: MARH (2005); Besbes *et al* (2007).

volume is strongly dependent on climatic variations. However, water budgets that result from foodstuffs trade balance are very favorable to Tunisia. The coverage rate of foodstuffs balance, more or less balanced in value, represents less than 30% in equivalent-water (Figure 9).

The current deficit of the equivalent-water related to foodstuffs exchanges might grow in the future because of the change that might occur in food demand, due to eating habits and to the increasing pressure on water resources. Indeed, food needs can change rapidly due to the expected improvements in living standards. The implications of these developments on agriculture and on foodstuffs trade balance are important, especially as changes in trade condition could occur in relation to market developments, to the modification of agricultural policies and to trade liberalization at the international level.

Table 1. *Blue Water* resources potential. After *Eau 21* (Khanfir *et al.*, 1998).

	Potential resources (Mm^3/yr)				Exploitable res. (Mm^3/yr)			
Year	1996	2010	2020	2030	1996	2010	2020	2030
Surface water	2,700	2,700	2,700	2,700	930	1,220	1,173	1,182
Water table aquifer	720	720	720	720	720	720	620	550
Deep aquifer	1,250	1,250	1,250	1,250	997	1,150	1,000	1,000
Total conventional resources	4,670	4,670	4,670	4,670	2,647	3,090	2,793	2,732
Treated sewerage water	250	400	420	440	120	200	290	340
Desalinization	0	0	0	0	0	10	24	49
Total non conventional resources	250	400	420	440	120	210	314	389
Total Resources	4,920	5,070	5,090	5,110	2,767	3,300	3,107	3,121

4 WATER SUPPLY AND DEMAND PROJECTIONS

4.1 *Tunisian water resources*

During the last decades, the Tunisian authorities have implemented a pragmatic policy of systematic development of water resources, in order to sustain the socio-economic development of the country and promote a modern agriculture, so as to increase local food production and promote its export.

Irrigation has been a crucial factor in the increase of agricultural production. Irrigated agriculture is essentially practiced with highly valued products (market vegetable, fruit production). With governmental encouragements for dairy production, irrigation is also practiced in complement for cereals and fodder crops production.

As to the production from rainfed agriculture, it strongly depends on climatic conditions. Nevertheless and despite its variability, rainfed agriculture production plays an essential role in food safety. It plays a double role in the food balance: first, production from rainfed agriculture represents a significant part of the food exports; second, the objective of importing basic food products, particularly cereals which constitute a significant component of the food imports, is to meet the deficit from local production by rainfed agriculture which depends strongly on climatic conditions.

Competition for limited water resources is rapidly increasing and will require more adapted modes of resource management and water allocation. All the studies carried out during the last decades (Hamdane, 1993; Besbes *et al.*, 2002) propose global responses in terms of administration methods and resource management principles, and introduce technical and economic solutions (resource protection, water loss reduction, improvement of water use efficiency, suitable water pricing). With regard to indicators for water resources conservation, results from water sector reforms are already remarkable both in urban systems (Limam, 2007) and in agriculture (Hamdane, 2007): i.e. reduction of water losses and the improvement on the efficiency of water uses.

4.2 *Demand projections*

The official Tunisian study on water resources *Eau 21*, has estimated water resource potential up to 2030 (Table 1, Khanfir *et al.*, 1998). It indicates that the amount of regularly produced conventional *Blue Water* will be almost 2,700 Mm^3 after 2010 (Table 1). The official long-term projections of *Eau 21* predict very rigorous water resource management. The supply-demand adequacy suggested in this study assumes a moderate evolution of direct water demand (domestic, industrial and tourism) and a drastic management of all water uses. According to *Eau 21*, domestic water consumption per capita will remain at 100 L/day in 2030 and water allowance for irrigation will be strongly reduced, by generalizing water saving irrigation systems, from an average allocation of 6,320 m^3/ha/yr in 1996 to 4,335 m^3/ha/yr by 2030. Irrigated surfaces for the same period would go from 335,000 to

Table 2. Initial *Blue Water* demand and projections (in Mm^3/yr). After *Eau 21* (Khanfir *et al.*, 1998).

Year	1996	2010	2015	2020	2025	2030
Drinking Water Demand	290	381	410	438	464	491
Industrial Water Demand	104	136	150	164	183	203
Tourism Water Demand	19	31	33	36	39	41
Total Direct Water Demand	413	548	593	638	686	735
Irrigation Water Demand	2,115	2,141	2,115	2,082	2,058	2,035
Total Water Demand	2,528	2,689	2,708	2,720	2,744	2,770

Table 3. Overall water demand of Tunisia (average values for 1990–1997).

Sector	Water Demand (Mm^3/yr)
Irrigation	2,100
Rainfed agriculture [Green Water]	6,000
Deficit of food balance [Imported Virtual Water]	3,700
Urban [Cities, tourism]	400
Industry	100
Forests and Rangelands	5,500
Water Bank [Storage in dams for droughts]	600
Environment [Conservation of humid areas]	100
Total Water Demand	18,500

467,000 ha. The annual water allocation to irrigation would thus be re-examined, from 2,100 Mm^3 in 1996 to 2,030 Mm^3 in 2030 (Table 2).

5 OVERALL WATER BALANCE OF TUNISIA

The future growth of irrigation is mainly limited by the availability of water resources. On the other hand, rainfed agriculture mobilizes a significant part of water resources and contributes to more than 2/3 of food production (cereals, olives, livestock, . . .).

The total available resources resulting from surface and underground runoff *Blue Water*, represents a relatively small part (about one quarter) of the total water resources of the country.

National accounting of water resources for an average year indicates that the overall volume of water resources effectively used is around 18,500 Mm^3/yr (approximately 2,000 m^3/yr per capita) (Table 3). Direct water demand (urban water-use, tourism, industry) is about 50 m^3/yr per capita, and the equivalent-water of food demand reaches 1,300 m^3/yr per capita. As mentioned before, part of this demand is provided by imported foodstuffs products (mainly cereals and vegetable oils) in the form of *Virtual Water*. The deficit of food balance expressed as Equivalent-Water represents 3,700 Mm^3 (about 400 m^3/yr per capita). The rainfed agriculture represents a significant part of water resources in the form of *Green Water* estimated at 6,000 Mm^3. Table 3 shows that rainfed agriculture production (especially cereals, leguminous plants and olives) represents, in terms of Equivalent-Water, a significant contribution (more than half the food demand and about three times the volume of water allocated to irrigated areas). It also appears that food imports represent a considerable contribution, which serves to meet the local production deficit especially that of rainfed agriculture, very variable because of rainfall variability. The agro-alimentary balance deficit represents on average an *Equivalent-Water* contribution of roughly a third of the total food demand expressed as Equivalent-Water.

The overall water balance model of Tunisia highlights some important aspects related to water resource management (Chahed *et al.*, 2008). This model assumes that the allowance in *Blue Water* to irrigation (*IW*) must adjust to the available water once the direct needs (*DD*) insured. This water balance is expressed as:

$$IW = EWR - (1 - RI)DD - ENV \tag{1}$$

$$VW = FD - GW - \lambda EWR + \lambda(1 - RI)DD - ENV \tag{2}$$

Equation 1 expresses the irrigation water volume (*IW*) as the difference between exploitable water resources (*EWR*) and the quantities consumed by the direct water demand (*DD*) or allocated to meet environmental demand, the coefficient (*RI*) representing the rate of water recycling. *Equation 2* expresses the water deficit (*VW*, *Virtual Water*) volume as the difference between the equivalent water demand for food (*FD*) and the equivalent-water of agricultural production in both rainfed (*GW*, *Green Water*) and irrigated contributions; the coefficient λ expresses the global irrigation factor which converts irrigation volumes into equivalent water; this factor integrates irrigation efficiency and rainfall contribution. The adjustment of these balances at the national level by comparing them with the data for the two years 1996 and 2004 led to the validation of this formulation. It appears that the rate of coverage of the demand in terms of equivalent-water is mainly controlled by the production of rainfed agriculture and to a lesser extent by the contribution of irrigated area, while direct water demand has a relatively weak effect on the overall balance, even though, as it could be expected, their socio-economic effects at the sectorial level are crucial.

6 AGRICULTURE STRATEGY AND WATER POLICY IN TUNISIA

Tunisia has implemented, relatively rapidly, a pragmatic policy of systematic development of water resources. After engaged in efforts over several decades, Tunisia has managed most of its conventional water resources. This policy also relied on large water transfers and played a crucial role in the development of different sectors of the economy at the national level, including irrigated agriculture. Irrigation has developed significantly and has played a pivotal role in development policies: e.g. the total area equipped for irrigation has increased sevenfold in the past four decades.

On the other hand, rainfed agriculture mobilizes a significant part of water resources and contributes to more than 2/3 of food production (cereals, olives, livestock . . .). Despite the variability of production, rainfed crops also play a key role in food security. Firstly, the production of rainfed agriculture accounts for a significant part of food products exports; and secondly, imports of staple foods including cereals, are mainly intended to cover the deficit in local rainfed agriculture production whose performance is highly dependent on climatic conditions. The two dominant rainfed cropping systems in Tunisia (cereals and olive oil), occupy more than two thirds of land on various soils and in weather conditions more or less favorable to intensification. The average yields of cereals per hectare have been multiplied by three over the past thirty years, but they remain relatively low. Cereal production is still insufficient; it only covers less than the half of the national demand and the difference is made up by imports, whose volume is increasing. The government tries to enhance cereals production promoting irrigation and new farming methods to increase crop productivity (seed selection, fertilizer use, etc.)

Irrigated areas represent 10% of the total cultivated area. Nowadays its future growth is mainly limited by the availability of water resources. The irrigation sector accounts for the largest water demand (80% of exploitable resources). With this potential, irrigated agriculture provides 35% of agricultural production in value and contributes to 25% of agricultural products exports. The direct water demand (urban, tourism, industry) has the advantage of having an absolute priority in the allocation of good quality water resources. This demand remains moderate, but the prospect of development of these different economic sectors will be accompanied by an increase in demand.

On the other hand, the almost complete exploitation of the available resources is accompanied by a number of risks and constraints, particularly with regard to environmental concerns (water quality, variability of resources, environmental restrictions, and overexploitation of underground resources...). The prospect for sustainable management of water resources should lead us to consider all factors and constraints that could affect the resource. In particular, the environmental demand in terms of direct allocation of water resources for lakes, wetlands and groundwater recharge is now considered as a primary factor of sustainable water management. For example, the overall releases from Sejnane, Joumine and Ghezala dams for feeding Ichkeul Lake totaled 750 Mm^3 between 1998 and 2005, corresponding to an inter-annual average volume of about 110 Mm^3.

7 PROSPECTIVE STUDY OF THE OVERALL WATER BALANCE

Since it is predicted that water supply at the national scale is going to stabilize, the pressure on the resource will result in inevitable reallocations that will be operated at the expense of the agricultural sector which, during periods of drought, is the first to suffer from the effects of water scarcity. The reduction of water availability for agriculture is a key issue for water problems in the future. It constitutes the ultimate challenge for current water policy which tries to find the most appropriate ways to meet the growing needs of the different economic sectors without burdening the agricultural sector. In these circumstances this approach leads to pursue the development of the supply to furnish sufficient water increasing the pressure on water resources. This classic hydraulic vision of water resource is incomplete. It does not take into account the important production potential of *Green Water* and does not consider the global food demand. The production of rainfed agriculture reflects an exploitation of non-productive rainfall resources which account for a large part of the overall water balance in the country and thus may effectively participate, as irrigated agriculture, in the national effort to increase agricultural production to meet food challenges.

Significant changes of dietary patterns have occurred in Tunisia in recent decades bringing the water equivalent of food consumption under 700 m^3/yr per capita in the 1960s to a level exceeding 1,400 m^3/yr in 2000 (Figure 10). The fact remains that this level of consumption is relatively low compared with the equivalent-water of developed countries food consumption (1,600 m^3/yr per capita in Europe, 2,200 m^3/yr in California) which corresponds to a diet richer in animal products (proteins).

The prospect of sustained growth in the purchasing power in Tunisia, will result in an increase of the equivalent water for food requirements due to the increase in proteins in the diet. It is indisputable that such changes will have important implications for the water balance and for agriculture. On the other hand, the development of the tourism sector will be coming with additional demand for food products. During 2007, the number of tourists reached almost 7 million with approximately 40 million tourist nights. Even if demand from the tourism sector may represent a relatively small

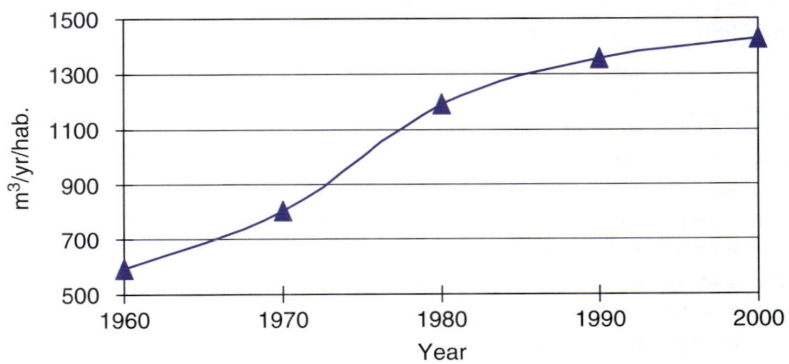

Figure 10. Evolution of per capita water equivalent of the Tunisian food demand during the last 40 years.

part of global food demand, the prospect of growth indicates that its long-term effect cannot be neglected.

Demographic change, economic growth and the aspiration for a better life will result an increased water demand with suitable quality but also quantities increasingly important of foodstuffs. These perspectives point to the need to study patterns of water development and their implications on the national water balance. The simplified formulation of the water balance is therefore useful for simulating the future of water resources in order to prospect different scenarios for water resource management and exploitation. The purpose of these simulations is to assess the impacts of different assumptions on water demand, including the demand for food production. The second objective is to compare different planning and development options for all water resources, including resources used in rainfed agriculture *Green Water*.

The simplified formulation of the water balance is useful for simulating the future of water resources in order to prospect different scenarios for water resource management and exploitation at the national level. Three scenarios are studied: the two first scenarios are called *trend scenarios*, based on the analysis of the actual trends in operations and management of the resource. The third scenario, called *sustainable scenario* is more conceptual in that it broadens the notion of resources to the potential of *Green Water*. This scenario stems from a vision of what we want to see happen in the future in order to ensure sustainable management of all water resources. All of the scenarios suppose the continuation and amplification of the efforts made in recent decades to ensure resource conservation. Taking into account the current effort to increase reuse and recycling of wastewater, the scenarios admit that the rate of water recycling will reach a maximum of 50% by 2025.

7.1 *Trend scenarios*

Trend scenarios are based on a classic vision of water resources planning which considers only *Blue Water* and tries to match the available resource with the water demand. In recent decades, the rainfed cultivated lands have not really evolved and these scenarios will extend these trends considering that the contribution of *Green Water* to the overall water balance will remain constant.

The first simulation (simulation 1) builds on a drastic control of all water uses and on a moderate growth in the direct water demand (potable, industrial, and tourism). The assumptions of this simulation assume that water demand for direct uses is supposed to remain relatively low (60 m^3/yr per capita by 2025 and 70 m^3/yr by 2050) and the equivalent water of food demand is expected to remain largely below the equivalent water of food demand of developed countries (1,600 m^3/yr per capita by 2025 and 1700 m^3/yr by 2050). The second simulation (simulation 2) goes in the direction of more *realism* assuming that water needs will increase reasonably. The assumptions of (simulation 2) consider that the direct water demand will reach 70 m^3/yr per capita by 2025 and 80 m^3/yr by 2050. The simulation also admits that the structure of the diet will be improved, so that the water equivalent of the food demand is expected to reach 1,700 m^3/yr per capita by 2025 and 1,800 m^3/yr by 2050.

7.2 *Sustainable scenario*

The sustainable scenario attempts to build a new water vision, at least in regard to agricultural sector. This scenario corresponds to a desire to benefit from all water and soil resources in order to strengthen the capacity of agricultural production to meet increasing food demand. This simulation implies that efforts will be made to strengthen rainfed agriculture production by developing cultivated areas and by improving crops productivities so that, in an average year, the production of the whole sector will be increased by 25% by 2025 and 40% by 2050. Furthermore, we adopt the same assumptions as (simulation 2) for direct water demand and for the equivalent water demand for food. The results of the simulations for the horizons 2025 and 2050 are summarized in Table 4.

Table 4. Adjustment and prospecting scenarios of the global water balance of Tunisia.

		Trend Scenario Simulation 1		Trend Scenario Simulation 2		Sustainable Scenario	
	2004	2025	2050	2025	2050	2025	2050
Population (10^6 habitants)	9.93	12.15	13	12.15	13	12.15	13
Exploitable Resource (EWR) (Mm^3/yr)	2,500	2,700	2,700	2,700	2,700	2,700	2,700
Food demand, Eq. water (m^3/yr per capita)	1,450	1,600	1,700	1,700	1,800	1,700	1,800
Total food demand (FD), Eq. water (Mm^3/yr)	14,399	19,440	22,100	20,655	23,400	20,655	23,400
Direct water demand (m^3/yr per capita)	55	60	70	70	80	70	80
Total direct water demand (DD) (Mm^3/yr)	545	729	910	851	1,040	851	1,040
Reuse rate (RI)	0.1	0.5	0.5	0.5	0.5	0.5	0.5
Total irrigation allocation (IW) (Mm^3/yr)	2,008	2,336	2,245	2,275	2,180	2,275	2,180
Rainfed agriculture (GW), Eq. water (Mm^3/yr)	8,000	8,000	8,000	8,000	8,000	10,000	11,200
Conversion factor, λ	0.9	0.9	0.9	0.9	0.9	0.9	0.9
Deficit of food balance (VW) Eq. water (Mm^3/yr)	4,591	9,338	12,080	10,608	13,438	8,608	10,238
Total water demand	14,945	20,169	23,010	21,506	24,440	21,506	24,440
Rate of dependency	31%	46%	52%	49%	55%	40%	42%

8 RESULTS, DISCUSSIONS AND FURTHER EXPECTED DEVELOPMENTS

Tunisia is an arid area where water resources are almost entirely mobilized. The country is therefore obliged to apply new concepts and paradigms to optimize the use of different types of water resources, and to change the behavior of some parts of the population (Chahed *et al.*, 2007; 2008; Besbes *et al.*, 2007). Global water resource assessment indicates that the improvement of food safety will depend on the capacity of the country to manage and optimize the use of all kinds of water resources. Expressed in term of Equivalent-Water, rainfed agriculture *Green Water* and the contribution of the agro-alimentary trade balance *Virtual Water* appear to be very important. On the other hand, food dependence is likely to develop with the change in diets.

Considering the current state of water resource management and exploitation in Tunisia, the improvement of the agro-alimentary balance requires a better understanding of the important relation between water and agricultural production in order to optimize the use of all water resources. This corresponds to a global vision of water resource which goes beyond the traditional concept of withdrawal water to cover all kinds of water resources and all water uses including those for food demand.

The simulations of the trend scenarios indicate that the increase in population, relatively well controlled, will induce an increase in foodstuffs imports (expressed in water equivalent), even if the consumption level is maintained relatively low (simulation 1). In this best case, the volume of *Virtual Water* will be multiplied by a factor 2 by 2025. Irrigated agriculture is unable by itself to meet changing demand and the coverage rate of the food balance expressed in Equivalent-Water decreases significantly.

The second trend scenario (simulation 2) based on a realistic increase in water uses supposes that the direct water demand will increase significantly. It follows that the residual resource to be allocated to the agricultural sector will be reduced. This reduction will have little effect on the agricultural water allocation because of the prospects for increased reuse and recycling of water resources allocated to urban, industrial and tourist sectors.

The simulation of the sustainable scenario shows the essential role of rainfed agriculture in the overall water balance. The increase in rainfed agriculture production will bring the averaged Equivalent-Water of rainfed agriculture to 10,000 Mm^3 by 2025 and to 11,200 Mm^3 by 2050 against 8,000 Mm^3 in the initial period. This would maintain the coverage rate of foodstuffs production expressed in water equivalent to a much more favorable level facing the increase of all water demands, including the demand for food production.

The integral vision of water resource leads to a global water balance which confronts the real possibilities of water resources to the whole water demand. In such conditions, it becomes possible to value and optimize all kinds of water resources (improvement of water use efficiency, inclusion of rainfed agriculture, development of alternative water resources, optimization of *Virtual Water* fluxes, etc.). The question of water resources scarcity is obviously related to the safety of direct water demand supply (urban, industrial, tourism), but and above all to food security.

In a context of water resource limitation, this large perception of water resources is most favorable to the development of new ideas and strategies aiming at reducing the deficit of the agro-alimentary trade balance and improving food security. Some progress has to be sought in: (a) improvement of water use efficiency, in particular for irrigation; (b) inclusion of all water resources in particular that involved in rainfed agriculture production; (c) intensive use of water and soil conservation techniques; (d) improvement of plant varieties with regard to their adaptation to aridity; (e) development of suitable policies for management and control of food demand evolution; (f) optimization of agro-alimentary exchanges by taking into account their effects on the water resource balance. This last point should result in agro-alimentary trade that promotes import of food (and more generally products and services) requiring a lot of water (water-intensive products) and export of products requiring less water (water-extensive products).

The formulation of the overall balance of water resources is relevant in that it highlights the order of magnitude of the different elements of the water balance. However, the overall balance established at the national level hides huge disparities affecting surface and groundwater resource *Blue Water* as well as water resources mobilized directly by the crops *Green Water*. On the other hand, the regions where water resources are available do not always correspond to regions of high water consumption. Thus water policy in Tunisia has drawn heavily on the massive transfer of resources from areas where the resource is abundant toward zones with high water consumption in the coastal regions. These transfers, which involve water with very good quality, represent 25% of the total exploitable resources.

These considerations lead to regionalize the water balance in order to clarify these differences and to measure their impacts on the overall water balance. Built on the same assumptions as for the national overall water balance, the water balance applied to the region (i) supposes also that water resources are divided according to a classification of quality (quality criterion noted k). This balance is written in the form:

$$IW_{ki} = \sum_k \left[EWR_{ki} + \sum_j [I_{kij}] - (1 - RI_{ki})DD_{ki} - ENV_{ki} \right] \tag{3}$$

$$VW = FD - \sum_i \left[GW_i + \sum_k \left[\lambda_{ki} \left[EWR_{ki} + \sum_j [I_{kij}] - DD_{ki} - ENV_{ki} \right] \right. \right.$$
$$\left. \left. + \lambda_{REU_{ki}} (1 - TR_{ki}) ED_{ki} \right] \right] \tag{4}$$

Where I_{ij} is the volume transferred from region (j) to region (i) that verifies $I_{ij} = -I_{ji}$ and $I_{ii} = 0$ so that $\sum_i \sum_j I_{ij} = 0$

The parameters of *Equations 3 and 4* are the same as those expressed in the national water balance (*Equations 1 and 2*). They are associated here to the region (i) and to the class of water quality (k). λ_{REU} denotes the irrigation factor related to reused treated water.

This regionalization leads to more detailed informations; not only on the availability of the resources but also on their quality. This regionalization will also pose the issue *Green Water* assessment at the regional scale; knowing that there can also be *Green Water* transfers between the resources involved in the maintenance of rangelands and in logging and those employed in agricultural production. As part of ongoing research, we are developing the evaluation, at regional scales, of the *Green Water* potential on the basis of statistical data of the regional agricultural production.

REFERENCES

Allan, J.A. (1998). Virtual water: a strategic resource. Global solutions to regional deficits. *Ground Water*, 36(4): 545–546.

Besbes, M. (1998). *Gestion durable des ressources en eau: éléments pour l'élaboration d'une stratégie*. Institut Tunisien d'Etudes Stratégiques, Tunis.

Besbes, M.; Chahed, J.; Bargaoui, Z.; Mattoussi, M.S. & Mhiri, A. (2002). *L'avenir de l'eau, un nouveau challenge pour la Tunisie*. Int. Rep. Tunisian Institute for Strategic Studies (ITES), Tunis.

Besbes, M.; Chahed, J. & Hamdane, A. (2007). Bilan intégral des ressources en eau, sécurité hydrique et sécurité alimentaire. Communication at Académie d'Agriculture. Paris, session of April, 4th 2007 (l'Avenir des Eaux Continentales).

Besbes, M.; Chahed, J.; Hamdane, A. & De Marsily, G. (2010). Changing Water Resources and Food Supply in Arid Zones: Tunisia. In: G. Schneider-Madanes & M.F Courel (eds.), *Water and Sustainability in Arid Regions*. Springer.

Chahed, J.; Besbes, M. & Hamdane, A. (2007). Stress hydrique et sécurité alimentaire, une vision intégrale des ressources en eau. *La Houille Blanche*, no. 3.

Chahed J.; Hamdane, A. & Besbes, M. (2008). A comprehensive water balance of Tunisia: blue water, green water and virtual water. *Water International*, 33(4): 415–424.

Chapagain, A.K. & Hoekstra, A.Y. (2004). Water footprints of nations, Value of Water Research Report Series, 16. UNESCO-IHE, Delft, the Netherlands.

De Marsily, G. (2006). Les eaux continentales. Rapport sur la Science et la Technologie, 25. Académie des Sciences, Institut de France. Paris: EDP Sciences.

Ennabli, N. (1993). Les aménagements hydrauliques et hydro-agricoles en Tunisie. Imprimerie Officielle, Tunis.

Frigui, H.L. (2005). Rapport National sur la mise en valeur des ressources en eau en Tunisie. Nations Unies, Commission Economique pour l'Afrique ; CEA-AN/PUB/EAU/TN/7.

Gleick, P.H. (1996). Basic water requirements for human activities: Meeting basic needs. *Water International*, 21(2): 83–92.

Hamdane, A. (1993). La gestion de l'eau en Tunisie. Rapport de synthèse. Tunis. Ministère de l'Agriculture, Tunisie.

Hamdane, A. (2007). Tunisia National report on monitoring progress and promotion of water demand management policies. Water demand management in the Mediterranean, progress and policies. Blue Plan, Zaragoza, Spain.

Hamza, M. (2006). Ressources en eau de la Tunisie. DGRE, January. Tunis.

Hoekstra, A.Y. (2003). Virtual water trade: Proceedings of the International Expert Meeting on Virtual Water Trade. Delft, The Netherlands, December 12–13, 2002. Value of Water Research Report Series, 12. IHE, Delft, the Netherlands.

Hoekstra, A.Y. & Chapagain, A.K. (2007). Water footprints of nations: water use by people as a function of their consumption pattern. *Water Resources Management*, 21(1): 35–48.

Hoekstra, A.Y. & Hung, P.Q. (2002). Virtual water trade: a quantification of virtual water flows between nations in relation to international crop trade. Value of Water Research Report Series, 11. UNESCO-IHE, Delft, the Netherlands.

Horriche, F. & Besbes, M. (2006). Analyse du Réseau Piézométrique National Tunisien. *Revue des Sciences de l'Eau*, 19(4): 347–363.

INS (Institut National des Statistiques) (2005). Prévisions Démographiques Nationales, Tunis, August.

Kallel, R. (1994). Caractéristiques morphologiques et hydrologiques du réseau hydrographique Tunisien. DGRE, Tunisia.

Khanfir, R.; El Echi, M.L.; Louati, M.; Marzouk, A.; Frigui, H.L. & Alouini, A. (1998). EAU21: Stratégie du secteur de l'eau en Tunisie à long terme 2030.

King, L.; Nasr, Z.; Almohamad, H. & Maag, C.C. (2007). Le Climat. In: MARH–GTZ–Gopa–EXAConsult, *Stratégie nationale d'adaptation de l'agriculture tunisienne et des écosystèmes aux changements climatiques*. Chapter 7.2. Tunis.

Lahache Gafrej, R. (2007a). Les Ressources en eau. In: MARH–GTZ–Gopa–EXAConsult, *Stratégie nationale d'adaptation de l'agriculture tunisienne et des écosystèmes aux changements climatiques*. Chapter 7.3. Tunis.

Lahache Gafrej, R. (2007b). Comment réduire la consommation d'Eau dans le secteur touristique en Tunisie, approche et stratégie. Water demand management in the Mediterranean, progress and policies. Blue Plan, Zaragoza, Spain.

Limam, A. (2007). La tarification progressive: un outil de gestion de la demande; cas de l'eau potable en Tunisie. Water demand management in the Mediterranean, progress and policies. Blue Plan, Zaragoza, Spain.

Liu, J. (2007). Modelling global water and food relations; development and application of a GIS-based EPIC model. Dissertation ETH No. 17069. Swiss Federal Institute of Technology, Zürich, Switzerland.

MARH (2005). *Annuaire des statistiques agricoles*. Ministère de l'Agriculture et des Ressources Hydrauliques, Tunisia.

MARH–GTZ–Gopa–EXAConsult (2007). Stratégie nationale d'adaptation de l'agriculture tunisienne et des écosystèmes aux changements climatiques, Tunis.

Oki, T.; Sato, M.; Kawamura, A.; Miyake, M.; Kanae, S. & Musiake, K. (2003). Virtual water trade to Japan and in the world. In: Hoekstra A.Y. (ed.), *Virtual water trade*. Proceedings of the International Expert Meeting on Virtual Water Trade. Delft, the Netherlands, 12–13 December 2002. Value of Water Research Report Series, 12. IHE, Delft, the Netherlands.

Renault, D. & Wallender, W.W. (2000). Nutritional Water Productivity and Diets: From crop per drop towards nutrition per drop. *Agricultural Water Management*, 45: 275–296.

UNESCO (2009). *The UN Water Development Report 3. Cases studies volume: Tunisia*: 12–15.

Warner, D.B. (1995). Water needs and demands: Trends and opportunities from a domestic water supply, sanitation and health perspective. *Workshop on Scenarios and Water Futures*, Stockholm Environment Institute, Boston, Massachusetts, USA, 28–30 September 1995.

II
Searching for *water intelligence* solutions

CHAPTER 4

Can human ingenuity, Science and Technology help solve the world's problems of water and food security?

Elena López-Gunn
Water Observatory of the Marcelino Botin Foundation, Spain; Complutense University of Madrid, Spain; London School of Economics, UK

M. Ramón Llamas
Water Observatory of the Marcelino Botin Foundation, Spain; Complutense University of Madrid, Spain; Royal Academy of Sciences, Spain

ABSTRACT: This chapter argues that advances in human ingenuity (*how we think about problems*), science (*what we know and not know about problems*) and technology (*what we can technically do about problems*) produced in the last decades contribute to *solving* problems of water and food security (or insecurity and associated conflicts), with methods that were unthinkable only two decades ago. This chapter only considers four of these advances. Obviously there are other promising advances but what is presented here refers mainly to those that are cheap and easily available. These are: 1) communication technology (chiefly the Internet) that facilitates, participation, public awareness and education, as well and transparency, which can contribute to reduce corruption; 2) the spectacular increase in virtual water trade due to advances in transportation; 3) desalination thanks to membrane chemical technology that is able to desalt seawater or brackish groundwater or to regenerate waste water at decreasing costs, affordable for many industrial and urban uses; and 4) the silent revolution of irrigation groundwater development that has produced stupendous social and economic benefits although in some countries this has also induced a colossal anarchy and ecological problems. All these aspects are intertwined. However, technology alone will not solve the main water problems. It is necessary to achieve an equilibrium between their utilitarian and symbolic or cultural values. Solutions to water conflicts require considering not only social ethics but also environmental ethics.

Keywords: science policy interface, extended water footprint, virtual water, desalination, groundwater governance, transparency and accountability

> *Paul Polak (2008: 218) writes that, when he returned home tired and discouraged from his trips trying to end poverty through the International Development Enterprises, his two daughters played the theme song of musical The Man of La Mancha.*
> *The main song in this musical is the IMPOSSIBLE DREAM.*
> *This paper argues that impossible dreams are in fact, possible.*

1 INTRODUCTION

Despite current predictions on doom and gloom, this chapter will argue that there is enough knowledge, technology and resources to address many predictions on water and food scarcity. The solutions to global water problems however do not lie in the water arena but rather in policy decisions currently being taken *out of the water box*, in current negotiations over trade, aid, development and energy policies. In this sense this chapter is a follow-up of an article already published (López-Gunn & Llamas, 2008). It was argued that at present there are opportunities in terms of available scientific knowledge and technology that are relatively cheap and accessible to address water problems of around 80 to 90% of the global population. Rogers *et al.* (2006) and Rogers (2008) present similar conclusions.

The chapter is structured in the following way: the section below highlights the key importance of changing the way we *frame the problem*. This in itself can have the highest impact: i.e. innovation in policy making by looking at old problems from a new perspective. The second section focuses on the gap between science and policy by focusing on the gap itself, i.e. on how research and knowledge has be to opened up through processes of inclusion, transparency and accountability. The third and fourth sections then focus on two examples of new ideas (virtual water and the water footprint) that can help re-frame old problems, and how new technologies (desalination) can open up the room for policy manoeuvre (i.e. new water). The final section addresses areas that are often neglected despite their significance (the silent revolution in groundwater uses) and the challenge in bringing them into the spotlight.

In the last two years we have received feedback that have shifted our thinking in ways that we are integrating into this chapter. Feedback to our 2008 paper seems to indicate that we were over-optimistic on the ability of science and technology to solve current global water problems. We believe that we probably did not give enough weight to a third pillar: *thinking out of the (water) box*, i.e. new ways of thinking about seemingly intractable problems. Also a fourth pillar, the fact that decision making is inherently political, and that *rational* decision making based on new evidence is not an automatic and quick process, and knowledge is by-and-of itself not sufficient to trigger change.

In fact, in previous publications one of the authors of this chapter (Llamas & Delli Priscoli, 2001; Delli Priscoli *et al.*, 2004; Llamas *et al.*, 2009) has long sustained the position that –in order to solve water conflict– it is necessary to achieve an equilibrium between the utilitarian (economic) value of water (i.e. irrigation, production of energy, and others) and the intangible (cultural, social, religious, and other values). This view was also adopted by a working group on the Ethics of Freshwater Uses and approved by the UNESCO Commission on the Ethics of Science and Technology (Selborne, 2001). These guiding principles are particularly relevant for policy making, giving clear directions for decision makers. The *ethics* that should underpin global policy making in the field of water echo other policy debates on the right to water, the right to sanitation, and the right to food, i.e. governments and other actors like corporations have clear *moral* duties and responsibilities *vis a vis* their citizens to provide access to affordable water and food. Water is an economic good, but it also has clear social and cultural values (Delli Priscoli *et al.*, 2004). The underlying concept is that science and technology are not by themselves sufficient to solve current water problems. The crux of the potential solutions demands that greater attention is paid to both social and environmental ethics. Social Ethics deals with the relations between human beings, and is reflected in a rights based approach to development (United Nations, 1998), based on human dignity and inalienable rights; meanwhile environmental ethics acknowledges the new capacity of humanity to dramatically alter the environment in ways unprecedented in human history. This links up with new theories on solidarity rights which encompass the right to a healthy environment.

The debate on the positive or negative effects of scientific and technological advances on the future of humanity and nature is alive (as well as on affluence, population growth and other factors). A recent article by Goklany (2009) provides evidence on the general positive impact of advances in technology on human welfare.

This chapter will discuss the following four advances in new ideas, scientific evidence and technology:

a) First, the revolution in Information Technology, which can *democratise* an inherently technical sector and can open up decision making towards deliberative processes in water governance (information, participation, education, accountability and data transparency).
b) Second, how globalization and cheap transport has facilitated virtual water trade. New information and evidence generated by the use of the *extended (economic and hydrologic) water footprint* provides an easily understandable and intuitive tool to showcase options *en route* to Integrated Water Resources Management.

c) Third, thanks to technical advances like membrane technology (Bernat *et al.*, this volume), the possibility opens to produce *new water* due to the decreased cost of desalination from seawater, brackish groundwater, or from water reuse from urban and industrial polluted water.
d) Fourth, the recent spectacular increase in groundwater abstraction (according to Shah, 2008) from 100 to 1,000 km^3/yr –occurring in some of the most heavily populated countries in the world– is facilitating fast socio-economic change, thanks to the massive uptake of drilling rigs and turbine pumps. This silent revolution has developed spontaneously, largely due to farmer initiative, in many cases by millions of modest farmers and often outside the planning and control by Water Authorities. There is evidence that this has generated social change, although there are some pending questions over this anarchical model of development.

In relation to the common denominator of the ideas discussed in this chapter, the key criteria is that these are *cheap* and *accessible* in the sense that these are affordable by private individuals, companies or governments and where the pay-off is greater than the cost.

For example, in developed countries investment in desalination and reuse in e.g. coastal areas, can free up water resources for other sectors. In middle income countries, or emerging economies like the BRIC countries (Brazil, Russia, India and China), virtual water trade and the *extended* water footprint can help identify where resources will deliver greater socio-economic welfare. In developing countries, groundwater offers a relatively easily accessible resource that requires small investment, and provides security in the context of e.g. rainfall variability, which is at the root of its silent revolution. Meanwhile issues related to advances in information technology are probably beneficial to all, in terms of data transparency, and these offer the potential to open up decision making processes and thus increase accountability, e.g. of all stakeholders: users, of water authorities, and of businesses (Transparency International, 2008).

2 WATER AND FOOD SECURITY: THINKING OUT OF THE *WATER BOX*

This chapter is mainly concerned with the generally positive influence of the main advances in science and technology to address global water and food security. For the purposes of this chapter, water security and food security are defined as having access to *sufficient, safe, acceptable, physically accessible, and affordable water and food, in the context of sustainable development and livelihoods*. There are differences to the specific way to secure water and food. For example, in the context of *water security* often this has been related to security from damage from extreme events, and linked to the capacity to adapt to these extreme events. Equally, food security also refers not only to the food quantity but also to its nutritious value.

The major driving forces for decision making in relation to water resources allocation and management however are located outside the water arena, i.e. decisions that have the most important effect on water use are often not taken by persons or institutions in Water Authorities. For example, decisions currently taken in the field of agriculture, energy (e.g. bio-fuels) and trade liberalisation are ultimately the key driving forces of water use at both the national and the river basin local level (United Nations, 2009: XIX); and *viceversa*, e.g. the 2008 food crises was also related to factors exogenous to agriculture (Lamo de Espinosa, 2008; 2009). This is exemplified in other chapters in this book, which highlight that the water footprint of any given country is mainly driven by its agricultural policy, since agriculture accounts globally for 70–80% of global water use and 85–95% of global water consumption (United Nations, 2009; Kuylenstierna *et al.*, 2008). This number is even higher in arid and semi-arid countries, and in developing countries and emerging economies (including BRIC countries).

Therefore there is a paradox in water policy (like in other policy areas) in the large gap between policy and scientific evidence, and where additional effort is needed to bridge the gap between scientific knowledge, and the world of policy and politics. It is now accepted that the assumption that *good* ideas and new evidence would filter automatically into policy is not guaranteed. Reality is much messier and more complex compared to this *assumed* rational decision making policy process.

The study of the interaction between science and policy is arriving at the same basic conclusion: good ideas can indeed change policy but good ideas by themselves are not always a sufficient condition. A number of factors are at play like: timing, communication and increasingly the role of policy entrepreneurs pushing the idea through. This is a topic previously described in a note by Margaret Catley-Carlson addressed to all the participants which can be read in Botin Foundation website [http://www.fundacionmbotin.org/agua] and as a Postscript in this book (Catley-Carlson, this volume). It was one of the most interesting *hot issues* debated in the Workshop brain storming.

In most countries there seems to be a *communication gap* among persons and/or institutions taking decision that significantly affect the water policy of that country. A good road map of these processes would help to bridge this gap. One of the stumbling blocks is the assumption that the transfer from knowledge to policy is simply a matter of improved communication. However, there is increased realisation that this linear-rational model is at best idealistic, and that *sound science* can in fact fail to have any meaningful policy impacts.

In relation to water policy and scientific knowledge, there are a number of issues in the science/policy interface:

First, the scope of valid knowledge (and its legitimacy) has expanded: different kinds of knowledge (expert, lay, scientific, quantitative and qualitative) are now perceived to fit different purposes.

Second, the aim of science/policy interaction cannot be idealised: knowledge can be also used for instrumental reasons, to fit pre-existing policy frames, and be discarded if it does not fit these frames. This situation has occurred throughout history and also in modern times. Scientists who dissent with what is *politically correct* can be ignored or stamped out by hidden or open lobbies (see for instance B. Martin in Newsweek, 1993). The recent debate on climate change, including the stalemate at COP 15, the *climategate*, and the Himalayas glaciers report may be other examples.

Third, issues of relevant timing. In many cases the adoption of new ideas capable of changing policy and shift political inertias can take anything from 10 to 25 years. As the saying goes today's philosophy 20 years later is common sense. This means that scientists should be aware that ideas, scientific advances and technologies currently being developed, could be in operation 10–25 years hence. Other times the uptake of research findings is quick and effective. This is because there are sometimes critical points in time, so called *tipping points* and policy windows which open and when the policy environment is receptive to new ideas and evidence. However, in other cases new research can be uncomfortably ahead of its time, and be *politically inconvenient*, if it does not suit pre-conceived agendas or give answers to pre-conceived problems (Owens *et al.*, 2007).

There are a number of scenarios, e.g. where there is little evidence and no policy, or when there is policy without evidence (Davoudi, 2006), or as in the case of water, where there is evidence in the water field, but most relevant policy decisions lie out of the *water box*. The challenge then is to align science and innovation in water with powerful processes and inertias taking place out of the *water box*. This process of *translation* and *adoption* may be *speeded up* or *fast forwarded*, with scientists acting as policy entrepreneurs (Kingdon, 2003; Jarvis, 2008) or knowledge producers (Catley-Carlson, this volume). As Lorenz (1992: 181) notes, according to Samuel Popkin a *political entrepreneur* is an "innovator who solves collective action problems not by offering selective incentives, but by persuasion and changing beliefs, beliefs about the value of the collective good and expectations about the behaviour of others". According to Schneider & Teske (1992), political entrepreneurs are individuals whose actions produce unexpected results and induce policy changes. Yet not all good ideas and innovations will be taken up, and the challenge is to analyse and understand key factors that increase the chances of an idea being adopted, and anticipate and characterize the likely barriers to enacting new policies and strategies for overcoming these barriers. If one adopts a non-linear and complex model of understanding the interaction between science and policy, i.e. a less purely rational approach to science, this can help identify supposed barriers or constraints which in a *rational* policy frame are often perceived to be in the cultural, social and political arenas. A complex perspective on policy in fact can help perceive these supposed barriers in a different way, to shift our understanding on these supposed barriers to policy change, which in fact can provide early indicators and powerful signals about deeper processes of socio-economic change. Carefully studying the politics of resistance and opposition e.g. to policy change, may assist in bridging knowledge and policy through processes of public engagement (Eden & Tunstall, 2006).

Fourth, influencing policy can happen in different timeframes: in some cases the most powerful impact might be as stated earlier the re-framing of the problem itself, which then makes it more open to *knowledge creep* (Radaelli, 1995) or to *enlightenment* (Weiss, 1977). In other cases a distinction has to be made between research, which might be very useful in the long term, yet it is not very usable in the short term (Owens, 2005), and knowledge. Yet science and policy interaction covers this whole spectrum from immediate impact to the long term re-framing of policy paradigms.

This happy consensus between politics and science however can be difficult, and in fact sometimes undesirable. For example, Malthus's catastrophic predictions or prophecies on the grim future of humanity if population continued to expand have not materialised. In fact, since Malthus time (1766–1834) the world's population has substantially increased by a factor of more than six[1], on track to reach 7,000 million in 2011, only 12 years after reaching 6,000 million in 1999. Virtually all of the growth is in developing countries. Also the growth of the world's youth population (ages 15 to 24) is shifting into the poorest of those countries (Population Reference Bureau's 2009 World Population Data Sheet). Food production –Malthus's main concern– has increased at a higher rhythm than population, and the average food kcal/day per person has increased from 1,700 to 2,500. Equally, since the 1950s, the area of the world under cultivation has increased by roughly 11%, while yields per hectare have increased by 120% (Goklany, 2009; Kuylenstierna *et al.*, 2008; CAWMA: Comprehensive Assessment of Water Management in Agriculture, 2007).

Box 1. Examples of increased water productivity in agriculture (*Source*: Allan, 2009).

Farmers as *big water* managers key to increased water productivity.

1) North West Europe:
   Farmers in North West Europe trebled rain-fed wheat yields from 3 to 9 t/ha [t = tonne = 1,000 kg] between 1961 and 1990. This is compared to the yield of 1 t/ha in 1800 using the same volume of water.
2) India and China:
   Indian and Chinese farmers increased wheat, rice and maize production between 1961 and 1990 by 4 and 5 times. Water productivity was not at this level but it was still in the order of three times higher.
3) Vietnam:
   In Vietnam after the end of hostilities in 1975 Vietnamese farmers moved rice yields from 2.5 t/ha to 7 t/ha and increased to double and triple cropping so that some tracts were raising 20 t/ha/yr. Vietnam is now the second rice exporter. It was importing in 1975.
4) Egypt:
   Egypt has increased its wheat yields from 2 t/ha in 1960 to 7 t/ha today and yields and returns to water are still increasing. In the case of Egypt the increased returns have been due to blue water.

From this point of view, it is interesting to reproduce most of the written statement of A. Allan (2009) on occasion of the Workshop:

> "Farmers are the main water managers and they manage the big water –that is the water used in agriculture which is more than 80% of all water in the world's economies of which 70% is

[1] [http://belfercenter.ksg.harvard.edu/publication/1734/dont_count_out_malthus.html].

green water. If farmers were to double water productivity –that is double returns to water– they would make a very important contribution to global water security. [In fact] more than any other community.

Yet farmers compete with the environment for water more than any other community, because of their key role in managing *big water*. Therefore we need to understand what motivates farmers. Society needs to value what they do more than it has until now. We need to encourage farmers to further increase water productivity and at the same time be environmentally aware. In addition to the above, consumption and international trade asymmetries need to be understood and where necessary remedied."

Malthus's oversight was to ignore human ingenuity and its capacity to solve problems. For instance, one hundred years ago the typical yield in staple food (e.g. wheat) was about 1,500 kg/ha/yr. Today in industrialized countries yields are around 8,000 kg/ha/yr, whilst in most developing countries, it is about 3,000 kg/ha/yr (Goklany, 2009; Kuylenstierna *et al.*, 2008).There is no reason to believe that human ingenuity to cope with new problems in water, food and other issues is now finished. Nevertheless, as The Economist (1997, December 20th) pointed out, there will be always forecasters of scarcity, the so called *limits to growth debate*: "they are invariably wrong but they think that being wrong proves them right". As the Director of the World Water Week wrote "the natural resources are limited but human ingenuity is boundless".

There are different opinions on whether it would be possible to feed the 3,000 million human beings who will probably inhabit earth within half a century. To cope with this problem different solutions are proposed but these are mainly classified in three groups:

a) An increase arable land in rain-fed agriculture.
b) An increase in irrigated land.
c) To increase agricultural productivity per m^3 of green or blue water used.

For instance, according to CAWMA (2007), the additional necessary food could be obtained if productivity in developing countries increased by about 30% from their current productivity; this also requires an increase in irrigated surface which will demand an additional 500 km^3/yr of blue water. However, Kuylenstierna *et al.* (2008) present a more optimistic perspective, because they consider that technology improvements are sufficient to cope with the increasing need for food due to a higher standard of life and population growth. Statement by Allan (2009) on occasion of the Workshop is quite hopeful.

Goklany (2009) shows that the situation is just the opposite: for example, the variation in life expectancy in the last hundred years has increased substantially in all the countries (industrialized and developing), which many authors consider as one of the best indicators for social well-being. These ideas are in close agreement with those presented decades ago by Simon (1996) and Clark (1968).

3 BRIDGING THE GAP BETWEEN SCIENCE AND POLICY: TRANSPARENCY IN DECISION MAKING AND THE ROLE OF THE INTERNET

The previous sections highlighted that more attention has to be given to the *gap* between policy and science and innovation. This gap sits mainly between the political economy of water and the scientific and technical contexts. There is an increased tension to open up decision making processes to society, and the potential powerful role (rarely realized in practice) of *meaningful* processes of participation, accountability and transparency. The call for greater civil society participation is redundantly mentioned in most documents on water governance, starting with the Dublin Principles (United Nations, 2009; CAWMA, 2007; Transparency International, 2008; Salman Salman, 2004). The Dublin Principles which *instituted* public participation as one of the cornerstones for water policy was re-instated in Agenda 21 and Rio. Nevertheless, the achievement of this goal, as adopted in Article 14 of the EU Water Framework Directive, is far from being a reality in most EU

Member States. The situation outside of the EU and the USA is possibly even more fraught with problems.

The onset of many social movements and contestation of key issues in relation to water (e.g. dams, human right to water, Doha round) is possibly a proxy indicator of the exclusion of different perspectives and types of knowledge and opinion. The challenge therefore for participation is to move beyond rhetoric, and this is where democratising knowledge, through science and innovation itself can contribute.

For example, Transparency International (2008) estimated that up to a 25% of the funds theoretically used for water infrastructures are lost in bribes and other types of corruption. Although the correlation between transparency and less corruption is still not fully proven, what is clear is that there is increased evidence between the link between transparency and economic growth. This relates to addressing the problem of information asymmetry, which is a classic example of market failure. Transparency is a core component of the so called second generation institutional reform, and it is increasingly associated with better socioeconomic development, as well as with higher competitiveness and lower corruption, which ultimately can improve policy outcomes (Bellver & Kaufmann, 2005). Transparency for example, can facilitate participation and collective action by stakeholders. Transparency is at the heart of water governance and a fair allocation to users and sound incentives for efficient water use (WEF, 2008). The authors of this chapter are currently involved in a project with Transparency International (Spanish chapter) to develop transparency indicators on water management that will be applied to all river basins in Spain.

Without an equivalent progress in their implementation, the value of international declarations to solve drinking water and malnourishment decreases as the number of declarations increase. The effect of their overall impact and eventual legitimacy wilts if it is not matched by their effectiveness and implementation. For example, paying lip service to participation can in fact backfire and be seen as hypocritical, eventually losing value and meaning. Capacity building, participation, bottom-up initiatives, the empowerment of women, transparency, accountability, and so on can become once more elements of social engineering, if these are not translated into results and real change to people on the ground.

The use of Internet and of GIS (Geographic Information System) can be a great aid to achieve the goals of transparency and participation, and may even indirectly be a great means to fight corruption in the water sector. Technology can now provide the means at a relatively affordable prize to provide better monitoring and data collection and transparency. Reform and change is inherently a political, negotiated process with potential winners and losers when, for example, water is subject to being re-allocated between e.g. agricultural and urban users. Data transparency can be a catalyst against the inherent resistance or inertia in the system against change. For example, transparency can shed light on false or inaccurate pre-conceptions on cost and benefits, identify data and knowledge gaps, and help shift to other criteria like e.g. cost effectiveness and full cost recovery (as demanded in Europe by the EU Water Framework Directive).

The case of virtual water is a case in hand, e.g. the *true* cost and benefits of physically transferring water resources, *vs.* the cost effectiveness of transferring virtual water. For example, the farmer lobby has proved very successful framing of what is *politically correct*, and in political terms the farming lobby is a feared enemy by any government having to tackle the re-allocation of water resources e.g. from agriculture to other uses. Transparency can help to assess trade-offs (water accounting, social and environmental impacts) and open up decision making through sharing information and knowledge in a more equitable way.

Policy making is a messy process with many drivers in constant flux, therefore new information has to be constantly produced and updated to help steer direction, and in the particular case of water, achieve the idealised model of adaptive management. For example, the problem of managing expectations in public consultation processes. As Catley-Carlson (this volume) points out: "Consultation is dangerous unless there is a consultation strategy –in terms of a realistic assessment of the key stakeholders". Here science and innovation can help democratise decision making processes, whilst encapsulating the complexity of the issues at hand, the use for example of techniques like Participatory Bayesian networks, participatory modeling and participatory GIS have provided powerful new tools for decision support (Zorrilla, 2009). New accounting tools like the water footprint can also help shift debates, e.g. by questioning established assumptions, e.g. in the case of Spain on the economic productivity of water by different uses (Garrido *et al.*, 2010).

4 VIRTUAL WATER TRADE AND THE WATER FOOTPRINT

For current global water use, agriculture is the key sector not only for food security but also for water security, since about 85 to 90% of all (blue and green) water consumed by humans is used for agricultural activities (Hoekstra & Chapagain, 2008). In this context, there are some areas which – due to a combination of geographic location, population pressure, and rapid economic growth – face key decisions in relation to water: Indus-Ganges, Northern China Plain, and the North America High Plains, also called *hot spots*.

The volume of water consumed by evapotranspiration in agriculture seems to be in the order of 7,000 km^3/yr (CAWMA, 2007; Hoekstra, 2009). A study by FAO-IFAD indicates that there are 1,500 million hectare (Mha) of cultivated land. Out of this 1,500 Mha, about 300 Mha are irrigated cropland. Out of this 7,000 km^3/yr, around 80% is used domestically (inside the country where the agricultural goods are produced) (Hoekstra & Chapagain, 2008). From this, 20% of virtual water traded in food only about 15–16% according to Yang & Zehnder (2008), is imported by arid and semiarid countries. In other words, it seems that the virtual water traded to solve water and food security is smaller than 4% of the total amount of water consumed currently in the planet to feed the 7,000 million persons who inhabit it. However, Liu *et al.* (2009) consider that the previous numbers are too high and that the real consumptive use is significantly smaller.

Nevertheless, this small percentage is very important to solve the problems in a good number of arid and semiarid countries. One typical example is the MENA (Middle East and North Africa region) where the volume of virtual water imported is greater than the Nile river average annual stream flow (Yang *et al.*, 2007; Kuylenstierna *et al.*, 2008). Increasingly more *water-poor* countries will be looking outside their borders to secure access to virtual water, or in some cases, for land to grow their food, as is the case e.g. with Saudi Arabia currently in the process of buying land in Sudan (Cotula *et al.*, 2009).

Virtual water trade has advantages and disadvantages. For example, it could be argued that the current land grab is an unintended consequence of the increased awareness on the key strategic role of virtual water. Equally, there is increased awareness of the potential *blue* water savings from international food trade for *water poor* countries. Overall the estimate on global water savings due to trade is around 300–500 km^3 (Hoekstra & Chapagain, 2008). This is between 4–8% of the total global agricultural footprint (around 7,000 km^3/yr). On the other hand, the cascade of errors in many estimations of evapotranspirative consumption (scale factor, type of soil ignored, and others) is probably much greater than 5%. Therefore, we have to be very cautious on the relevance or scale of global water savings due to virtual water trade. The main advantage might be that virtual water trade could represent strategically significant outside options for intensively developed river basins. This is a way to secure access to additional (virtual) water resources, thus keeping real green and blue water for key uses like public water supply and e.g. high added value industrial uses like solar thermal energy. For instance, preliminary assessments indicate that in Spain the economic productivity for water in thermo solar is between 100 and 200 greater that the economic productivity of water for irrigation for cereals or cotton (Garrido *et al.*, 2010). It seems obvious that cereals and cotton can be imported from other water rich countries and would be therefore meaningless to prioritize water uses for irrigation over thermo solar production.

However, this raises some key issues related to regulatory structures on both national and global food trade, as well as national and global regional structures in order to adapt them and make them resilient to these fast changes in trade and land ownership. Trade has the potential to help countries manage water security in a globalised world system. However, the global trade system is outdated and in need of deep reform (WEF, 2008). In the context of the current stall in World Trade Organization (WTO) negotiations, and the growth in bilateral negotiations, questions are centred on how food trade fits into inherently political decisions. If water scarcity or its increased opportunity cost is internalised into global trade structures, there are some key areas that should be addressed: e.g. could food and fibre international trade be regulated by policies derived from virtual water? Is livestock a sector that merits special attention due to changes in diet, and its implications in terms of increased water demand? Should regulation be introduced on clean and dirty energy and incorporated in the goods that are traded internationally? Would there be additional problems if virtual water and dirty and clean energy are used to design international trade policies?

There are also some pertinent questions raised on whether implications are different if this trade refers to *domestic trade* (e.g. India and China) compared to *global trade*. To achieve the potential global water savings from trade, it is important to consider that water has not traditionally been a direct determining factor for trade, as compared to other production factors like land, labour, or capital.

Global food trade is not driven by-an-large by water scarcity. For example, if Canada imports bananas or flowers from Central America, it is not because Canada is water scarce, in fact Canada is one of the most global *water rich* countries in terms of water resources. International commodity trade mainly depends on factors such as availability of land, labour, technology, the costs of engaging in trade, national food policies and international trade agreements. In this sense, what was previously mentioned is a logical development: and just a small amount of international virtual water trade is due to water scarcity. However, some countries like Saudi Arabia as a *cash rich, water poor* country have taken the policy decision to secure food and water from outside their border (see above). Meanwhile other *water poor* countries will be unable to adopt this strategy because they are *cash poor*. This is also the case for some countries of the MENA region (Yang *et al*., 2007) and in other countries (Yang & Zehnder, 2008; Liu *et al*., 2009). Preconditions for trade are a minimum of wealth in terms of GNP and a fair trading system. The virtual water *trade* is fraught with diverse problems and possibly the most urgent is the lack of fair international trade regulations, and where attempts to solve this impasse by the WTO have failed up to now. The main obstacle seems to be the position adopted by some industrialized countries.

Another key associated topic to the regulatory reform of the current trade system is the issue of perverse incentives and in particular the current subsidy structure in agriculture by most OECD countries. This probably deserves deeper analysis because of some media grabbing *moral hazards* in current European subsidies to agriculture, where e.g. it appears that each cow receives about 1 US$ per day, which is higher than the income of more than 1,000 million human beings (Development Policy Forum, 2009). In the context of recent large subsidies to other sectors (e.g. the bail-out of the financial system and banks), key questions need to be posed on what are the principles that define which subsidies are permissible and which are detrimental. The WTO agreements on subsidies and countervailing measures have established a legal definition of subsidy. However, in reality there is no clear over arching consensus and analytical framework to help evaluate the use of public budgets, in relation to subsidies to energy (electricity), subsidies to agricultural prices, and cross subsidies, e.g. for public water supply or to water infrastructure.

The analysis of the flare up of the price of staple food two years ago has already identified a number of drivers (biofuels, speculation by some food international corporations, oil price increase, lack of enough storage for staple foods, economic and financial crisis, and others) (Lamo de Espinosa, 2008; 2009; Paarlberg, 2008a; 2008b). One of the main potential reasons is that millions of people in BRIC countries are starting to eat meat, and this will drive cereals prices up for the long run. Therefore, the issue of incentives (and subsidies) –in the context of regulatory reform for trade liberalization– is the key complementary side to trade, and questions on the criteria used to assess the *kindness* of subsidies. For example, do subsidies lock in inappropriate technologies; lead to inefficient use of water, or remove incentives to invest on infrastructures? It is also necessary to identify where, how and when to reduce perverse incentives (including subsidies), and where subsidies are in fact needed.

A number of authors (Falkenmark & Rockström, 2004) consider that rain-fed agriculture should play a more important role in providing food for a growing world population. Nevertheless, some authors call attention to the issue of rainfall variability, which increasingly is perceived as more urgently pressing in terms of adaptation in the short term than climate change itself. This is because addressing variability is a win-win strategy since adapting to rainfall variability is a good adaptation measure to prepare for the much larger variability range of potential climate change. However, this variability in the production of staple foods is not global since up to now droughts have been regional; a *global drought* has never occurred. It may be appropriate to assess the global staple food storage capacity because lack of storage might have been one of the key factors that explains the increase in the prices of staple food two years ago.

Climate change will be mainly mediated through water, i.e. most climate change impacts are related to water (sea-level rise, rainfall variability, melting of glaciers, river flow and groundwater recharge, availability of water for crops production, frequency and magnitude of floods and droughts, ...). Water management is therefore key in adaptation to both climate variability and climate change and to amplifying the resilience of the global economy (López-Gunn, 2009). Innovation which incorporates both cutting edge research and scientific and technological advances as well as indigenous/local knowledge on e.g. good husbandry of green and blue water management, e.g. age-old techniques like terracing, fallow periods, etc. represent valid and valuable traditional knowledge. Recycling this knowledge can be a relatively cheap investment with potential high pay offs. For example, the European Union could consider the potential for incorporating green water credits, i.e. a mechanism to transfer cash to rural people in return for water management activities that determine the supply of all fresh water at source – activities that are presently unrecognised and un-rewarded. Direct payment could enable better management of the resource[2] in rural development policy (Dent & Kauffman, 2007; Dent & Kauffman, 2008), and their linkage with payments for environmental services. In addition, green water credits might become increasingly seen as part of climate justice demanded by African countries and which have to be incorporated into the reform of foreign and development policies in the EU. Both these examples highlight how green water credits and payments for environmental services are intimately linked with the need to reform agricultural policies like the Common Agricultural Policy, away from price support, and towards rural diversification and land use and good husbandry policies.

Many predictions or assessments state the need for more water or arable land to cope with the extra food needed for the growing world population (2,000 or 3,000 million persons within 25 to 50 years). These consider that the agricultural productivity of staple foods in developing and/or emerging countries will continue to be small in comparison with productivity in industrialized countries. For example, today the yield of rainfed cereal across humid Europe might be up to 8–10 t/ha, but in developing countries productivity is assumed to rarely reach one third of this amount. This is equivalent to assuming that most farmers in the Third World are not going to have access to technology, institutions, transportation, markets, or are incapable of innovating. Yet data on the yearly economic growth (8 to 10%) in a good number of developing countries, like India and China, seems to reject such a pessimistic assumption. As was previously mentioned, the papers by Goklany (2009) and Kuylenstierna *et al.* (2008) in addition to the communication of Prof. Allan (2009) supports the idea of a potential increase in agricultural productivity in the near future based on previous precedents in now developed countries. To conclude this section, virtual water and the water footprint, represent a new take on an old idea (embedded water and comparative advantage). It shows that increasing the hydrological and economic productivity of water used for agriculture offers an entry point for relatively simple, but high impact changes in national and global water policy (WEF, 2008). The extended water footprint may become a good tool to achieve Integrated Water Resources Management (IWRM). Furthermore, in many countries like Spain it may facilitate the transition from the motto ***more crops and jobs per drop*** to the new motto ***more cash and care of nature per drop*** as required by the EU Water Framework Directive.

---

[2] [http://www.isric.org/UK/About+ISRIC/Projects/Current+Projects/Green+Water+Credits.htm].

5 NEW AND RECYCLED WATER: DESALINATION AND REUSE[3]

The global water footprint at the moment is mainly agricultural (i.e. about 90%), and this is likely to decrease in the coming thirty to fifty years (WEF, 2008). Therefore there are key questions related to this transition in the global economy, towards a service based global economy and the re-allocation between sectors on the existing water resources and uses. In this changing global water scenario, certain areas and sectors become particularly crucial. For example, at present more than half the world's population lives in urban areas, and of these, about a third are located in coastal areas. This opens a number of opportunities, e.g. related to desalination and reuse of urban waste water for agricultural uses. Economies of scale exist, however there remain some constraining factors like access to financial resources and the cost of energy. Recent events related to lack of water resources for urban water supply, e.g. due to drought in cities like Barcelona, Atlanta and Istanbul are highlighting the problems of guaranteeing water supply and of making these cities water secure. This will be reflected in the political decisions on trade-offs, for example, whether to e.g. transfer water resources physically, opt for desalination or instead for re-allocation between existing dominant uses (mainly agriculture). All these options have pros and cons and require a great dose of transparency and education to the public at large. For example, transparency on prices paid per m^3, and the willingness to pay by different economic sectors (e.g. agriculture *vs.* urban users) and whether desalination plants fulfill their projections (e.g. time span before they operate at full capacity). However, reallocation of water used to irrigate low value crops to new water demands like public water supply may be the best way of solving water supply problems in the ever growing cities. This, of course, requires that the staple food supply for developing countries is guaranteed through better international regulation of food trade.

There is little doubt about the relevance of recent advances in chemical Engineering to provide *new water*[4]. Nevertheless, in Spain and other countries the clear advantages of this *new water* technology have to be assessed realistically, to dispel the possibility that sometimes this potential might have been overstated for political or other reasons. There is also the potential for water reuse. However data in this area are not easily accessible or detailed enough (including costs), to be able to assess easily the scale and full potential of reuse in terms of timing, possibilities, costs e.g. based on sewage systems in operation[5].

In addition, there is dispersed data at the global and national level on the current uses, costs and prices of desalinated sea water, desalinated brackish groundwater and reused waste water. In some specific local circumstances desalination may be a useful solution. This could be the case for example in the Middle East, where thanks to advances in desalination, water does not have

---

[3] This topic will be dealt with in detail by other participants at the Workshop. Therefore, here we only comment on some general aspects.

[4] *New Water* refers to "additional volumes of fresh water that can be introduced to the hydrological cycle, through various means" such as desalination, wastewater reuse, or bulk importation of fresh water (Phillips *et al.*, 2008: 11).

[5] Water harvesting, supplemental irrigation, managed aquifer recharge, field water conservation and deficit irrigation offer relatively low-cost technologies that can have real impacts at the local scale. For example, the decentralized recharge movement was an almost spontaneous response to India's groundwater depletion to help water tables rebound to pre-development levels at the end of the monsoon season in pockets of intensive use (Polak, 2008). This is an example of the contrast in perception between *popular hydrogeology* and formal *hydrogeology*, e.g. scientists argue hard rock have too little storage and advocate recharge, meanwhile the prolific growth of recharge structures is based on the value people attach to a check dam even if these wells would only provide 1,000 m^3. This 1,000 m^3 is nevertheless crucial for livelihoods irrigation in times of delayed rain. However, it seems important to obtain better data on the practical relevance of these methods in countries with a medium or high economic standard. Possibly, the real significance in the water balance of the country is irrelevant, but they can be vital in specific points in time, and furthermore can contribute to create a better public awareness on the duty of care for nature. At a bigger scale, new ideas like virtual water trade and the extended water footprint can offer better national and global strategic water policy choices in water allocation.

to be a cause for conflict between Israel and Palestine. The fact that discourses on water conflict continue despite available options highlights that water can become an excuse or smokescreen for other conflicts. The main conflict over water centers on a volume of water that is less than 400 Mm^3/yr of groundwater in the Judean aquifer. Today the cost of this volume as desalinated seawater is less than US$ 300 millions per year. This is significantly less than 1% of the GDP of Israel's (Llamas, 2008). Provided there is political will, water can be a *positive sum outcome*. This becomes obvious when one can show through a theoretical economic model, like WAS (Fisher & Huber-Lee, 2008) that water is mainly valuable where it is scarce. In fact, the generation of *new water* in terms of quantity is a trivial amount for governments. As Fisher & Huber-Lee (2008: xiv) stress: "thinking about water values rather than water quantities can lead to useful and surprising results".

6 DEEP BLUE WATER: GROUNDWATER DEVELOPMENT SILENT REVOLUTION[6]

Groundwater is the largest volume of global freshwater resources in this planet, and yet it is largely invisible for civil society and in the mind of many water planners. Yet in the last half century groundwater use has increased ten times –from 100×10^3 Mm^3/yr to more than $1{,}000 \times 10^3$ Mm^3/yr (Shah, 2008). This spectacular increase has been generally undertaken by millions of modest farmers, a true example of human innovation and initiative. Yet it has been undertaken with scarce or no planning and control by conventional governmental water authorities, and the scale of this silent revolution are now becoming more startling thanks to new technologies.

This silent revolution has produced large economic and social benefits, particularly since it has been an emergent, spontaneous process that has enabled socioeconomic transition and change (Llamas & Custodio, 2003; Shah, 2008; Mukherji *et al.*, 2009a; Giordano & Villholth, 2007). Groundwater irrigation is considered by many as a driver for important social changes. However, it is less clear –since it is a relatively new global phenomenon (half a century of groundwater irrigation compared to 5,000 years of surface water irrigation)– what the long term consequences of this intensive groundwater use might be. What is clear however is that there can be and have been substantial environmental externalities to this groundwater growth model, which some authors think might follow the boom and bust model of mining and non-renewable resource use (Hotelling rule[7]). There is increased awareness on the scale of ecological impacts, such as the problems created by overdraft of aquifers in Southwest and Central USA, China, India, Australia, Spain, Mexico. Nevertheless, it is appropriate to know that this concern has grown especially among experts in surface water resources who usually are non-experts in groundwater hydrology. As a matter of fact the UNESCO World Commission on the Ethics of Science and Technology (COMEST), published a report (Selborne, 2001) that gives recommendations on this issue[8].

Nevertheless, we are not aware of practically any case where the ecological, economic and political externalities caused by the use of non-renewable groundwater have been analyzed in adequate detail from multiple perspectives. In general the concept of sustainability has been defined as ecological impacts, the emphasis predominant in a Northern and Central European approach, which only takes socio-economic aspects as a secondary aspect. The WFD of the European Union is

---

[6] This topic is dealt in detail in other chapters in the book, therefore we will only emphasize some aspects.

[7] Hotelling's rule is a 1931 economic model of non-renewable resource management by Harold Hotelling. It shows that the efficient exploitation of a non-renewable and non-augmentable resource would, under otherwise stable economic conditions, lead to a depletion of the resource. The rule states that this would lead to a net price or *Hotelling rent* for it that rose annually at a rate equal to the rate of interest, reflecting the increasing scarcity of the resources (http://en.wikipedia.org/wiki/Non-renewable_resource).

[8] This report is largely based on a previous monograph by Llamas & Delli Priscolli (2001).

the epitome of this perspective. This is rather narrow and sometimes highly unrealistic when looking at the reality of what is happening e.g. in developing countries. This was already commented by the UNESCO Working Group on the Ethics of Freshwater Uses and the COMEST (Llamas & Delli Priscoli, 2001; Selborne, 2001; Delli Priscoli *et al.*, 2004). In Llamas *et al.* (2008) some cases are described where it is the economic and political circumstances that are the main drivers in groundwater development and not ecological motives. Some may think that this situation is typical of developing countries: this is not the case.

For example, in Spain the Upper Guadiana basin has experienced an intensive groundwater development for irrigation since the 1960s. These groundwater abstractions have generated important socio-economic benefits to the farmers and to the region but has also caused an important ecological degradation to the UNESCO Biosphere Reserve *La Mancha Húmeda*, to the Tablas de Daimiel National Park and to several Ramsar sites located in the aquifer. There has been –and continues– a strong conflict between the conservation lobbies and the agricultural lobbies. Nevertheless, up to now the Central and Regional Governments have not been able to enforce the Spanish Water Code that theoretically has tools to solve this situation. Apparently the power of the farmers is higher than that of the conservationists (Llamas *et al.*, 2010).

In the case of groundwater security, there are a number of issues that need further analysis, because of particular characteristics inherent to groundwater as a resource. This relate to (and will be discussed in turn): first, the incentive structure for pumping groundwater; second, the inherent resilience of groundwater to dry spells; third, the individualistic nature and high level of agency inherent in groundwater use as a common pool resource; and fourth, the need in groundwater governance to devise ways to incentivise collective action, seeking win-win scenarios.

Llamas & Martinez-Santos (2005) described the evolution of a series of phases related to groundwater development in most arid and semi-arid countries. This chapter revisits this analysis to add an additional stage. Figure 1 presents an idealised overview of the different water policy stages that

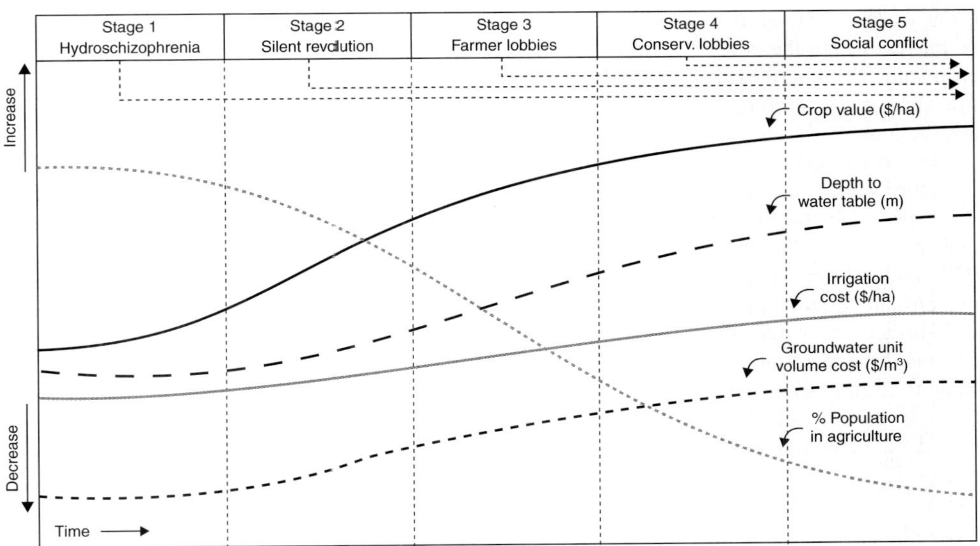

Examples				
California (1920) Texas (1930) Arizona (1950) Spain (1960) India (1960) Mexico (1960)	California (1930) Texas (1940) Arizona (1960) Spain (1970) India (1970) Mexico (1970)	California (1950) Texas (1970) Arizona (1970) Spain (1980) India (1990) Mexico (1990)	California (1980) Arizona (1980) Spain (1990)	Spain (Ebro Transfer, 2000) California (Bay-Delta Plan, 1999) India (Energy Subsidies, 2004)

Figure 1. Idealised stages in groundwater-related development in arid and semiarid regions (after Llamas & Martínez-Santos, 2005).

many arid and semiarid countries experience due to intensive groundwater use. Each of the five stages is roughly equivalent to one generation (about 15–25 years).

The five stages that can be distinguished are: a first stage when groundwater is treated as a *pillar of sand* compared to surface water and therefore is either neglected, ignored or both in water planning by conventional hydrologists or water planners (Stage 1. Hydroschizophrenia). The second stage is mainly due to the inherent *resource blessing* characteristics that groundwater has (namely resilience to drought, low investment costs, agency i.e. control by individuals –e.g. in decision making). This triggers a *silent revolution* in a vacuum of conventional water authorities, often incentivized or encouraged by other authorities (normally agricultural departments). In the third stage the stream of economic benefits, where normally farmers tend to be the main beneficiaries, stimulates the creation and/or strengthening of farmer lobbies to defend groundwater use. In the fourth stage this often comes head to head with conservation lobbies, who campaign to reduce abstractions due to environmental externalities, which become an issue for public policy and a cause for social conflict between competing users and where there is difficulty in changing the *status quo* and inertias in the system. The beginning of these conflicts can generally be traced back to the moment when intensive groundwater development begins. In the later stages (particularly 3–5) there might be some overlapping between stages.

One of the most significant aspects of this *Silent Revolution* is the manner in which farmers, as they grow richer and more educated, move from low value crops to cash crops. This is mainly due to the intrinsic reliability of groundwater, where encouraged by the expectation of enhanced revenues, farmers decide to invest in better irrigation technology and, in turn, shift to higher value crops. As crop value is related to the type of crop, climatic and other natural and social variables particular to each site, and also subject to trade constraints, crop value ranges widely. In Europe for instance crop value ranges between US$ 500–800 per hectare (e.g. cereals) to more than US$ 60,000 per hectare for tomatoes, cucumbers and other greenhouse crops. Frequently in Spain, and probably in other industrialized countries, the ratio between crop value and groundwater irrigation cost is greater than 20; in other words, the cost of water abstraction is usually in the order of 5% (Llamas & Custodio, 2003). This ratio may be radically different in developing countries, where the cost of groundwater abstraction may be in the order of almost 50% of the crop value (Mukherji *et al.*, 2009b). Obviously this is a very different scenario, mainly due to the low value of the crop (usually rice and other cereals) and to the use of poor technology in water abstractions.

The *more crops and jobs per drop* motto is considered one of the most effective policy options to secure both water and food. This is because of the large share irrigation has in global blue water use, and the often low efficiency in irrigation. However, few water experts or decision-makers are aware that the goal behind such motto is now often achieved by groundwater irrigation (Hernandez-Mora *et al.*, 2001). Groundwater irrigation will play a key role and useful testing ground because of its resilience to climate variability in contrast with rain-fed agriculture and surface water irrigation systems. In the industrialized or rich countries in arid and semiarid regions the new motto should be *more cash and care of nature per drop*, which adds a new 6th stage to groundwater intensive use, since it becomes a win-win for both farmers (and new users like e.g. thermo-solar groundwater use in Spain) and nature. For instance, as previously mentioned, the economic productivity of 1 m^3 used for a thermo-solar plant is about 100 to 200 times higher than if this m^3 was used to produce cereals or cotton. This efficiency gains in both water use and increased productivity can release water for other uses like e.g. water needed by local wetlands.

Examples like this are particularly relevant for *water stressed regions*, since these *water poor* regions can buy staple food from *water rich* regions, normally with abundant (usually green) water and save their scarce blue water for cash crops or other beneficial uses, like wetlands. Green water and surface water irrigation systems cannot guarantee crops due to climate variability, and this is why growers of cash crops generally opt for groundwater irrigation.

The analysis of the *extended* water footprint considers not only water embedded in a product, but also the economic value obtained from every m^3 assigned to each use. For instance in Spain about

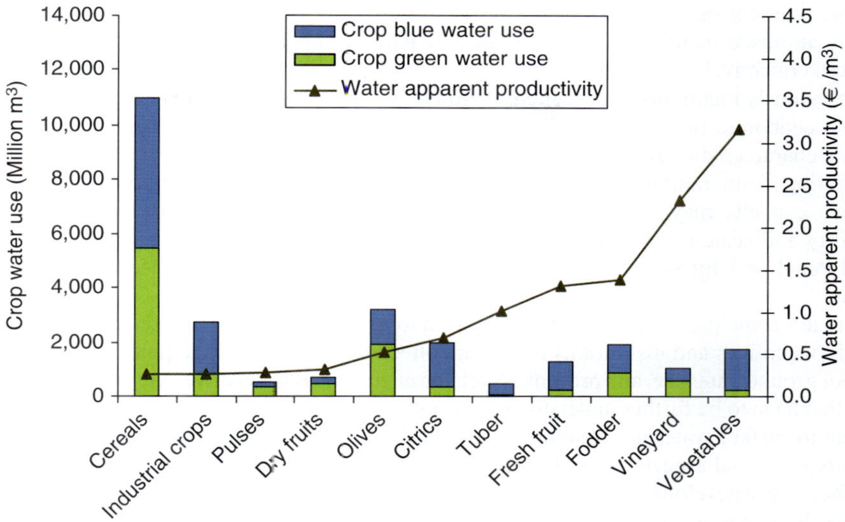

Figure 2. Apparent water productivity and blue and green water footprint for different crops in Spanish agriculture (2002).
Source: Aldaya *et al.* (2009a; 2009b).

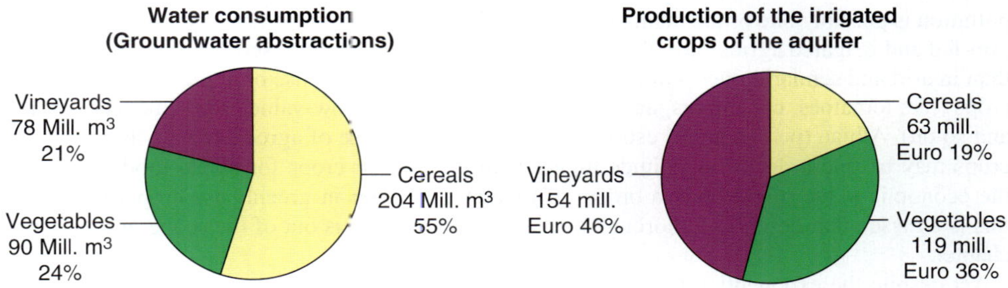

Figure 3. Water consumption and irrigated crops in the Western Mancha aquifer (Upper Guadiana Basin, Spain).
Source: López-Gunn & Zorrilla (2010).

90% of green and blue water is used for agricultural production; yet in Spain the driest country in the European Union, 10% from this 80–90% of water used for agriculture produces almost 90% of all the economic agricultural value (Aldaya *et al.* 2009a; 2009b) (see Figure 2). This means that almost 70% of water used in Spain is applied to produce *low value crops*. It might be better to buy these low value agricultural goods in countries with abundant (green) water and use the theoretical scarce water for cash crops, tourism, industry or meeting the new regulatory requirements for ecological flows under the EU Water Framework Directive.

For example, in the Upper Guadiana basin, in 2006 cereals represented 55% of the water volume abstracted; yet, in terms of economic productivity, cereals only accounted for 19% of the economic added value generated (see Figure 3). Meanwhile, vineyards, which accounted for only 21% of the water abstracted, generated 46% of the economic added value. This situation is even clearer in the Mediterranean regions of Spain where it is possible to grow crops with much higher economic returns than in the Upper Guadiana basin which has a continental climate.

At the moment the dominant incentive structures drive the use of groundwater intensively in a highly atomised manner, where there are few perceived benefits from managing groundwater resources collectively[9].

As previously mentioned, the economic productivity of groundwater irrigation is usually higher than the economic productivity of surface water irrigation. This can be explained because of inherent characteristics in groundwater resources which make it a *resource blessing*, in terms of buffering from rainfall variability, low investment costs, control in time and space. The buffering capacity may become also a crucial point for the adaptation to the potential greater variability and scarcity of precipitation generally predicted by some models of the Intergovernmental Panel on Climate Change (IPCC) for the Mediterranean region, identified as a potential hotspot.

There are some pressing questions in relation to groundwater governance and how to generate the right incentives and institutions to favour collective action, avoid the potential boom and bust model of groundwater use, and prevent associated environmental negative externalities. Some of the issues that have to be further understood are *whether* and *if*, groundwater governance is inherently different to surface water governance because of the nature of the resource and the individualised and entrepreneurial aspects of groundwater use. This also set in the context of almost 5,000 years of surface water development as compared to a short 50 year span of groundwater use.

In previous sections we have mentioned the new motto *more cash and care of nature per drop* as one goal to achieve in the near future in global water policy, beginning in the industrialized and emergent economies. At a later stage this might be applicable to developing countries. The first part of the motto *more cash per drop* seems clearly feasible, mainly if more just and equitable international trade regulations are achieved. However, we cannot ignore that the second part of the motto: *more care of nature per drop* may be fraught with difficulties. Agricultural diffuse pollution is possibly one of the main problems in global water policy. It is a common issue to both rain-fed and irrigated agriculture. Probably this problem is today more serious in humid countries than in arid and semiarid ones. On the other hand, it seems that the use of agrochemicals for cash crops (like tomatoes, cucumbers, and so on) is higher than for low-value crops (like cereals, rice and so on). Albiac (pers. comm.) estimates that in Spain the use of agrochemicals in greenhouse crops may be one order of magnitude higher than in open air crops for staple foods. However, the economic value might be two orders of magnitude higher in greenhouse crops. Obviously, this is an issue that deserves a thorough assessment although it is out of the scope of the present chapter.

Yet despite the exponential growth in groundwater use and its increasing strategic value, e.g. for public water supply, in coastal areas and for some of the most heavily populated countries in the world (e.g. China, India, and Brazil), groundwater governance is in its infancy. The approach has to be very different to surface water irrigation governance. In order to cope with the current groundwater development anarchy, it is increasingly important to analyse institutional aspects of water use and allocation for groundwater, since much groundwater use in many countries operates under *informal* rules. Yet the key issue is the same in surface and subsurface water: institutions for cooperation among stakeholders to achieve collective action.

Groundwater offers unique opportunities for testing out the first part of the motto *more cash and care for nature per drop* because groundwater development is resilient to climate variability, which

[9] This is also known as the Gisser-Sánchez effect, i.e. an economic model that demonstrates theoretically what many people observe empirically, that farmers experience great incentives to use groundwater intensively and have little economic benefit from controlling use (see Koundouri, 2004, for a detailed explanation), i.e. the *no-management* (competitive) dynamic solution of groundwater exploitation is almost identical (in terms of derived social welfare) to the *efficient management* (optimal control) solution (Koundouri, 2004: 706). This is a management paradox because the serious depletion of aquifers is a major risk to many freshwater ecosystems yet the social benefits from managing collectively groundwater abstractions are numerically insignificant. This also has significant implication for water managers because it is a severe constraint to the effectiveness of policy options since implementing reduced extractions is not socially, economically or politically costless.

is the Achilles heel of green water (rain-fed) agriculture as to most surface water irrigation[10]. If the pessimistic predictions of the IPCC concerning the decrease and greater irregularity of precipitation in some regions materialize, the role of groundwater to achieve an adaptation to these changes will increase as a natural insurance and buffering system to climate variability (Hetzel *et al.*, 2008; López-Gunn, 2009). Groundwater has higher hydrologic and economic productivity in comparison with surface water irrigation. This advantageous economic aspect of groundwater irrigation may compensate the requirement of *more care of nature* (using less agrochemicals) and the result means a win-win for the farmer lobby to then facilitate farmers acceptance of *more care for nature*.

7 CONCLUSION: 10 SUGGESTIONS TO HELP RE-FRAME GLOBAL WATER POLICY

1) Freshwater and food are not a scarce in this blue planet. The existing security problems are local and affect to a small proportion of humanity (less than 15%).
2) The increasing need for food because of the improvement in quality of life and population growth can be met without a significant increase in either the use of blue water or area of arable land. However, this requires a greater effort from industrialized countries to transfer agricultural technology and know-how to developing countries.
3) The scarcity of green and blue water in arid and semiarid regions can be partly compensated through virtual water trade. In reality, virtual water trade is already functioning and solving the problems in some countries in the MENA region, like Jordan, Libya and Israel.
4) Virtual water trade is not yet a panacea to solve water and food security problems. The main obstacle is the lack of international regulations that guarantee the security of poor countries against political pressures, embargos, and oligopolies by international food corporations. The efforts currently underway under the World Trade Organization should be improved and strengthened.
5) Most virtual water traded is green water from humid countries, like Canada or Argentina. One potential problem is the great variability of precipitation and its impact on corresponding crops. Up to now there has never been a global drought. Nevertheless, it seems that the global staple food storage should be assessed in order to mitigate this potential contingency.
6) While irrigated agriculture is more resilient to precipitation variability, surface water irrigation systems can also fail in long dry spells. Meanwhile groundwater irrigation is strongly resilient to dry spells. Groundwater farmers obtain greater profit during droughts, when irrigation with surface water systems are vulnerable to failure.
7) The groundwater *silent revolution* is a fact in most arid and semiarid regions. It has increased ten fold in the last half century. It has produced stupendous social and economic benefits but also colossal groundwater management anarchy.
8) It seems relevant and urgent that governments in arid and semiarid regions obtain better information on the situation of groundwater irrigation as a preliminary stage to correct this current general global anarchy.
9) Probably the most serious concern in water resources policy is how to cope with diffuse pollution due to rain-fed and irrigated agriculture.
10) The new motto in industrialized and emergent countries should be *more cash and care of nature per drop*. Developing countries probably should still operate in the paradigm of *more crops and jobs per drop*.

---

[10] Pressures and increased competition on water resources, and in particular groundwater which is particularly attractive because of its high productivity, means that the *low hanging fruit* (i.e. quickest and easiest gains) lies in increasing productivity, for which groundwater is particularly well adapted to, and where generating *more cash and care of nature per drop* is a win-win for farmers and the environment.

ACKNOWLEDGEMENTS

Many of the ideas used in this chapter have been the result of work and discussions by the authors with their colleagues at the FMB Water Observatory. We have benefitted greatly from these group discussions.

REFERENCES

Aldaya, M.M.; Garrido, A.; Llamas, M.R.; Varela-Ortega, C.; Novo, P. & Rodriguez, R. (2009a). Water footprint and Virtual Water Trade in Spain. In: A. Garrido & M.R. Llamas (eds.), *Water Policy in Spain*. CRC Press: 49–61. ISBN-10: 0415554114.

Aldaya, M.M.; Martínez-Santos, P. & Llamas, M.R. (2009b). Incorporating the water footprint and virtual water into policy: reflections from the Mancha Occidental region, Spain, *Water Resources Management*, Springer. Published on-line, July 15th, 2009.

Allan, J.A. (2009). Written communication addressed to the Fourth Botin Foundation Water Workshop participants on September 2nd, 2009.

Bellver, A. & Kaufmann D. (2005). *Transparenting Transparency: Initial Empirics and Policy Applications*. Preliminary draft. World Bank. Available at: [http://siteresources.worldbank.org/INTWBIGOVANTCOR/Resources/Transparenting_Transparency171005.pdf].

CAWMA (Comprehensive Assessment of Water Management in Agriculture) (2007). *Water for Food, Water for Life: A Comprehensive Assessment of Water Management in Agriculture*. D. Molden (ed.). Earthscan, London, UK; and International Water Management Institute, Colombo, Sri Lanka.

Clark, C. (1968). *Population Growth and Land Use* [Spanish version: Crecimiento Demográfico y Uso de la Tierra]. Alianza Editorial, Madrid, Spain: 469 pp.

Cotula, L.; Vermeulen, S.; Leonard, R. & Keeley, J. (2009). *Land grab or development opportunity? Agricultural investment and international land deals in Africa*. IIED/FAO/IFAD, London/Rome. ISBN: 978-1-84369-741-1.

Davoudi, S. (2006). Evidence-Based Planning. *Rhetoric and Reality*, 42(165): 14–24.

Delli Priscoli, J.; Dooge, J. & Llamas, M.R. (2004). Overview. Series on Water and Ethics, Essay 1. UNESCO, Paris, France. 31 pp. ISBN 92-9220-016-X.

Dent, D. & Kauffman, S. (2007). *The spark has jumped the gap. Green Water Credits Proof-of-Concept*. October 2007. Available at: [http://www.isric.org/isric/webdocs/docs/GWC_Report_7.pdf].

Dent, D. & Kauffman, S. (2008). *Green Water Credits*. ISRIC – World Soil Information Policy Brief. Available at: [http://www.isric.org/isric/webdocs/docs/PolicyBrief11web.pdf].

Development Policy Forum (2009). *Saving the MDGs from Global Recession*. Report of the Fourth Development Policy Forum (DPF). Roundtable organised by Friends of Europe in partnership with UN, World Bank, AFD, DFID and GTZ.

Eden, S. & Tunstall, S. (2006). Ecological versus social restoration? How urban river restoration challenges but also fails to challenge the science-policy nexus in the United Kingdom. *Environment and Planning C: Government and Policy*, 24: 661–680.

Falkenmark, M. & Rockström, J. (2004). *Balancing water for humans and nature: The new approach in ecohydrology*. Earthscan, London, UK.

Fisher, F. & Huber-Lee, A. (2008). WAS-Guided cooperation in water management. MIT Department of Economics. Available at: [http://econ-www.mit.edu/faculty/ffisher/water]. Accessed September 15th 2009.

Garrido, A.; Llamas, M.R.; Varela-Ortega, C.; Novo, P.; Rodríguez Casado, R. & Aldaya, M.M. (2010). *Water Footprint and virtual water trade of Spain: policy implications*. Springer Publishers.

Giordano, M. & Villholth, K.G. (eds.) (2007). *The Agricultural Groundwater Revolution: Opportunities and Threats to Development*. Volume 3, CABI Publication: Wallingford, UK; and Cambridge, Massachusetts, USA.

Goklany, I.M. (2009). Have increases in population, affluence and technology worsened human and environmental well-being? *The Electronic Journal of Sustainable Development* no. I/3.

Hernández-Mora, N.; Llamas, M.R. & Martínez-Cortina, L. (2001). Misconceptions in Aquifer Over-Exploitation. Implications for Water Policy in Southern Europe. In: C. Dosi (ed.), *Agricultural Use of Groundwater. Towards Integration between Agricultural Policy and Water Resources Management*. Kluwer Academic Publishers: 107–125. ISBN 0-7923-6805-3.

Hetzel, F.; Vaessen, V.; Himmelsbach, T.; Struckmeier, W. & Villholth, K.G. (2008). *Groundwater and climate change: challenges and possibilities*. BGR, Hannover, Germany. 20 pp.

Hoekstra, A.Y. (2009). Human appropriation of natural capital: A comparison of ecological footprint and water footprint analysis. *Ecological Economics*: 1963–1974. Available at: [http://www.sciencedirect.com].

Hoekstra, A.Y. & Chapagain, A.K. (2008). *Globalization of water: Sharing the planet's freshwater resources*. Blackwell Publishing, Oxford, UK.

Jarvis, T. (2008). Strategies for groundwater resources conflict resolution and management. Paper presented in the Conference of the NATO Advanced Study Institute. Varna, Bulgaria, October 1–12, 2006. In: C.J.G. Darnault (ed.), *Overexploitation and Contamination of Shared Groundwater Resources*. Springer Verlag Science + Business Media: 393–414.

Kingdon, J. (2003). *Agenda, alternatives and public policy*. 3rd Edition. Longman, New York, USA.

Koundouri, P. (2004). Current issues in the economics of groundwater resource management. *Journal of Economic Surveys*, 18(5): 703–740.

Kuylenstierna, J.; Destouni, G. & Lundqvist, J. (2008). Feeding the future world – securing enough food for 10 billion people. In: *Water for Food*, the Swedish Research Council Formas: 9–21. ISBN: 978-91-540-6019-1.

Lamo de Espinosa, J. (2008). *El regadío en el mundo* [Irrigation in the World]. Congreso Internacional de Riego [National Irrigation Congress]. Zaragoza, Spain. June 2008, (manuscript in press).

Lamo de Espinosa, J. (2009). La crisis económica global y la crisis agraria española [The global economic crisis and the agricultural Spanish crisis]. *Mediterráneo Económico*, Madrid, Spain.

Liu, J.; Zehnder, A.J.B. & Yang, H. (2009). Global consumptive water use for crop production: the importance of green water and virtual water. *Water Resources Research*, 45, W05428: 15 pp.

Llamas, M.R. (2008). Some suggestions on the summary of discussion. Presented to the CIT-PAX Meeting on *Water in the Middle East: Source of Confrontation or Cooperation*. Madrid, Spain. September 13th, 2008.

Llamas, M.R. & Custodio, E. (eds.) (2003). *Intensive Use of Groundwater: Challenges and Opportunities*. Balkema Publishers, Lisse, the Netherlands. 478 pp. ISBN: 9058013905.

Llamas, M.R. & Delli Priscoli, J. (2001). *Water and Ethics*. Papeles del Proyecto Aguas Subterráneas, Serie A (5), 99 pp. Fundación Marcelino Botín, Santander, Spain. ISBN 84-95516-14-4.

Llamas, M.R. & Martínez-Santos, P. (2005). Intensive Groundwater Use: Silent Revolution and Potential Source of Social Conflicts. *Journal of Water Resources Planning and Management*, September/October 2005: 337–341. American Society of Civil Engineers.

Llamas, M.R.; Martínez-Santos, P. & de la Hera, A. (2008). Hydropolitics and Hydroeconomics of Shared Groundwater Resources: Experience in arid and Semiarid Regions. Paper presented in the Conference of the NATO Advanced Study Institute. Varna, Bulgaria, October 1–12, 2006. In: C.J.G. Darnault (ed.), *Overexploitation and Contamination of Shared Groundwater Resources*. Springer Verlag Science + Business Media: 415–431.

Llamas, M.R.; Martinez-Cortina, L. & Mukherji, A. (eds.) (2009). *Water Ethics*. Marcelino Botín Water Forum 2007. Taylor and Francis Group, London, UK. 353 pp.

Llamas, M.R.; Varela-Ortega, C.; de la Hera, A.; Aldaya, M.M.; Villarroya, F.; Martínez-Santos, P.; Blanco-Gutiérrez, I.; Carmona-García, G. Esteve-Bengoechea, P.; de Stefano, L.; Hernández-Mora, N. & Zorrilla, P. (2010). The Guadiana Basin. In: Mysiak, J. *et al.* (eds.), The Adaptive Water Resource Management Handbook. Earthscan, London, UK: 103–114. ISBN: 978-1-84407-792-2.

López-Gunn, E. (2009). Spain, Water and Climate Change in COP 15 and Beyond: Aligning Mitigation and Adaptation through Innovation. Working Paper: Royal Elcano Institute for International Affairs. Area: International Economy and Trade, Working Paper 65/2009. Available at: [http://www.realinstitutoelcano.org/wps/portal/rielcano_eng/Content?WCM_GLOBAL_CONTEXT=/elcano/Elcano_in/Zonas_in/DT65-2009].

López-Gunn, E & Llamas, M.R. (2008). Re-thinking water scarcity: can science and technology solve the global water crisis? *Natural Resources Forum*, 32, September: 228–238.

López-Gunn, E. & Zorrilla, P. (2010). The Western Mancha aquifer: data mapping to provide transparency to aid stakeholder participation and decision making. *Journal of Maps*, 2010: 1–9.

Lorenz, E.H. (1992). Trust, Community and Cooperation. In: M. Storper & A.J. Scott (eds.), *Pathways to Industrialization and Regional Development*. Routledge, London, UK: 175–182.

Mukherji, A.; Facon, T.; Burke, J.; de Fraiture, C.; Faurès, J.M.; Füleki, B.; Giordano, M.; Molden, D. & Shah, T. (2009a). *Revitalizing Asia's irrigation: to sustainably meet tomorrow's feed needs*. International Water Management Institute, Colombo, Sri Lanka; and FAO, Rome, Italy.

Mukherji, A.; Villholth, K.G.; Sharma, B.R. & Wang, J. (eds.) (2009b). *Groundwater governance in the Indo-Gangetic and Yellow River basins: realities and challenges*. IAH Selected Papers on Hydrogeology, 15. CRC Press, London, UK. 325 pp.

Newsweek (1993). Stamping out dissent. Too often unconventional or unpopular scientific views are simply suppressed. By B. Martin, *Newsweek*, April 26th: 49–50.

Owens, S. (2005). Making a difference? Some reflections on environmental research and policy. *Transactions of the Institute of British Geographers New Series:* 287–292.

Owens, S.; Petts, J. & Bulkeley, H. (2007). Boundary work: knowledge, policy and the urban environment. *Environment and Planning C: Government and Policy*, 24: 633–643.

Paarlberg, R. (2008a). Rises in world food prices may hurt city-dwellers, but they do not increase global hunger. *Prospect*, June 29th, Issue 147. Available at: [http://www.prospectmagazine.co.uk/2008/06/thetruthaboutfood/].

Paarlberg, R. (2008b). It's not the Price that causes Hunger. *International Herald Tribune*, April 22nd. Available at: [http://www.nytimes.com/2008/04/22/opinion/22iht-edpaarlberg.1.12230340.html?_r=1].

Phillips, D.J.H.; Allan, J.A.; Claassen, M.; Granit, J.; Jägerskog, A.; Kistin, E.; Patrick, M. & Turton, A. (2008). *The TWO Analysis – Introducing a Methodology for the Transboundary Waters Opportunity Analysis*. Report No. 23: 5–35. Stockholm International Water Institute, Sweden.

Polak, P. (2008). *Out of Poverty, what works when traditional approaches fail*? Berrett-Koehler Publishers, Inc.

Radaelli, C.M. (1995). The role of knowledge in the policy process. *Journal of European Public Policy*, 2: 159–183.

Rogers, P. (2008). Facing the Freshwater Crisis. *Scientific American*, August 2008 Issue. Available at: [http://www.scientificamerican.com/article.cfm?id=facing-the-freshwater-crisis].

Rogers, P.; Llamas, M.R. & Martínez Cortina, L. (eds.) (2006). *Water Crisis: Myth or Reality*? Marcelino Botin Water Forum 2004. Taylor and Francis Group, London, UK. 333 pp.

Salman Salman, M.A. (2004). From Marrakech through The Hague to Kyoto: has the global debate on water reached a dead end? Part I and Part II. *Water International*, Vol. 27 and 28.

Schneider, M. & Teske, P. (1992). Toward a Theory of the Political Entrepreneur: Evidence from Local Government. *The American Political Science Review*, 86(3): 737–747, September.

Selborne, J. (2001). *UNESCO World Commission on the Ethics of Science and Technology*. 62 pp.

Shah, T. (2008). *Taming the Anarchy? Groundwater Governance in South Asia*. Resources for Future Press, Washington, D.C., USA.

Simon, J. (1996). *The Ultimate Resource*. Princeton University Press, Princeton, New Jersey, USA.

The Economist (1997). Plenty of gloom. December 20th: 21–22.

Transparency International (2008). *Global Corruption Report: Corruption in the Water Sector*. Cambridge University Press, Cambridge, UK. 365 pp.

United Nations (1998). *Annual Report of the Secretary-General* 173, cf. Chapter V.

United Nations (2009). *World Water Development Program-3*. UNESCO Publishing, Paris, France; and Earthscan, London, UK. 318 pp.

WEF (World Economic Forum) (2008). *Managing our Future Water Needs for Agriculture, Industry, Human Health and the Environment*. Discussion Document for the World Economic Forum. Annual Meeting 2008.

Weiss, C. (1977). Research for policy's sake: the enlightenment function of social research. *Policy Analysis*, 3: 531–545.

Yang, H. & Zehnder, A.J.B. (2008). *Globalization of Water Resources through Virtual Water Trade*. Manuscript presented in the Sixth Biennial Rosenberg International Forum on Water Policy, Zaragoza, Spain. 19 pp.

Yang, H.; Wang, L. & Zehnder, A.J.B. (2007). Water Scarcity and food trade in the Southern and Eastern Mediterranean countries. *Food Policy*, 32: 585–605.

Zorrilla, P. (2009). *Análisis de la gestión del agua en el acuífero de la Mancha Occidental: construcción de una red bayesiana mediante procesos de participación pública*. Doctoral Thesis. Autonomous University of Madrid, Spain. 355 pp.

CHAPTER 5

Producing more or wasting less? Bracing the food security challenge of unpredictable rainfall

Jan Lundqvist
Stockholm International Water Institute (SIWI), Stockholm, Sweden

ABSTRACT: A dramatic jump in the number of undernourished people around 2007/08 and conclusion at the World Food Summit, November 2009, to increase food production by 70% to 2050 are put in context. By the time decisions about a massive increase in production were in print, figures revealed that global food production had never been more impressive. Paradoxically, undernourishment increased alongside increases in production for an extensive period.

It makes little sense to increase production if the produce is not beneficially used. However, production, storage but also transport and marketing need to improve in areas where poverty and undernourishment co-exist. Despite an adverse climate with a highly unpredictable rainfall pattern in these areas, farmers occasionally produce bumper harvests. If the produce can be secured, especially during the good years, and if part of it could be transported and sold also outside local markets, devastating consequences of climatic variation could be reduced.

Food insecurity is a harsh reality for 1,000 million people. At the other end of the spectrum, overeating and food waste is common among 1,000 million plus people. Waste of food is waste of water and other resources. Availability of empirical data is one constraint to analyze the low efficiency in the food chain. Poor conceptualization is another.

Keywords: food losses, food waste, stochastic rainfall, carryover stocks of food, food chain, value chain

1 THE FOOD SECURITY CHALLENGE: CONTRADICTING TENDENCIES

Between 2007 and 2008, world cereal supply increased by more than 5% according to FAO (2009a). This may be compared with the current annual global population increase of about 1.2%. A record supply of $2{,}242 \times 10^6$ t [t = tonne = 1,000 kg] was due to a sharp increase in world cereal production and not the result of changes in the stocks of cereals[1]. Never before had so much food, in terms of cereals, been available on the world market.

[1] A word of caution is warranted about statistical information and terminology. Apart from a slight difference between tons and tonnes, information on food production, stocks, utilization, supply and consumption must be interpreted with care. It is often based on Food Balance Sheets compiled by FAO on data provided by countries. Many of the variables are difficult to define and measure and important data are derived from certain assumptions. In literature and in policy debates, differences between production and supply are seldom made. Losses in the field or during transport and storage are not recorded in any systematic statistical data base. Similarly, consumption, which is a key concept with regard to food security, is often estimated with reference to the amount of food produced in a country/other unit, plus import, minus export, non-food uses and assumed losses. Food waste is, however, not considered although it could be significant. The calculations are used as a proxy for food consumption, which invariably is quite different from food intake. See further below under: 5.5 *Food Security versus food supply and beneficial consumption.*

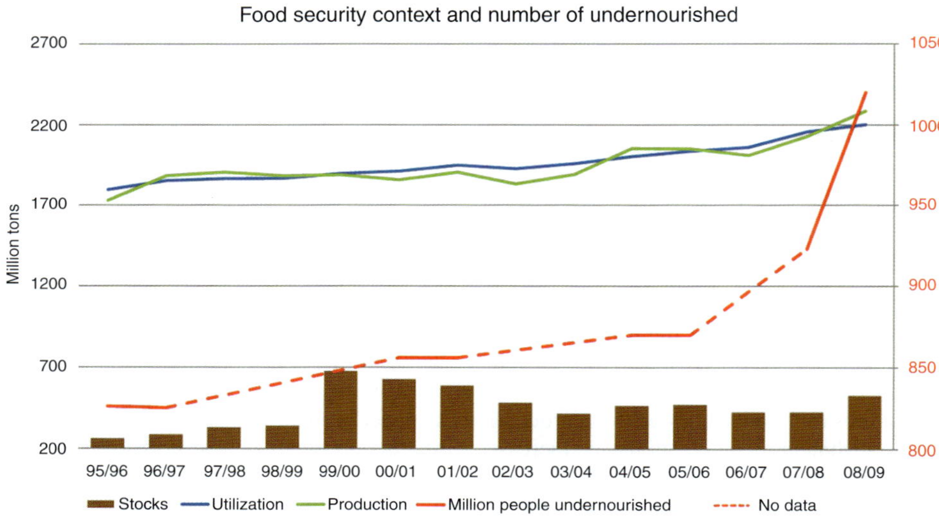

Figure 1. Global trends and variations in food production, supply and stocks (in 10^6 t), and number of people suffering from undernourishment (million).
Source: FAO: Food Outlook (FAO, 2009a); FAOSTAT.

Parallel with these optimistic calculations, dire figures about a dramatic increase in the number of people who were undernourished made headlines in the media. Human sufferings and discontent were demonstrated in the streets. Naturally, the areas and farmers who contributed to the record supply were far away from those who took to the streets. In food deficit areas, the political legitimacy and the capacity to handle riots were tested in countries across the globe, e.g. in Haiti, Egypt and Bangladesh. Export restrictions were rapidly imposed in more than thirty countries and in others, subsidies and price controls were tested (Evans, 2009). But this situation was not entirely new. The absolute and relative number of people who were undernourished had gradually slowed down since the end of the 1960s, when about 1,000 million people went hungry till the beginning of the 1990s. In 1995/97 the downward trend was broken and the number of undernourished started to increase. But it was the sudden and significant jump in 2008 from an estimated 850 million to about 1,000 million (FAO, 2009b) within months, i.e. an increase by some 15%, that renewed and augmented the concern (Figure 1).

It is relevant to underline that production and supply are very different from access and beneficial consumption. In literature and in policy, concepts and statistical information in these regards are invariably mixed as discussed later in this chapter. Poverty, conflict and adverse weather conditions are the main causes that prevent hundreds of millions of people to get access to food or to grow the food that they need. Recent price hikes on food for consumers have received much attention. Equally significant but much less discussed are the substantial increases on the price of inputs in agriculture, e.g. energy and fertilizers that the farmers need, and price on transport to the market. The prices on important inputs in food production are likely to remain high, which may be regarded as a new scourge that is added in the 21st century. Subsidies could distort the picture.

The international food price index has come down from a peak in 2008, but not for millions of people in Nigeria, Brazil, India and to some extent China, that is, for a major part of the world's population. They desperately need food at a price that they can afford, but the prices on the local markets have not come down (UN, 2009).

With the grave situation at hand, it is surprising that so little attention is paid to what happens to the food once it is produced. The losses and waste of the food[2] are largely ignored. The statistical information is quite limited but it is plausible to argue that the aggregate and cumulative losses and waste are in the order of 50% of the food that is produced albeit with different combinations and magnitudes of losses and waste in different countries and in different periods (FAO, 1981; Kantor *et al.*, 1997; Smil, 2000; Kader, 2005; Lundqvist *et al.*, 2008; Stuart, 2009). Losses and waste could be quite limited among small farmers. Leftovers, for instance, are used for feed. Paradoxically, losses could be high during good years (see Section 4.4). The estimate of 50% does not include overeating of food which may be perceived as another dimension of food waste. In terms of the number of people involved and affected, overeating is much more prevalent compared undernourishment. Similar to the throwing away of food, it implies a pressure on natural resources over and above what may be regarded as *sound* for food security.

Global figures and trends are deceptive. However, in a globalised world it is noteworthy that a gradual deterioration in food security over several years has occurred parallel with an augmentation in the production and supply of cereals and other food items.

The lingering mismatch between trends in production and supply, and the number of undernourished is worrisome. In view of the high costs and stakes involved, it jeopardizes a stable and sound development. Increasing production to the level envisaged, i.e. by 70% till 2050 will most probably involve an expansion of agricultural land and a gradual increase in the pressure on finite and vulnerable water resources. High output agriculture presumes high inputs of fossil energy and nutrients, some of which are finite and cannot be substituted, e.g. phosphorous (Cordell, 2010). In spite of the additional efforts required, few well-informed analysts doubt that increases in yields are possible. They are conceivable. But without investments, additional inputs and well-functioning institutional arrangements, they will not be realized in areas and among people who most urgently need such improvements.

2 CONTENTS AND PURPOSE OF THE CHAPTER

The conventional approach to improve food security through an exclusive focus on production and supply is put in context. The purpose is to show that it is rational, even smart, to combine strategies that ensure that a larger fraction of the food produced will be secured and beneficially used parallel with efforts to increase production.

Reductions in losses and waste of food, as well as other agricultural commodities, will ease the pressure on water, land and other resources for a given amount of food that is available for beneficial consumption. The argument is not to totally eliminate losses and waste, since such a goal is not realistic. But losses can be reduced, for instance, through improved storage, transport and marketing arrangements of the food produced. The waste of food, i.e. the throwing away of food that is *perfectly fit for consumption* requires different policies as commented in footnote 2.

Reducing losses and waste improves the efficiency in the food chain, and is likely to generate significant and multiple gains. Foremost among these is an urgently needed opportunity to improve

---

[2] From an analytical point of view, it is useful to make a distinction between causes for and remedies to combat reductions in the amount of food that is beneficially used. In this chapter, losses refer to a reduction in the amount of food during the first segments of the food chain, i.e. during harvest, storage and transport, i.e. *from field to market*. Spoilage is used with reference to losses in quantity and quality during storage. Conversion is still another important dimension in the agricultural commodity and value chain, e.g. the use of cereals for feed. Waste refers to the throwing away of food that is "perfectly fit for eating" (WRAP, 2009a; 2009b), i.e. a dubious use of food in the latter segment of the food chain. The accumulated losses and waste refer to the entire food chain, *from field to fork* (Lundqvist *et al.*, 2008). The causes for losses and waste, respectively, differ and so do the remedies to deal with them. Reducing losses implies investments, technical and institutional arrangements. Reduction of waste can, in principle, be achieved at no cost, but with multiple benefits, i.e. it is a win-win-win option.

food security for a growing world population *without* a corresponding increase in resource pressure and environmental damage. Better storage and transport would lessen the risk for price collapses at local markets, enhance rural livelihoods and provide opportunities for a diversification of agricultural production and rural livelihoods. In addition to resource and environmental benefits, waste reduction is, arguably, the most cost effective and sensible effort with several winners. It is hard to envisage losers.

Some of the questions touched upon above will be discussed:

- In an era when considerable attention is given to climate change, it is relevant to discuss how the assumed increase of extreme climatic events and higher temperature will affect prospects for increased production and supply of food. To what extent and where will an erratic and unpredictable rainfall pose constraints on agriculture and what measures could be used to counter the consequences of an adverse climate?
- What are the resource implications from demographic trends and what seems to be a continuous improvement in purchasing power of large segments of the world's population, including a rapid growth in the urban population?
- What strategies are realistic and effective to achieve an improvement in food security, that is, in addition to efforts to increase production and supply?
- The discussion below will follow the food chain. The need to develop a systems perspective from *field to fork* and to pay more attention to what is happening to the food that is produced is emphasized.

3 REVIEWING THE EXCLUSIVE FOCUS ON INCREASED FOOD PRODUCTION

When the Millennium Development Goals (MDGs) were decided in 2000, the prospects for a continuous reduction in the number of hungry people seemed promising. The aggregate value of world agriculture including all food and non-food crops and livestock commodities had grown by 2.1–2.3% annually on average during the last four decades (Alexandratos *et al.*, 2006; Bruinsma, 2009), i.e. faster than population increase. The number of undernourished had steadily been pushed back over several decades although the pace was not sufficient to reach the first MDG, i.e. to reduce the number of hungry people by half till 2015, from about 840 to 420 million. As shown in Figure 1, the positive trend in terms of a decline in the number of food insecure people turned upwards already in 1995/97.

With the recent dramatic acceleration in the number of people who are food insecure, the predominant response from policy makers and analysts has been to suggest increases in food production. At a high-level expert forum organized by FAO as an input to the World Food Summit in November 2009, about 300 leading experts from academic, non-governmental and private sector institutions from developed and developing countries, concluded that food production will have to increase by 70% till 2050 in order to feed another 2,300 million people, an increase from the current 6,800 to 9,100 million. A year earlier, the World Bank in its World Development Report argued that cereal production needs to increase by nearly 50% and meat production by 85% to meet the expected growth in demand by 2030. It was stressed that a renewed emphasis on agriculture would be a driver for development and poverty reduction (World Bank, 2008). Similar suggestions have been aired by highly esteemed colleagues. Norman Borlaug, Nobel Prize Laureate in 1970, argued that we must learn to produce three times as much food for the more populous and more prosperous world of 2050 (Borlaug, 2002).

Optimistic visions are important. Many of them have unfortunately not been achieved as solemnly envisaged. At the first World Food Conference, in 1974, Mr. Henry Kissinger, who was then Secretary of State in the USA, declared that no child would go to bed hungry within ten years (Economist, 2009). Few, if any, circumstances have changed in a direction to make this goal easier to accomplish today. Even if production and supply targets were achieved what is it that tells us that this would reduce the number of people who are undernourished (cf. Figure 1)?

However, as many colleagues emphasize, there is a significant yield gap in areas where food insecurity prevails, i.e. the yields are below the level that is realistic (Molden, 2007; World Bank, 2008; Adesina, 2009; Godfray *et al.*, 2009). This is argued to be the case also in areas that are generally perceived as having a low potential, e.g. the Guinea zone in Africa, stretching over 400 million hectares of land, which is characterized as a *sleeping giant* (Morris *et al.*, 2009), i.e. the potential is much higher as compared to the current output. The discussion about a yield gap raises an important question: if the potential is real and the most basic needs are not satisfied in the areas where it exists, why is it not realized more often?

4 VARIATION IN YIELDS AND PER CAPITA FOOD SUPPLY

The trends in yields during the past decades are disappointing in parts of the world where food security is already a serious challenge. Sub-Saharan Africa has experienced a declining trend in per capita cereal production since the 1960s. For many countries in the region, the level (in 2003–2005) is below 100 kg of cereal produced on a per capita basis, compared with e.g. India, 219 kg per capita; China, 313 kg per capita; Argentina, 914 kg per capita; the USA, 1,253 kg per capita (World Bank, 2008).

The combination of low yields and high rates of poverty implies a low food supply, and a high risk for food insecurity. In countries and for people with an economic strength and/or political influence, a low domestic food production can be compensated through imports and a reliance on stocks, which *smoothen* the trends as shown in Figure 1. Several distinguishing differences between the food supply in Western countries and NICs, the *Newly Industrialized Countries*, on the one hand; and many developing countries on the other, are shown in Figure 3. The fraction of animal calories in the supply, for instance, varies from a few percent to about a fifth. Another worrying difference is the steady increase in food supply for the former group of countries whereas the fluctuations over time are considerable in many developing countries.

The graphs illustrate that the food supply, at national level, varies from about 1,800 to about 3,800 kcal/day per person in 2005.

The trends shown in Figure 2 refer to the aggregate regional situation and the columns in Figure 3 show the situation at national level. Both Figures conceal the fact that the situation is often more dramatic at sub-national levels and during individual years and seasons.

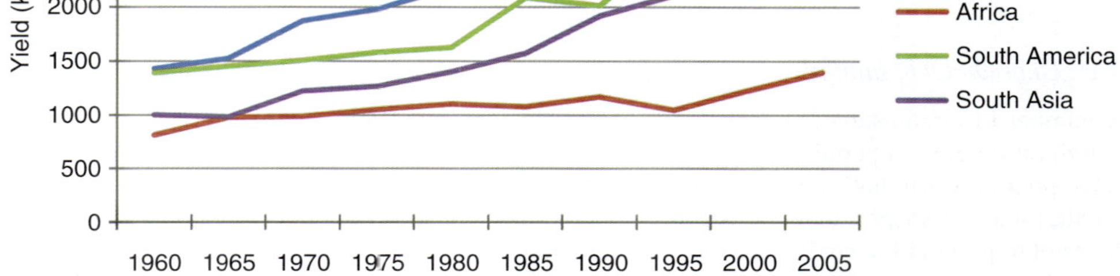

Figure 2. Trends in yields for major world regions 1960–2005.
Source: FAOSTAT.

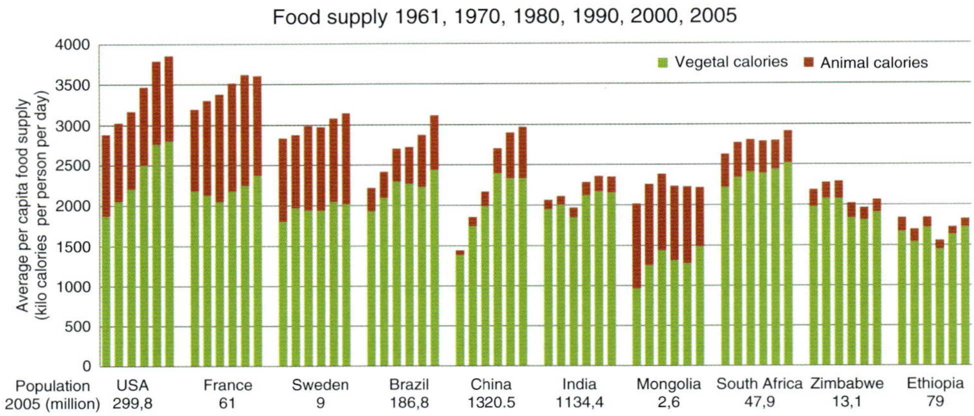

Figure 3. Per capita food supply per day, 1961–2005, separated into vegetal and animal calories.
Source: FAOSTAT.

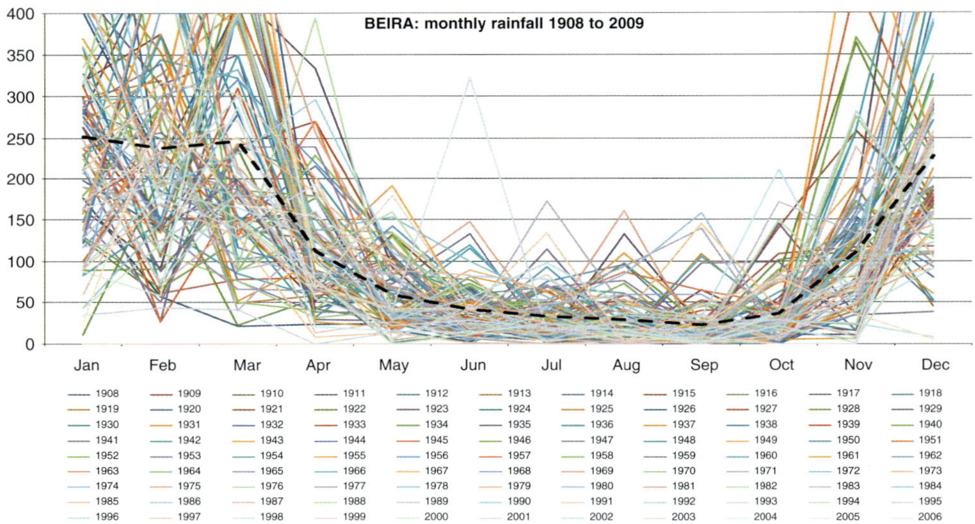

Figure 4. Monthly rainfall in Beira, Mozambique, 1908–2009.
Reproduced by kind permission from a presentation made by Jean-Marc Faures, FAO, at World Water Week in Stockholm, August 2009.
Source: FAO (2009d).

4.1 *Unpredictable rainfall*

A number of circumstances contribute to low yields and insufficient domestic food production: a high ratio between population size and resource endowment, adverse climate and impoverished soils, poor access to high quality seeds and sources of nutrients, expensive and limited agricultural credit, inferior transport to markets and small holdings with uncertain tenure. For this chapter, it is relevant to point at the challenges that climate imposes in terms of water scarcity and particularly the risks and implications of a highly erratic rainfall pattern.

Figure 4 provides a visual presentation of the tremendous variability in monthly amounts of rainfall measured at a station in Beira, Mozambique, over a hundred year period. The challenges

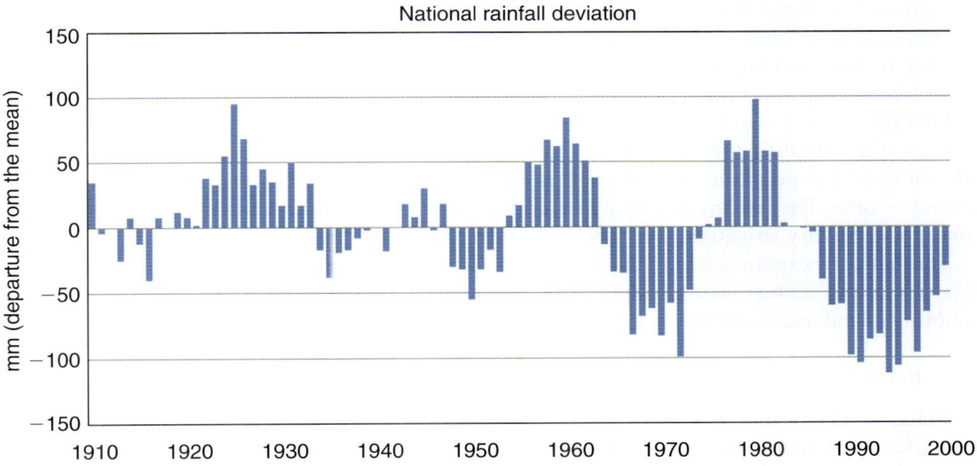

Figure 5. National rainfall deviations in millimeters from the long-term mean, 1910–2000, Zimbabwe. *Source*: GRID-A based on data from Zimbabwe Department of Meteorological Service. [www.grida.no/zw/climate/vitalafrica/english/02.htm].

are often compounded by a high intensity in rainfall and the associated rapid flows of water, overland and back to atmosphere. Figure 5 gives an indication of an increased amplitude in the deviation from the mean national rainfall value in Zimbabwe. The Figure also shows that positive and negative deviations, respectively, tend to last for a number of consecutive years.

Torrential rainfall during short periods has devastating consequences in arid and semi-arid areas. For vulnerable communities in the Sahel region, the situation is described as *ground zero*, and the need for effective adaptation measures is obvious (UN-IRIN, 2008; 2010).

The significant variation in the amount of precipitation between seasons and years and the deviations from an average, as illuminated in Figures 4 and 5, are the reality for farmers in many parts of the world. This is, for example, the case in monsoon areas where an unpredictable rainfall is part of human history. Lannerstad (2009) compiled information of incidences of failures of the monsoon and the related human sufferings in Southern India during a hundred year period from the beginning of the 19th century. ". . . scarcity, desolation and disease" were rampant at 18 occasions with a duration of one or a couple of consecutive years. For individual years, like 1808, failure of both monsoons "carried off half the population". With the huge population in the region today, the scale of the challenge is mindboggling. For the 540 million people living in the Ganges-Brahmaputra-Meghna Basin, 50% of the annual precipitation falls in 15 days and 90% of runoff is in 4 months (Grey & Sadoff, 2009). Yet, the magnitude of the challenge is increasing.

The association between a *difficult hydrologic legacy*, including a highly unpredictable rainfall and insecurity, is found in many parts of the world (Grey & Sadoff, 2007; Brown & Lall, 2006). The situation was quite bad in Africa around 1910 with droughts in both East and West Africa. The beginning of the 1980s was the culmination of a few decades of diminishing rainfall. The magnitude of the problem in Africa is less in quantitative terms when compared to Asia but devastating for those who are affected. In semi-arid and dry sub-humid Sub-Saharan Africa, about 250 million people today depend on rain fed crop farming as their primary source for food and income (Enfors, 2009).

4.2 *Lack of water where and when it is needed*

The rainfall pattern and the high temperatures are together resulting in a high risk for agricultural droughts, i.e. moisture deficits in the root zone during critical stages of the cultivation season.

The amount of water that is available to the roots of the crops, i.e. the size of the green water resource, depends on the amount of rainwater that infiltrates through the soil surface. In areas with an impermeable layer on top of the soil, farmers must break up the crusts through ripping or some other land management practice (Enfors, 2009; Molden *et al.*, 2010; Rockström *et al.*, 2009).

Otherwise, there will be no water where it is needed, i.e. in the root zone. Vulnerability and risk are often amplified by a low water holding capacity of the soil in tropical areas, the faction of rainwater that permeates into the soil is accessible to the roots of the plants only for a short period of time. The risk associated with rain fed systems in semi-arid parts of the world without any supplementary irrigation is very high (Molden, 2007).

Through their experience from this harsh reality and in the absence of insurances and *risk capital*, farmers in this context naturally apply a precautionary strategy with the result that experiments, innovative land and water use and new crops are seldom tried. The yield gap remains.

4.3 *Rainfall pattern decisive for lean and fat years across the globe*

Extreme weather conditions are experienced also in major food producing and export oriented countries. Starting in 2002, Australia experienced the worst multi-year drought in a century (Evans, 2009). For part of the country, e.g. in the Murray Darling basin, a dramatic reduction of the inflow into the river was observed and the water situation is still problematic, especially in the Murray river (Michael Moore, Stockholm International Water Institute, 2010, pers. comm.). More or less during the same period, droughts and other adverse weather conditions affected other important export countries, e.g. Russia, Ukraine and Canada, with the result that world cereal production fell by 2.1% in 2006 (Evans, 2009) and cereal stocks diminished (cf. Figure 1).

A reduction in world cereal production during one or a couple of years will hit various countries quite differently, primarily because adaptation capacity differs. Stimulated by the wave of food export restrictions, which distort international trade, the leasing of land in other countries has recently become a state driven strategy to secure access to food on a long-term basis. Individual farmers have always bought or leased lands in other countries. However, no poor country with a large fraction of its population being undernourished, or facing such a risk, is in a position to enter into such lease contracts. Nor can they access adequate amounts of food through trade. Equally problematic, food aid tends to undercut prospects for domestic farmers to sell their produce since food aid is *free* and the price on imported food is artificially low due to subsidies in exporting countries. Conversely, for wealthy people and politically powerful countries access to food is not an issue that cannot be overcome.

For local communities in most parts of Africa, the occurrence of bad years dominates. Enfors (2009) found that rainfall is highly unpredictable in semi-arid parts of Sub-Saharan Africa. In areas with an average annual rainfall below 600 mm, the farmers experienced critical dry spells (defined as no rain during 21 consecutive days, or more) during 60% of the seasons or more. Farmers in areas with an annual rainfall of 400 mm, or less, experienced dry spells almost every season. Even without dry spells, 400 mm is close to the margin for the successful cultivation of many crops in this climatic context. In northern Tanzania the conditions have become worse with a doubling in the frequency of dry spells during the last 60 years (*ibid.*).

The dynamic links between adverse weather conditions, agriculture and limited options for alternative livelihoods for an increasing population pose significant challenges. A low and uncertain return on hard work under a burning sun and a high incidence of food insecurity is common for a large number of the about 1,500 million people who depend on smallholder agriculture (World Bank, 2008). With low and uncertain yields the *one-acre dilemma* is a huge challenge, socially and in terms of resource pressure (Lannerstad, 2009). Is the exclusive focus on increases in production the most promising strategy among this group of food producers?

4.4 *Losses increase during good years*

Occasionally, the weather and other conditions are favorable. Like elsewhere, yields in the semi-arid area in Northern Tanzania are sometimes much higher than the prevailing low output. In 2006, yields increased to about 3,000 kg/ha, and with conservation tillage yields could reach 4,800 kg/ha (Enfors, 2009). When farmers are able to produce high yields, they are left with a couple of choices. They can try to sell the surplus or they can store it. Some may use it for feed. Selling the surplus during a good year is however likely to be frustrating. With a lack of transport and poor market access, it is only local markets that are accessible. When many farmers want to sell, the result is invariably market shocks and the collapse of prices as noted for different parts of Africa (Adesina, 2009). In the communities studied in Northern Tanzania, farmers choose to store the surplus for later consumption. Unfortunately, pests destroyed most of the crops that were stored (Enfors, 2009). Drastically falling prices at local markets and inferior storage are not peculiar to semi-arid parts in Northern Tanzania and Africa, but is also a significant problem, for instance, in China (Dinghuan Hu, Chinese Academy of Agricultural Sciences, 2010, pers. comm.).

A paradoxical situation is experienced by millions of farmers caught up in the context just described. Farmers face considerable risks related to the vagaries of an adverse climate. Yields are low and so are total production and income. When they are fortunate and produce high yields, there is an imminent risk that a large part of the surplus will be lost, or alternatively, that the price at local markets collapses. Conversely, with low yields, farmers will, of course, use the harvest judiciously and losses may be insignificant.

In this kind of situation, which is common for poor households in smallholder communities and in areas where food insecurity is rampant and where a stable level of production from year to year cannot be expected, efforts to increase production must be combined with investments in post harvest technologies and marketing arrangements. As aptly formulated in an FAO report (1981:2): "It is distressing to note that so much time is being devoted to the culture of the plant, so much money spent on irrigation, fertilization and crop protection measures, only to be wasted about a week after harvest".

4.5 *Carry-over stocks of water or food?*

This sensible argument has been left without attention for decades. To overcome the consequences of unpredictable rainfall on food production and rural livelihoods, it is necessary to balance the deficiencies during bad years with the surplus during good years. Water storage in reservoirs together with lifting of groundwater has been one important strategy to reduce the risk of water deficits during cultivation seasons. The strategy is certainly important but it is increasingly costly, financially and in energy and environmental terms. There are few sites left where carry-over stocks of water from rainy seasons or years can be impounded in surface reservoirs at reasonable expense. Similarly, the energy cost to lift water must be covered by the user or society. Storing rain water in the soil is very important but presumes better coordination between land and water management (Falkenmark & Rockström, 2004). In areas with frequent dry spells, supplementary irrigation from various types and sizes of water storages is important to reduce the high risk associated with rain fed cultivation.

In this kind of situation it seems rational to make arrangements for storing food as a complement to storing water from the good to the bad years, i.e. facilities that could help the farmers to save part of the agricultural produce when conditions are favorable for use at a later occasion or for a staggered sale (Lundqvist, 2009a).

Improved storage, transport and marketing arrangements for food, and other agricultural commodities, are likely to bring multiple benefits. In addition to a reduction of losses of the crops currently cultivated and more stable prices and, hence, income, these kinds of improvements will enable the farmers to produce other crops for which there is a demand outside the local market. Local food processing could generate opportunities for farmers to grow sensitive crops that cannot be transported over long distances in an unprocessed form. This is, for example, the case with fruits which can successfully be grown in Northern India and sold in cities far away or even abroad (Paul Appasamy, Karunya University, India, 2009, pers. comm.; Narendra Kumar Tyagi, Department of Agricultural Research and Education –ICAR–, New Delhi, 2009, pers. comm.).

Table 1. Comparison of population increase, growth of GDP and water withdrawals, 1800–2000 with projections for 2050. Population figures from UN population statistics.

	Population (million people)	GDP (10^3 million US$, 2005 PPP)	Water Withdrawals (km^3)
1800	<1,000	913[1]	–
1900	1,650		700
1950	2,500	7,006	–
2000	6,000	56,593	3,500
2050	9,100[2]	193,318[3]	Increase by 10–15%[4]

[1] Figure refers to 1820.
[2] UN medium projection.
[3] Trend growth projection (Hillebrand, 2009). PPP refers to Purchasing Power Parity.
[4] Various projections have been made. Increase of this magnitude is used in FAO reports.

5 INCREASED PURCHASING POWER AND URBANIZATION

Low yields and missed opportunities to increase and stabilize the income for the farmer and deliver food and other agricultural produce to the market are obstacles to a sound development. It also implies a low efficiency in resource use and in the food supply chain (Nellemann *et al.*, 2009). A high efficiency in resource use in connection with production will be of limited value if the food produced is lost or wasted as eloquently formulated in the 1981 FAO report (quoted above). A glance at the demand side dynamics reveals that strategies that link production with demand and beneficial consumption must be developed.

For the future, the significant change in the food and water linkages will be related to the combination of an urban expansion and an increase in purchasing power rather than to population size *per se*. Socioeconomic and demographic dynamics is concentrated in urban settings, but the implications are, to a very large extent, in rural areas. For example, the demand for food and many other goods and services is largely driven by urban interests and it is in urban centers that a larger and larger fraction of consumption, and waste, takes place.

The figures in Table 1 show that GDP has historically grown faster than population. But a new situation is unfolding. Population increase has slowed down and may reach a replacement level in a generation or two, whereas economic growth is projected to continue unabated. In the contemporary setting and for the future, the growth of GDP is significantly faster than population increase. Between 2000 and 2050, world population increase is estimated to be about 50%, a staggering 3,000 million, while the global GDP is expected to expand by some 400% (Hillebrand, 2009). Purchasing power will thus be relatively more important than population size. For the poor and undernourished, an improved purchasing power is certainly desirable. For society as a whole and with regard to the natural resources and environmental situation the combined effect of demography and economics in terms of the character of the increased demand is important. If the demand for water intensive products increases heavily, which it currently does, and if a large share of the procured products are lost and wasted, the future and the prospects for sustainability appear bleak.

However, the number of people living under various levels of poverty is still staggering. About 880 million people had less than 1 US$ a day to spend in 2005, estimated in terms of Purchasing Power Parity (PPP), and 2,600 million had less than 2 US$ a day, 13% and 38% respectively of the world's population. It is also mindboggling that 5,150 million, i.e. 77% of the world population, have less than 10 US$ a day to spend (Shah, 2009). The vast majority of the world's population is thus far from wealthy. But it is reasonable to assume that all people, even those who are better off, have an ambition to earn more. And spend more, with implications for resource pressure.

5.1 *Disposable income, dietary preferences and water pressure*

An increase in disposable income will increase the demand for food among other things. In combination with urbanization the composition in food preferences is changing, sometimes referred to as a *nutrition transition* (Schmidhuber & Shetty, 2005). As noted above, the demand for meat products may increase by something like 85% till 2030 (World Bank, 2008). Some of the increasing demand will be for meat. However it is important to recognize that the composition in the demand for various meat and animal products varies and that the consequences and benefits are quite different between various categories. More cattle mean an increased demand for feed, which translates into a demand for land and water. Instead of direct human consumption of grain products, about 36% (738×10^6 t in 2005 according to FAOSTAT) are used for feed to cattle. This fact has, however, received much less attention as compared to the significantly much smaller use of cereals for bio-energy production, which is about 5% or some 120×10^6 t in 2008/09 (FAO, 2009c).

One of the most rapid transformations in socioeconomic terms takes place in China with significant repercussions on water pressure and the environment. After the cultural revolution (1966–1976) and in pace with the rapid economic growth after the economic reforms at the end of 1970s and together with the demographic dynamics, the demand for food items has increased rapidly, totally and on a per capita basis. Farmers have been stimulated to increase production to meet the growing demand. Initially, the demand for cereals grew, but from around the mid 1980s a shift in the demand pattern towards high quality food items became obvious, i.e. an increased demand for animal products, vegetable oils, vegetables and fruits (Liu & Savenije, 2008). This is an important development in terms of nutritional improvements and quality of life for hundreds of millions. It is also in line with the progress to reach the MDGs. However, it also meant that the pressure on water resources increased rapidly as illustrated in Figure 6.

5.2 *Nuances in the meat debate*

Most colleagues would agree that animal products have a much higher impact on water resources and the environment as compared to cereals and other food items as illustrated in Figure 6. However, it is important to recognize a few points.

The virtual water content in different meat products varies. Generally, it is much higher for red meat as compared to white meat, whereas pork is somewhere in between. The circumstances under which meat products are produced, also need to be considered. Grazing of cattle in areas that cannot be used for crops or other agricultural production and where no irrigation water is used is quite different compared to raising cattle in feedlots. Parts of Southern Africa, for instance, are suitable for wild animals, which produce high quality meat. These areas are not suitable for crop production. Cattle raised in feedlots with fodder coming from irrigated fields or feed grown in areas where alternative land use is possible naturally have a different impact on resource systems.

A third important feature refers to the role of animal husbandry in rural livelihoods in many poor countries. Rearing of different animals alongside with other agricultural activities helps millions of farmers to diversify agricultural production and to obtain their income from various sources (Peden *et al.*, 2007). For many communities, cattle represent an important part of their capital as well as an integral part of their culture and livelihood. On the demand side, this corresponds to the fact that people obtain a range of high quality food and other items from animals, e.g. milk and other dairy products, blood and fat apart from meat. Multipurpose benefits are thus derived from animals especially for many poor communities.

However, it is relevant to observe that a high or very high per capita animal food supply tends to be associated with an intake of animal based food items in excess of what is motivated from a medical and nutritional point of view. Similarly, the environmental implications of livestock are considerable (Smil, 2000; McMichael *et al.*, 2007; Steinfeld *et al.*, 2006). In addition to these circumstances, we should also realize that animal based food items are quickly and easily degraded in quality, which increases the risk of waste particularly if supply is excessive (Lundqvist, 2009b) as discussed in next section.

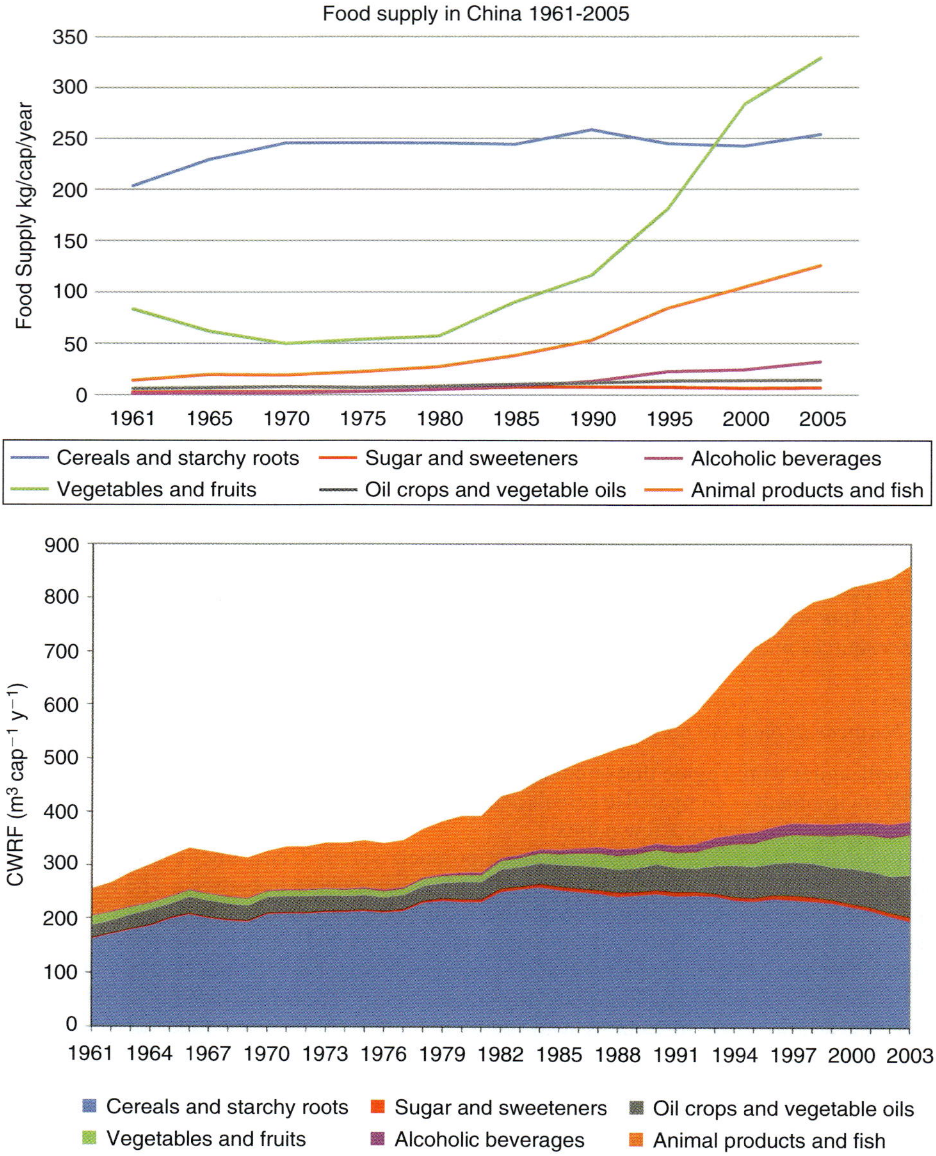

Figure 6. The top diagram illustrates the increasing supply of different food items and the diagram below shows the water footprint associated with food supply in China, 1961–2005.
Source: Modified from: Liu & Savenije, 2008; Liu *et al.*, 2008; FAOSTAT.

5.3 *Increased vulnerability and impacts of food systems*

Many of the food items for which demand is increasing with growing prosperity are both water intensive, environmentally dubious and vulnerable, that is, they will easily and quickly degrade in quality. Without proper handling, storage, cold or chilled transport, animal based products, fruits and vegetables will start to deteriorate after days or even hours. As a comparison, cereals can be kept for long periods and presume less sophisticated treatment during storage and transport. In urban societies and perhaps particularly among people who are far away from food production, physically and in terms of comprehension, there is a tendency to perceive blemishes and cosmetic appearance of food e.g. on fruits and vegetables as a sign of poor quality (Stuart, 2009; Godfray *et al.*, 2009). Safety standards and a fear among consumers about health risks associated with eating food that is proven to be –but often rather suspected to be– of inferior quality adds to the vulnerability of the food systems. Reports reveal that other types of food waste can be significant but largely unnoticed. This is e.g. the case with supermarkets and restaurants (Stuart, 2009). Finally, the recall of huge quantities of beef in the wholesale segment of the food chain is documented (Rano, 2008; Lundqvist *et al.*, 2008).

5.4 *The food chain is also a value chain*

Since expenditure and efforts are made in different segments of the food chain, in farms, in transport and storage, processing, marketing, etc., the accumulated investments and value are higher towards the end of the food chain. Similarly, the accumulated energy input increases in the food chain per unit of food. Of the total energy used from production to consumption of food items, a large part of it refers to the last part of the food chain, i.e. in households where especially cooking means a high energy use per food unit. Situation varies between countries, but the general picture seems to be that the economic value and inputs increase along the food chain. Consequently, wasting food and especially food that is prepared and ready to eat incurs higher accumulated costs as compared to the loss of food at the beginning of the food and value chain.

An additional perspective is relevant: when food that is prepared and ready to eat is tossed, a range of actors who have been involved in various activities in the food chain have already been paid for their efforts. The GDP has increased while, ironically, the benefits of the product itself are foregone. When food is lost in the field or at the beginning of the food chain, on the other hand, it is primarily the producer who is deprived of an income, and society, at large, will not get access to these commodities. In both cases, water and other resources have been used to produce the commodities that are used as well as those that are lost and wasted.

Based upon the assumption that about half of the food available *in the field* is lost or wasted and thus not beneficially consumed, a conservative estimate of the water that is used in vain is about 1,350 km^3 (Lundqvist *et al.*, 2008). This is half of the water that is estimated to be withdrawn from rivers, lakes and aquifers and used for irrigation, i.e. it does not include any reference to the water used in rain fed systems. Other relevant calculations are made about e.g. impacts of food waste in terms of green house emissions (Nellemann *et al.*, 2009; WRAP, 2009a) and the "Livestock's Long Shadow" (Steinfeld *et al.*, 2006).

5.5 *Food security versus food supply and beneficial consumption*

The laudable increases in global food production and supply have resulted in a situation where food security should not be a problem. On a per capita basis, much more food is produced and also supplied, as compared to what is required for the world population to lead an *active and healthy life*. But as we all know, what *could be* is different from *what is*. And it is different from commitments made at various meetings. Hundreds of millions of children still go to bed hungry. This is not to say that commitments made at various meetings are not seriously intended. It rather illustrates that the reality is extremely complex.

It is relevant to relate the discussion above to the definition of food security: "food security exists when all people at all times, have physical and economic access to *sufficient*, safe and nutritious food to meet their dietary needs and *food preferences* for an active and healthy life" (FAO, 1996, *italics* added). In connection with the World Food Summit, November 2009, the following sentences were added: "The four pillars of food security are availability, access, utilization and stability. The nutritional dimension is integral to the concept of food security". The most widely used definition on food security is thus about the combination of availability, i.e. production and supply, and access. It is also about the use of food. The notion of *sufficient* is one key word in the definition. For about 1,000 million people the harsh reality is that they do not have access to what is sufficient. Paradoxically, the number of people who overeat and suffer from overweight and obesity is considerably higher. Obviously, *food preferences* and human behavior deviate from what is sufficient.

It is interesting to compare the bars in Figure 3 with figures on food intake requirements for *an active and healthy life*. Food intake requirements usually refer to dietary energy contents of food, expressed in kcal or Joule per person and day. For good nutrition, proteins, vitamins, fat, etc. are required in addition to the energy contents of food. The average energy intake requirements, which allow light activity, is estimated to range between 1,720 to 1,840 kcal/day per person, for different developing countries with the current age structure and increase to 1,820 to 1,980 kcal/day per person by 2050 as a result of changes in age composition (Alexandratos *et al.*, 2006). With increasing physical activity, the requirement is naturally increasing. In medical and nutritional literature, an average intake of 2,000–2,200 kcal/day per person is considered adequate (Schäfer-Elinder, 2005; Smil, 2000). Lower intake levels are seen as acceptable (MSSRF, 2002). Interestingly the intake is comparatively low for some groups in many different societies, e.g. the USA, Canada and in Africa (Smil, 2000). For food security, a common international norm is that food supply at a *national level* should be in the order of 2,700 to 2,800 kcal/day per person (Molden, 2007). This level of supply is supposed to be adequate to ensure that the *individuals* may have a food intake in line with the definition quoted above.

As shown in Figure 7, the food supply is much higher in many countries as compared to the international norm and substantially much higher as compared to the food intake requirements. If a comparison is made between the amount of food produced in the field and food intake requirements, the gap is even wider since losses and conversions are substantial. These kinds of distinctions are rarely, if ever, made in literature and policy. The notion of food consumption, for instance, is not used with reference to food intake requirements, which would be natural from a semantic point of view. National average food consumption is a proxy that is based on figures in National Food Balance Sheets (FBS). The FBS include figures on production, trade and estimates of losses, but not waste at household level. As noted by Alexandratos *et al.* (2006), waste can be significant. In other words, figures on food consumption, sometimes called *apparent consumption* (*ibid.*), convey deceptive information about food intake requirements and they mask considerable waste of food.

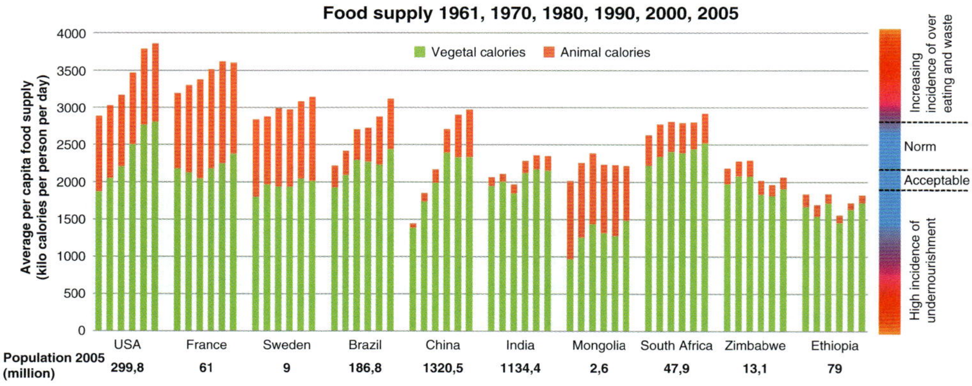

Figure 7. Comparison between the variation and trends in food supply and norms and recommendations to reach food security and national and individual levels.
The columns in green and red show daily per capita food supply in selected countries (cf. Figure 3). The column to the right illustrates interpretations of food security in terms of average energy intake requirements at the individual level and food supply at the national level. An intake of 1,980 kcal/day per person corresponds to the minimum requirement, and 2,200 kcal/day per person to an *acceptable level*. For food supply at national level, reference is made to the international norm, which is usually set at 2,700/2,800 kcal/day per person. For details, see comments in text above.

6 THE MULTIPLE COST OF LOSSES AND WASTE OF FOOD

It may be discussed whether the definition on food security should only refer to what is *sufficient* in terms of energy and nutrient contents, or whether it should also refer to the implications when supply and intake increase. Keeping in mind that the objective of food security is to ensure that *all people at all times* may lead *an active and healthy life*, it is logical to recognize that overeating is associated with overweight and obesity and thus curtails the ability to lead an active and healthy life. In an interesting study recently published, the relationships between food supply, intake and food waste in America have been analyzed. Calculations imply a steady increase in body weight among USA adults over the past 30 years and a progressive increase in food waste, from 900 to 1,400 kcal/day per person between 1974 and 2003 (Hall *et al.*, 2009). When supply of food increases and food is perceived as relatively cheap and easily accessible, the risk for a dual problem increases: the public health situation deteriorates and the waste increases with negative repercussions on resource pressure, environment and productivity in society.

The waste of food is significant in affluent societies with multiple costs. Detailed studies in the UK show that households throw away about a third of the food that they buy and that some 60% of this is perfectly fit for consumption. The annual avoidable waste is about 70 kg per person. A similar level of waste is found for Scotland (WRAP, 2009 a; 2009b). Studies made in other OECD countries show partly similar figures, but also that the magnitude of waste varies significantly. Norway has about the same level of waste as the UK, i.e. 71 kg/yr per person (Hanssen & Olsen, 2008). For Holland, Thoenissen (2009) reports that 8–11% of edible food is wasted (43–60 kg) and that this corresponds to US$ 364–539 per household, annually. Ca. 20% of all the food bought is thrown away. In the USA, the estimates vary significantly from about 25 to 50% (Jones, 2006; Schiller, 2009) and in Australia an average household annually throws out an estimated AUD$ 239 per person, or US$ 222 (Baker *et al.*, 2009); the corresponding figure for UK is £ 430 or US$ 659 (WRAP, 2009a). The surprisingly large differences between societies that seemingly have similar socioeconomic and cultural characteristics may partly depend on methodological differences and difficulties to define the boundaries for the measurements.

A common finding is that most people are not aware of the magnitude of the waste and they have vague and sometimes faulty notions of the consequences of food waste for resource pressure and environmental damage. For instance, the generation of methane from sites where food waste is improperly disposed is not known among people in the UK (WRAP, 2009a). Convenience and time constraints in relation to expenditure on food presumably play a major role in the level of waste. Cultural norms and social habit play their part. Household expenditure on food is typically about 10% or slightly more of household budgets in many OECD countries, which, however, is a misleading figure. An illusion of an abundance of cheap food is perpetuated in many societies. To the price paid in the shop, people pay for their food via their tax bills and indirectly in terms of environmental and other social costs. The expression that "there is no such thing as a free lunch" is more valid than ever but not internalized in the minds and habits of most people.

7 CONCLUSIONS

A dramatic and sudden jump in the number of people who are undernourished triggered vivid debates and activity around 2007–2009. Many countries were severely affected. Protests and grief from those who were affected were displayed in real time through the media. Export restrictions were hastily introduced in some thirty countries. Distortions in trade and access to food were the result of, or resulted in, price hikes on inputs in production as well as price on food in the market. This chapter has elaborated on the conclusion issued at high level meetings that food production must increase substantially. Reports and decisions talk about the need to increase cereal production by 50% to 2030 and 70% by 2050. For animal based food products, the corresponding figures are higher.

The focus on increased production is noteworthy. By the time these analyses and decisions were in print, figures started to show that global food production and supply had never been more impressive. Comparing the global trends in food production and supply reveals, strikingly, that increases on the supply side are no guarantee for a desirable improvement among those who need improvements. For individual societies or countries, a worrying association is rather between a progressively increasing food supply and an increasing number of overweight and obese people. In addition, high supply levels seem to be linked to high waste rates with associated costs in resource and environmental terms. Most of these costs are *unnecessary* and function as hindrances to sustainability.

What happens with the food once it is produced and how much of it will reach various social groups in society and how much will be beneficially used is, unfortunately, not an issue in the policy discussion nor in literature.

With a highly unpredictable rainfall as a result of an adverse climate and other problematic preconditions, increasing production *to the extent proposed* is a tremendous challenge. To stabilize production is even tougher. If production is increased in other areas than where food and improved income are needed, the prospects for reduction of undernourishment and poverty are limited. However, sometimes farmers are fortunate and produce bumper harvests also in areas that, generally, show low yields. If the produce can be secured, and especially during the good years, for years with low production and/or if some of it could be transported and sold also at places and to customers outside local markets, the devastating consequences of seasonal and annual climatic variation in many poor communities could be offset to some degree.

This chapter has elaborated on a few questions: to what extent may investments and institutional arrangements in post-harvest segments of the food and value chain improve food chain efficiency? To what extent can losses and waste of food be reduced and in what ways and to what extent would they help in feeding a growing world population, keeping in mind that some are more and more wealthy and some are trapped in extreme poverty?

Efforts to reduce losses and waste can be motivated from the point of view of water and other resources, environmental and public health. But reducing losses and waste is not easy. On the other hand, increasing production by 70% is not easy either. A mixture of production and supply and demand side strategies seems most promising. What other options would be easier and better to pursue if the objective is to feed the world without an undue pressure on natural resources and impact on the environment?

In an era when food insecurity is a tough reality for 1,000 million people and when resource scarcity and rainfall unpredictability together with environmental impacts cause widespread political and public concern, attention needs to be devoted to the food chain dynamics with its inherent complexities and opportunities. The relation between food security, our life support system, including scarce and unpredictable water resources, and human wellbeing is a key issue in this regard. Unfortunately, the availability of empirical data is an absolute constraint in analyses and policy development. Combined with a deceptive terminology, and a less-than-strict use of key concepts, the efforts to understand and effectively deal with this huge complex are thwarted.

ACKNOWLEDGEMENTS

Several colleagues have given important comments on previous drafts of the article, including Claudia Sadoff, Rosemary Stanton, Peter Rogers, Junguo Liu, David Molden and Roland Thönissen. Britt-Louise Andersson has drawn the Figures, except Figure 4. Thanks to all of you!

REFERENCES

Adesina, A. (2009). *Africa's Food Crisis: Conditioning Trends and Global Development Policy*. Plenary keynote paper presented at the International Association of Agricultural Economists Conference, August. Beijing, China. [http://ageconsearch.umn.edu/bitstream/53199/2/Adesina%20final.pdf].

Alexandratos, N.; Bruinsma, J.; Bödeker, G.; Schmidhuber, J.; Broca, S.; Shetty, P. & Ottaviani, M.G. (2006). *World agriculture: towards 2030/2050*. Interim report. FAO, Rome, Italy. [http://www.fao.org/es/esd/AT2050web.pdf].

Baker, D.; Fear, J. & Denniss, R. (2009). *What a waste. An analysis of household expenditure on food.* Policy brief, 6. November. The Australian Institute. [https://www.tai.org.au].

Borlaug, N. (2002). Can we feed the World? Here's How. *Wall Street Journal*, May 13.

Brown, C. & Lall, U. (2006). Water and economic development: The role of variability and a framework for resilience. *Natural Resources Forum*, 30: 306–317.

Bruinsma, J. (2009). The Resource Outlook to 2050. By How Much do Land, Water Use and Crop Yields Need to Increase by 2050? Expert Meeting on: *How to Feed the World in 2050*, 24–26 June. FAO, Rome, Italy. [ftp://ftp.fao.org/docrep/fao/012/ak971e/ak971e00.pdf].

Cordell, D. (2010). *The Story of Phosphorous. Sustainability implications of global phosphorous scarcity for food security*. Department of Water and Environmental Studies, Linköping University. Linköping Studies in Arts and Science, 509. PhD Thesis. Linköping, Sweden.

Economist (2009). *How to feed the world*. November 19. [http://www.economist.com/PrinterFriendly.cfm?story_id=14915144].

Enfors, E. (2009). *Traps and Transformations. Exploring the Potential of Water System Innovations in dryland sub-Saharan Africa*. Doctoral Thesis in Natural Resource Management. Department of Systems Ecology. Stockholm University, Sweden.

Evans, A. (2009). *The Feeding of the Nine Billion: Global Food Security for the 21st Century*. Chatham House Report. January. [http://www.chathamhouse.org.uk/files/13179_r0109food.pdf].

Falkenmark, M. & Rockström, J. (2004). *Balancing Water for Humans and Nature*. Earthscan. London, UK.

FAO (1981). *Food loss prevention in perishable crops*. Agricultural Services Bulletin, 43. FAO, Rome, Italy. [http://www.fao.org/docrep/011/ai481e04.htm].

FAO (1996). *Rome Declaration on World Food Security*. World Food Summit, 13–17 November. FAO, Rome, Italy. [http://www.fao.org/docrep/003/w3613e00.htm].

FAO (2009a). Food Outlook. [http://www.fao.org/docrep/011/ai474e/ai474e02.htm].

FAO (2009b). The State of Food Insecurity in the World 2009. Economic crises – impacts and lessons learned. [http://www.fao.org/docrep/012/i0876e/i0876e00.htm].

FAO (2009c). *Crops Prospects and Food Situation*. No. 2. [http://www.fao.org/docrep/011/ai481e/ai481e04.htm].

FAO (2009d). *Mozambique: PRO-IRRI Farming systems project. Farming systems assessment and perspectives in Sofala and Manica*. Draft v1. FAO, Rome, Italy.

Godfray, H.C.J.; Beddington, J.R.; Crute, I.R.; Haddad, L.; Muir, J.F.; Pretty, J.; Robinson, S.; Thomas, S.M. & Toulmin, C. (2009). Food Security: The Challenge of Feeding 9 Billion People. *Science*, 327: 812. (DOI: 10.1126/science.1185383).

Grey, D. & Sadoff, C. (2007). Sink or Swim? Water security for growth and development. *Water Policy*. 9: 545–571.

Grey, D. & Sadoff, C. (2009). Beyond the river: A practitioner perspective. *Stockholm Water Front*. No 1, May. [http://www.siwi.org/documents/Resources/Water_Front/Water_Front_1_lores.pdf].

Hall, K.D.; Guo, J.; Dore, M. & Chow, C.C. (2009). The Progressive Increase of Food Waste in America and its Environmental Impact. PLoS ONE 4(11): e7940. (DOI:10.1371/journal.pone.0007940).

Hanssen, O.J. & Olsen, A. (2008). Kartlegging av matavfall. Forprosjekt for NorgesGruppen. Östfoldforskning. Norway. (Mimeo).

Hillebrand, E. (2009). *Poverty, Growth, and Inequality over the next 50 years*. Expert Meeting on How to feed the World in 2050. FAO, Rome, Italy. 24–26 June. [ftp://ftp.fao.org/docrep/fao/012/ak968e/ak968e00.pdf]

Jones, T. (2006). Addressing food wastage in the US. Interview: *The Science Show*, 8 April. [http://www.abc.netau/m/scienceshow/stories/2006/1608131.htm].

Kader, A.A. (2005). Increasing food availability by reducing post harvest losses of fresh produce. V International Postharvest Symposium. *Acta Horticulturae*, 682.

Kantor, L.; Lipton, K.; Manchester, A. & Oliveira, V. (1997). Estimating and addressing America's food losses. *Food Review*, Jan-Apr: 2–12.

Lannerstad, M. (2009). *Water Realities and Development Trajectories. Global and Local Agricultural Production Dynamics*. PhD Thesis. Linköping Studies in Arts and Science. No. 475.

Liu, J. & Savenije, H.H.G. (2008). Food consumption patterns and their effect on water requirement in China. *Hydrology and Earth System Sciences*, 12: 887–898.

Liu, J.; Yang, H. & Savenije, H.H.G. (2008). China's move to higher-diet hits water security. *Nature*, 454 (7203): 397.

Lundqvist, J. (2009a). Unpredictable and significant variability of rainfall: carryover stocks of water and food necessary. *Reviews in Environmental Science and Bio-Technology*. (DOI 10.1007/s11157-009-9167-x).

Lundqvist, J. (2009b). The DNA of Waste. There is no such thing as a free loss of food. *Forum CSR International*. [http://www.nachhaltigwirtschaften.net/scripts/basics/forumcsrE/basics.prg?session=3eb5f8054bbc5ac6_146629&a_no=289&suchbegriff=The%20DNA%20of%20waste].

Lundqvist, J.; de Fraiture, C.; Molden, D.; Berndes, G.; Falkenmark, M.; Holmen, H.; Karlberg, L. & Lannerstad, M. (2008). *Saving Water: From Field to Fork. Curbing Losses and Wastage in the Food Chain*. SIWI Policy Brief. SIWI, Stockholm, Sweden. [http://www.siwi.org/].

McMichael, A.J.; Powels, J.W.; Butler, C.D. & Uauy, R. (2007). Food, livestock production, energy, climate change, and health. *The Lancet*, 370. No 9594. 6 Oct.: 1253–1263.

Molden. D. (ed.). (2007). *Water for food, water for life: a comprehensive assessment of water management in agriculture*. Earthscan Publications. London, UK; and IWMI, Colombo, Sri Lanka.

Molden, D.; Oweis, T.; Steduto, P.; Bindraban, P.; Hanjra, M. & Kijne, J. (2010). Improving agricultural water productivity: Between optimism and caution. *Agricultural Water Management*, 97(4): 528–535.

Morris, M.; Binswanger-Mhkize, H.P. & Byerlee, D. (2009). *Awakening Africa's Sleeping Giant: Prospects for Commercial Agriculture in the Guinea Savannah Zone and Beyond*. The World Bank. Washington, D.C., USA.

MSSRF [M.S. Swaminathan Research Foundation] (2002). *Food insecurity atlas of urban India*. M.S. Swaminathan Research Foundation and World Food Programme. MSSRF, Chennai, India.

Nellemann, C.; MacDevette, M.; Manders, T.; Eikhout, B.; Svihus, B.; Prins, A.G. & Kaltenborn, B.P. (eds.) (2009). *The environmental food crisis. The environment's role in averting future food crisis*. UNEP.

Peden, D.; Tadesse, G. & Misra, A.K. (2007). Water and livestock for human development. In: D. Molden (ed.), *Water for food, water for life: a comprehensive assessment of water management in agriculture*. Earthscan Publications. London, UK; and IWMI, Colombo, Sri Lanka: 485–514.

Rano, L. (2008). Industry concerns follow massive beef recall. *Food Production Daily*, February 19. [http://www.foodproductiondaily.com/news/ng.asp?id=83362].

Rockström, J.; Kaumbutho, P.; Mwalley, J.; Nzabi, A.W.; Temesgen, M.; Mawenya, L.; Barron, J.; Mutua, J. & Damgaard-Larsen, S. (2009). Conservation farming strategies in East and Southern Africa: Yields and rainwater productivity from on-farm action research. *Soil and Tillage Research*, 103: 23–32.

Schäfer-Elinder, L. (2005). Obesity, Hunger, and Agriculture. The damaging role of subsidies. *British Medical Journal*, 331: 1333–1336.

Schiller, M. (2009). *Dear EarthTalk: Food waste has become major issue in United States*. [http://www.pasadenastarnews.com/ci_14343450 Posted: 02/05/2010].

Schmidhuber, J. & Shetty, P. (2005). *The nutrition transition to 2030. Why developing countries are likely to bear the major burden*. Plenary paper presented at the 97th Seminar of the European Association of Agricultural Economists, Reading England, 21–22 April. [http://www.informaworld.com/smpp/content~content=a743906480].

Shah, A. (2009). *Poverty Facts and Stats*. [http://www.globalissues.org/article/26/poverty-facts-and-stats]. Last updated, March 28.

Smil, V. (2000). *Feeding the world. A Challenge for the Twenty-First Century*. The MIT Press, Cambridge.

Steinfeld, H.; Gerber, P.; Wassenaar, T.; Castel, V.; Rosales, M. & de Haan, C. (2006). *Livestock's long shadow: environmental issues and options*. FAO, Rome, Italy.

Stuart, T. (2009). *Waste. Uncovering the global food scandal*. Penguin Books, London, UK.

Thoenissen, R. (2009). *Fact Sheet: Food Waste in the Netherlands*. Ministry of Agriculture, Nature and Food Quality. November. [http://www.se2009.eu/polopoly_fs/1.24471!menu/standard/file/Roland%20Th%C3%B6nissen.pdf].

UN (2009). *The Millennium Development Goals. Report 2009*. United Nations, New York, USA. [http://unstats.un.org/unsd/mdg/Resources/Static/Products/Progress2009/MDG_Report_2009_En.pdf]

UN-IRIN (2008). *SAHEL: Region is ground zero for climate change*. June 2. [http://www.irinnews.org/Report.aspx?ReportId=78515].

UN-IRIN (2010). *Burkina Faso: Farmers act on climate change*. March 11. [http://www.irinnews.org/Report.aspx?Reportid=87806].

World Bank (2008). *World Development Report*. Washington, D.C., USA.

WRAP (Waste & Resources Action Programme) (2009a). *Household Food and Drink Waste in the UK*. [http://www.wrap.org.uk/retail/case_studies_research/report_household.html].

WRAP (Waste & Resources Action Programme) (2009b). *The food we waste in Scotland*. September. [http://www.wrapscotland.org.uk/the_food_we_waste_in.html].

CHAPTER 6

Prioritising the processes beyond the water sector that will secure water for society – farmers, fair international trade and food consumption and waste

J.A. [Tony] Allan
King's College, London, UK; SOAS, London, UK

ABSTRACT: Most of the water used in our economies is invisible. Invisible *green water* in soil profiles accounts for over 70% of the water used in food production. Also invisible is the water embedded in commodities traded internationally – know as *virtual water*. The role of farmers in achieving water security is also invisible. They manage all of the green water and with engineers most of the blue water worldwide. By doubling and trebling water productivity in both rainfed and irrigated agriculture farmers have met the water and food needs of the global population, which trebled in the second half of the 20th century. The international food commodity trading corporations are also pivotal in achieving global water security. It is argued that it is important that they exercise influence on the supply chain in proportion to this invisible pivotal role. There are other invisibles. Consumers unwittingly determine the volumes of water used worldwide. Diet is key. A non-vegetarian has a daily footprint of about 5 m^3 of water per day. The daily footprint of the vegetarian is only 2.5 m^3. Food waste is another invisible factor. Consumers in industrialized countries waste over 30% of the food –and embedded water– that they purchase. The same proportion is wasted between the farm gate and the market in developing countries. Reductions in these numbers are strategic and accounting for them will be essential in the achievement of water security.

Keywords: water security, green water, virtual water trade, consumers, trading corporations, food waste

> *It is on world's farms that the challenge must be faced and success achieved. Farmers are the world's major water managers*

1 INTRODUCTION

Water security is a vast topic. Scientists who deconstruct the topic find it daunting. Water professionals find it daunting because there is so much there that is beyond the water sector. The private sector finds it daunting because there is so much that is beyond the market that is of elemental importance. Even politicians who can think in a language of social and economic priorities find it daunting. They find it hard to allow water security onto the public agenda because the political price that has to be paid to change the approach of society to water use and management is so high. We all find it daunting because there are so many interacting uncertainties. Politics is the only trade that engages day in and day out with uncertainty and *wicked problems* (Conklin, 2006). And the ways of politics are very rough and ready. We must nevertheless learn to engage with this rough domain in which urgent and wicked problems are either addressed or ignored.

A wicked problem is not only urgent it is also extremely uncertain. Wicked problems are normally the outcome of previous attempts to deal with an earlier version of a similar problem (Conklin *et al.*, 2007). For example the big-oil/government nexus responded to the oil and gas crisis of the 1970s by racing to develop North Sea and North Slope oil and gas resources. In this they did the wrong thing extremely well. Oil prices stayed low for another three decades with devastating environmental consequences. An element of this rush to do the wrong thing extremely well was to crush precious and appropriate initiatives to mobilise renewable energy technologies[1]. There are less iconic examples in the water sector in the excesses of the era of the hydraulic mission in industrialised economies and still in train in the BRICS economies (Brazil, Russia, India, China) and in some other developing economies.

The purpose of the chapter is to focus on four elements of the intellectual challenges and of the water policy challenges associated with this vast topic. *First*, it will be shown that it is farming communities that manage the big water used and consumed by society. They manage about 80% of the water used in our economies – about 70% by volume of this water is green water and 30% is blue water. Unless society reduces its unnecessary food consumption we shall have to rely on farmers worldwide to raise the productivity of green and blue water to meet the food consumption of a future global population of about 9,000 millions. It will be shown that they have the propensity to do this. The *second* element of a more secure water managing scenario is one where the prices of food commodities send signals to farmers that they should produce crops with less water and with practices that do not impair the services of the water environment. A number of examples will show of how farmers have achieved spectacular increases in returns to water even in our own lifetimes. Price incentives work when they are reliable and long term and not misguided attempts to address wicked problems. The *third* element of fundamental importance is *international trade*. International food commodity trade has also been spectacularly successful in meeting the needs of at least 180 economies that have run out of water. But global trade is not fair. International trade in agricultural commodities has been severely distorted for decades. Its terms punish the weak economies of, for example, Africa. Food production is the major livelihood of most of the peoples of Sub-Saharan Africa. But the water productivity of Sub-Saharan farmers is the lowest worldwide. Millions of African farming families are repeatedly driven back into poverty by the level of international food prices kept low by the governments and food producing interests in Europe and the USA.

Fourthly, it will be emphasised that farmers compete on our behalf with the environment for water and tend to impair the water services of the environment. A joint effort is needed – both via increases in water productivity on the *farms* and *changed* patterns of *food consumption* by *society*. The latter will both *enhance human* and *environmental* health and reduce the consumptive use of water.

Farmers everywhere must be nurtured and encouraged to prosper by increasing – in some cases doubling or more– their water productivity. Doubled rainfed water productivity in Africa would be of global significance with respect to future global water security.

The issues of wasteful practices in the ways society handles the food supply chain in advanced economies is treated in another chapter (Lundqvist, this volume). Food transportation, storage and marketing –between the field and the market in developing countries accounts for even larger proportion of perishable food losses. In this case it is not the perverse behaviour of consumers. The waste of food, and therefore of consumptive water use, occurs because of the absence of effective transportation and food storage technologies (Biswas, 2009, pers. comm.).

The chapter will not address water security issues related to the small volumes of water needed for drinking, domestic use and for non-farming livelihoods. This form of water security is rarely constrained by the availability of water resources. Very few economies do not have the volumes of freshwater needed to supply enough water for their domestic water and water for non-agricultural jobs. Some economies in the Gulf and Yemen with its 23 million population living mainly between 1,200 and 2,000 metres above sea level, and many small island states are in difficulties with water

[1] The controversial example of energy is used because everyone can engage with this very topical issue. It especially highlights the dangers of *government/private sector alliances* where those at the heart of this political system are dependent for their continued exercise of power on very poorly informed *voters* or for their continued leadership of prosperous companies on badly informed *customers*. The *voters* and *customers* are very often manipulated to have appetites for types and levels of consumption that are lethal for society and its long term survival. The political economy is a wholly owned subsidiary of the environment.

security for industry and domestic uses. However, all those economies that have a sea coast or lie on major rivers can now desalinate water at costs that are similar to providing potable water by any other means.

2 GLOBAL WATER SECURITY

We have been trying for some decades to answer the question – *will there be enough water to meet the needs of a bigger population of between 8,000 and 9,000 million by mid-century in a warmer world with some different rainfall patterns?* The purpose of this study will be to show that the key factor relevant to answering this question and related policy issues is crop productivity on the farms of the world. On the subsistence farms of Africa, on the family farms of India and Europe, as well as on the corporate farms of North America and Brazil.

Since 1950 we have placed unprecedented pressures on our water resources – especially on our blue water. But we have also reached deeper and deeper into tracts with green soil water by turning natural vegetation into cropland. Over the past half century the response of farmers almost everywhere has been extraordinary[2]. Even more important than the mobilisation of new land and its green water and the expansion of irrigation has been the increase in crop and livestock yields. Increased yields on rainfed farms mean that the productivity of water has been increased. These increases in the yield of green water are easy to track. Increases in the returns to blue water have also been achieved but these are more difficult to evaluate.

These increases in returns to water have made it possible for the world to remain water and food secure. There will have to be further increases in the productivity of the big water deployed in food production if we are to remain water and food secure by the time we reach peak population in the second half of this century. At the same time society will have to re-think its food consumption habits and preferences if the water environment is to remain in good health and the health of society improved.

The second half of the 20th century has been a story of what we would now call *adaptiveness*. This *adaptiveness* has been unprecedented because it has been associated – in many parts of the world, particularly in the temperate latitudes – with the industrialisation of farming. The major feature of this industrialisation has been the unprecedented increases in the use of hydrocarbon-based energy in transport, cultivation, fertilizers and other inputs. The water/energy link is important but it is not the topic of this study.

Since 1950 the world's population has almost trebled. The use of freshwater has almost doubled. Food production has increased about fourfold and the only continent that now has a high proportion of persistently undernourished people is Africa. South Asia also has undernourished people but it

[2] The farmer, whether a family farmer or a global corporation, has in their hands the decision making power to gain high or low returns to water by combining a vast array of endowments and inputs. These endowments include good or poor soils, which may be well watered or not. The climate may enable single or multiple crops and provide optimum growing temperatures of 30°C or not. Droughts and floods have to be taken into account. The progressive and extraordinary increases in water productivity have mainly been the consequence of technological advances – equipment to transport and cultivate as well as fertilizer, herbicides and pesticides – since the beginning of the early industrial revolution in about 1800. The technological advances would have had little impact without a suite of economic and social adaptations. Technology helped the farmer to cope with the environmental circumstances and occasional extreme events. In parallel, urbanised industrial society evolved markets with demands for food commodities at prices that stimulated production and investment. Farmers could accumulate surpluses that enabled assets to be capitalised. A suite of other conditions have also evolved with complicated impacts. These include farm policies that protect and subsidise farmers. It will be shown that these range from those that are essential to protect the weak to those that wreak havoc internationally across over 100 poor mainly agricultural economies. The global market will be identified as the infrastructure that has both the capacity to remedy the water scarcity of most of the world's economies via virtual water *trade* as well as the potential means to ensure that the incentives for poor farmers in weak economies to have the incentives to improve their water productivity.

has managed to remain a net food exporter despite its very rapid rate of population increase. As a consequence India's groundwater resources have been very heavily overdrawn in some regions. But as in many regions water is still being moved via internal food trade from relatively water scarce regions such as Northwest India to the water rich regions of Eastern India (Verma, 2009). The same phenomenon occurs in China where the water scarce North China Plains are exporting food commodities to the water rich south of the country (Chapagain, 2007). California performs the same remarkable role in the USA and Andalucia in Spain. Industrialised irrigated farming achieves water miracles if the environmental costs are ignored.

This analysis will focus on food production because it is the demand for food which has exerted pressure on the world's water resources. Every additional individual, whether on a farm in a remote rural region of Africa, a new immigrant to the USA or another baby in a rich European city needs about 1,000 m^3/yr of water to provide its food. Food consumption accounts for about 80% of the water needs of a society. A 40% increase in the productivity of green and blue water used consumptively in crop and livestock production will meet the future water and food needs of the world. A lesser increase will meet the needs of a human population which might decide to consume food of types and in lower volumes and in different mixes that will be good for human and environmental health.

We also need to use and dispose of the water used for domestic and industrial purposes effectively as this water is costly to mobilise and dispose of. But any economies in these water-using sectors will be of minor significance with respect to the goal of achieving water security. It is the volume of water that is important for global water security and it is in food production that the big volumes are mobilised.

It is on world's farms that the challenge must be faced and success achieved. Farmers are the world's major water managers.

3 DEFINING WATER SECURITY

The important message is that water security is achieved outside the water sector in economies that are diverse and strong. A strong economy can afford to import water intensive commodities.

The important question is –is there enough water in the global system to meet the demands of the water scarce in a world with 25% more people?

To be water secure an individual needs about 1,200 m^3/yr. The volume of water needed for food per head appears to range from about 600 m^3/yr to 1,500 m^3/yr [3]. The water needed for the non-agricultural commodities consumed each year varies much more and these numbers are even more difficult to model than those for food consumption[4].

The current 6,800 million global population *consumes* about 8,000 Mm^3/yr of water in meeting its food needs. This volume happens to be equivalent to the annual flow of the biggest river in the world –the Amazon. The Amazon Basin's vast water resources are scarcely touched. Perversely the vast majority of the green and blue water *consumed* is in relatively water scarce regions and not in the regions of the world that have high levels of rainfall. Green and blue water in less well water endowed regions is used to produce food commodities that stay for the most part within the economies to feed their own populations.

By 2000 most of the 210 economies of the world had become net food importers. The volume of virtual water *trade* in food commodities is over 1,000 km^3/yr. This 15% or so of global green and blue water *consumption* is equivalent to the flow of any one of the next four biggest rivers – the Orinoco, the Congo, the Ganges or the Yangtse. Interestingly only one of these tropical/equatorial rivers – the Ganges – is utilised to provide significant volumes of consumptive use in food production. The

[3] These numbers are based on the modelling of Arjen Hoekstra.

[4] The assumptions used in the water footprint modelling to date.

vast blue water resources in these mighty tropical/equatorial surface water systems remain safely in the environment for the moment.

It is the blue water in the south to north and north to south flowing rivers – and their associated groundwaters – that have supported the majority of the crop production that has secured at least half of the extraordinary four-fold increase in food production in the second half of the 20th century. These are minor rivers in global terms normally with less than 10% of the flow of the major tropical rivers. These rivers – the Mississippi, the Nile, the Amu and Syr Darya, the Indus, the Irrawaddy, the Mekong and the wandering Yellow river have been raided for their blue water. As a consequence, some of them, the Nile, the Aral Basin rivers and the Yellow scarcely reach the sea.

When an economy runs out of water it will be evident in its inability to feed itself. The small volumes of water needed to provide potable water for drinking and high quality water for other domestic use and for industry and services account only for between 10% and 20% of the water utilised. High proportions of the water used at home and in industry returns to the environment and where treated this returning water can be re-used by society and the economy. Water used by farmers, by contrast, is utilised in a system where the water transpires to the atmosphere. There it is inaccessible to farmers and to the economy of which they are a part.

Moving food around the world in international trade is the main way that water security is achieved. It requires about 1,000 m^3 (tonnes) of water to produce a tonne of wheat [tonne (t) = 1,000 kg]. Economies that cannot find the 1,000 m^3 of water avoid all the economic and political stress of being unable to mobilise the water in their own economy by importing wheat and other water intensive commodities. The practice is as old as civilization as no city can survive without a rural hinterland able to provide embedded water in food. The embedded water is known as virtual water (Allan, 2001; 2003; Hoekstra & Chapagain, 2008). Most water – both green and blue water – stays in the economies where it is able to be used. Only about 15% of the water used to produce crops and livestock enters international trade. This proportion has been extremely successful in meeting the water needs of the water deficit economies. In practice about one third of the water traded is in exchanges between highly industrialised economies. The important message here is that water security is achieved outside the water sector in economies that have become diverse and strong. A strong economy can afford to import water intensive commodities. Strong and diverse economies are not dependent on water resource endowments. Strong and diverse economies evolve in well-governed political economies where human resources have been nurtured.

Understanding international crop and livestock trade is an important part of understanding current and future global water security. Some of big virtual water net *exporter* economies are advanced industrialised economies[5] – the USA, Canada and Australia. Other major net virtual water *exporters* are BRICS[6] economies – Brazil and India, or potential major food exporters such as Russia.

Almost all of the major virtual water *importers* are industrialised countries. In East Asia, Japan and Korea are in this group. Only France of Europe's 27 economies is a major net virtual water *exporter*. The Middle East and North African economies engage in virtual water *import trade* more than any other non-advanced economies and they will increase their participation in such *trade* as their populations will double by mid-century.

At least half of the additional food needed to meet the increased global demands of the second half of the 20th century has come from green water in temperate and semi-arid tracts in both the Northern and the Southern hemispheres; in North America and Europe and in South America and Australasia. Grain yields of such crops as wheat have been doubled or trebled since 1950 and amazingly they continue to increase in these regions. These globally strategic increases in

---

[5] There are about 35 advanced industrialised economies. They include the economies of the EU, the USA, Canada, the economies of Australasia and some of the economies of East Asia.

[6] BRICS – Brazil, Russia, India, China and South Africa. The BRICS economies comprise over 40% of the global population. They include the water tower of the world – Brazil and two pivotal economies: India and China that have either not come on to the world market at all (India), or only to a minor extent (China). Russia has vast underused water resources.

returns to water, mainly to green water, are the result of farmers combining their inputs – including water – ever more effectively.

They have been enabled to do this because they have been given incentives to produce food commodities in secure protected economies especially in the EU and the USA. In the process, rural societies have been transformed and it is this complex adaptiveness of advanced economies that has enabled the increases in returns to water. These advances in water productivity have brought us the version of global water security that we currently enjoy.

4 FARMERS, RURAL TRANSFORMATION AND FAIR INTERNATIONAL TRADE

4.1 *Farmers*

> *Farmers do not consciously address global water security but collectively they are key to its achievement.*
>
> *What is water security in volumetric terms if water use efficiency can be increased almost tenfold?*

Farmers have shown that it is possible to achieve tenfold increases in water productivity. In North-west Europe the yields of wheat per hectare in 1800 were about 1 t/ha. By 1900 they were 2 t. By 1950 they had reached 3 t/ha. Remarkably, by 1990 yields had reached 9 t/ha. All of these gains were achieved with the same water, as annual rainfall had not changed. It was not that three times the water had been mobilised between 1800 and 1950, nor three times more by 1990. After mobilising a range of inputs and enjoying – at least in recent history – reliable prices and helpful market circumstances water was being used nine times more effectively. What is water security in volumetric terms if water use efficiency can be increased almost tenfold?

The local impact of these changes on the farms of the UK was to transform the 60% dependence on imported food in the 1930s to a 40% dependence by the 1990s despite a 15% increase in population. A significant proportion of the global food exports have for the past four centuries been commodities exported from the tropics that could not be grown in Northern Europe. By the 1980s UK was even exporting grain for animal feed.

In Southeast Asia there has been another example of extraordinary increases in water productivity. In this case in the productivity of blue water. After the end of hostilities in Vietnam in 1975, farmers were able to operate in a secure environment. They could increase the rate of cropping from one to two, and in some cases, to three crops per year. Yields per crop also increased – they sometimes doubled. With double and treble cropping yields per hectare increased from 3 to 20 t/ha/yr in some regions. Vietnam changed within a decade from being a minor rice importer in 1975 to being a rice exporter by end of the 1980s (FAOSTAT, 2009). It became the second after Thailand in rice exports in the world by the mid-1990s. No one has completed any calculations of the increases in returns to green and blue water. One of the seasonal rice crops – that in the monsoon season – doubled yield to mainly green water. This increase reflected a significant increase in green water productivity. The calculation of the increased yield to blue water in the second and third rice crops are much more difficult to model. But in the firmament of the unknown unknowns on global water security this gap in knowledge is minor.

Farmers do not consciously address global water security but collectively they are key to its achievement. Every increment of improved yield, that is not just the consequence of increased blue water applications, contributes to local water security directly. And indirectly to global water security as water productivity increases can be fed through into the global system via *trade* in virtual water.

It is not being argued that the installation of water infrastructures created by engineers, and improved seeds, fertilizers, pesticides and herbicides are unimportant. They are important but the extent to which they are effectively combined with scarce water is in the hands of farmers.

It is being emphasised that if we are serious about addressing global water security one of the main ways we can promote this is by making the lives of farmers easier by protecting them from the uncertainties of markets and international trade. In the next section it will be emphasised that farmers combine available inputs most effectively if the rural economy and rural social infrastructures are favourable.

Farmers operate in pre-modern modes of food production with low levels of returns to water until political and economic processes allow them to operate modern systems with different options and the associated very much better returns to water. The importance of the transformative context will be considered next.

4.2 *Rural transformation*

Over the next half century, farmers will continue to adapt their capacities and will respond to the increased demand for food in all regions if they are given the economic and social circumstances in which to do so.

Farmers have achieved doubled and even tenfold increases in crop productivity and returns to water over the past two centuries when their economic and social conditions changed and they were able to take advantage of new circumstances.

Environmental hazards have not varied much during the past two centuries. In the next century in some areas they will worsen with climate change and in other regions crop and livestock production might be easier. The calculus of the worsening and improving has yet to be completed. Two things are certain. First, farmers have delivered adaptive solutions to the increased demands for water to meet demographically imposed demands for food during the last century. These increased demands have been much greater than anything climate change will bring about in the current century in the availability of global water resources[7]. Secondly, over the next half century farmers will continue to adapt their capacities and will respond to the increased demand for food in all regions if they are given the economic and social incentives and favoured circumstances in which to operate.

They will do these things quickly or slowly to the extent that commodity prices are attractive and other incentives are intelligent and sound. If we are serious about global water security we need to bring in measures – *at the pace that is politically feasible* – to ensure a number of conditions. First, commodity prices should capture the value of water. Secondly, incentives put in place should take into account the needs of the farmers to be able to cope with environmental shocks. In addition, in advanced economies immediately, and at a politically feasible rate in less developed economies, incentives should also be introduced which promote practices that reduce the environmental impacts of intensified crop and livestock production on water and atmospheric services.

Rural transformation that enables farmers to achieve higher returns to water is not just about the economics of production and the adoption of sustainable practices. The past half century has shown that it is a complex mix of corporatisation at one extreme and a wide range of changes in the structure of rural family economies at the other. Financial services are essential. The features of rural transformation of fundamental importance to farmers in poor and non-diverse economies also include improvements in the material infrastructure. These improvements include village electrification, improved roads and bus services and reliable water services. And in the social infrastructure these improvements include education and health services. Just as important is the integration of the rural economies into regional, national and international economies in circumstances that are fair. Markets that favour the advanced economies will do just the reverse.

The integration of poor economies into global trade can be very dangerous and is only positive when the international terms of trade are fair. I guess there are many readers who at this point feel that fair international trade is a fanciful element of global water security. I would argue that it is amongst the top four factors preventing Sub-Sharan African farmers – the farming community that has been unable to keep pace with demographically driven demands – to match increased local

[7] The comparison of the demographic and climate change impacts is deliberately stated in the way it is in the text at the level of hyperbole in order to get attention.

demand for food and water. The other three are unreliable rainfall, poor soil fertility and the socio-economic context. The next five decades will answer the question – have rainfall reliability and soil fertility been determining in sub-Saharan Africa? Or is it the political and economic contexts in which Sub-Saharan farmers operate?

Rural transformation has some other important features that help a farmer to survive environmental and market shocks and go on to increase crop and livestock productivity. Farmers across the world both survive and increase elements of their productivity by engaging in off-farm employment. Everywhere, as soon as rural transformation begins, the proportion of off-farm income that supports the farm enterprise increases. Subsistence farmers are full-time farmers. At the other extreme corporate farms employ full time professionals. At all points, in between these two extremes, the proportion of part-time and off-farm activity is universal although with different degrees of importance. A feature of off-farm employment of all or some of the farming family is that it hedges the farm enterprise against the brutal uncertainties of the environment and the market. This hedging is not trivial. It is a life and death issue. The suicide rate amongst farmers is above the society average on subsistence farms, as well as on farms in industrialised countries. We need farming communities to survive to be food and water secure.

Numerous conditions of industrialisation and of modernity have made it possible for farmers to increase their returns to water. Notwithstanding these advances farmers still have to cope with the environment. With rainfall variability, with multi-year droughts and seasons of excessive rainfall and flooding. But the extreme hazards of the market place which are just as lethal, and the deathly circumstances of no market at all, are nowhere experienced in the rural societies of advanced economies. Droughts are painful but survivable albeit with new patterns of farm ownership and adapted cropping.

A feature of modernising rural political economies is their juxtaposition with urbanising and industrialising economies. Rural depopulation is closely associated with a developing, industrialising and diversifying economy. I shall never be the same after spending an afternoon on a Canadian prairie farm at the height of the Saskatchewan harvest of lentils and durum wheat a few days before drafting this chapter. These commodities were all destined for the Mediterranean and the Middle East via the extraordinary global commodity trading corporations – such as Cargill. One could only be impressed by the way that the specific commodity needs of distant economies with very serious water deficits are supplied with their deeply preferred and essential food commodities grown on an almost completely depopulated landscape. The farm operations which achieve very high and increasing returns to only 500 mm of annual rainfall are dependent on *huge inputs of energy*. These energy inputs mobilise the tiny in number – but massively capitalized – human skills and equipment focused on the vital harvesting, transportation, storage and marketing infrastructures.

One was curious to know why the green water productivity had improved over the past decade. As usual it was a mix of agronomic and technological inputs and changes. Yields of the grain crops, and therefore returns to green water, had increased by about 20% in the past decade. The main reason for this increase was the adoption of low-till methods of cultivation and seeding with the parallel development of herbicide technologies. Low-till cultivation and seeding could have been adopted a century ago. But it was not adopted until over half a century ago in Australia and since then the approach has been widely adopted. The point being made is that even already high returns to green water can be improved with significant impacts on global water security.

4.3 *International trade and fair international trade*

The purpose of this section is to highlight the major forces associated with international trade that bring about global water and food security on the one hand and impair it on the other. Figure 1 illustrates the volumes of virtual water associated food commodity trade. It shows how the industrialised regions dominate the exports as they have in some cases substantial water resources – especially green water resources in North America, South America and Australasia. They are the *water losing* economies. The *water saving* economies are in East and South Asia. Figures 2 and 3 show that the majority of trade in food commodities is between the industrialised economies.

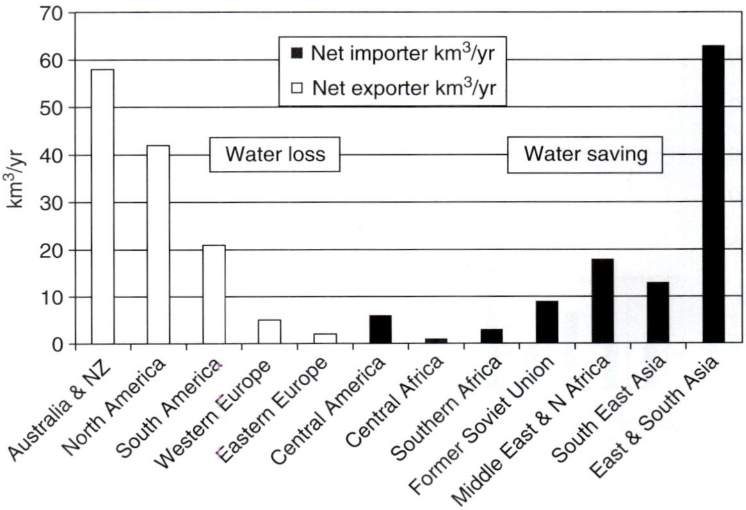

Figure 1. International virtual water *trade* showing the predominant *exporting* regions and the major *importing* regions.
Source: Hoekstra & Chapagain (2007)

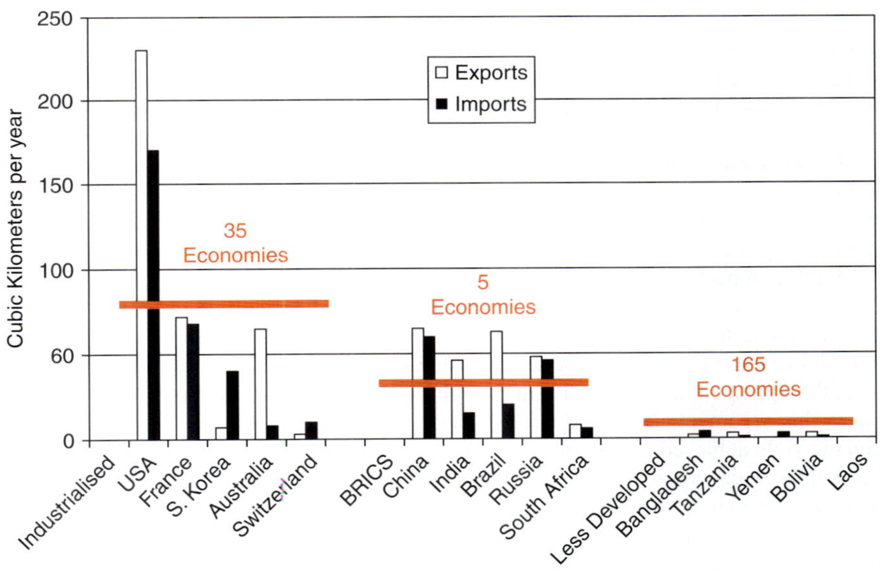

Figure 2. The volumes of virtual water *imports* and *exports* of some industrialised, BRICS and developing economies.
Source: Hoekstra & Chapagain (2007)

Trade in virtual water is accepted as the major means of remedying the water deficits of water scarce economies. Trade in food commodities is mainly between the industrialised economies. The BRICS economies are also significantly involved. The developing economies are scarcely involved and for the poorest developing economies, for example those of Sub-Saharan Africa, international food commodity trade can be very dangerously negative. For these economies virtual water *trade* associated with food commodity trade has a downside.

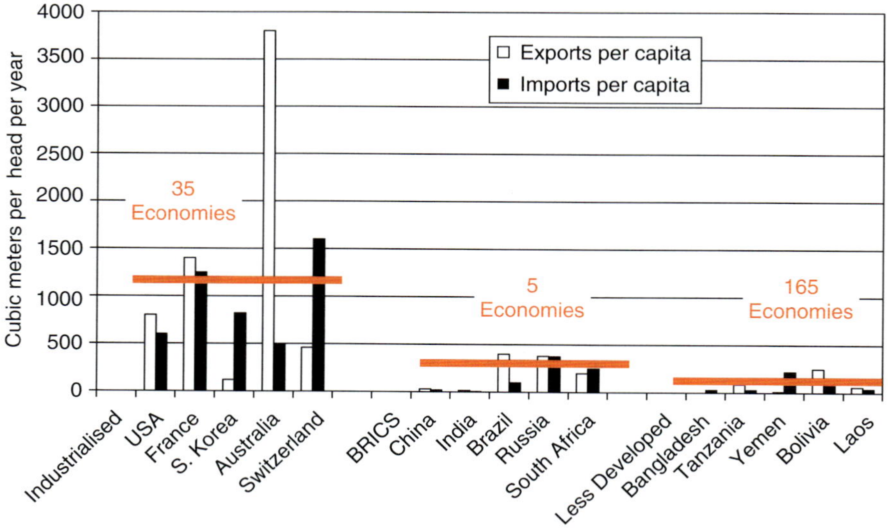

Figure 3. The volumes of virtual water *imports* and *exports per head* of some industrialised, BRICS and developing economies.
Source: Hoekstra & Chapagain (2007)

International trade in water intensive food commodities is mainly about meeting the food preferences of consumers in rich regions. But a substantial proportion of virtual water *trade* does address the water and food deficits of two BRICS and very many developing economies. Especially 20 or so economies in the Middle East where they are dependent for over one third of their water on imported food commodities. And when their populations double they will be 60% dependent.

The international food trade and emergency food assistance are currently important in Sub-Saharan Africa. The water intensive food imports are very necessary for humanitarian reasons during periods of drought. But at other times the terms of international food commodity trade have a very negative impact on the productivity of water in African economies. FAO data does show that there have been minor increases in crop yields in rainfed agriculture in East Africa. But crop yields of 1 t/ha or less are the norm in Sub-Saharan Africa. FAO agronomists argue that these levels of yield cannot be improved because rainfall is unreliable and soil fertility is low (Burke & Steduto, 2007, pers. comm.). The current drought in Kenya and the Southern Horn of Africa lend support to this argument. But there is other evidence that the yield gap – that is the difference between the yield achieved in experimental farm circumstances and that achieved by farmers who lack the technical, economic and social infrastructures of the experimental farm – can be reduced. Dorosh *et al.* (2008) have shown that there is strong relationship between crop production and crop productivity in rainfed tracts of Sub-Saharan Africa and the availability of transportation[8].

It is difficult to identify the relative impacts of the three factors determining crop yields: first, the environmental endowments; secondly the socio-economic infrastructures that can transform returns to water as demonstrated in industrialised economies; and thirdly, the terms of engagement in international markets.

[8] "In terms of agronomic potential, there is substantial scope for increasing agricultural production in sub-Saharan Africa, particularly in more remote areas. Total crop production relative to potential production is approximately 45% for areas within 4 hours travel time from a city of 100,000 population. In contrast, total crop production relative to potential production is only about 5% for areas more than 8 hours travel time from a city of 100,000 population. These differences in actual versus potential production arise mainly because of the relatively small share of land cultivated out of total arable land in more remote areas." (Dorosh *et al.* 2008).

The relative impacts of the third factor – the unfair terms of international trade – are especially hard to quantify. They are certainly a very significant factor in reducing the incentives for poor farmers on rainfed African farms to increase their crop yields and the returns to green water. The disincentives arise because the very productive farmers in the USA and Europe – enjoying powerful technical, market and social infrastructures as well as additional protection and close relations with USA based global trading corporations – have been putting staple grains on the world market for half a century at about half cost. There have been occasional price-spikes but the price of wheat has been falling for 1,000 years and especially since the 200 years of the industrial era. Staple food commodity prices have neither reflected the true costs of many inputs such as energy, nor the real costs of water –if environmental impacts were to be internalised.

The history of crop yield improvement in industrialised countries has been very long and arduous. The early advances were the result of agronomic improvements associated with investment. Farmers can only invest when they have financial surpluses and surpluses accumulate when prices are favourable. Prices tend to be favourable when there is scarcity associated with natural events such as droughts.

Sub-Saharan farmers endure droughts. But every time there is a drought and evident scarcity African governments with too many urgent problems including large and very poor urban populations – understandably – cannot resist the temptation of importing the under-priced grain on the world market. There has been an endless cycle of punishing rural poverty perpetuated by engagement of Sub-Saharan economies in international commodity trade during the second half of the 20th century. The experience has been damaging for the farmers and for Sub-Saharan economies. The current progress in the World Trade Organisation negotiations do not augur well.

It is suggested here that international trade in food commodities should be reformed to enable the farmers of Sub-Saharan Africa to be able to enjoy prices for their output that reflects their inputs and further enable them to invest in improved cultivation and harvesting systems and in modern and sustainable agronomic practices.

The visit to Saskatchewan at the height of record breaking harvest made one aware of what can be achieved in an environment that is *extremely hostile* most of the year. Farming has evolved from a life of non-mechanised poverty and cycles of failure (Waiser, 2005) to one where extreme environmental and market shocks can be accommodated. Last year's productive vigour on the grain farms of North America is a consequence of reasonable although late rains and the vital stimulus of two years of high commodity prices. The previous year's price spikes came after over a decade since the last spikes in 1995. In 2008 grain prices were exceptionally high as they tend to be when there is an unwelcome spike in global energy prices.

Canadian farmers – like African farmers – have to survive in a global market dominated by the distortions of EU and USDA subsidies. For decades the EU and the USA have competed down the global price for some of the main grain staples. The potential of the Canadian farm sector has been revealed by the price spike of 2008 when it was briefly released from the numerous disincentives and the lack of investment associated with low commodity prices. One does not have to go very far back, however, to encounter a period when the farmers of the prairies of Canada faced the inevitable crisis which follows a number of years of low rainfall. The recent extreme year was 2002 when some farms did not take their combine harvesters on to the fields in August or September. Canadian farms do fail but the whole industry is kept in place by the vast suite of hedging processes which relatively prosperous farmers can engage in themselves as well as those of the banks and in extremis of government.

5 COMPETING WITH THE ENVIRONMENTAL SERVICES OF WATER

Farmers compete on our behalf for water resources with the environmental services of water. We have a voice. The environment can only be given a voice.

Sustainable intensification is sound. Unsustainable intensification is not.

Farmers are also key players with reference to the environmental services of water. In delivering food and water security to society they compete – on our behalf – more significantly than any other water resources users with the environment for water. Agriculture tends to impair the water services of the environment. In many hotspots world-wide irrigation has made extensive tracts of land unusable.

When incentives are being put in place to bring about increased returns to water, measures should be put in place that incentivise consideration of the environmental services of water.

6 CONSUMPTION – WASTE, DIET AND DEMOGRAPHIC POLICY

> *A joint effort is needed – both via increases in water productivity on the farms and changed patterns of food consumption by society.*
>
> *China has taken 5% out of global water demand. Their population policy is the biggest water demand management measure in modern history.*

The analysis so far has shown that global water security has been achieved by increasing the productivity of green and blue water in food production. Where the intensification is sustainable there is a virtuous circle.

Water consumption for food production is such a dominant and pivotal use that many ways of reducing levels of agricultural demand for water resources must be deployed. Levels of food consumption are determined by consumers. Consumers in rich economies tend to overconsume food and associated embedded water. And they also waste food. Other chapters draw attention to the waste of water associated with the waste of food by society from harvest to table and beyond.

The relationship between the water consumption of individuals and societies and their diet has been emphasised by a number of water scientists (Hoekstra & Chapagain, 2008; Waterwise, 2008; Allan 2010). The water demand of the individual vegetarian South Asian has a light daily water footprint of 2.5 m^3/day of water. The heavy beef consumers of Europe and the USA require about 5.0 m^3/day of water. Since the consumption of a lot of red meat is bad for human health we have the remarkable situation in the industrialised world that people are eating their way to poor health and at the same time unnecessarily depleting and misusing scarce water resources. About one in five of the world's population live in industrialised countries and tend to have these bad consumption habits. If some were prepared to abandon these bad diets and others were prepared to eat more sensibly the demand for water in industrialised economies could be reduced by up to 30%. The challenge in the BRICS and developing economies is that the majority vegetarians in the 1,000 million population of India should remain vegetarian. The Chinese should remain committed to their fine cuisine and resist the industrialisation of their food consumption. And those in developing economies should not emulate blindly the bad consumption habits of the industrialised world. The difference between the water demand of a sensible food consumption scenario and one similar to that of the industrialised world will make at least a 20% difference to the level of future global water demand.

Global water demand is not only driven by diet. China has shown that it is possible to gain a substantial demographic dividend for the water environment by addressing the problem of rising demands on natural resources by introducing a population policy. The one-child policy has taken 300 millions out of China's and the global economy. Global water resource demands have been reduced by a volume equivalent to the water demand of the USA, or Old Europe or the Middle East. While the populations of the USA and Europe will not rise significantly, the population of the Middle East will double. The global calculus of water security has been impacted positively by China and will be impacted negatively by the Middle East. China has taken 5% out of global water demand. Their population policy is the biggest water demand management measure in modern history.

A joint effort is needed –both via increases in water productivity on the *farms* and *changed* patterns of *food consumption* by *society*. The latter will have the dual benefit of enhancing human and environmental health and reducing the consumptive use of water.

7 CONCLUDING COMMENTS

Water security has been shown to be an elusive concept. Most of the key factors have been addressed – invisible water resources, invisible solutions to water security beyond the water sector, virtual water and international commodity *trade*, water productivity, socio-economic contexts consumption and waste. But there are some others that compound the problem of conceptualizing water security. Water provides many services from food production to environmental amenity. New demands on water resources come along. Brazil has been producing bio-ethanol since the oil price shocks of the 1970s. Many other economies, notably the USA, have followed suit in the recent era of unstable global energy prices. Both Brazil and USA use food crops – mainly sugar cane, soya and maize – to produce bio-ethanol. The volumes of green water associated with this supposedly renewable energy production are very negatively significant with respect to global water security. Brazil is the only economy that is so well endowed with water that it can engage in bio-energy production without much reflection. Using food crops for energy production has a very large water footprint of strategic significance with respect to global water security. This example of society's latest way of testing the capacity of the global water environment is just another complicating factor in any attempt to conceptualise the already elusive concept of global water security.

It has been emphasised in this chapter that the volume of water is important but what we do with water is more important. Farmers do more with water than any other agent and have the potential to increase the productivity more than any other user because they handle 80% of water resources. Societies must nurture farmers and incentivise them to gain even higher returns – sustainably – from green and blue water. They can achieve such higher returns, in some cases much higher returns, if society enables them to operate in conditions that protect them from environmental and especially market uncertainties. They must be incentivized to achieve sustainable intensification of water resource use.

Mobilising new water has been important in achieving the contemporary version of global water security. But managing water, and especially allocating it effectively and *trading* a crucial proportion of it embedded in water intensive commodities have been other means that have been just as important. They will be even more important in future.

Global water security will be achieved with a wide variety of improved engineering, improved water management on farms and with an even wider variety of improved economic and social infrastructures. New blue water could increase food production by between 20 and 40% but with significant negative environmental costs. Increased green and blue water productivity could achieve even higher increases in global food production. The ways we consume food could have even bigger impacts. Diets of the one in five who live in industrialized economies could be changed to advance human health and reduce their water consumption by at least 20%. Future demand for water in the BRICS economies and the developing economies is very hard to predict and will be determined by demographic trends and even by demographic policies. The latter have been shown to be very significant in reducing water demand in China. In addition society itself has the solution in its own hands as it can modify its demands for water resources by reducing waste, by modifying diets and by levelling off the rate of population increase either as a consequence of socio-economic advancement or demographic policy.

We need to look beyond the water sector to achieve water security.

REFERENCES

Allan, J.A. (2001). *The Middle East water question: hydropolitics and the global economy*. I.B. Tauris, London, UK.

Allan, J.A. (2003). Virtual water – the water, food, and trade nexus: useful concept or misleading metaphor? *Water International*, 28: 1–11.

Allan, J.A. (2010). *Virtual water*. I.B. Tauris, London, UK.

Chapagain, A.K. (2007). *Globalisation of water: opportunities and threats of virtual water trade*. PhD Thesis, UNESCO-IHE, Delft, the Netherlands.

Conklin, J. (2006). Wicked problems & social complexity. In: J. Conkin (ed.), *Dialogue mapping: Building Shared Understanding of Wicked Problems*. Wiley, Chichester, UK.

Conklin, J.; Basadur, M. & VanPatter, G.K. (2007). *Rethinking Wicked Problems: Unpacking Paradigms, Bridging Universes*. NextDesign Leadership Institute Journal, New York, USA. 30 pp. Available at: [http://www.humantific.com/downloads/NextD_10_1.pdf].

Dorosh, P.; Wang, H-G. & You, L. (2008). *Crop production and road connectivity in Sub-Saharan Africa: preliminary findings of a spatial analysis*. World Bank, Washington, D.C., USA.

FAOSTAT (2009). FAO production and trade statistics. [www.fao.org].

Hoekstra, A. & Chapagain, A. (2008). *Globalisation of water: sharing the planet's freshwater resources*. Blackwell Publishing, Oxford, UK.

Verma, S. (2009). Securing India's water future. Stockholm World Water Week. Stockholm International Water Institute (SIWI), Sweden. Available at: [http://www.worldwaterweek.org/documents/WWW_PDF/2009/thursday/T3/Shilp_Verma_2009-08-20_Workshop_6.pdf]

Waiser, B. (2005). *Saskatchewan: a new history*. Fifth House, Calgary, Canada.

Waterwise (2008). *Hidden waters*. Waterwise, London, UK.

III
Economic development, the role of the private sector and ethical aspects

CHAPTER 7

Present and future roles of water and food trade in achieving food security, reducing poverty and water use

Juan A. Sagardoy
Mediterranean Agronomic Institute of Bari, Italy (MAI–B); International Center for Advanced Mediterranean Agronomic Studies (CIHEAM)

Consuelo Varela-Ortega
Technical University of Madrid, Spain (UPM)

ABSTRACT: The development of scarce water resources in the Mediterranean Region has been very important to satisfy the high demands for food, drinking and industrial water of a population that at regional levels has increased substantially. The chapter analyses, in the first part, the past trend of the irrigation development and the future demand. It is evident that irrigation development has contributed substantially to increase the food security. From the point of view of the water resources irrigation development has been and continues to be the largest user. Although irrigation areas have been growing substantially over the last 25 years the possibilities for continue such trend appear small, exception made of few countries, due to scarcity of the water resources. The chapter also reviews the interrelations between poverty, hunger, gender with the development of the water resources and the increased agricultural trade.

The scarcity of the water resources has reached alarming levels in several countries of the Region that made a call to reconsider the future options. One of such option is to use agricultural trade as a source to supplement the national agricultural production capacity. But here several options open to countries in terms of deciding what to produce and what to import/export. The concept of virtual water developed in recent years has helped to understand better the implications of trade in terms of water. However, this concept has not yet penetrated the sphere of the trade decision makers. The chapter intends to look further into this possibility by analyzing historical statistical data (production, irrigation areas, net trade and virtual water) of two selected countries (Egypt and Syria) to identify feasible possibilities of modifying the production cropping pattern – and consequently the trade– so that the water resources are used more efficiently.

The second part of the chapter focuses on a specific country (Syria) to assess the implications that would have changing the cropping pattern of main crops (wheat, cotton, maize) so that the water resources are used in a more sustainable manner while the income of the farmers do not deteriorate. This analysis is made at farm level and shows that, when attempting such changes, other aspects (labor, environmental consequences and incomes) are also important considerations. The consequences of this change at farm level are extrapolated at the national level to analyze the impact that such changes will have in the agricultural trade and the consequences for the government.

The chapter concludes that analyzing trade and production data through *water policy lenses* provides useful orientations regarding possible changes that can be attempted in the production and trade. However political decisions require consideration of several other factors that affect the production patterns and this may call for detailed studies. Ultimately governments have to make trade off choices between the trade policies and the use of the production factors.

Keywords: water resources, irrigation development, virtual water, agricultural trade, food security, poverty, Mediterranean Region

1 INTRODUCTION

The development of the water resources has been mainly geared to satisfy the food and domestic demand of the increasing population by expanding the irrigation areas and extending the distribution of domestic water supply. However this policy is approaching the limits of the available water resources in many countries and new ideas and policies are emerging that are influencing the sector. These new trends are reflected by the evolution of the different paradigms that have been proposed in the last decades. The first one was *more food per drop* that soon evolved to *more crops and jobs per drop* and now recently to *more cash and nature per drop.*

It is clear that several countries of the Mediterranean Region have not sufficient water resources to produce all the food they need and will have to rely on the import of food. The number of countries in this situation is likely to increase in the future. Consequently, it is relevant to look into the likely impact of the agricultural trade in the use of water resources. This has been done through the work on *Virtual Water* (VW) and *Foot print Water* by several authors (Chapagain & Hoekstra, 2004). The concept of virtual water has contributed to understand the flows of water involved in food trade, however rarely are used as a policy decision instrument. The chapter, in its second part, will simulate with a country example what are the issues that policy makers will have to face when reorienting production having into consideration the agricultural trade

The geographical focus of this chapter is the Mediterranean Region. The reason for this selection is related to aridity conditions of this region that makes the role of available water resources very essential, and the second reason to the fact that agricultural trade among all Mediterranean countries is already intensive and is likely to increase in the future. This second point is particularly relevant for the scope of the chapter.

2 THE PRESENT SITUATION OF THE WATER RESOURCES DEVELOPMENT AND MANAGEMENT

The available natural water resources per capita in the countries of the Mediterranean Region are the lowest among other regions of the world. For this reason their development and management constitute one of the leading sectors in the government actions. So far governments have tried to respond to the increased demand by spending sizeable sums of their budgets in the development of the resources (regulating dams, distribution networks, etc.). However, it has become evident that for several countries this policy is no longer possible due to the scarcity of the water resources and other alternatives need to be considered including demand management. There is however a great variability of situations and a closer analysis is needed and is attempted in the following sections.

2.1 *A Region of limited natural renewable water resources and high demand*

In the Mediterranean countries there is considerable variability of available natural water resources and their use by the populations of the respective countries. Figure 1 illustrates this variability. The scarcity index (the ratio of total water withdrawals/average natural renewable resources flow) already exceeds 50% in eight Mediterranean countries. When such rates reach or even exceed 100%, they indicate that there is a deliberate, but unsustainable use of non-renewable resource (Libya) or indeed reuse of part of the non-conventional resource such as treated wastewater or drainage water as in Egypt and Israel. However, the scarcity index refers to the totality of renewable water resources; the situation would appear even more critical if the indicator is referred to those resources considered exploitable (Hamdy & Lacirignola, 2005). In this case, a more dramatic picture will emerge, but the criteria used for determining the exploitable resources are quite variable and the figures are hardly comparable and hence the analysis is restricted to the natural renewable water resources.

However not only the availability of water resources is limited but also the variability of their occurrence is very high. Thus the figures of the natural available water resources can be misleading since only a small fraction of them is available under natural conditions. This makes necessary the

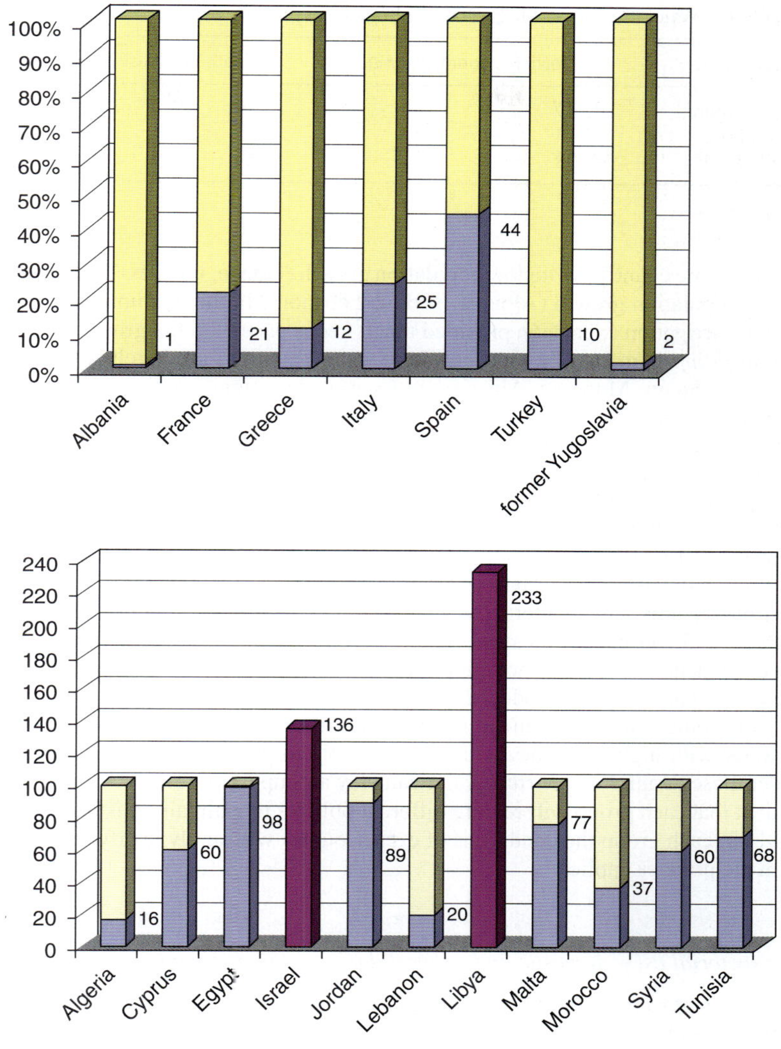

Figure 1. Ratio between the total water withdrawal and average natural renewable water resources (in percentage).
Source: Hamdy & Lacirignola (2005).

construction of regulating dams and in fact the Mediterranean region contains the largest proportion (80%) in the word of available water resources regulated by dams. Although dams have been the object of heated debates particularly from environmentalists, recognizing that the construction of some would be difficult to justify, the large majority has brought considerable benefits to the Region.

From the present high levels of regulation: 80% according to the World Bank (2006), is implicit that the opportunities for more regulating dams are very small and surely they are highly expensive and difficult to justify in economic terms. Here again another limit is being approached.

Most of the studies carried out about the water resources in the Mediterranean agree that there are three groups of countries that can be established:

The first group consists of countries where water availability will remain adequate up to 2025 and beyond, and where there is even a fairly comfortable margin for increased per capita draw-offs. This

Table 1. Water uses in the Mediterranean Region (in percentage).

Water Uses	Northern countries (%)	South and East countries (%)
Agriculture	49	79
Domestic	13	13
Industrial	38	8

group includes some countries with low population growth (France, Italy, ex-Yugoslavia) and some with stronger population growth (Albania, Turkey, Lebanon). In these countries there are limited possibilities for irrigation expansion provided that it keeps pace with the growth of population.

The second group is made of countries where water resources are barely adequate at present and this include: Spain, Morocco, Algeria and Cyprus. Any significant growth in the per capita draw-off or further increase of the irrigation areas will put these countries quite quickly in the critical situation being faced by the countries in the next group. Demand management is absolutely essential in these countries.

The third group is made by the countries where current water availability is already limited or negligible and includes: Malta, Egypt, Syria, Libya, Israel, Jordan, Tunisia and Palestinian Territories. These countries will probably have to face per capita draw-offs of conventional resources, or compensate it through the import of virtual water or increase dramatically the use of non-conventional water resources. As this second alternative may be too expensive for agricultural uses it is evident that food trade, and consequently virtual water, will play a major role in the future of these countries. The consequences of being a net importer of food are that the country has to dedicate a good part of the economic resources produced in the country to purchase agricultural goods produced elsewhere. It becomes therefore critical in these countries to allocate water to the most beneficial uses, use water with high efficiency and when possible to produce agriculture products of high value. Under these assumptions virtual water can play an important role.

It is evident that each group will require different policies to maintain satisfactory levels of use, and even within each group the conditions of every country will surely justify somewhat different emphasis in the policies applied.

2.2 *Water sectorial uses: domestic and industrial uses gaining grounds*

The sectorial uses of water in the Mediterranean Region are quite different in the North and South-East countries as indicated in Table 1. This clearly indicates the different economic development trends in both regions and in particular the different development of the industrial sector. It remains the fact that irrigation uses are very high in the South-East Mediterranean Region (SEMR) as agriculture remains an important sector of the economy while in the Northern countries the contribution to the GDP is less than 5%.

The figures of Table 1 do not reflect an important trend that is taking place in the domestic uses of the SEMR where the growth of the population is still high and where future demand is expected to increase several times the present values. Furthermore the coverage of the water supply in the EU countries is practically 100% while in some Mediterranean countries (Morocco, Algeria, Syria, Jordan and Egypt) still the coverage, particularly in the rural areas, is often far from complete as shown in Figure 2.

The growth of tourism in the Mediterranean countries exerts a strong seasonal increase of the domestic demand considering that the estimated number of tourists exceeds 250 million with an estimated demand of 1,000 Mm^3/yr. The basic question remains if these additional needs can be met from existing water resources or unconventional water resources will have to be developed. Another feasible alternative would be to reduce the agriculture use (using more virtual water) and transferring resources to the domestic and industrial sectors.

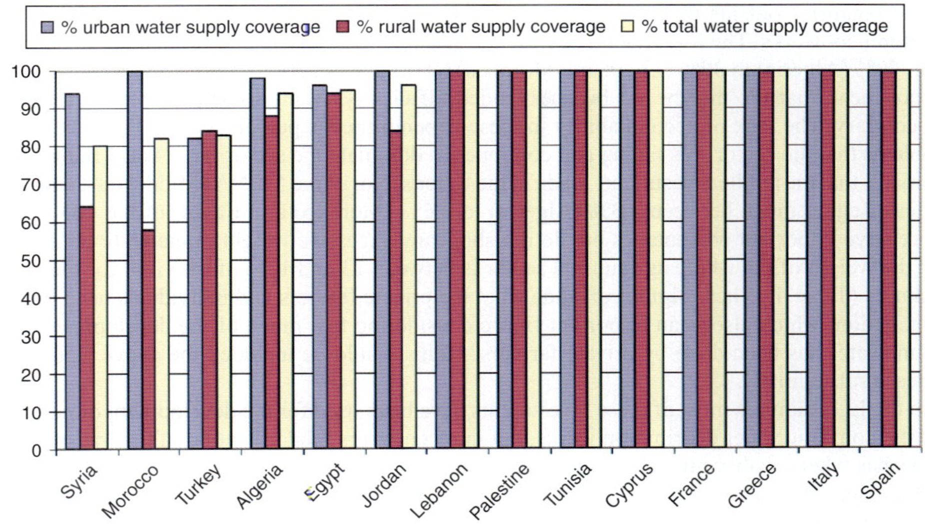

Figure 2. Percentage of coverage of the domestic water supply.

The use of non-conventional water resources can be an important component of coping with the future demand, particularly in the domestic and industrial sectors. Their use for agriculture is very limited as the costs are incompatible with the production of most crops. Countries of the MENA (Middle East and North Africa) Region are increasingly producing water for municipal and industrial use by removing salt from sea or brackish water. The region has 60% of the world's capacity and has been using this technology to supply more than half of all municipal water needs since 1990, producing 2,377 Mm^3/yr (World Bank, 2006). It is therefore foreseeable that they will continue to play an important role in satisfying the demand of the domestic and industrial sectors. Therefore it is foreseeable that much of the demand of the urban-industrial sector will be met from these non-conventional resources and will not add additional pressure over the existing conventional water resources.

2.3 *Investing in water resources development. The EU structural support policies and other financial resources*

The structural policies were established in the European Community (EC) to support the development of regions that have lower indexes of economic development than the averages. Essentially they are funds that do not need to be returned and constitute a part of the investments made. The main sources are the FEDER (European Fund for Regional Development), the Cohesion Funds, and the FEOGA (European Fund of Agricultural Orientation and Guarantee). In general, the subsidy ranges between 10 and 25% of the total investments, and the requesting regions must comply with the eligibility criteria. These European funds (FEOGA and Cohesion Fund) have been widely used in the National Plan for the Modernization of Irrigation in Spain, but still the larger part of the total investments is shared in nearly equal parts by the government and the users.

To some extent this policy has greatly contributed to the irrigation development in Spain which is one of the countries that have been using these European funds intensively in the modernization plan and in the provision of domestic water supply (FEDER). However, not all European Union (EU) countries of the Mediterranean Region has been using these funds effectively, as for example Italy and Greece, where there is a real contraction of the irrigated area (Table 1) particularly in the period 2000–2009.

The countries outside of the EU in the Mediterranean region have heavily invested in water development works. The investments represent a sizeable share of the public expenditures ranging from 20% (Algeria) to 30% (Egypt and Morocco) (World Bank, 2006). Large public expenditures do not mean that the money has been expended efficiently. Many local experiences document the contrary. The lack of accountability in these expenditures is perhaps one of the most serious problems that the sector was facing and continues to dominate the sector. The World Bank (2006) estimates that the public investments made in the water sector of the MENA (Middle East and North Africa) Region range between 1.3% (Algeria) to 3.6% (Morocco and Egypt) of the GDP, which are relative important percentages but do not show any excessive weight in the GDP considering that the weight of the whole agricultural sector is between 13 and 15% of the GDP for the SEMR countries (Table 4). These figures exclude what the private farmers and individuals have invested in irrigation, which also is an important figure considering that much of the development of the groundwater has been done by private farmers. If this would be included the above figures will increase substantially.

In the EU countries most of these investments have been used for the development of new areas and the domestic water distribution works, but in recent years modernization of the irrigation networks has taken a substantial share, particularly in Spain. For the SEMR countries, the expansion of the irrigated area has continued and constitutes the greater part of the investments. The modernization of irrigation systems is limited to some modest government contribution to the farmers willing to modernize their irrigation methods. As traditional surface water distribution methods (mostly using rotations of different kinds) are incompatible with the requirements of localized irrigation, this type of modernization has been largely restricted to groundwater areas where the water availability is not constrained by rotation rules.

3 IRRIGATION DEVELOPMENT. APPROACHING THE LIMITS

Governments and private individuals have invested large sums to provide irrigation water to their farms. The main reason for governments to support this development is to increase the food security of the country, provide employment and additional rents for farmers. Individual farmers are mostly geared by the aim of improving their financial situations but they contribute to increase food security directly or indirectly (by the exports). In the following sections the past trends are reviewed and the future possibilities discussed.

3.1 *The past irrigation expansion. Was population growth a real driver?*

Table 2 provides an overview of the irrigation development in the last 25 years for selected EU and SEMR countries. The growth patterns of the EU countries are quite high although they tend to be lower than those of the SEMR, exception made of France where the irrigated area is mostly made of supplemental irrigation where the additional irrigation water provided to crops are lower than for the rest of the countries. However this has proved to be economically attractive (largely due to the relatively low cost of the available water) and has produced a large expansion of the irrigated area in France. However in nearly all countries a slowdown of the growth during the last period reported is evident.

For the SEMR countries, the data available show the impressive growth made by all of them. Even Egypt that has extremely limited water resources has been able to expand the irrigated area in a substantial manner. In the case of Egypt much of this expansion is due to the reuse of drainage water making the overall water use efficiency of the country rather high. Another interesting case is Syria where a large expansion has taken place but the country has reached the ceiling of the available water resources and recent information (Sagardoy & Varela-Ortega, 2007) shows a decline in the irrigated area. In general the SEMR countries have made enormous effort to cope with the demand but the level of efficiency in the management of the resources is low.

Table 2. Irrigated land and its share in arable land and permanent crops for selected countries of the Mediterranean region.

	Irrigated land					Share in arable land and permanent crops			
Countries	79–81 (10^3 ha)	89–91 (10^3 ha)	99–01 (10^3 ha)	02–03 (10^3 ha)	% of increase	79–81 (%)	89–91 (%)	9–01 (%)	03 (%)
EU Countries									
France	1,369	1,980	2,628	2,600	89.87	7.24	10.33	13.44	13.28
Greece	950	1,200	1,441	1,431	50.58	24.16	30.33	37.34	37.93
Italy	2,400	2,615	2,699	2,750	14.58	19.30	21.92	23.93	25.71
Spain	3,028	3,387	3,719	3,780	24.83	14.77	16.77	20.31	20.20
Subtotal	7,748	9,183	10,487	10,561	36.31				
SEMR Countries									
Egypt	2,453	2,621	3,310	3,422	39.48	100.00	100.00	98.19	99.94
Morocco	1,208	1,258	1,397	1,445	19.59	14.96	13.45	14.60	15.41
Syrian Arab Rep.	548	717	1,221	1,333	143.10	9.60	12.88	22.47	24.59
Tunisia	232	328	393	394	69.83	4.83	6.75	7.90	7.99
Turkey	2,712	4,024	4,743	5,215	92.32	9.48	14.50	17.89	20.05
Subtotal	7,154	8,948	11,064	11,809	65.08				
Total	14,901	18,130	21,551	22,370	50.12				

Source: FAOSTAT (FAO, 2009).

The other important point that Table 2 illustrates is that the share of the irrigated area in the arable land and permanent crops has also increased considerably. In most cases the percentage of the increase exceeds the 100% reaching 200% and more (Turkey and Syria). In any case the higher the percentage the higher is the dependency of food security on irrigated agriculture. But even more important is the fact that even when the share is around 20% the value of the corresponding agricultural production is often larger than 50%. For this reason most countries in the region have seen irrigation development as the main strategy to ensure food security.

One relevant question is to what extend the above growth of the irrigated areas correspond or is parallel to the growth of the population to ensure their food needs. A simple way to estimate this relationship is to compare the growth of the irrigated area with the growth of the population. Table 3 provides the estimation of the growth of the population for selected countries and the growth of the irrigated area for the years 1980 and 2003. An indicator has been developed to relate irrigated area with population by dividing the two factors. It can be observed that the indicator of the number of persons to the irrigated hectares has decreased in general for most countries and therefore the situation has generally improved. However, when the percentages of increase in the number of person per irrigated hectare are compared with the percentages of increase in the population, the picture changes considerably. Most of the European countries show that the percentage of increase in persons per irrigated hectare is greater than the percentage of growth of the population while it is the opposite for most of the SEMR countries during the period considered. In other words, the capacity of the irrigated lands to contribute to food security has improved for most countries, but not sufficiently to compensate the growth of the population in most of the SEMR countries and for some, like Egypt and Tunisia, it has deteriorated. However for most of the European countries the situation has improved except for Spain.

The above consideration explains why the agricultural trade has increased substantially in the period considered. In particular the imports have increased much more than the exports (see Table 5) for most countries of the region and particularly for the SEMR countries. In a way, the SEMR countries have decreased its food security in relative terms and this is compensated by the increase in trade.

Table 3. Irrigated area per capita of selected European countries (1980–2003).

Countries	Year 1980 Persons (10^3)	Irrigated (10^3 ha)	Persons per irrigated ha	Year 2003 Persons (10^3)	Irrigated (10^3 ha)	Persons per irrigated ha	% decrease [increase –] of the persons per ha	% population growth
EU Countries								
France	53,950	1,369	39	61,013	2,600	23	40	13
Greece	9,643	950	10	11,064	1,431	8	24	15
Italy	56,307	2,400	23	58,645	2,750	21	9	4
Spain	37,527	3,028	12	43,060	3,780	11	8	15
Subtotal	159,407	7,748	21	175,785	10,561	17	19	10
SEMR Countries								
Egypt	44,433	2,453	18	77,154	3,422	23	−24	74
Morocco	19,567	1,208	16	30,495	1,445	21	−30	56
Syria	8,971	548	16	19,121	1,333	14	12	113
Tunisia	6,457	232	28	9,878	394	25	10	53
Turkey	46,161	2,712	17	71,169	5,215	14	20	54
Subtotal	125,589	7,154	18	207,817	11,809	18	0	65

Source: UN, 2008; FAOSTAT (FAO, 2009).

3.2 *What future for irrigation development?*

Several institutions (FAO, INWI, World Bank, Plan Bleu and others) have developed scenarios of the possible future development of irrigation in the next 25–50 years at the global, regional and national scales. Most of them agree that the future development of irrigation will be selected and limited to few countries in the world that will be the main drivers of this development.

At the Mediterranean level the Plan Bleu has been particularly active in this area, and Figure 3 contains the projections made for the future water demand up to year 2025 (Benoit & Comeau, 2005). Only 3 countries, namely Turkey, Egypt and Syria show an important growth of the demand with regard to the present situation. For most countries of the Mediterranean, the foreseeable expansion of the water demand is small. For some European countries like Italy and France even a decline is foreseen. Consequently for most of the Mediterranean countries the expansion of the irrigated area will be modest partly due to the scarcity of water but also due to the fact that the demand is not increasing in a substantial manner. For those countries where the demand will grow significantly, like Syria and Egypt, and do not have enough water resources to satisfy the future demand, much will depend on the international agreements with neighbor countries to obtain additional water resources or they will have to increase the imports of agricultural products. Turkey has enough water resources to meet the increasing demand so it is foreseeable that the expansion of irrigation will continue for some time.

Certainly, the demand management represents an additional possibility to increase food production with a more efficient management of the existing water resources (*more crops per drop*). This search for a greater efficiency will continue in most countries of the region. While some countries, like Spain, are betting strongly on this option, most countries of the Region move slowly in this direction limiting their support to some economic incentives for those farmers willing to invest in the modernization of their irrigation systems.

4 FOOD SECURITY, POVERTY AND HUNGER

The basic equation of food security is that what is produced internally plus the net balance of imports and exports should be adequate to cover the needs of the country populations, provided that the people have the financial resources to buy the necessary food. It is precisely this last *proviso*

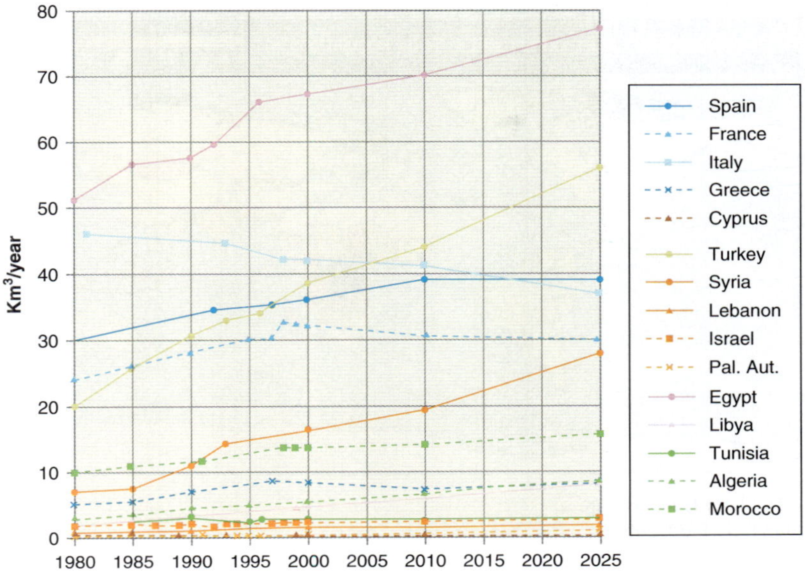

Figure 3. Total water demand by country in the Mediterranean (1980–2025).
Source: Plan Bleu, baseline scenario (Benoit & Comeau, 2005).

that during the last few years has been missing in many countries and that has produced an increase of 75 million people of undernourished people. According to the FAO estimates, in 2003–2004 the number of undernourished people was 848 million while it has increased to 923 in 2007 (FAO, 2008). This is largely attributed to the increases in food prices during this period. As the trend of high prices has continued in 2008 the figure of hungry people has increased further. In fact, the estimations for 2009 are of 1,000 million of undernourished people.

It is remarkable that hunger has increased as the world has grown richer and produced more food than ever in the last decade. This shows that in reality the levels of under nourishment are more connected to poverty than to the food production capacity. It also shows that when such circumstances arise, those that suffer most are the poorer strata of the society: landless workers and females-headed households. On the other hand high prices are also an opportunity for small farmers if the increases in food prices are accompanied for higher prices of inputs that may write off any potential gain.

While the above arguments are of fundamental importance at world level, when we look at the Mediterranean Region the situation is less dramatic because the levels of food insecurity in most countries of the Region are low. Figure 4 shows the *Hunger Map* prepared by FAO and it is evident that for all countries of the Mediterranean the under nourishment levels are below 5%, exception made of the Palestinian Territories where the level is estimated at 34%.

At country level there are large differences in the Region. The percentage of poor people in Syria is 61%, out of which approximately 19% of people are estimated as food insecure; in Egypt the population still falling below the poverty line is 23%; in Jordan this percentage varies between 15–20% due to the country's very limited resources and high dependency on imported food and agricultural products. In the Palestinian Territories 34% of the population is food insecure and the proportion of households living under poverty conditions reaches 60%, a number that gets to 73% in households headed by women[1].

[1] Most of the poverty and malnutrition figures mentioned here are coming from national reports where the criteria used to define poverty and under nourishment were different from those used internationally.

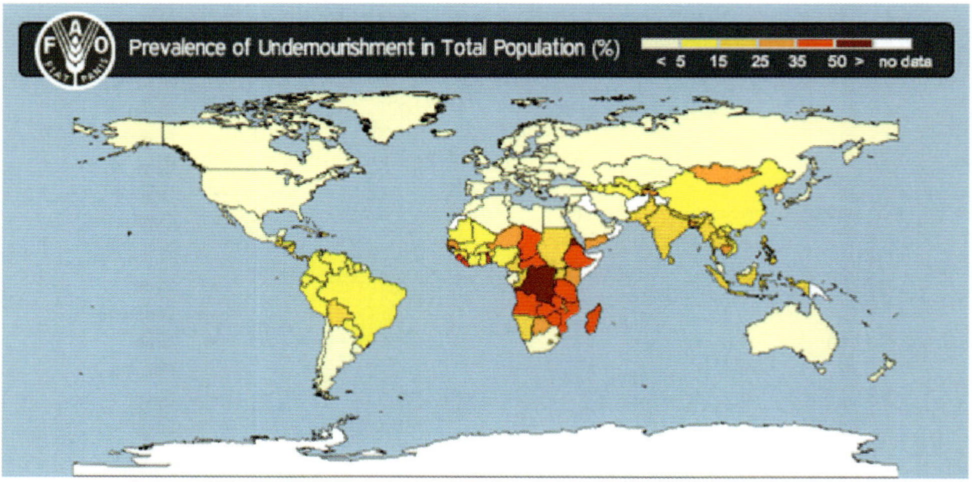

Figure 4. The FAO Hunger Map.
Source: FAOSTAT (FAO, 2009).

Often the concepts of poverty and hunger are used in similar contexts but in reality there is a considerable difference. Poverty is basically an economic concept that establishes levels of income below which acceptable living standards are not possible in a national context, while hunger is a physiological concept related to the caloric intake. Up to recent years the poverty level was internationally established in 1 US$/day but recently have been increased to 1.25 US$/day. However, often countries use different thresholds to assess poverty within their boundaries.

At the level of 1.25 US$/day the number of poor people in the world is estimated at 1,374 millions in 2005 (World Bank, 2008) and 11 million for the countries of the Middle East and North Africa (which include most of the countries of the SEMR). If the poverty line is raised to 2 US$/day the figures become respectively 2,564 million and 51 million, which indicate the large extent of poverty in the world (nearly 50% of the active population) and even in the Mediterranean countries it appears quite relevant. The tragedy associated with these figures is that much of this poverty (70%) live in the rural environments.

The most striking consequence of poverty is that if you are a poor person you may have the capacity to buy some of the local food, but imported food will be far too expensive. If the water scarcity is compensated with food imports, the poor will not have much access to such food as it may be too expensive for them. Here the role of governments may be fundamental to guarantee that basic food remains accessible to all levels of the population.

4.1 *The role of irrigation development in alleviating poverty*

The role of irrigation development to reduce poverty appears quite significant. Most of the irrigated farms in the Mediterranean Region are small. For instance in Egypt the average size of farms is 1.7 ha and the revenue generated by this average farm is 1,705 E£ (Egyptian pound)[2] (IFPRI, 2000), i.e. slightly over 300 US$, but these figures do not account for the internal farm food consumption and therefore this income is somewhat beyond the poverty line earlier mentioned. In Morocco the average size farm is 5 ha but those under irrigation are much smaller. Similarly, in Spain the number of farms with less than 5 ha is 647,134, representing 53% of the total (INE, 1997) and those under irrigation are even smaller. In summary, the prevalence of the small farms in the Mediterranean irrigated agriculture has provided the opportunity to millions of farmers to earn revenues that place

[2] 100 E£ (EGP) = 13.54 US$ (USD) = 18.03 € (EUR) (April 27, 2010).

them beyond the poverty lines. Certainly, national averages may hide the fact that there are also large irrigation farms which owners are far from poor. However, even in these large farms owners employ seasonal and permanent workers earning salaries that may place them also beyond the poverty lines.

If the expansion of irrigation development in the coming years will reduce, or grow little, the consequence will be that its contribution to reduce poverty will be also much smaller. So far, countries have been reluctant to finance *social irrigation* and it is unlikely that this may change in the future. The general response to the poverty problems has been the implementation of rural development programs that not only reduce poverty but enhance rural life. Such programs have gained considerable recognition and support and they are widely used in the Mediterranean countries, not only to reduce poverty but also to improve the living conditions of the rural communities. The EU has specific programs (The Co-operation Specific Programme and the Sustainable Development Strategy –SDS– in the Seventh Framework Programme –FP7) dedicated to support these programs and many of the bilateral projects have focused in these types of projects. Also governments of the SEMR are quite supportive of the rural development programs, but the resources allocated to them are still modest and need to be increase to improve the living conditions of the rural poor.

4.2 *The role of food aid*

Food trade is playing an increasingly important role in the economies and in the agriculture sector of many countries. However, it is not at all clear that the poor sections of the population are benefiting from this increase in trade (FAO, 2005). For many, the trade restrictions are increasing poverty and hunger and the need for reforming trade rules are strongly felt. Trade reforms must be accompanied by government policies to enhance the capacity of the poor to share the gains from trade and to compensate those who lose from the process, perhaps through social safety net programs. The overall domestic policy environment is just as important as trade policies, and must be conducive to private investment and private activity.

The role of food trade in alleviating poverty can be highly relevant when food imports are used to alleviate the hunger of the poorest people through the food emergency programs and small farmers have a better access to markets. It may be worth noting that the recent increase rate of natural disasters has lead to an increase in the number of food emergency programs in recent years. In fact, several of the SEMR countries have been receiving food aid in a consistent manner up to very recently. As an example, in 2009 the EC has approved emergency food aid for 6 million € to the Palestinian Territories and Syria to assist in the problems originated by the persistent droughts and the effect of the recent war conflicts. In addition to these emergency situations governments provide food for children in many public schools and also in many pro-poor institutions.

In crisis situations food aid is of critical importance, but in the short term. The key to sustainability, however, is to ensure that the aid provided does not create dependency or harm the communities and stakeholders it hopes to assist. Since women are typically responsible for food production, preparation, storage, and marketing, it is crucial to include them in emergency-related food security planning and decision-making as potential agents of change and decision-makers, rather than as the *victims* they are often portrayed to be. Traditional forms of food aid have largely failed to recognize and enhance the productive capacity of women, and this in turn means that food aid has been ineffective in contributing to lasting solutions to poverty. While short-term emergency food aid is often essential, it must be balanced with longer-term assistance and more comprehensive programs for agricultural development that are designed to support men and women crucial contributions to agricultural production.

5 AGRICULTURAL TRADE. THE NEW FRONTIER OF FOOD SECURITY

5.1 *A general overview of agriculture and trade in the Mediterranean Region*

Before entering into the analysis of what role is now playing agricultural trade, it will be relevant to address a more general question related to what role is now playing the agricultural sector in

Table 4. General economic data for selected Mediterranean countries.

	GDP per capita (US$, 1990)			Agriculture (as % of GDP)		Agriculture value added per worker (US$, 2000)	
	1990	2004 *	% increase	1990	2004	1990–92	2002–04
SEMR countries							
Egypt	1,185	1,600	35.0	19	15	1,575	2,007
Morocco	1,057	1,150	8.8	18	16	1,275	1,582
Syria	1,129	1,650	46.1	28	23	2,356	2,977
Tunisia	1,485	2,300	54.9	16	13	2,365	2,415
Turkey	2,563	3,100	21.0	18	13	1,772	1,793
EU countries							
France	21,321	25,000	17.3	4	3	24,724	40,521
Greece	8,360	9,950	19.0	11	7	8,315	9,303
Italy	19,401	22,450	15.7	4	3	13,672	21,553
Spain	12,928	17,750	37.3	7	4	12,611	19,132

* Values are estimated for the year 2004
Source: [http://www.un.org/esa/population/publications/countryprofile/]

the economies of countries and how has been evolving in the last few years. Table 4 provides an overview of the GDP and other economic data for some selected Mediterranean countries (EU and SEMR) and some of the following observations emerge:

- The first point that emerges is that in spite of the efforts made to increase the agricultural production, its impact in the GDP has decreased, for the period considered, by at least 2–5 percentage points and this applies equally to all countries analyzed.
- The increase in the GDP for countries of the SEMR is generally greater than for those of the EU. However, as the impact of the agricultural sector in the GDP has decreased in 2004 it is evident that there are other sectors of the economy that have grown more dynamically than agriculture.
- The agriculture added value per worker is much higher in the EU countries than in the SEMR countries. This is a clear indication that most of the agricultural products in the SEMR countries are subject to little processing and probably exported as raw materials. This is an indication that the opportunities for rural industries and food processing in these countries are still ample.

The overall conclusion is that all Mediterranean countries are making a substantial economic progress but the agriculture sector is reducing its importance in the national economies.

Table 5 provides the national data regarding the agricultural net trade (imports minus exports) for the years 1990 and 2004. Also the virtual water traded is included. The analysis of these data provides the bases for the following relevant observations:

- All SEMR countries are net importers of agricultural products, except Turkey that is exporter. The fact that they are mostly importers seems to have some correlation with the limited availability of water resources. During the period considered only Tunisia and Egypt have managed to reduce the net trade. Morocco and Syria have largely increased the net agricultural trade during the period (see percentage of increase per capita). Syria has increased its exporting capacity to the expense of a critical exploitation of the water resources.
- The European countries show a dichotomy that is interesting. France and Spain are net exporters of agricultural products. On the contrary, Greece and Italy are net importers. It is interesting to note that France and Spain have greatly increased their respective irrigated areas in the period considered while Greece and Italy have not. It seems that the expansion of the irrigated area has led to an increase of the agricultural exports in these countries.

Table 5. Agricultural net trade balance (imports – exports).

Country	Agricultural Net trade (imports – exports) value					% of net trade over GDP per capita		Net balance of virtual water (m^3 per capita) *	Value of m^3 (2004) **
	Year 1990		Year 2004		% of increase per cap.	Year 1990	Year 2004		
	10^6 US$	Per capita	10^6 US$	Per capita					
SEMR countries									
Egypt	2,649	46	1,700	22	−51	4	1	−250	−0.09
Morocco	153	6	1,094	36	491	1	3	−1,050	−0.03
Syria	28	2	123	7	200	0	0	−200	−0.03
Tunisia	371	45	207	21	−53	3	1	−1,250	−0.02
Turkey	−858	−15	−1,298	−19	21	−1	−1	−100	0.19
EU countries									
France	−10,819	−190	−12,026	−211	11	−1	−1	200	−1.06
Greece	564	56	2,632	238	329	1	2	−600	−0.40
Italy	12,517	220	7,270	132	−40	1	1	−300	−0.44
Spain	213	5	−4,495	−112	−2,140	0	−1	−90	1.25

* *Source*: Fernandez, 2007
** Net trade value/virtual water per capita.
Source: FAOSTAT (FAO, 2009).

- The percentage of the net agricultural trade per capita over the GDP per capita shows some interesting features. For all countries and the two years (1990 and 2004) the impact remains low (1 to 4%) indicating that trade remains a minor component in the GDP per capita. The per capita values are generally higher for 2004 than in 1990 indicating a growing trend in the net trade. As the impact of the increases on the GDP per capita remains small it is apparent that they are not affecting the purchasing power of the individuals in a substantial manner.
- The net balance of the virtual water values per capita, taken from Fernandez (2007), has been included in Table 5 to examine the possible correlation with the net trade values per capita. If the net values of trade for the year 2004 are divided by the net balance of virtual water per capita one finds the *net trade value of the* m^3. The resulting figures present reasonable values for most of the countries, indicating a certain consistency among the two parameters. The fact that the higher values correspond to Spain and France also makes sense since both countries are net exporters of predominantly high value crops. This value of the m^3 can be useful in the economic evaluations related to virtual water trade.

In summary, the analysis of this information at national level permits to assess the financial impact of agricultural trade on the consumer and the trends that emerge. It does also provide an assessment of how much the consumer is paying for the m^3 of virtual water trade imported providing some warning when unusual values emerge.

5.2 *Analyzing agricultural production, trade and virtual water trade to reorient national crop production. A methodological approach*

The essential question that we are addressing here is to what extent countries should change their internal production crop patterns to reduce the net crop trade to the lowest possible level, and at the same time maintain the water abstractions within reasonable limits or even reduce them. To try to respond to this important question the authors propose an outline of a methodology consisting in analyzing 4 sets of data. The first set of data contains the production data (area, quantity and value) for the main crops in the period chosen. The second set of data includes the irrigation water used

Table 6. Evolution of the surface, production and value of main crops in Egypt (1990–2004).

Main crops	Year 1990 Surface (10^3 ha)	%	Production (10^3 t)	%	Price (US$/t)	Value (10^6 US$)	%
Cotton	417	13	838	3	91	76	2
Sugar cane	111	3	8,895	34	18	164	4
Rice	436	13	3,167	12	127	404	9
Wheat	821	25	4,268	16	159	678	16
Maize	830	26	4,799	18	141	674	16
Other crops	616	19	4,101	16	Not avail.	2,281	53
Total	3,232	100	26,069	100		4,277	100

Main crops	Year 2004 Surface (10^3 ha)	%	Product. (10^3 t)	%	Price (US$/t)	Value (10^6 US$)	%	Increase 1990–2004 Surf. (%)	Prod. (%)	Value (%)
Cotton	300	8	785	2	93	73	1	−28	−6	−5
Sugar cane	135	3	11,230	27	18	205	4	−22	26	24
Rice	646	17	6,352	15	186	1,183	20	−48	101	193
Wheat	1,095	28	7,178	17	132	945	16	−33	68	39
Maize	789	20	6,236	15	121	754	13	−5	30	12
Other crops	903	23	9,314	23	Not av.	2,619	45	−46	127	15
Total	3,867	100	41,096	100		5,778	100	−20	58	35

Source: FAOSTAT (FAO, 2009).
[t = tonne = 1,000 kg]

in the production of the selected crops. The third one refers to the export and import of the same products and the net trade. The fourth estimates the blue virtual water traded for the selected crops. The examination of these sets of information provides some interesting indications regarding the convenience or not to modify the production patterns at national level. Although these indications are very useful they are not sufficient to make policy decisions. Any intend to modify the existing national cropping pattern will require a more detailed analysis of local information. This exercise will be attempted in the last section of this chapter through the simulation of a number of scenarios at farm level and analyzing their impact at national level.

The two countries selected for this exercise are Egypt and Syria. The reason for selecting Egypt is because practically all agriculture is irrigated and this facilitates the interrelation with the water side. Syria has been selected because the authors have access to detailed farm information resulting from earlier work. The analysis of data from Egypt follows but stops at the level of the analysis of the national information. The one of Syria is included in a separate section since it contains a more complete analysis covering from the national to the farm level.

5.2.1 *Analysis of the agricultural production, trade and virtual water. A case study for Egypt.* This analysis is based in the following tables: Table 6, Evolution of the surface, production and value of main crops in Egypt (1990–2004); Table 7, Irrigation water use for the main crops of Egypt (2004); Table 8, Net trade of the crops selected (import–exports); and Table 9, Estimation of the blue virtual water traded in 2004. Nearly all the information presented is derived from the FAOSTAT data bases.

The crops selected for this analysis are cotton, sugar cane, rice, maize and wheat. In the year 2004 they represent 78% of the food production in tonnes (t = 1,000 kg) and 53% in value (Table 6). Hence the group chosen represents a large portion of the agriculture in Egypt. As in Egypt all these crops are irrigated no differentiation is made between rainfed and irrigated production.

Table 7. Irrigation water use (IWU) for the main crops of Egypt (2004).

	Irrigation water use (IWU) at farm				Total IWU (2004)		Irrigation water per tonne		
Main crops	Irrigation req. (m^3/ha)	Mm^3 1990	Mm^3 2004	Increase (%)	Distribut. efficiency	Total IWU (Mm^3)	Main crop	Yield (t/ha)	m^3 of total IWU per tonne
Cotton	9,000	3,755	2,703	−28	0.70	3,861	Cotton seed	2.61	4,926
Sugar cane	10,000	1,106	1,353	22	0.70	1,933	Sugar cane	83.00	172
Rice	11,000	4,795	7,102	48	0.70	10,146	Rice	9.84	1,597
Wheat	4,000	3,285	4,379	33	0.70	6,256	Wheat	6.56	872
Maize	8,000	6,641	6,308	−5	0.70	9,012	Maize	7.91	1,445
Other crops	6,000	3,698	5,416	46	0.70	7,737	Cotton lint	0.87	14,778
Total						38,945			
Total per inhabitant						505			

Source: elaborated data.

Table 8. Net trade of the crops selected (import – exports).

	Net trade (import – exports)			
	1990		2004	
Main crops	Quantity (t)	Value (10^6 US$)	Quantity (t)	Value (10^6 US$)
Cotton	20,562	−66	−101,549	−389.66
Maize	1,899,937	249	2,427,724	364.45
Rice	−63,924	−15	−804,730	−222.97
Sugar cane	−548	0	−4,227	−0.69
Wheat	5,400,000	853	4,366,462	727.57
Total Merchandise Trade	No data	6,617	No data	7,522.70

Table 9. Estimation of the blue virtual water traded in 2004.

Blue virtual water traded (year 2004)		
Main crops	m^3/t	Mm^3(*)
Cotton lint	14,778	−1,500.72
Maize	1,587	3,853.53
Rice	1,597	−1,285.34
Sugar cane	172	−0.73
Wheat	872	3,805.47
Net Total		4,872.20
m^3 per inhabitant		63.15

* [Quantity 2004 (Table 8) × m^3/t (Table 9)]

For each crop the mentioned set of tables are examined and conclusion drawn. They are reflected in the following sections:

a) Rice

Rice is a crop that has been expanding in area (48.12%) during the last 14 years; and the production (101%) and its value (193%) have also increased considerably (Table 6). Much of this expansion is due to the fact that an important part of the drainage water is being re-used to irrigate the predominantly saline soils of the *new lands* located in the delta. The increase in the production indicates that the productivity for unit of land has increase considerably. In fact the yields of rice in Egypt are high.

Table 7 indicates that the irrigation water used for this crop is quite high: 10,146 Mm^3/yr (in 2004), which represent nearly 26% of the total irrigation water used in the country.

Already in 1990, Egypt was a modest net exporter of rice with 63,924 t, but in 2004 it has grown to 804,730 t, which is about the 12% of the total rice production. Although this represents a modest share of the production, the trade value is significant: 223 million US$ (Table 8). In 2004 rice was the second export crop (in value) after cotton. The value of the crop has benefited from the high international prices of this commodity in recent years.

The quantity of blue virtual water exported through this crop amounts to 1,285 Mm^3/yr (Table 9). This represents about the 26% of the total net blue virtual water traded. In spite of the revenue generated by the export of this crop remains questionable if this export policy is sustainable from the point of view of the water resources.

In summary, the analysis of the data suggests not to expand further the production of rice and where possible to reduce the area and replace it for another crop. However, as rice is mostly produced in saline soils with drainage water of medium level of salinity, the replacement of rice for another crop does not seem feasible in many areas. This is a clear example where the analyses of data suggest replacement of a crop for a less consuming water crop, but the environmental conditions where rice is grown will limit the possible change. To quantify the magnitude of the areas that could be subject to the cultivation of other crops soils information and other local data are needed. This points out that a reduction of the rice area cannot be implemented without a closer analysis at the local level.

b) Cotton

Cotton has been an important traditional crop in Egypt and its production has been under government control. The cotton area was almost halved between 1952 and 1987, due to the government policy requiring farmers to sell all their cotton output to the government at fixed prices that were kept below world market prices thus rendering its cultivation not very attractive. During the period considered the area of cotton crop has further declined (28%) and logically also the total production (−6%) and value of the production (−6%) (Table 6). Although the area has reduced by 28%, the production has only reduced by 6%, indicating that there has been some modest increase in the productivity. Cotton has high labor requirements and since wages have increased in the period considered the net returns for farmer may have decreased.

As the area dedicated to cotton has reduced, the use of the irrigation water has also declined in the period considered by about 28%. The total irrigation water used in 2004 (Table 7) was 3,861 Mm^3 which represents 10% of the total irrigation water use (38,945 Mm^3). Therefore it has not a great impact in the irrigation water used.

Table 8 shows that Egypt has evolved from being an importer of cotton in 1990 to be a large exporter with 101,549 t of cotton lint exported and a value of 389.6 million US$. Cotton is the main export crop of Egypt. As the relation between cotton seed and cotton lint is approximately 3:1, it can be estimated that nearly 39% of the production was exported in 2004.

Regarding the blue virtual water (Table 9) the amount exported by the cotton lint is 1,500 Mm^3 which is quite a substantial figure (35% of the total net blue virtual water).

Given the above observations it seems reasonable to conclude that there is margin for a further modest reduction of the cotton area cultivated. This may happen if Government policy prices are maintained at the same time that production cost increase due to labor increasing wages. This may free some 10–20% of the area for other crops and reduce the water use.

c) Sugar cane

The area dedicated to this crop has increased by 22.36% and production by 31.64% indicating a modest increase in the productivity due to new varieties and improved cultivation practices. The price of the crop has remained nearly constant (18 US$/t) and therefore the increase in the value of the crop (30.6%) (Table 6) is basically due to the increase in area. This increase in area probably goes in parallel with the increasing internal demand for sugar.

The amount of irrigation water used in the production in 2004 was 1,933 Mm^3. Therefore it accounts for a modest share (4.9%) of the total irrigation water (38,945 Mm^3).

Sugar has been exported in very modest quantities: 4,227 t in 2004 (see Table 9) that have very little impact in the total trade. For practical purposes it can be said that is a crop grown for internal consumption. Consequently the blue virtual water traded is also very small (−0.73 Mm^3) and therefore with small significance in the total traded.

The above observations do not point for the need of any substantial change in the planted areas of this crop.

d) Wheat

Wheat is now the main crop of Egypt with more than 1 million ha cultivated in the year 2004. The area has been expanding considerably (33%), as well as the production (68%) and the market value (39%) (Table 6). The yields are high and there have been a further increase in the productivity during the considered period.

Although wheat has low irrigation requirements, the large area cultivated makes that the total irrigation water used be very high with 6,256 Mm^3 in 2004 (Table 7) which represents 16% of the total irrigation water use.

The national production of wheat is not sufficient to satisfy the demand, and Egypt is a large importer of wheat with 4.3×10^6 t imported in 2004. Interestingly, the imports of wheat have been reducing from 5.4×10^6 t in 1990 (Table 8). Egypt subsidizes the price of flour and therefore the imported wheat represents an important lost to the Government. This may be the reason behind the expansion of the production area. However if the subsidies are removed, the expansion of the wheat may no longer be attractive.

From the point of view of the blue virtual water trade the imported amount is very high: 3,805 Mm^3, because the amount of wheat traded is also high. In a way, this imported virtual water serves to compensate the scarcity of water resources in the country.

Consequently, on one hand it seems advisable to expand the area cultivated of wheat, since this will contribute to reduce the gap between production and demand. Furthermore this will contribute to reduce the imports and the financial loses associated with the subsidy of imported flour. On the other hand, increasing the national production of wheat will create additional pressure on the limited water resources. Adopting one or the other solution not only depends on economic considerations but on the social acceptance of the alternatives.

e) Maize

Maize is an important crop for the diet of the rural populations. For this reason maize was as important as wheat in 1990 in terms of area cultivated but it has declined moderately (−5%) in 2004. However the production has increased by 30% indicating substantial gains in productivity. The value of the crop has only increased by 15% due to the fact that the international price of maize has declined in the period. Maize is part of the traditional rotation of Egyptian farmers that includes: wheat, clover and maize. In terms of farmers' revenue it has very similar values than those of wheat. Maize is a competitive crop for cotton because the revenue is also similar and requires much less labor.

The total irrigation water used is 9,012 for the year 2004, i.e. 23.3% of the irrigation water used at national level. Therefore is the second largest consuming crop from those selected.

Egypt has been a traditional large importer of maize. The net imports in 2004 were 2.42×10^6 t with an increase of 27.8% with respect to the values of 1990 (Table 8). Consequently the values of these imports have increased by 46%.

The blue virtual water imported has a substantial value of 3,853 Mm^3.

In summary maize appears as a stable crop with some potential for expanding and replace some of the crops earlier mentioned (cotton and rice mainly).

The analysis made above for the different crops indicates some desirable changes in the production and in the trade of the crops considered. This analysis is based on information that is easily accessible from international organizations (mostly FAO) and known references through Internet. The emerging suggestions for change have therefore a preliminary nature. Before attempting the suggested changes a much more detailed examination is required. This is attempted in the case of study of Syria that follows in the next section of the chapter.

A final consideration is that the suggested methodological approach can also be used to make projections following the existing national crop patterns.

6 THE ROLE OF IRRIGATION, TRADE AND VIRTUAL WATER FOR DESIGNING PRODUCTION POLICIES: A CASE STUDY IN SYRIA

This country case study is intended to assess the impact at farm and national levels of the application of certain policies related to water use in agriculture in Syria. These policies emerge from the analysis of the general data of production and trade in Syria, one of the countries in the Mediterranean region that faces a long-trend increase in water demand towards 2025 (see Figure 3, in Section 3.2) and that, consequently, is prone to severe water deficits in some regions.

In its first section, the Syria case study follows a similar pattern than the previous section dedicated to Egypt and it is based on the analysis of the evolution, at national level, of agricultural production, irrigation trends and trade balance along several selected years (Sections 6.1 to 6.6). Emerging from this analysis, it is possible to foresee a set of potential policies that will seek to reduce the negative agricultural trade balance, and at the same time, will try to preserve the scarce water resources in Syria. These policies could be translated into specific policy scenarios, of which three have been selected to illustrate the methodological approach used in this case study. These are, on the one side, a land use policy aimed at encouraging a cropping shift towards less water consuming crops and, on the other side, an irrigation modernization policy aimed at reducing water use in the farms (Section 6.7). Both policies have the common objective of reducing water use in the farms and at national level but the land use police makes use of agronomic instruments whereas the irrigation modernization policy is based on technological instruments, defined by the adoption of new irrigation technologies. The center of this case study is the analysis of the impact of these two policies on statistically-based representative farms that characterize irrigation agriculture in Syria and that consider various regions, cropping and irrigation technologies. The farm-level results are then scaled-up to the country's national level (Section 6.8). Throughout the analysis, we compare the implementation of both types of policies with the purpose of deriving policy relevant conclusions for the Syrian agricultural sector and for different types of farms as well (Sections 6.9 and 6.10). The implementation of both policies is defined by different environmental and socio-economic indicators, such as crop mix, water use for irrigation, farmers' income and overall farm employment.

6.1 *Main features of the agricultural production in Syria*

From the 18.5 million ha (Mha) of total lands of the Syrian Arab Republic, cultivated land extends over an area of 5.48 Mha of which 1.40 Mha are irrigated land (22%), 3.47 Mha are rainfed (63%) and 0.62 Mha are fallow land (11%). Of the total irrigated area, 551,000 ha are irrigated from surface waters while the remaining 851,000 ha use groundwater resources (Table 10).

As Table 10 indicates, the main driver for the expansion of irrigation in Syria has been predominantly the use of groundwater resources to the point that the area has nearly tripled with respect to 1985. As a consequence, most of the aquifers of the country have been overexploited and in some areas the decline of the water table has even surpassed 50 m, placing some farmers in the difficult position of having to close some of their wells. It is worth noting that in 2006, for the first time in the last two decades, the area irrigated from groundwater sources has diminished by 10,000 ha which is a clear sign that for some farmers this type of water source is no longer feasible. This

Table 10. Evolution of the irrigated agriculture in Syria by type of water resources (modified from Somi *et al*., 2002).

	Irrigated area				Total irrigated area	
	From surface water		From groundwater			% increase with respect to the area in 1985
Year	10^3 ha	% of the total	10^3 ha	% of the total	10^3 ha	
1985	334	51	318	49	652	
1990	351	51	342	49	693	6
1995	388	36	694	64	1,082	66
2000	512	42	698	58	1,210	86
2002	583	43	764	57	1,347	107
2004*	624	43	815	57	1,439	121
2005	561	39	865	61	1,426	121
2006	551	39	851	61	1,402	115

Source: Sagardoy & Varela-Ortega (2007).
**Source*: MAAR (2004).

Table 11. Available Water Resources in Syria (in Mm^3 for year 2000).

	Barada Awag	Yarmouk	Al Badia	Orontos (Al Asi)	Coastal	Al Khabour	Euphrates and Tigris	Total
Total available for use	1,277	272	70	1,831	1,257	1,371	9,981	16,058
Water use								
Irrigation water use	1,207	360	43	2,306	433	4,283	7,228	15,860
Domestic water use	298	69	8	185	134	49	300	1,042
Industry water use	77	18	2	48	35	13	78	315
Evaporation	5	31	15	148	16	132	1,643	1,990
Total uses	1,588	478	68	2,687	617	4,477	9,249	19,162
Water balance	−311	−206	2	−856	640	−3,105	732	−3,104

Source: Own elaboration based on Varela-Ortega & Sagardoy (2003).

situation is a clear sign that the exploitation of groundwater resources has reached the point of no recovery and therefore the reduction in irrigated area is likely to continue for several years to come.

6.2 *Available water resources of Syria and main river basins*

Water resources in Syria are very limited compared to the needs of the country. Estimations of the available water resources in Syria vary considerably depending on the source of information. One of the major problems for getting reliable estimations arises from the difficulty to obtain realistic data of the natural flows of the Euphrates River. Since the construction of the large dam of the Great Anatolia Project (GAP) in Turkey the data related to the flows of the Euphrates River have not been published and are not available for the international community. Nevertheless in this study it has been assumed that the flow share of Syria from the Euphrates River is 210 m^3/sec, an equivalent to 6,818 Mm^3/yr, which is the most conservative hypothesis. Estimations of available water resources tend to be rather constant unless severe climate changes take place, and therefore the estimations that we have made for this study can be considered a good approximation to the actual available resources in year 2007.

Water uses are estimated in Table 11 for the year 2000 for each of the Syrian basins and for the nation's total. Irrigation water use represents 82% of total uses whereas domestic and industrial uses are only 7% of the total.

Based on the values shown in Table 11, at country level, Syria has a negative water balance that results in a structural water deficit of 3,104 Mm^3/yr. In this case, the total annual water resources for the country are 16,058 Mm^3 while the total annual uses are 19,162 Mm^3. Although this estimation is for the year 2000, it indicates the gravity of the water situation in Syria which has certainly deteriorated to the current year.

The water balance per basin shows that only three basins, namely Euphrates, Coastal and Al Badia have positive balances. The remaining ones have considerable negative water balances. The case of the Al Khabour basin with an annual deficit of 3,104 Mm^3 is of extreme gravity as evidenced by the persistent annual increase in pumping depth due to the depletion of the aquifers. This is followed by the Orontes basin with an annual deficit of 856 Mm^3. The magnitude of the deficit of the Al Khabour basin indicates that it will be difficult to correct it without special and severe measures. Putting to a halt the drilling of wells has shown limited effectiveness to control the problem. Changing the cropping pattern in favor of less water consuming crops has often been advocated as one potential solution to the excessive water mining and will be analyzed later in this chapter.

6.3 *Evolution of the cultivated area, production and value for the main crops of Syria*

Table 12 shows the evolution of the main crops in Syria in terms of area, production and value. The crops selected for this analysis covered 85% of the total cultivated area in 1999 while in 2004 it diminished to 61% indicating that the cropping pattern of the country has evolved considerably over the period considered. Nevertheless the selected crops still represent the largest part of the agricultural production in Syria.

The analysis of the data and of some complementary information allows for the following observations:

a) Wheat is a highly relevant crop in Syria and object of food security policies by the government. Prices are high compared to international standards and thus make the crop very attractive to farmers. Table 12 shows that the area cultivated has increased by 37%, but the production has more than duplicated in the period considered. This represents a large increase in productivity probably due to the increase of irrigated areas where wheat is planted. The present production level seems stabilized around 4.5×10^6 t and the area planted around 1.7 Mha. The price per tonne has improved in the period considered and is slightly over the international price. Favorable prices have surely contributed to increase the interest in the cultivation of the crop.
b) Cotton production in Syria is controlled to a great extent by the Cotton Bureau of the Syrian Ministry of Agriculture and Agrarian Reform. The Cotton Bureau sets the total planted areas and encourages early planting and harvesting of seed cotton. When irrigation water is not a constraint, farmers exceed the licensed areas and the crop exceeds 10^6 t of seed cotton as in the year 2004 (see Table 12). Cotton production has been encouraged by the Government and in the period considered the production has duplicated (133% increase). The area planted in 2004 was 234,181 ha, but Government of Syria plans to reduce its cotton planted areas to 225,000 ha and produce 900,000 t of seed cotton in 2006/2007. Actual cotton production exceeds production plans by about 20% in order to maintain higher prices and reduce water use.
c) Barley is the most important feed grain grown in Syria and is used principally as sheep and cattle feed, but sometimes replaces corn in poultry feed rations. Barley is a non irrigated crop in Syria and the variability of the production is very high depending on climate conditions. Table 12 shows the large reduction in cultivated area (53%) for the year 2004 due to shortage in rainfall. Year 2000 was a record drought and the production of barley fall to less than 150,000 t of production. Normally the annual production ranges in the order of 700,000–800,000 t but the yields are very low compared to international standards.
d) Sugar beet appears as a stable crop in the agricultural production. The area cultivated has increased from 21,400 ha to 27,500 ha in the period considered. The considerable increase in the price (more than 100%) has stimulated the growth of the area and the production. The area cultivated represents less than 4% of the total and therefore is a crop of modest importance at national level.

Table 12. Evolution of cultivated area, production and value for the main crops of Syria.

		Wheat		Cotton Seed		Barley	
		Quantity	%	Quantity	%	Quantity	%
Year 1990	Surface (ha)	1,341,000	24	156,400	3	2,729,000	49
	Production (t)	2,070,000	25	441,200	5	846,000	10
	Price (S£/t) *	8,500		2,070		6,250	
	Value (10^6 S£)	17,595	63	913	3	5,288	19
	Value (US$)	391	63	20	3	118	19
Year 2004	Surface (ha)	1,831,226	37	234,181	5	1,290,570	26
	Production (t)	4,537,459	32	1,029,232	7	527,193	4
	Price (S£/t)	12,588		5,673		7,484	
	Value (10^6S£)	57,118	77	5,839	8	3,946	5
	Value (US$)	1,071	77	109	8	74	5
% of increase in the period	Surface	37%		50%		−53%	
	Production	119 %		133%		−38%	
	Value	174 %		439%		−37%	

		Sugar beet		Maize		Other crops		
		Quantity	%	Quantity	%	Quantity	%	Total
Year 1990	Surface (ha)	21,400	0	60,200	1	1,313,713	23	5,621,789
	Production (t)	421,800	5	180,000	2	4,194,559	51	8,153,605
	Price (S£/t) *	1,400		7,300				
	Value (10^6 S£)	591	2	1,314	5	2,373	8	28,073
	Value (US$)	13	2	29	5	53	8	624
Year 2004	Surface (ha)	27,590	1	56,516	1	2,632,023	53	4,965,885
	Production (t)	1,217,658	9	210,166	1	6,792,339	47	14,314,098
	Price (S£/t)	2,424		8,619				
	Value (10^6S£)	2,952	4	1,811	2	2,073	3	73,738
	Value (US$)	55	4	34	2	39	3	1,382
% of increase in the period	Surface	29%		−6%		100%		
	Production	189%		17%		62%		
	Value	322%		16%		−26%		

Exchange rate: 1 USD (US$) = 45.5 SYP (S£, Syrian pound) (1990); 1 USD = 53.7 SYP (2004)
*Data from 1991
Source: FAOSTAT (FAO, 2009).

e) Maize is mainly an irrigated crop with relatively low yield in Syria. In general, the agro-meteorological conditions are not favorable for this crop. Table 12 shows a slight reduction in the cultivated area (6%) from year 1990 to 2004 but there is a modest increase (17%) in the production. Maize represents only 1% of the national agricultural production (in t) and its price has increased moderately.

In summary, wheat has some potential for expansion –as long as the high prices are maintained– but cotton should reduce the area. Barley and sugar beet are stable crops and maize has a tendency to reduce the area.

6.4 *Irrigation water use in Syria*

Table 13 provides an overview of the irrigation cropping pattern of the country for the main crops and gives grounds for the following observations.

Table 13. Irrigated crop distribution in Syria.

	Area (ha)	Percentage of irrigated area (%)	On farm irrigat. water use (m^3/ha)	Irrigation distribution efficiency	Total irrig. water use (m^3/ha)	Total irrig. water use (Mm^3)
Wheat	689,868	57	4,000	0.7	5,714	3,942
Maize	72,627	6	5,000	0.7	7,143	519
Cotton	274,585	23	11,000	0.7	15,714	4,315
Sugar beet	28,667	2	6,000	0.7	8,571	246
Potato	21,668	2	5,500	0.7	7,857	170
Tomato	9,743	1	9,300	0.7	13,286	129
Others	109,715	10	6,000	0.7	8,571	940
TOTAL	1,206,873	100				10,262
Total m^3 per inhabitant						537

Source: Area of crops from Agricultural Census of 1998 (Central Bureau of Statistics, 1998).

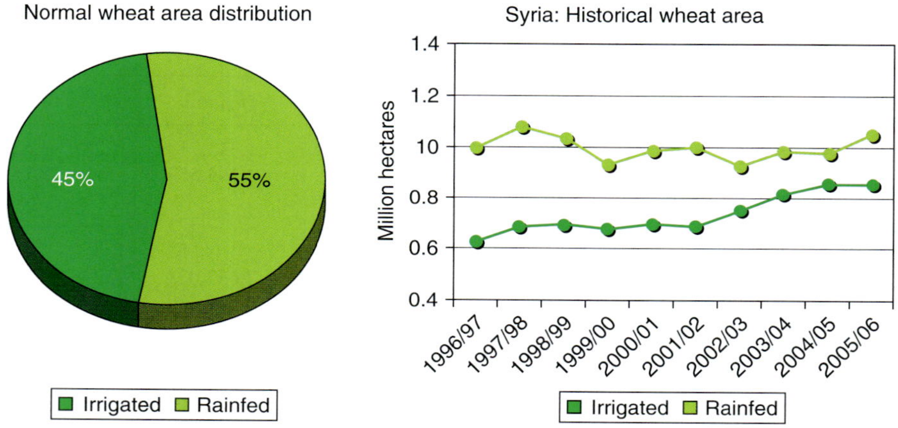

Figure 5. Relation between irrigated and rainfed wheat in Syria.
Source: USDA (2008).

Wheat and cotton together represent 80% of the total irrigated area in Syria and therefore they are by far the main irrigated crops of the country. In the Governorate of Al-Hasakah these two crops cover 73.5% and 23.6% of the surface respectively, which amounts to 98.3% of the total irrigated area in the region.

a) Irrigated wheat with 689,868 ha represented 37% of the total cultivated area, but this proportion has increased nowadays. It is by far the largest irrigated crop in terms of area. It is also the largest consumer of irrigation water with 3,942 Mm^3 (Table 13), which represents nearly 40% of the total. Figure 5 shows the relations between irrigated and non-irrigated wheat in Syria and shows that the irrigated area has been growing during the period while rainfed areas have remained nearly constant. Most of the irrigated wheat (nearly 70%) is cultivated in Al-Hasakah Governorate where water deficits are very high.
b) Cotton is the second largest consumer of irrigation water in the country with nearly 40% of all irrigation water used. More than 50% of this water is used for exporting the production. As earlier indicated, a substantial part (25%) of the cotton production takes place in Al-Hasakah where is mostly irrigated by groundwater wells.

Table 14. Evolution of net trade for main crops of Syria.

	Net trade (import – exports)			
	1990		2004	
Main crops	Quantity (t)	Value (10^6 US$)	Quantity (t)	Value (10^6 US$)
Barley	96,089	17.51	426,326	42.45
Cotton	−66,193	−152.90	−113,601	−164.05
Maize	249,332	60.10	854,841	118.56
Sugar beet	0	0	−1	0
Wheat	934,844	148.20	− 558,471	− 105.34
Total merchandise	No data	− 1,812.57	No data	880.36

Source: FAOSTAT (FAO, 2009).

c) Most of the area dedicated to maize is irrigated and it represents only 5% of the total irrigated area in the country and of a small share of the nation's total use of irrigation water. Maize is very sensitive to water stress in certain critical phases of the crop's growing period, and therefore timing of irrigation is very critical. In general, yields are below international standards.
d) The remaining crops represent small percentages of the irrigated area and of the total irrigation water.

In sum, the two crops that determine the high use of irrigation water in Syria are wheat and cotton. In consequence, reducing the cultivated area of these two crops could be considered a reasonable objective for the conservation of the water resources.

6.5 *Agricultural trade in Syria*

Table 14 summarizes the evolution of net trade for the considered period (1990–2004). For the purpose of the analysis, net trade has been defined by the difference between imports and exports which reflects the actual water flow balance of the country through the trade of agricultural commodities in the two years considered. This is discussed in the next section.

From Table 14 we can draw the following comments.

a) The efforts made in expanding the wheat production in Syria have resulted in the surprising change to be an importer of wheat in 1990 (with 934,844 t) to be a net exporter in 2004 with 558,471 t. However this change is highly dependent on rainfall. Low rainfall years may change completely the situation. Nevertheless, it appears questionable to export part of the production in normal rainfall years since water is a valuable asset. Furthermore, the production costs are higher that the international ones. While a self sufficient policy could be justified the export of wheat needs to be analyzed closely.
b) The government sets the prices for buying cotton seeds from the farmers. The crop, estimated at 1,029,252 t of seed cotton (Table 12), is ginned to produce approximately 350,000 t of cotton lint. Spinning facilities are not sufficient to process the whole crop. Only 150,000 t of cotton lint are utilized locally for yarn production. The balance of the crop, 200,000 t of cotton lint, is exported. Syria needs more than double its current spinning facilities to process all the cotton lint production and make use of the value added in exporting yarn and textiles instead of cotton lint. The Syrian Government does not officially subsidize cotton lint exports. However, the Cotton Market Organization (CMO) faces a difficult situation because international prices for cotton lint are below the cost of production. This is another reason for the government to limit the national production (US Syrian Embassy, 2006).

Table 15. Virtual water traded in Syria. Net trade balance (imports – exports, year 2004).

Crop	Yield (t/ha)	Water use (m^3/ha)	Water use (m^3/t)	Virtual water traded (year 2004) Quantity traded (t)	Virtual water (Mm^3)
Wheat	6	5,714	952	−558,471	−531.85
Cotton lint	1.5	15,714	10,476	−113,601	−1,190.08
Maize	7	7,142	1,020	854,841	872.18
Sugar beet	60	8,571	143	−1	0
Barley	0.4	2,500	6,250	426,326	2,664.54
Net total of virtual water traded					1,814.78
m^3 per inhabitant				94.91	

c) Table 14 shows that cotton is the main export crop of Syria (164 million US$ of net trade) contributing substantially to the agricultural sector. However is evident that the production cost need to be reduced and the area kept under control.
d) The country is a net importer of maize and the demand has more than tripled in the period considered. The value of the imported maize has increased from 60.1 million US$ to 118.5 million US$. This shows that there is considerable space for increasing the maize production and reduce imports.
e) Syria is a net importer of barley but the amounts imported are highly dependent of the rainfed production. Table 14 shows that the net imported quantity in 2004 was nearly 5 times the value of 1990, but this is precisely due to the low rainfed production of year 2004.

Table 14 shows that there is practically no import or export of sugar beet. Hence, it is used mostly for internal consumption and the area is surely proportional to the national needs.

In sum, Table 14 points out that it does not seem reasonable that Syria has become an exporter of wheat and that therefore wheat production should be reduced. Although cotton is an attractive export crop, also its production should be reduced. On the contrary the production of maize should be expanded to reduce the large imports. Sugar beet should be maintained at its current production level. Barley imports are largely depending of climatic conditions but it appears that the production area should be expanded.

6.6 *Virtual water trade in Syria*

Table 15 shows a summary of the blue virtual water traded in Syria in the year 2004 for the main crops considered.

Table 15 illustrates that the export of cotton lint represent the highest value in terms of virtual water exported. It is followed by wheat with 531 Mm^3 exported. This is largely off set by the imports of maize and barley. The final virtual water trade balance shows that Syria is a net importer of water for the total crops considered. On average, Syria is importing annually 94 m^3 per person, which does not appear as an alarming figure considering the limited water resources of the country. This figure is lower than the one estimated by Fernandez (2007), because this study includes all crops in Syria while in our analysis we have considered only the main five crops.

The analysis of the virtual water flows in Syria through agricultural trade points out clearly in the direction of reducing the exports of cotton and wheat and, where feasible, to increase the imports of maize and barley. In conclusion, from the above analysis it is apparent that there are good opportunities to reduce the area allocated to cotton. Wheat under irrigation has been expanding as prices have remained lucrative for the farmers but, according to the recent trends, it is likely that the areas of rainfed wheat will be reduced and, where possible, will be dedicated to maize.

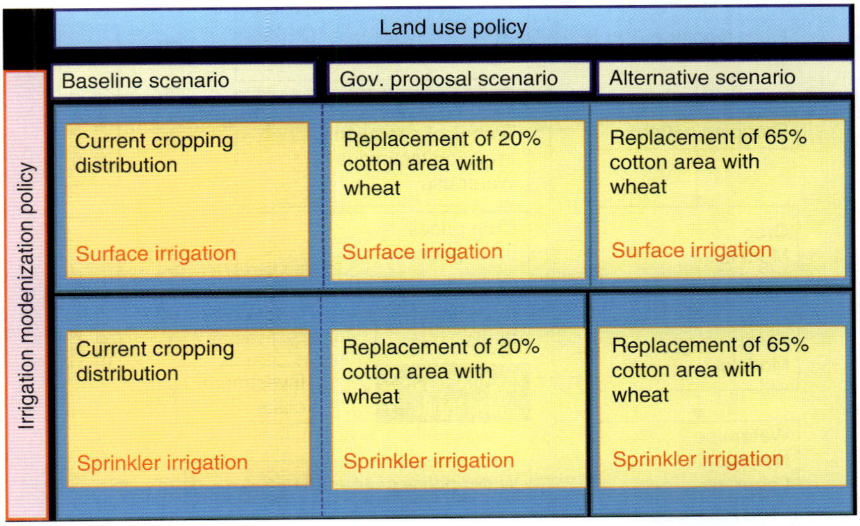

Figure 6. Simulated policy scenarios for Syrian irrigated agriculture.

6.7 *Policy simulations: Comparison between land use and irrigation modernization policies in Syria*

Based on the analysis of the national figures in Syria for the main cultivated crops, agricultural trade and virtual water trade, this section analyzes the comparative effects of the implementation of selected policies aimed to reduce the nation's structural water deficit. The two policies selected respond to the water related problems that have emerged from the analysis carried out in the previous sections. Figure 6 summarizes the two policy options that can be defined as follows:

a) *Land use policy* that seeks to modify the current cropping pattern in the main irrigated areas in Syria towards low water consuming crops. It is represented by two policy alternatives based on the reduction of the area cropped with cotton and the equivalent increase in the area cultivated with wheat. Groundwater irrigation is key in Syria, especially in the cotton growing regions of the water-scarce northeastern plains (region of Al-Hasakah). Within the cotton areas, the main limitation is the availability of groundwater for irrigation that is causing a strong depletion of the aquifers. For this reason the usual relation between wheat and cotton is 2 to 1 for medium size farms. Therefore, given the intentions of the Syrian government to reduce the cotton area by 20%, the challenge is to replace part of the cotton area by some other crop. Wheat is already grown in the areas where cotton is planted and is a technically feasible alternative.
   i. The first land use policy scenario is defined by the reduction of 20% in the cotton growing area and the equivalent increase in the wheat acreage.
   ii. The alternative policy scenario will be a more drastic option defined by a reduction of 65% in the cotton cultivated area and an equivalent increase in the area cropped with wheat. Although this latter option will be difficult to implement it has been chosen to represent the maximum possible level of water use reduction

b) *Irrigation modernization policy* that seeks to attain the same policy goal of reducing water use as in the former land use policy but making use of technical instruments. Modernization of the irrigation systems has been supported widely by the Syrian government during the last decades with the purpose of saving water by increasing technical efficiency. In spite of the recent advances, gravity irrigation continues to be predominant throughout the country, especially in the cereal and cotton growing areas and covers more than 85% of all irrigated lands. Localized drip irrigation is still limited and is mostly concentrated along the coastal zones for the production of higher value-added crops such as fruits and vegetables.

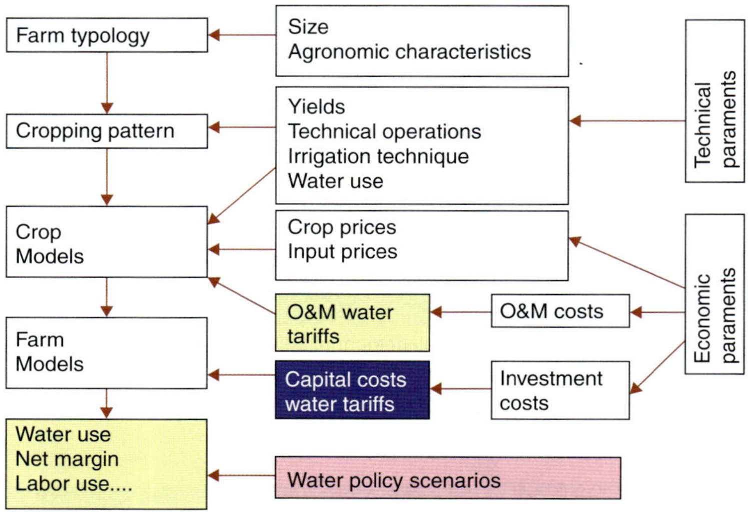

Figure 7. Methodology for assessing the effects of water policies at farm level.
Source: Own elaboration based on Varela-Ortega & Sagardoy (2003).

Implementing water conservation policies has environmental effects as well as economic and social effects for the rural areas. Therefore, these policies have been assessed taking into account their impacts on the use of water resources for agriculture, on farmer's income and on agricultural employment.

6.7.1 *Down-scaling the policy options to the farm level*

However, these global nation-wide policies have distinct effects at regional and local scales, which underline the necessity to downscale the analysis at farm level. This permits to capture the effects of the different policy options on various types of farms and local settings, and evaluate the local actions that could be undertaken. For this reason our analysis focuses on the effects of these policy scenarios at farm level, taking into account different types of farms, water sources and irrigation techniques. In the case of the land use policy the main questions that arise in our analysis are the following: Can water savings be met by changing the crop mix? Is it socially and financially sustainable? Are other factors, besides water use reduction, relevant for the rural livelihoods? In the case of the modernization policy the questions that arise are: Can water savings be met across farm types by changing to more efficient irrigation techniques? What are the economic and social impacts of these technical options? Which of the two policy options is more effective for attaining the desired water reduction objectives at the lowest social cost? All these questions are the object of our analysis that is presented in the following sections.

6.8 *A methodology for assessing the effects of water conservation policies*

Figure 7 summarizes the methodology for the disaggregated farm-level analysis that is required for the evaluation of the different policy options selected. Based on an ample field work, a farm model typology was constructed based on the combination of three statistically-based representative farms, two types of water sources (surface and ground water) and three types of irrigation techniques (gravity, sprinkler and drip). The representative farms are intended to represent the Syrian irrigated agriculture in different regional environments and the selection was made according to farm size, production potential, factor allocation, cropping pattern and type of water source. The characteristics of these farms appear in Table 16 and their location in the different Syrian regions in

Table 16. Selection of representative farms for irrigated agriculture in Syria.

Farm typology	Size (ha)	Cropping pattern		Region
		Crop	% of surface	
Large farm	14	Wheat	70	Al-Hasakah
		Cotton	30	
Medium farm	5	Wheat	50	Hama
		Cotton	20	
		Potato	15	
		Sugar beet	15	
Small farm	1.5	Tomato	50	Lattakia
		Potato	25	
		Oranges	25	

Source: Varela-Ortega & Sagardoy (2003).

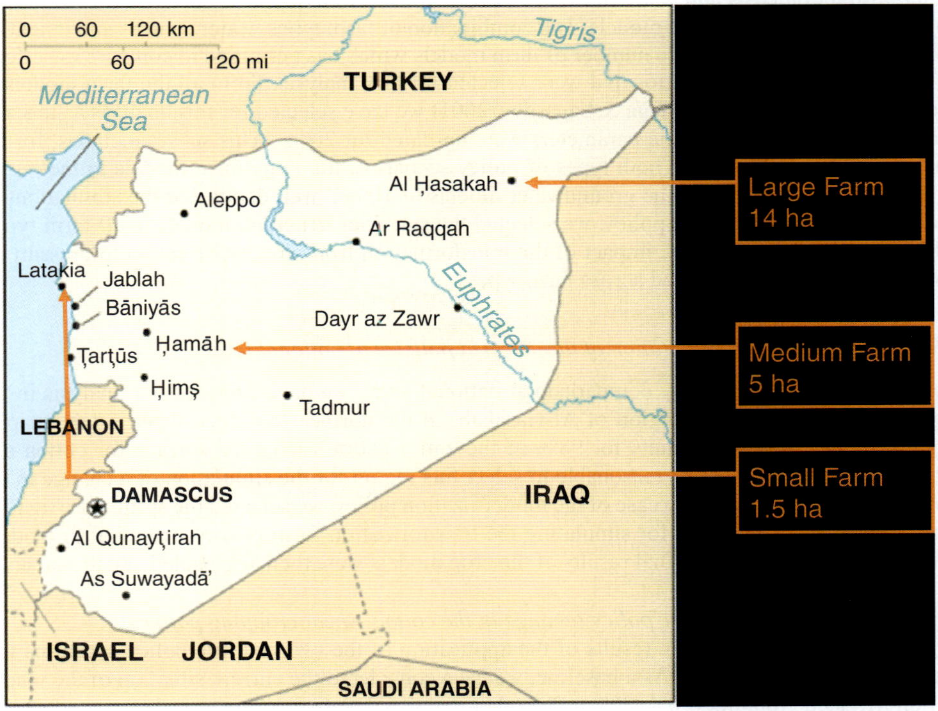

Figure 8. Location of the representative farms in the selected regions in Syria.
Source: The World Factbook 2002 [www.faqs.org/docs/factbook/print/sy.html]

shown in Figure 8. These are namely a large extensive farm of 14 ha, a medium size semi-intensive farm of 5 ha and a small intensive farm of 1.5 ha, which together represent 77% of all irrigated holdings and 64% of the irrigated surface (Varela-Ortega & Sagardoy, 2001; Varela-Ortega & Sagardoy, 2003).

Table 17. Effects at national level of the land use policy on water use, farm income and employment (year 2004).

Land Use Policy Comparisons (Gravity Irrigation)

	Water use (Mm^3)			Gross Margin (10^6 US$)			Labor Use (10^6 working days)		
Policy Scenarios	Total	%	Balance	Total	%	Balance	Total	%	Balance
Current situation	5,764	100		497	100		31	100	
Cotton/wheat 20%	5,386	93	−378	473	95	−24	27	85	−4.63
Cotton/wheat 65%	4,536	79	−1,228	419	84	−78	16	52	−15.04

6.8.1 *Farm models for the current situation and simulated scenarios*

Water sources in the different farm models are shallow water diverted from rivers, and groundwater extracted from wells (at different levels of well depth). The technical, agronomic and economic parameters of the farms were obtained from the field survey as well as all irrigation variables. These are, for instance, water use volumes and technical efficiency for each irrigation technique, water abstraction costs and water delivery costs, as well as water application standards for all the irrigation options in the farm models. The combination of farm types, water sources and irrigation techniques results in an ample number of farm models which has permitted to compare the effects of the different policy scenarios that were selected for this analysis. From all the farm models of the original study (Varela-Ortega & Sagardoy, 2001) we have selected for this analysis a subset of models in which the economic parameters were up-dated for 2004[3]. This subset includes, for the analysis at national level for both types of policy scenarios, the large-farm models in the region of Al-Hasakah, specifically the groundwater models of 100 m. well depth. For the simulations of the irrigation modernization policy, we selected the sprinkler irrigation models in all farm types, which permitted to assess the impact of the transformation from gravity irrigation to pressurized sprinklers at national level and across farm types.

6.9 *Effects of the implementation of the selected policies at national level*

For the purpose of the policy comparison at national level, we have centered our analysis in the cereal and cotton growing region of Al-Hasakah, in the northeastern part of Syria (see map in Figure 8). This region accounts for 98% of the entire nation's irrigated surface for cotton and wheat and therefore it seems reasonable to select this region for the simulation of a change in the cotton-wheat crop mix. In the case of the modernization policy, we selected the sprinkler irrigation farm models in Al-Hasakah for simulating the transformation form gravity irrigation to modern pressurized sprinklers. The final results of the crop models were then aggregated at national level.

6.9.1 *Results of the land use policy (change in the cotton-wheat cropping pattern)*

Table 17 shows the aggregate results of the application of the two land use policy scenarios that were defined in Section 6.7. As a baseline, we have considered the current situation of the wheat-cotton irrigated crop mix (70%–30%). The other two simulated scenarios included, respectively, a reduction of 20% and 65% of the cotton surface and the equivalent increase in the wheat panting area.

From the results we observe that in the case that the cotton growing area will be reduced by 20%, the volume of water used is a mere 7% less of what was used in the current crop mix situation. In total, water savings reach 378 Mm^3, well below the nation's annual shortage. A more drastic reduction of 65% in the area planted with cotton will reduce water use by 20%, still less than

[3] This year was chosen for consistency with the 2004 data presented in the tables of the previous sections. Data were updated considering the annual inflation rate in Syria for the period 2001–2004, and converted into US$ using the official exchange rate 53.52 S£/US$.

Table 18. Effects at national level of irrigation modernization on water use and farm income.

Irrigation Modernization Comparisons

	Water use (Mm^3)			Gross Margin (10^6 US$)			Labor Use (10^6 working days)		
Policy Scenarios	Total	%	Balance	Total	%	Balance	Total	%	Balance
Current situation (gravity)	5,764	100		497	100		31	100	
Current situation (modernization sprinkler)	4,323	75	− 1,441	831	167	335	24	77	−7.05
Cotton/wheat 20%	5,386	93		473	95		27	85	
Cotton/wheat 20% (modernization sprinkler)	4,040	70	−1,347	786	158	313	20	64	−6.51
Cotton/wheat 65%	4,536	79		419	84		16	52	
Cotton/wheat 65% (modernization sprinkler)	3,402	59	−1,134	683	137	264	11	35	−5.29

proportional, and the volume of water saved will reach 1,228 m^3/yr, about one third of the total deficit in the country (see Table 11). Looking at the economic and social effects of this land use policy, we observe that the first policy option produces a negative impact on farm income and employment. A total 24 million US$ and 4.63 million working days will be lost, in the nation's aggregate, as a response to this cotton-reduction cropping change. Equivalently, the more drastic 65% reduction of the cotton growing area, in the alternative scenario, will likely diminish farm income by 78 million US$ as labor use will diminish by 15 million working days. According to these preliminary results, the social impact of this cotton-reduction policy is not negligible, being cotton a labor intensive crop. In fact, in its less drastic version of 20% cotton reduction, this policy brings about 7% reduction in water use, which is clearly insufficient to balance the overall annual water deficit, but, in turn, it provokes an aggregate loss of 15% working days, more than twofold the water saving percent. From an overall policy perspective, balancing environmental and social objectives is a desired policy goal and, in this respect, it remains questionable whether it will be possible for the Syrian agrarian economy, to provide sufficient employment opportunities in the cotton growing areas, to compensate for this potential negative social impact.

6.9.2 *Results of the irrigation modernization policy*

Table 18 shows the aggregate results at national level of the application of the irrigation modernization policy in Syria. The Table summarizes the effects on water use and farm income. Irrigation modernization is represented by the substitution of gravity irrigation by sprinkler irrigation in the farm models. This substitution entails a reduction in water losses in the irrigation system, and hence an increase in technical efficiency for irrigation applications that reduces the water volumes applied to the crops. It also entails an increase in investment costs in the farms related to the installation of the pressurized system, the pump set and the sprinklers.

We can observe that when irrigation modernization takes place (that is, gravity systems are substituted by sprinklers) water use is reduced in all types of land use policy scenarios. These are defined as before by a reduction of, respectively, 20% and 65% of the cotton surface and an equivalent increase in the whet growing surface. This trend is larger when modernization occurs in the current situation with a larger proportion of cotton surface with respect to the cotton-reduction options.

With respect to the *effects on water consumption*, when gravity systems are substituted by pressurized sprinklers, in the current scenario, water use diminishes by 25% that amounts to 1,441 Mm^3, about half of the nation's water deficit. In the case of the 20% cotton-reduction scenario, water use is reduced by a slightly smaller percent amount of 23%, which corresponds to a volume of 1,347 Mm^3. Equivalently, in the case of the 65% cotton-reduction scenario, water use is reduced by a lesser percent amount, 20%, that totals a volume saved of 1,134 Mm^3. In sum, as land use favors less water intensive crops, such as wheat, by increasing its surface, the modernization of irrigation systems has a smaller impact on water savings than in the case of more intensive cropping patterns with more cotton acreage.

If we compare the results of the modernization policy to the results of the land use policy presented in the previous section (Tables 17 and 18), we can observe that irrigation modernization is a much more effective policy for attaining water savings that the land use policy. In fact, in the current situation, irrigation modernization attains a water use reduction that is fourfold the volume saved by the land use policy (1,441 Mm^3 and 378 Mm^3 respectively).

Observing the *effects on farm income*, the irrigation modernization policy produces an increase in farm income across all policy scenarios. When sprinklers are installed substituting for gravity systems in the baseline scenario, the 20% cotton-reduction scenario and the 65% cotton-reduction scenario, farm income increases respectively by 335, 313 and 264 million US$. In average, modernization of irrigation systems results in income gains that range between 70 to 40% of their non-modernized original farm income. When comparing these results to those obtained by the land use policy, we observe, as in the case of water use discussed above, that the modernization policy is more effective and does not inflict income losses to the farmers. On the contrary, adopting sprinkler irrigation has clear incentives to the farmers for using less water in their crops. It saves water and therefore results in clear income gains to the farmers, as water is measured and priced by the volume consumed in the pressurized systems. Conversely, the land use policy that foresees a reduction of 20% and 65% in the cotton cultivation area provoked a farm income loss of 24 and 78 million US$ respectively.

Equivalently, the *effects on labor use* and farm employment are less drastic in the case of the modernization policy than in the case of the land use policy. In fact, as land use becomes more extensive and wheat occupies a larger proportion of the cropping land, the impact of the modernization policy on labor use is progressively effaced. Farm employment is reduced by about 7 million, 6 million and 5 million working days when modernization takes place respectively in the current cropping scenario and in the 20% and 60% cotton-reduction scenarios. By itself, the land use policy inflicted a total job loss of three times more (15 million working days) in the case of the more drastic cotton-reduction case than when irrigation systems are upgraded. However, it has to be noted that half of the labor used in the large Syrian farms is family labor and therefore the employment loss that comes along the installment of pressurized irrigation systems will be partially absorbed by the families' on-farm labor commitments.

6.10 *Impacts at farm level*

Figure 9 shows the water use comparative results in the case of the land use policy and the modernization policy for the different types of farms. Figure 10 shows the farm income equivalent results. In both figures, the numbers shown in the small rectangles refer, respectively, to the difference between the current scenario and the cotton-reduction scenario in the case of the land use policy, and in the case of the modernization policy, for the difference between the gravity systems and the modern pressurized sprinklers. Comparatively, we can state that the aggregate nation-wide policy effects are also apparent across farm types. Water savings are substantially larger in the case of the modernization policy than in the case of the cotton-reduction policy for all farm types. In fact, in the current situation, installing modern irrigation systems in the larger farm saves five times more water per ha than substituting wheat for cotton (1,520 m^3/ha and 344 m^3/ha respectively). In the medium-size farm, this difference is increased to seven times more. Economies of scale favor the land use policy, as large farms save more water (344 m^3/ha) than medium size farms (260 m^3/ha). However, the adoption of modern irrigation technologies favors inverse economies of scale, and water savings increase as farm size diminishes. That is, the smaller more intensive farms have a higher water-saving capacity than their larger less intensive counterparts (3,106, 1,544 and 1,520 m^3/ha for the small, medium and large farms respectively).

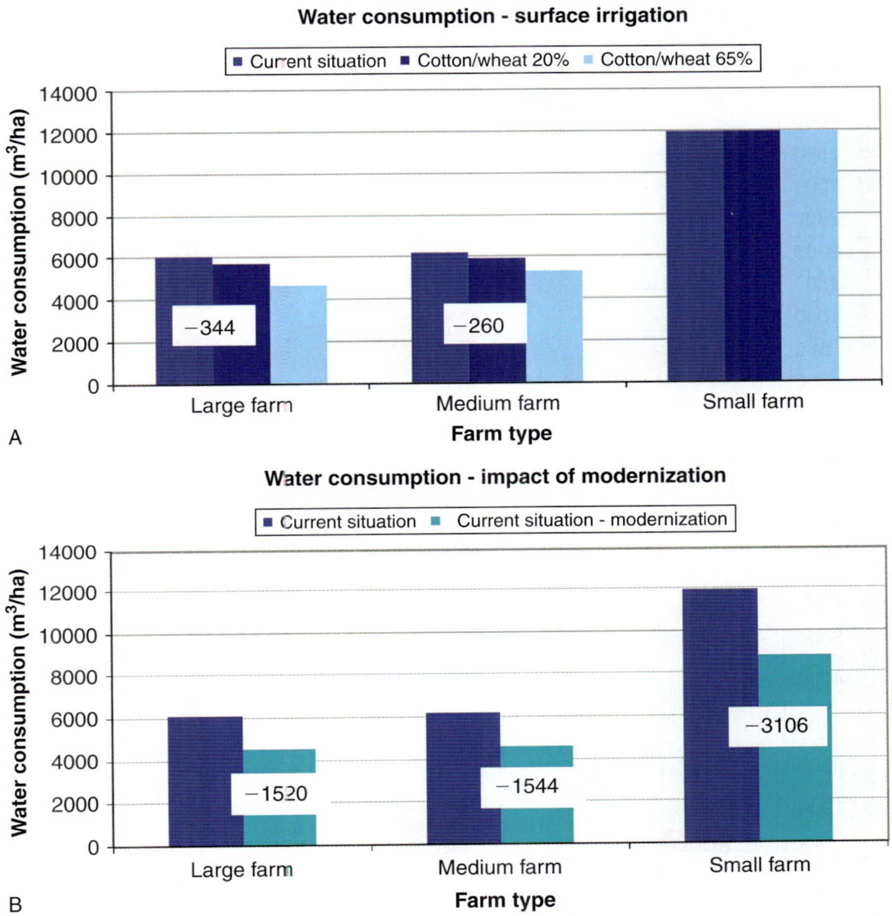

Figure 9. Comparative effects of land use and modernization policies across farm types: Impact on water consumption[4].

Adopting the cotton-reduction land use policy inflicts income losses to the farmers, almost equally in the large and medium-size farms (30 and 35 US$/ha respectively). However, adopting modern irrigation technologies produces an inverse effect and water savings come along with clear income gains for the farmers (around 50% for all farm types). Parallel to water use, modern irrigation technologies show inverse economies of scale with respect to farm income. Farmers' profits mount as farm size diminishes. The more intense small farms, that are capable of better adjusting to water-saving technologies, use less water and hence farm income is augmented (2,003, 548 and 362 US$/ha in the small, medium and large farm respectively) as water volumes and costs diminish in the pressurized systems. In sum, an irrigation modernization policy seems more effective and more socially acceptable for all farm types than the land use policy. However, even if both policies have the same goal of reducing water use, their impact varies for different types of farms and regions, and therefore, a local regionally-based vision is needed for a careful assessment of the effects of these policies.

---

[4]The land use policy has no effects on the small farm as this farm does not cultivate cotton or wheat and is therefore out of the crop-mix change defined in this policy (see Table 16).

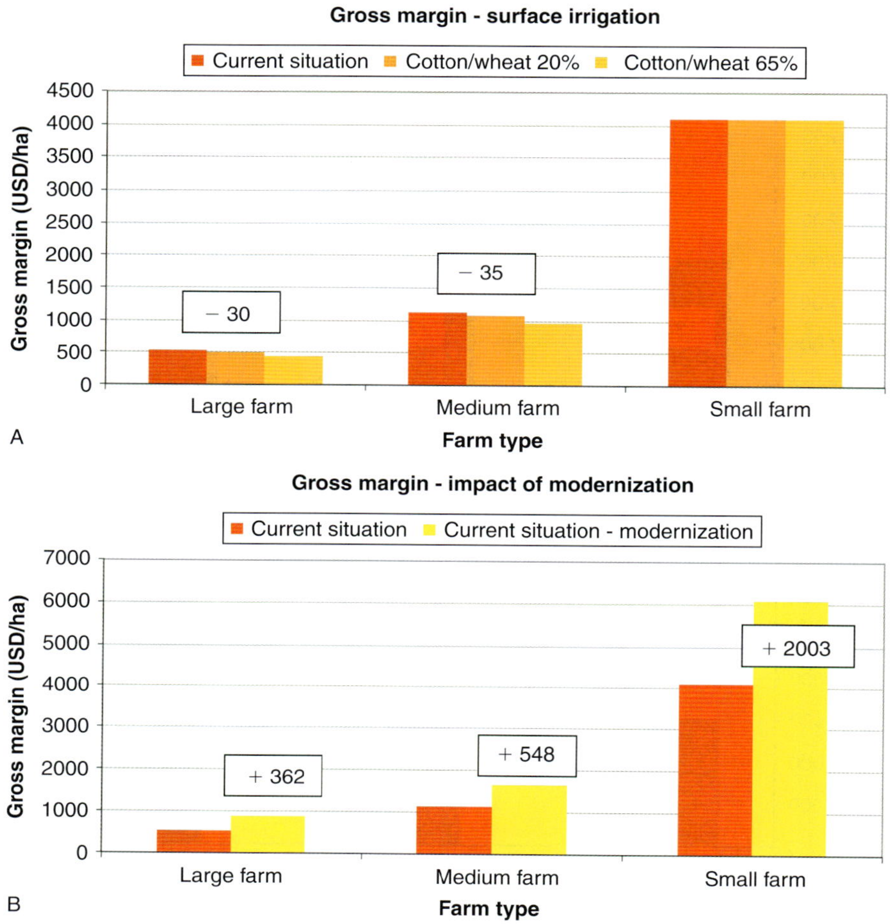

Figure 10. Comparative effects of land use and modernization policies across farm types: Impact on farm income.

Looking at the *combined application of the two types of policies*, the cotton-reduction scenarios do not provoke greater water savings when more efficient modern irrigation systems are installed (Table 17). Moreover, farm income gains are not augmented when modern irrigation systems are installed in conjunction with cotton-reduction options (Table 18). This evidences that both policies do not necessarily complement their water-saving objectives and that, for the Syrian economy, as well as for the economies of other Mediterranean arid countries, policy synergies have to be developed to attain commonly defined goals. These policies will have to interact with the aim of balancing the trade-offs between using water for conserving the environment and for maintaining the rural livelihoods. This is the case of the semi-arid regions in Spain in which the EU water and agricultural policies are seeking to converge, in specific rural settings, to balance the conservation of water resources and the irrigation-based agrarian economy (Varela-Ortega, 2010). In a regional perspective, the application of these two types of policies can have distinct effects across different types of farms. Hence, the regional and local effects of nation-wide policies have to be carefully explored to achieve a sound and socially-accepted implementation of water conservation policies in Syria.

7 CONCLUSIONS

The scarcity of water resources in the Mediterranean countries is reaching alarming levels. The scarcity index (the ratio of total water withdrawals to the average of natural renewable resources flow) already exceeds 50% in eight Mediterranean countries and this number will increase in the future.

The increasing water scarcity has led to high investments in regulating river flows. The region has the highest percentage of river regulation (80%) in the world. Also high investments have been made in irrigation development resulting in high increases in the irrigated areas during the last 25 years. The reuse of non-conventional waters has not escaped to these efforts and in the MENA (Middle East and North Africa) Region a high percentage (60% of the world capacity) of the domestic supply comes from this type of water resources. Still most of these efforts are not sufficient to compensate the growing demand.

Irrigation development has contributed to increase food security in most of the European countries. However in SEMR (South East Mediterranean Region) countries the situation has improved, but not sufficiently to compensate population growth. Some countries like Egypt and Morocco have reduced their food security standards in recent years, and this situation will worsen due to significant population growth rates.

The chapter also analyzes the interrelation between poverty and development and water scarcity. It concludes that the poor sections of the population suffer more from water scarcity. It also acknowledges that irrigation development has contributed substantially to reduce poverty. Food aid can be a useful instrument to alleviate emergency situations of hunger, but cannot be the remedy to poverty and hunger.

The SEMR countries, and to some extent the EU countries, have resorted to augment agricultural trade to meet the growing domestic demand that could not be satisfied by the national agricultural production. Net trade increases are moderate in all countries analyzed, mostly due to increased demand driven by population growth.

The increase in trade leads also to great transfers of virtual water. The use of this relatively new concept is becoming more common, but still rarely used for planning agricultural production in a given country. To contribute to this purpose, this chapter advances a methodology that through the systematic analysis of some basic information related to production, irrigation water use, agricultural trade and virtual water trade, provides some policy lines for redirecting the agricultural production. However these policies require careful analyses before attempting their implementation. A first attempt has been made with Egypt where the mentioned information has been analyzed for the main crops and a period of 15 years. Some interesting conclusions emerge, but it is obvious that in order to translate them into executable policies a deeper analysis is required and this was done in the case study of Syria were the authors had access to detailed field information.

The same methodology has been tested for Syria and has proven its usefulness for identifying the effects of the main policy actions that could be undertaken. However the implementation of these policies requires testing their feasibility through more detailed studies. Some of the emerging conclusions state that, given the structural water deficit in Syria, policies that aim to conserve water resources can have very different effects on the overall economy and on the rural livelihoods depending on the instruments chosen to attain the desired objectives.

A nationwide land use policy that foresees the reduction of water intensive crops, such as cotton, and the equivalent increase in wheat, will only attain water savings up to a mere 10% of the nation's water deficit. From the social perspective, the economic and societal impacts of this cotton-reduction policy are not negligible, being cotton a labor intensive crop, as the overall farm income and employment are severely impaired at the nation's aggregate. Therefore, this policy will be neither environmentally sustainable nor socially acceptable. It remains questionable if the Syrian economy will be flexible enough to provide alternative employment opportunities in the agricultural sector, or in any other economic sector, to compensate for this potential social loss.

If Syria chooses to implement a technically-oriented water saving policy, such as the modernization of the existing irrigation systems, we can conclude, based on our preliminary findings, that this policy is much more effective and socially balanced than the land use policy for attaining water conservation objectives. Provided that irrigators are faced with proper volume-based and price-based incentives to use less water in the new pressurized systems, the total volume saved in the nation can be up to four times larger than in the case of the land use policy. Under this latter policy, water is diverted and charged by surface in the non-modernized gravity systems in the absence of technically-driven incentives. In turn, proper economic and technical incentives result in social gains. Farm income can rise substantially across all farm types and regions, family labor will have to be more technically oriented and trained, and employment opportunities will have to shift to the production of more labor intensive higher value-added crops.

In Syria, as in other Mediterranean arid countries, policy synergies have to be developed to attain commonly defined goals in the entire nation and in the different regions as well. These policies need to balance the environmental objectives of water conservation and the socio-economic objectives of maintaining the rural livelihoods. However, it seems conclusive that the nation-wide policies that emerge from agricultural trade and production have distinct regional and local effects. Then, these implications have to be carefully explored to achieve a sound and socially-accepted implementation of water conservation policies in the water-scarce Mediterranean basin.

ACKNOWLEDGEMENTS

The authors would like to acknowledge the EU project SCENES ("Water Scenarios for Europe and Neighbouring States" FP6 GOCE, 036822) for supporting partially the data-base processing of this research at Universidad Politécnica de Madrid, Spain (UPM), and the Mediterranean Agronomic Institute of Bari, Italy (CIHEAM).

REFERENCES

Benoit, G. & Comeau, A. (eds.) (2005). Plan Bleu. *A Sustainable Future for the Mediterranean: The Blue Plan's Environment and Development Outlook*. UNEP-MAP-Blue Plan. Earthscan, London, UK: 464 pp.

Central Bureau of Statistics (1998). Agricultural census of 1998, Crop harvested areas. Damascus, Syria.

Chapagain, A.K. & Hoekstra, A.Y. (2004). *Water footprint of nations. Volume 1: Main Report*. Value of Water Research Report Series, 16. UNESCO-IHE, Delft, the Netherlands.

FAO (2005). The state of food and agriculture. Agricultural trade and poverty. Can trade work for the poor? Food and Agriculture Organization of the United Nations, Rome, Italy.

FAO (2008). *The state of food insecurity. High food prices and food security –threats and opportunities*. Food and Agriculture Organization of the United Nations, Rome, Italy.

FAO (2009). *FAOSTAT Database* [http://faostat.fao.org].

Fernandez, S. (2007). *L'eau virtuelle dans les pays méditerranéens: un indicateur pour contribuer à l'analyse des questions de gestion et de répartition de l'eau en situation de pénurie?* Rapport d'étude régionale du Plan Bleu [www.planbleu.org].

Hamdy, A. & Lacirignola, C. (2005). *Coping with water scarcity in the Mediterranean. What, Why and How?* Mediterranean Agronomic Institute of Bari, Italy: 17–45.

IFPRI (2000). *Wheat policy reform in Egypt: Adjustment of local markets and options for future reforms*. By M. Kherallah. Research Report, 115. Washington, D.C., USA.

INE (1997). *Encuesta sobre las estructuras agrarias de 1997*. Instituto Nacional de Estadística, Spain. [www.ine.es].

MAAR (Ministry of Agriculture and Agrarian Reform) (2004). Memorandum of Transfer to Modern Irrigation. Administration of Natural Resources. Research General Commission for Scientific Agriculture Research. Internal Communication. Damascus, Syria.

Sagardoy, J.A. & Varela-Ortega, C. (2007). *Sustainability of the exploitation of the water resources of Syria and its implications for the future development of the irrigated agriculture*. Paper presented at the Seminar: *The Contribution of Agriculture and Rural Development to the Process of Economic Reforms of Syria*. FAO Project GCP/SYR/006/ITA Phase III. Damascus, Syria.

Somi, G.; Zein, A.; Dawood, M. & Sayyed-Hassan, A. (2002). *Progress Report on the transformation to modern Irrigation Methods until the end of 2001*. Internal report. Ministry of Agriculture and Agrarian Reform. Damascus, Syria.

UN (2008). *World Population Prospects*. Population Division of the Department of Economic and Social Affairs of the United Nations Secretariat, New York, USA.

US Syrian Embassy (2006). *The Syria Cotton Report 2006. Executive Summary*. [http://damascus.usembassy.gov/uploads/].

USDA (2008). *Syria: wheat production in 2008/09 declines owing to season-long drought*. Foreign Agriculture Service. Commodity Intelligence Report, May 2008. Washington, D.C., USA.

Varela-Ortega C. (2010). The Water Policies in Spain: Balancing water for food and water for nature. In: A. Garrido & H. Ingram (eds.), *Water for Food in a Changing World*. Routledge, Abingdon, UK.

Varela-Ortega, C. & Sagardoy, J.A. (2001). *The utilization of water resources for agriculture: Analysis of current regime and policy*. Final Report. FAO project GCP/SYR/006/ITA. FAO-Italian Government Cooperative Program. Damascus, Syria.

Varela-Ortega C. & Sagardoy, J.A. (2003). Water Policies in Syria: Current Developments and Future Options. In: C. Fiorillo & J. Vercueil (eds.), *Syrian Agriculture and the Crossroads*. FAO Agricultural Policy and Economic Development Series, 8. Policy Assistance Division. Food and Agriculture Organization of the United Nations, Rome, Italy.

World Bank (2006). *Making the Most of Scarcity Accountability for Better Water Management Results in the Middle East and North Africa*. MENA Development Report. The World Bank, Washington, D.C., USA.

World Bank (2008). *World Development Indicators. Poverty Data Supplement to the 2008 World Development indicators*. The World Bank, Washington, D.C., USA.

CHAPTER 8

Economic aspects of virtual water trade: Lessons from the Spanish case

Alberto Garrido, Paula Novo, Roberto Rodríguez-Casado &
Consuelo Varela-Ortega
Technical University of Madrid, Spain

ABSTRACT: Virtual water is the volume of water that is used in the production process of a commodity (Allan, 2003). Agricultural trade is by far the largest vehicle to move water virtually around the world. The effects of water scarcity in many countries and regions can be reduced through open farm trade and specialization. Many world countries are now suffering water shortages and expect worsening conditions in the future due to climate change, economic development and food demand increases. We first review a number of key facts about world water scarcity and reflect on the role of virtual water trade. Observing that most countries import and export water embedded in the exchanged products, we review the example of our evaluation of water *trade* in Spain for the period 1997–2006. We differentiate between the green (soil water) and blue (surface water and/or groundwater) components of virtual water from a hydrological and economic perspective. The combination of spatial and time dimensions offers a unique empirical setting to determine whether virtual water *trade* can contribute to reduce water scarcity. The study reveals that Spain is a net and increasing *importer* of virtual water. By far the largest virtual water *imports* are linked to cereals and animal feed products whilst the virtual water *exports* are linked to exports of animal products, fruits and vegetables. Virtual water *trade* is one way to reduce the vulnerability of the agri-food sector to climate instability. It reinforces the competitive advantages of its natural endowments and capital investments in agriculture. The econometric analysis using provincial water exports for 10-years shows that water exports are invariable to cyclical water scarcity, and largely explained by fixed factors. Virtual water *trade* does not exacerbate water scarcity, though it is certainly a source of pressure for resource management. Adequate water pricing would make virtual water trade more efficient.

Keywords: virtual water trade, Spain, food production, agricultural trade

1 INTRODUCTION

Trading water virtually is a means to cope with water scarcity, but no trading partner exchanges goods based on the embedded quantity of a resource that is not properly priced. While labor, capital and other inputs have formal and tangible prices, water does not. This implies that virtual water trade occurs as an underlying process that is generally connected to differences in natural endowments and competitive advantage for food production, filtered out or influenced by commercial agreements, but not directly linked to the level of water scarcity or stress of the trading partners.

And yet, global trade establishes an *invisible* and indirect link between water demand and water consumption sites. The literature on virtual water *trade* has emphasized the options available to arid and semiarid countries to use international trade to deal with water resources scarcity (Allan, 2003; Chapagain *et al.*, 2006; CAWMA: Comprehensive Assessment of Water Management in Agriculture, 2007; Yang & Zehnder, 2007; Aldaya *et al.*, 2008; Novo *et al.*, 2009; Garrido *et al.*, 2010). However, determining whether this strategy is economically and environmentally efficient will depend on whether the real opportunity cost of water resources is properly internalized, and whether the trade is actually based on differences in competitive advantage among trading partners. It is also doubtful that virtual water *trade* should be termed a *strategy*, because up to now no government has been documented to pursue it directly. Rather, it is a process that is naturally linked to the trade and exchange of goods, with the main exception of arid and semi-arid countries in the Middle East and North Africa (MENA) region.

Since water is a limiting resource for many food importing countries, in most cases the observed patterns of trade are more consistent with relative differences of production costs, and highly dependent on trade regimes. Relative levels of water scarcity among trading partners do not seem to be an important explanatory factor. While global benefits can be associated with virtual water trade (Hoekstra & Chapagain, 2008; Garrido *et al.*, 2010), there are pitfalls in pursuing the idea to its most extreme format, which is to import virtually water *and* land. Furthermore, many countries engage in two-way flows of trade (Spain imports mainly cereals and feed and exports oranges, grapes, wine, etc.). While one can trace back the region or basin from which virtual water exports originate, it is almost impossible to record the imports and assign them to specific geographical zones.

Up until recently, virtual water *trade* was primarily an unintended consequence of farm commodities trade. When countries have deliberately sought to *import* water, very soon they have seen the advantages to *import* land and natural endowments as well. Recent massive land purchases in Africa through state negotiations are the genuine expression of virtual *trade* of natural resources. The ethical consequences of these exchanges have not been yet thought out in detail. Purchasing states provide capital, technology and know-how, land selling states offer abundant land and water in exchange for hard currency or some other compensation. It has been reported that between 15 and 20 million hectares of African farmland have been sold to food-importing or water-scarce countries, China leading the group.

Another potentially damaging effect of virtual water *trade* is the specialization of water *exporters* in goods that are water-intensive. This can exacerbate water scarcity for domestic users, and increase the price of food for the poorest. Ensuring food self-sufficiency was seen as an objective of massive costs for an importing country, but the costly reliance on international food markets has revisited the concept of self-sufficiency and the value of food sovereignty. Furthermore, pressures to develop water works so that more crops can be produced for the importing countries may bring the scarcity pressure to adjacent basins or catchments, from which water resources can be transferred. Foreign food demand may become thus an indirect source of pressure for exporting countries.

Underlying the ethical and political dimensions of land-purchases, there are clear and unambiguous economic signs of the gains from trade and the efficiencies that can be achieved by expanding the capital base of many African agricultural areas. As von Braun & Meinzen-Dick (2009: 4) state: "Foreign investment can provide key resources for agriculture, including development of needed infrastructure and expansion of livelihood options for local people". In a very different vain, Peter Brabeck-Letmathe, President of Nestlé, stated: "The purchases weren't about land, but water . . . they should be called the great water-grab" (The Economist, 2009). But at the same time, it may be a consequence of the repeated failure of industrialized countries to help African countries get away from their undeveloped situation.

In this chapter, we analyze the economic implications of virtual water *trade* with a double objective. On the one hand, we summarize the findings from a more extensive work carried out by Garrido *et al.* (2010), in which economic evaluations of *exported/imported* water of Spain were performed. On the other hand, we take the case of Spain to reflect on the economic meaning of the predictions of worldwide water scarcity, and its implications for food global demand and production.

We first review the major criticisms and reflections on virtual water *trade*, then synthesize the most recent literature on water scarcity and food demand worldwide. In the fourth section we review the Spanish virtual *trade*, summarizing and extending the main conclusions of Garrido *et al.* (2010). The fifth section highlights the main conclusions.

2 MAJOR ECONOMIC CRITICISMS ABOUT VIRTUAL WATER TRADE

International virtual water *trade* can be evaluated in terms of comparative advantage (first explicitly formulated by the British economist David Ricardo) (Rosegrant *et al.*, 2002), and the fact that natural resources are unevenly distributed over space and time. It is claimed that nations can profit from trading if they concentrate on, or specialize in, the production of goods and services for which they have a comparative advantage, while importing goods and services for which they have a comparative disadvantage.

Whether international trade actually helps alleviate global water stress is still an issue that has not been settled in the literature (Falkenmark & Rockström, in press; Hoff *et al.*, in press). Nevertheless, an increasing number of authors recognize this role (Aldaya *et al.*, 2008; Comprehensive Assessment of Water Management in Agriculture, 2007). Worldwide global water savings as a result of trade is estimated to have reached 450 km^3/yr (Oki & Kanae, 2004; Hoekstra & Chapagain, 2008). Most of these savings come in the form of international trade of cereals, protein crops and oil crops (Hoekstra & Chapagain, 2008).

Several conceptual and practical problems about virtual water *trade* remain relevant. Some of these are:

- Green and blue water components are crucial to determine whether observed exchanges contribute to a sustainable world economy.
- The virtual water metaphor addresses resource endowments but not production technologies. Hence, the metaphor does not include the concept of comparative advantage (Wichelns, 2004).
- Political and economic considerations often outweigh water scarcity concerns, limiting the potential of trade as a policy tool to mitigate water scarcity (Fraiture *et al.*, 2004). Very little of the calculated virtual water *trade* is due to water scarcity.
- Other factors, like land, nitrogen, phosphorous and potassium, should be added to water scarcity measurements.
- Emphasizing virtual water *imports* is not a neutral policy for a water-scarce country, since this affects, among other things, urbanization, rural-urban migration and income distribution (Roth & Warner, 2008)
- Expanding agricultural commodities trade generates overall welfare gains, but also winners and losers among trading partners (Berrittella *et al.*, 2007).
- Virtual water *trade* may be exacerbating water scarcity in water-stressed regions, as shown for the case of India by Verma *et al.* (2008). In explaining virtual water flows, these authors identify key explanatory factors other than water scarcity, including per capita gross cropped area (an indicator of land concentration and population density) and access to secure markets (an indicator of institutional performance).

None of these studies, except for Garrido *et al.* (2010), have evaluated the economic value of the imported or exported water.

3 UPDATE OF WORLDWIDE WATER SCARCITY EVALUATIONS

Concern about worldwide water scarcity and its relation to food production has inspired a number of recent studies. In addition to CAWMA (2007), other studies by UNESCO (2008), Formas (2008), Falkenmark & Rockstrom (in press) have looked in detail in the implications of water resources and food production.

For one thing, the linkages between water availability and economic development are weak for developed and developing countries. In Figure 1 (taken from FAO, 2009), per capita water resources are plotted against the percentage of undernourished people for a number of countries.

However, according to CAWMA (2007), achieving the Millennium Development Goals or erasing poverty by 2030 imply substantial increases of world water use, as shown in Figure 2. Very often the water scarcity problem is taken as a synonym of a food problem.

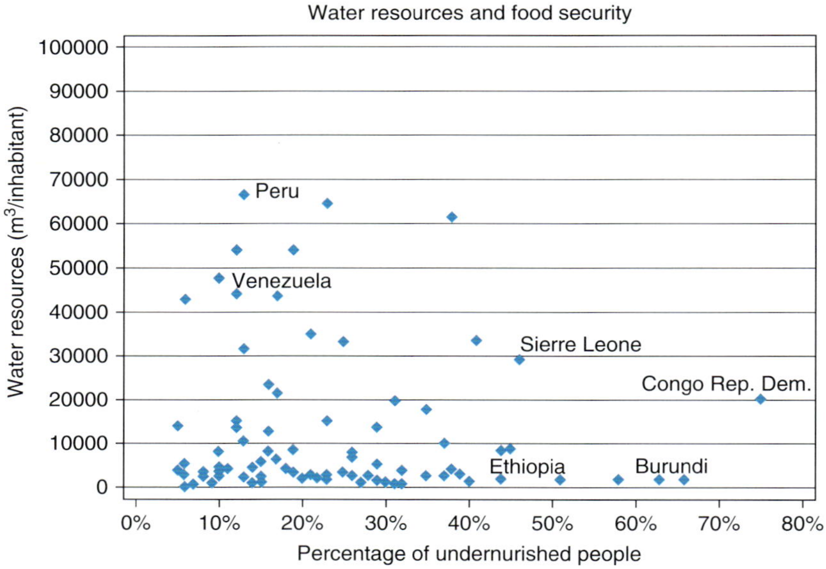

Figure 1. Water resources and food security in developing countries.
Source: FAO (2009).

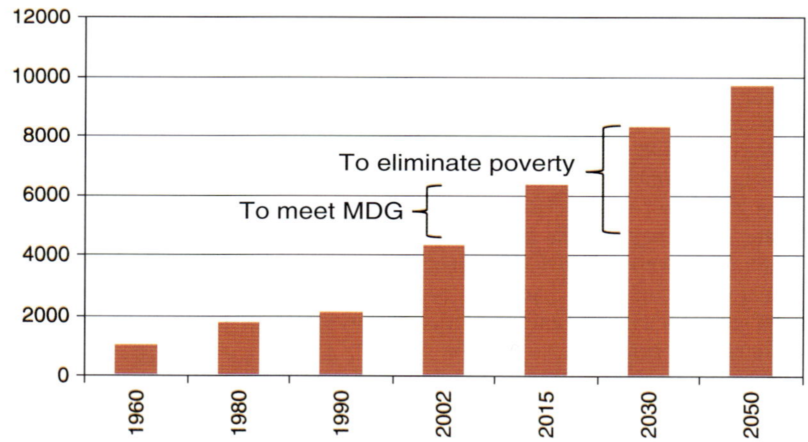

Figure 2. Projections of human water uses (in km^3) in different scenarios.
Source: CAWMA (2007).

World physical systems have also been modeled to project future impacts on precipitation, evapotranspiration, run-off and land-cover. It is impossible to summarize the findings of the most recent literature on the subject. Alcamo *et al.* (2003) concluded that by 2050 water stress will be increasing over most developing regions, but decreasing to a significant extent of industrialized regions. Bates *et al.* (2008) showed that the tropical and subtropical regions, together with those that have a Mediterranean-type climate (the Mediterranean itself, South Africa, large parts of the Southwest of North America), will experience lower and more unstable precipitation regimes. Nevertheless, Bates *et al.* (2008) already warn that the model used for water predictions are inadequate. Therefore, these projections are rather uncertain.

Table 1. Regional available resources and present utilization.

	IRWR*	Total volume of freshwater utilization km^3/yr	Freshwater utilization by purpose: Domestic use km^3/yr	%	Industrial use km^3/yr	%	Agricultural use km^3/yr	%	Utilization as % of resources
World	43,764	3,811.3	376.2	9.9	783.0	20.5	2,652.1	69.6	9
Latin America and Caribbean	13,570	265.1	50.4	19.0	27.4	10.3	187.3	70.7	2
Near East and N Africa	516	322.6	25.1	7.8	19.5	6.0	278.0	86.2	63
Sub-Saharan Africa	3,856	98.1	6.9	7.0	2.8	2.9	88.3	90.1	3
East and SE Asia	8,720	977.4	71.2	7.3	192.3	19.7	714.0	73.0	11
South Asia	1,761	917.8	58.7	6.4	39.6	4.3	819.6	89.3	52
Oceania developing	874	0.1	0.1	35.5	0.0	28.4	0.1	36.2	0

*Internal Renewable Water Resources
Source: CAWMA (2007).

As on May 2009, the list of *Low-Income Food-Deficit Countries* (LIFDC) included 77 countries, five less than in 2008. In April out of the 82, 44 are located in Africa, 25 in Asia, 6 in Oceania, 3 en Central America and 4 in Europe, comprising an urgent demand for 85×10^6 t of cereals [t = tonne = 1,000 kg]. Most analyses assume that a healthy diet requires about 1,700 m^3/yr per person (Kuylenstierna *et al.*, 2008).

Increasing food demand, global warming and more unstable precipitation regimes worldwide are the three major factors that back the somber projections for the future. Do they stand a closer scrutiny in light CAWMA's continental evaluations about available resources and current utilization rates? (see Table 1).

UNESCO (2008) reports that by 2025 there will be 1,800 million people living in regions with absolute water scarcity. Gleick (2009) discusses the concept of *peak water* as a parallel concept to *peak oil* (a point which is reached when half of existing stock of petroleum has been depleted and the rate of production peaks). Gleick, in discussing the differences between oil and water, claims that while oil can be transported economically water cannot. It certainly can, and much cheaper than oil, simply because the footprint of food production is the largest among all economic goods in the economy and have large abundant water embedded in it. So water can be transported embedded in commodities.

As Garrido & Dinar (2009) show, most world water problems materialize most severely in a number of known watersheds and basins, where water use and water pollution may have gone beyond any possible threshold of sustainability. Reversing these trends is a daunting and extremely expensive task, even for developed countries.

In this sense, the notion of *peak ecological water* proposed by Gleick (2009) and the transition from *more crops and cash per drop* to *more cash and care of nature per drop* are relevant means to think about water resources.

In any case, virtual water *trade* is by far the easiest and most economic way to save water resources without threatening the most pressing human needs.

In Table 2 we report various national data about water resources, arable land and irrigation for selected countries from Africa, South America and the European Union. The purpose of putting together basic data from quite disparate countries is to highlight the vast untapped resources of many world countries for food production. According to FAO, countries like Angola, Ethiopia, Mozambique, Brazil, Colombia, Nigeria or Bolivia, to cite just a few, can multiply their present irrigated acreage by a factor of 5 or more. Whereas more mature countries in terms of water utilization rates, like Egypt, Spain, Israel, USA, Mexico or Australia have very few opportunities to

Table 2. Basic water and agricultural land statistics of selected countries from Africa, South America and the European Union.

	Renewable resources (km^3)	Withdrawal (km^3)	Per capita withdrawal (m^3/yr)	Arable land (10^3 ha)	Irrigated land (10^3 ha)	Irrigation potential (10^3 ha)	Potent/actual
Africa							
Angola	184	6.07	22	3,000	60	3,700	61.7
Cameroon	285	0.99	61	5,960		290	
Ethiopia	110	5.56	72	10,000	300	2,700	9.0
Madagascar	337	14.96	804	2,900	899	1,500	1.7
Mali	100	6.55	484	4,634	232	566	2.4
Mozambique	216	0.63	32	3,900	117	3,070	26.2
Niger	33.7	2.18	156	14,483	145	270	1.9
Nigeria	286.2	8.01	61	28,200	282	2,330	8.3
Sudan	154	37.32	1,030	16,233	1,786	2,700	1.5
Zambia	105	1.74	149	5,260	158	523	3.3
South America							
Argentina	814	29.19	753	27,800	1,390	6,200	4.5
Bolivia	622	1.44	157	2,928	117	2,000	17.1
Brazil	8,233	59.3	318	57,640	2,306	29,000	12.6
Colombia	2,132	10.71	235	2,818	564	6,600	11.7
Paraguay	336	0.49	80	2,850	57		
Uruguay	139	3.15	910	1,373	206	1,700	8.3
Venezuela	1,233	8.37	313	2,595	441	1,700	3.9
EU							
France	189	33.16	548	18,440	2,397		
Germany	188	38.01	460	11,804	472		
Italy	175	41.98	723	8,479	2,035		
Spain	111	37.22	864	13,400	3,300		
UK	160	11.75	197	5,876	176		

Sources: AQUASTAT (FAO), FAOSTAT, Gleick (2009).

increase food production at faster rates than those enabled by technical and scientific developments (less than 1% per year).

In Table 3 we report the most recent data on production and yields of three basic commodities (rice, corn and wheat). Yields in less developed countries are still far from the largest potential. The case of Sudan is noteworthy: while wheat yields are comparable to those of Spain and Italy, total production is extremely low. Apart from its serious political and civil conflicts, it is clear that, from its resource base, there is an extraordinary potential to increase food production in Sudan. The fact that Sudan sold 690,000 ha to private entrepreneurs helped by the government of South Korea attests for the potential of the country.

4 THE CASE OF SPAIN[1]

The objective of this part of the study is to assess the virtual water *trade* in Spain, differentiating the green (soil water) and blue (surface water and/or groundwater) components of virtual water from both a hydrological and economic perspective. As Spain encompasses very diverse agricultural regions, the combination of spatial and time dimensions offers a unique empirical setting to

[1]This section borrows from Garrido *et al.* (2010).

Table 3. Rice, corn and wheat yields and production in selected countries from Africa, South America and the European Union.

	Yields (kg/ha)			Production (10^6 t)		
	Rice	Corn	Wheat	Rice	Corn	Wheat
Argentina	7,061.0	5,902.9	2,623.4	1.193	14.446	14.663
Bolivia	2,651.2	2,607.0	1,111.2	0.446	0.894	0.144
Brazil	3,879.8	3,382.3	1,592.6	11.527	42.662	2.485
Cameroon	1,300.0	1,888.8	1,333.3	0.052	0.850	0.000
Ethiopia	1,868.8	2,640.4	1,904.0	0.012	4.030	2.779
France	5,533.0	8,585.6	6,740.7	0.095	12.902	35.367
Germany		8,030.6	7,200.6		3.220	22.428
Italy	6,277.1	8,728.5	3,729.5	1.419	9.671	7.182
Madagascar	2,699.4	1,500.0	2,380.9	3.485	0.495	0.010
Mali	2,553.3	1,730.0	2,422.9	1.053	0.707	0.009
Mozambique	619.8	852.0	1,052.6	0.099	1.418	0.002
Niger	3,708.2	812.5	1,499.2	0.078	0.007	0.008
Nigeria	1,483.3	1,818.1	1,126.9	4.042	7.100	0.071
Paraguay	3,000.0	4,878.0	2,191.7	0.126	2.000	0.800
Spain	6,799.1	9,743.6	2,875.4	0.724	3.356	5.522
Sudan	3,439.1	1,046.3	3,831.9	0.026	0.109	0.669
UK			8,036.5			14.747
Venezuela	4,950.2	3,334.1	301.3	1.123	2.337	0.000
Zambia	972.5	1,815.6	5,494.6	0.014	1.424	0.094

Source: FAOstat (www.fao.org).

determine whether virtual water *trade* can contribute to reduce water scarcity. In short, the main contributions of our study to the literature are: 1) the spatial and temporal analysis, as the study covers all Spanish provinces for the period 1997–2006; and 2) the econometric analysis of virtual water *trade*, making use of spatial and temporal variations of water scarcity value and irrigated land productivity.

Spain is considered a semi-arid country, but nationally it is relatively well-endowed with blue water resources (more than 2,000 m^3 of renewable resources per capita). Geographically, resources are abundant in the less-populated basins and scarce in the most populated regions, especially the Mediterranean arc and the South.

The water footprint of Spain has been estimated at about 45,000 Mm^3/yr for the period 1997–2006 (Garrido *et al.*, 2010). This, on a per capita basis, is about 1,100 m^3/yr per capita, suggesting that, despite the fact that Spain is usually classified as an arid country, its average is close to the global average of around 1,300 m^3/yr per capita. This figure is close to the global average of around 1,700 m^3/yr per capita, which also corresponds to a food supply need of 3,000 kcal/day per person, out of which 20% are animal products (Kuylenstierna *et al.*, 2008).

The largest water user, in line with global trends (although not necessarily developed countries of a similar level), is the agricultural sector. Considering both green and blue crop consumption and livestock water use, agricultural water use represents about 85% of the total water use, while it contributes about 3% of the GDP (26,000 million €) and employs 5% of the economically active population (1 million jobs).

There is a growing body of literature focusing on virtual water and the water footprint. However, few of these studies deal with the economic valuation of virtual water. From an economic perspective, only blue water is valued. Green water has certainly an economic value both for agricultural production and natural ecosystems. However, it is complex to attach an opportunity cost to green water since it cannot be easily allocated to other uses.

Table 4. Water scarcity values and scarcity levels.

River Basin	Provinces	Scarcity level	Scarcity value €/m^3	Volume stored (s) (in % over total storage capacity)
Duero	Ávila, Burgos, León,	1	0	s > 75.2
	Palencia, Salamanca,	2	0.06	63.2 < s < 75.2
	Segovia, Soria,	3	0.12	56.4 < s < 63.2
	Valladolid, Zamora	4	0.361	s < 56.4
Ebro	Álava, La Rioja,	1	0.01	s > 80.2
	Navarra, Huesca,	2	0.06	71.7 < s < 80.2
	Lleida, Zaragoza,	3	0.09	71 < s < 71.7
	Tarragona, Teruel	4	0.15	s < 71
Guadalquivir	Cádiz, Córdoba,	1	0.005	s > 66.2
	Jaén, Sevilla	2	0.1	46.2 < s < 66.2
		3	0.25	18 < s < 46.2
		4	0.96	s < 18
Guadiana	Ciudad Real,	1	0.033	s > 65.8
	Badajoz, Huelva	2	0.058	57.5 < s < 65.8
		3	0.137	16.8 < s < 57.5
		4	0.678	s < 16.8
Júcar	Castellón, Alicante,	1	0.07	s > 33.3
	Cuenca, Valencia	2	0.19	23.2 < s < 33.3
		3	0.35	18.6 < s < 23.2
		4	0.52	s < 18.6
Segura	Murcia, Albacete	1	0.12	s > 22.5
		2	0.27	19.7 < s < 22.5
		3	0.52	12.1 < s < 19.7
		4	0.61	s < 12.1

Source: Garrido *et al.* (2010).

For the purpose of this study, the economic value of blue water is defined in terms of shadow prices or scarcity values. Using the shadow price of water to measure the economic value of blue water seems consistent with the analysis of virtual water *trade* in arid and semiarid countries, where the distinction between green and blue water is essential to relate land and water management to drought and climate variability.

The shadow prices or scarcity value of blue water, as reported in Table 4, have been selected based on a comprehensive literature review. Blue water values are defined for each river basin and scarcity level. In this framework, each Spanish province is identified with a specific river basin, although the administrative and basin boundaries do not perfectly overlap. Blue water scarcity value varies depending on the scarcity level, which in turns depends on the volume of water stored in each river basin. Scarcity levels are defined on a scale from 1 to 4, being 4 the scarcest level. For each river basin, storage thresholds are defined based on a percentile analysis for the period 1997–2006. Thus, when in a certain year the volume stored in May is higher than the 50th percentile the scarcity level is 1. Scarcity level 2 corresponds with a volume stored between 25th and 50th percentiles. Scarcity level 3 is defined between 10th and 25th percentiles and the scarcity level 4 occurs when the stored volume is lower than 10th percentile.

4.1 *Econometric analysis*

Our data generation process allows for testing the hypothesis that the blue virtual water *exports* are dependent on water scarcity and land productivity. Basic economic theory would suggest that as water and land become scarcer, users would be more efficient.

Making use of the spatial and temporal variations of both water scarcity and land productivity, we can pose the following model, only relevant for irrigated agriculture:

$$BVWE_{it} = \alpha + \beta_1 SV_{it} + \beta_2 LP_{it} + \varepsilon_{it} \quad (1)$$

$BVWE_{it}$ denotes blue virtual water *exports* expressed in volumetric terms, that, is in 1,000 m^3 of blue water of province i and year t; SV_{it} represents the water scarcity value in €/m^3, which varies across years and basins, using the parameterisation shown in Table 4; LP_{it} is the land productivity of irrigated production in province i and year t, measured in €/ha of crop value.

The time-series and panel structure of our database can be best estimated using *Feasible Generalised Least Squares*, assuming heterocedastic, but uncorrelated panels (provinces).

$$\hat{\beta}_{GLS} = (X'V^{-1}X)^{-1}X'V^{-1}y \quad (2)$$

Where matrix V, with n being the number of provinces, is as follows:

$$V = \begin{bmatrix} \sigma_1^2 I & \dots & 0 \\ \vdots & \ddots & \vdots \\ 0 & \dots & \sigma_n^2 I \end{bmatrix} \quad (3)$$

Model [1] hypothesises that coefficient β_1 could be negative. Parameter β_1 measures the effect that the SV_{it} has on the blue virtual water *exports*. Model [1] permits both general estimations as well as regional estimations. This strategy will be pursued by estimating the model for all provinces, only for Mediterranean provinces and only for the inland provinces. That is, if we control for the geographic provinces, one would expect that as water becomes scarcer, provinces would *export* less virtual water in the form of farm exports.

In terms of the model's variables and crops' demand, these major *regions* differ in two essential aspects: a) the lower percentage of irrigated land in the inland regions than in the Mediterranean regions; b) the fact that water is scarcer in economic terms in the Mediterranean regions than in the inland regions.

β_2 could be either positive or negative. Positive (negative) means that higher irrigated land productivity would increase (reduce) the volume of blue virtual water *exported*. Note that the direction of the causality would either assume that as land becomes more productive more water would be virtually *exported* in the form of farm products, or that higher land productivity could be caused by scarcer water for irrigation so that less water could be *exported* in farm production.

Since both water scarcity and land productivity values differ among river basins, a set of dummy variables is introduced to explain as much of these inter-basin differences.

$$BVWE_{it} = \alpha + \beta_1 SV_{it} + \beta_2 LP_{it} + \sum_i \beta_{3i} SR_i + \varepsilon_{it} \quad (4)$$

In which SR_t controls for each river basin coefficient. There are a total of 12 basin variables in the model. Once the geographical differences are controlled by coefficients β_{3i}, model [4] allows for testing the hypothesis of whether larger scarcity permits lower values of blue virtual water *exports*.

4.2 *Virtual water flows*

Spain is a net virtual water *importer* country. In terms of volume, net virtual water *imports* amount to an average of 12,800 Mm^3 for the period 1997–2006. International trade data reveals that Spain exports high value crops, such as fruits and vegetables, and imports less valuable crops, such as cereals and industrial crops. This fact shows the importance of considering both volume and economic value of the virtual water *exchanged*.

Virtual water *imports* totaled 20,147 Mm^3 in the year 1997 and increased up to 29,150 Mm^3 in the year 2006 (Garrido *et al.*, 2010). A maximum of 32,500 Mm^3 was reached in the year 2005, which in terms of precipitation was also the driest year of the series. Even though farm trade responds

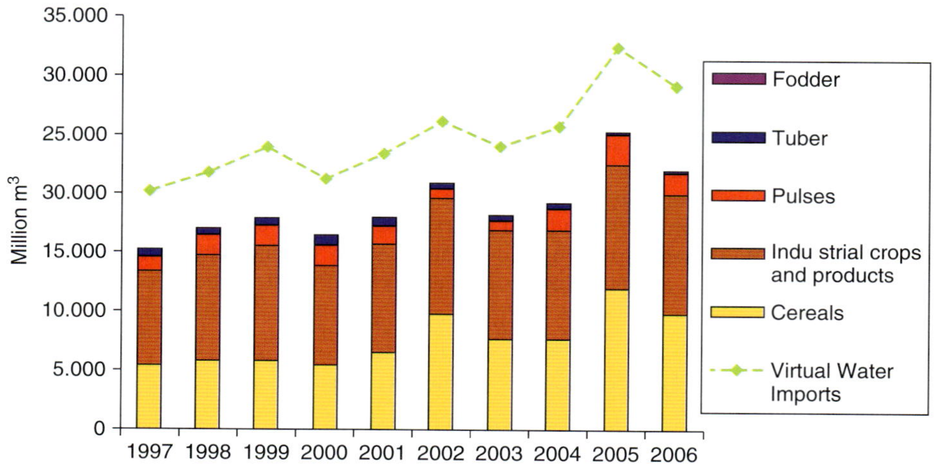

Figure 3. Virtual water *imports* related to livestock production.
Source: Own elaboration.

primarily to the relative prices and resources' productivity, variations in agricultural trade patterns might to some extent be explained by climatic variability.

The main groups of imported crops are cereals and industrial crops (and their products), which represent 70% of total virtual water *imports*. Major virtual water volumes are *imported* from France, Argentina, Brazil and USA, where primary crops are mainly cultivated under rainfed conditions. Therefore, their virtual water *exports* are predominantly green and consequently with a lower opportunity cost. The case of exports from France may be slightly different since maize is by far the most important irrigated crop in France. The Spanish imported maize could embed blue water resources that have a non-negligible cost.

Most virtual water *imports* are directly connected to the livestock sector (see Figure 3). Almost 100% of the soya cake consumption and 75% of cereals and pulses' consumption is used for animal feeding (MARM, 2008). Spanish meat production has grown from 3.6×10^6 to 5.8×10^6 t during the period 1992–2007 (*ibid.*).

According to official data, livestock direct water use is about 260 Mm^3 (MARM, 2008). However, Spain has virtually *exported* about 10,000 Mm^3/yr by means of animal product exports. During the 1997–2006 animal-related virtual water *exports* have experienced a steady growth, although *imports* have remained fairly stable. The swine sector expansion underscores the growth of *exports*, reaching a maximum of 4,500 Mm^3 in 2005. The bovine exports, second in importance, exhibit more variability. The sanitary and veterinary crisis experienced in the bovine sector explains its virtual water *trade* variability and its decline in the most recent years.

4.3 *Econometric analysis: Regression results*

Garrido *et al.* (2010) show that there are very large differences of green and blue water use across basins in agriculture. In addition, water scarcity varies also significantly across years, due to drought cycles. The question of whether virtual water *trade* increases or reduces water scarcity at regional level can be tested using a regression analysis with the cross-section and time-series data developed in this research.

We run a number of specifications using the styled model described above in equation [1]. Tables 5 and 6 summarize the main results. As hypothesized earlier, coefficient β_1 is significant and negative in Mediterranean regions, but non-significant and positive in the mainland provinces. Mediterranean blue virtual water *exports* are more responsive to changes in scarcity values than

Table 5. Blue virtual water *exports* in Mediterranean and Inland regions, period 1997–2006.

	Mainland regions			Inland regions		
	Coef.	Std. Err.	Elasticity ey/ex	Coef.	Std. Err.	Elasticity ey/ex
Scarcity value (β_1)	−226.4286*	39.9971	−0.3868	4.0493	10.7320	0.0096
Irrigated land productivity (β_2)	−6.2016*	0.7644	−0.3719	9.1597*	1.0612	0.8589
Constant α	201.5281*	10.5028	–	4.1305	2.8381	–
Number of obs.	190			220		
Number of groups	19			22		
Time periods	10			10		

$p < 0.01$*

Table 6. Blue virtual water *exports* by provinces, period 1997–2006.

	Coef.	Std. Err.	Elasticity ey/ex
Scarcity value (β_1)	3.7581	6.2427	0.0062
Irrigated land productivity (β_2)	−3.0540*	0.4018	−0.1832
Constant α	23.003*	1.254874	
Number of obs.	410		
Number of groups	41		
Time periods	10		

Basin (β_{3i})	Coef.	Std. Err.	Basin (β_{3i})	Coef.	Std. Err.
Ebro	55.8346*	3.4548	Tajo	8.5104*	2.0861
Guadalquivir	200.5525*	17.2891	Sur	115.9937*	7.7616
Guadiana	71.2109*	4.9076	Catalonia	1.1457	3.3966
Júcar	132.112*	7.1650	Canarias	44.5612*	5.0822
Segura	232.9477*	9.4501	Baleares	15.5284*	3.8057

$p < 0.01$*

inland regions, the elasticity being significant and different in both equations by more than one order of magnitude.

Our model also hypothesized that irrigated land productivity can have an impact on the blue virtual water *exports*. While this variable is significant in both models, the direction of the effect is negative in Mediterranean provinces and positive in the mainland provinces. This indicates that higher irrigated land productivity decreases the blue virtual water *exports* in the provinces where blue water is scarcer. In the inland regions, higher irrigated land productivity is generally explained by higher blue water availability and larger productions. In turn, more blue virtual water is *exported*. These findings suggest that the export-oriented Mediterranean provinces are generally more responsive to variations of water scarcity and land productivity than mainland provinces.

The estimation of model [3] provides a complementary interpretation. As we control for the basins, we indirectly control for the water scarcity levels. The resulting effect is that the scarcity value of water becomes insignificant, while the basins' controls become very significant, except for the internal basins of Catalonia. The geographical latent conditions –temperature and precipitation regimes– become more relevant than the time-variation of water availability and economic scarcity. This implies that these natural endowment factors have more explanatory power of the volumes of *exported* water than the scarcity conditions prevailing in each region and year. So one can conclude that virtual water *trade* does not aggravate water scarcity, which is in fact caused by the greater competitive advantage of those regions with better natural endowments. Furthermore, we see a higher response of blue water *exports* to changes in irrigated land productivity. This means that it is the allocation of land and water that influences more the amount of *exported* water in each province, than the water scarcity component.

5 CONCLUSIONS

The Spanish water economy, like most developed and many emergent and developing countries', has become increasingly globalised. About half of all imports come from the southern hemisphere, and its virtual water *trade* in terms of *exports* and *imports* actually accounts for a volume larger than the water actually used in the country. Exports have been effectively decoupled from the fluctuations in water availability for two main reasons: first, animal product exports rely primarily on imported products (animal feed from cereals and soy), which complement national production. Second, exports originating in Mediterranean provinces show little variation in total water use.

Not all traded m^3 are equally valuable: as a general rule, green ones are less *valuable* than blue ones. There are years when water in exactly the same location is twice as valuable as in other years because surface reservoirs storage levels are lower due to drought cycles. Nevertheless, these results will need to be reassessed when reliable data on the role of groundwater storage are available. This is a gap in Spain, as well as in most countries.

At a global scale, it is essential that water values are presented jointly with virtual water *trade* economic evaluations. Even if gathering this information is a significant challenge, existing water stress indicators that are currently recorded at national or basin levels (see CAWMA, 2007) can be used to help placing a monetary value to virtual water *traded* globally. Based on the work carried out by Hoekstra & Chapagain (2008) at the global level, it is likely that the value of virtual water *trade* is much more significant than originally estimated because water savings could translate into euro savings as well.

Water policies must be placed in a global context. Water users in semi-arid countries need a way to internalise the international crop prices, in competition with foreign producers. The EU policy has taken a bold step by pushing to integrate water policy as a requirement of the new *Health Check* of a revised Common Agricultural Policy. Spain has seen large increases in farm water productivity, especially in the provinces where land and water were once less productive. A main reason is that farmers have been able to open their farms increasingly to global competition, whilst simultaneously reducing their dependence on subsidies that targeted specific crops over others, therefore distorting market signals and isolating farmers from key market information of shifts due to increased globalisation. Spain has in fact saved billions of m^3 by opening its farming and livestock industries to world market opportunities, and has been able to offer foreign consumers competitively priced products. Farm trade has in turn helped the Spanish rural economy to cope better with droughts, and it will also help the country face the upcoming challenge: to change the now outdated motto *more crops and jobs per drop* to *more cash and care nature per drop*.

This chapter also examined the hypothesis of whether virtual water *trade* aggravates water scarcity in the most competitive and exporting regions. Instead of looking at nation-wide trade, we scaled down the analysis to examine the regional and time differences of virtual water *exports* based on the variations of both water scarcity and irrigated land productivity. The findings show that virtual water *exports* do not respond to changes in water scarcity, but essentially to the natural and capital endowments of the provinces. So we conclude that farm trade, and the virtual water *trade* that comes with it, adds a degree of latent pressure to the water resources of the exporting provinces. But farm exports show very little response to variations of economic water scarcity, and seem to evolve quite invariably to the variations of water availability and economic value. To ensure that virtual trade offer robust policy prescription, water should be adequately priced in the exporting countries. By taking into account the varying scarcity value of water, commodity exports would internalize its value and the trade regime would be consistent international trade postulates.

The economics of virtual water trade is still in its infant stage. Despite the comprehensive evaluations of Hoekstra & Chapagain and Hong & Zehnder, we know very little about the actual value of the water that is used in the exporting countries. While it is known that most of it is green water –certainly with less economic value than blue water–, there are initiatives in China and Turkey which are meant to use blue water for the production of low-value commodities.

What the virtual water *trade* literature has not yet come to grips is the risk that food crises, like the one of 2007 and 2008, represent for food importing countries. Possibly one of the main reasons is that an adequate analysis is still missing on the main causes of that crisis (corruption, oligopolies, biofuels, oil price increase, economic crisis and others). Those wealthy enough paid the prices of the commodities as asked in the international markets, and more recently resorted to purchase and lease suitable land with accessible water resources. The poorest among the poor did not and will not have that option. And yet, ironically those countries selling the land stand among the poorest in the world. This means that with capital and know-how they should be able to feed themselves. It is not the lack of water resources that explain the world's hunger and malnutrition.

REFERENCES

Alcamo, J.; Döll, P.; Henrichs, T.; Kaspar, F.; Lehner, B.; Rösch, T. & Siebert, S. (2003). Development and testing of the WaterGAP 2 global model of water use and availability. *Hydrological Sciences*, 48(3), 317–337.

Aldaya, M.M.; Hoekstra, A. & Allan, J.A. (2008). *Strategic importance of green water in international crop trade*. Value of Water Research Report Series No. 25, March. UNESCO-IHE; Institute for Water Education. Delft, the Netherlands.

Allan, J.A. (2003). Virtual Water – the water, food, and trade nexus useful concept or misleading metaphor? *Water International*, 28(1): 4–11.

Bates, B.C.; Kundzewicz, Z.W.; Wu, S. & Palutikof, J.P. (eds.) (2008). *Climate Change and Water*. Technical Paper of the Intergovernmental Panel on Climate Change, IPCC Secretariat, Geneva, Switzerland. 210 pp.

Berrittella, M.; Hoekstra, A.Y.; Rehdanz, K.; Roson, R. & Tol, R.S.J. (2007). The economic impact of restricted water supply: a computable general equilibrium analysis. *Water Research*, 42: 1799–1813.

CAWMA (Comprehensive Assessment of Water Management in Agriculture) (2007). *Water for Food, Water for Life: A Comprehensive Assessment of Water Management in Agriculture*. D. Molden (ed.). Earthscan, London, UK; and International Water Management Institute, Colombo, Sri Lanka.

Chapagain, A.K.; Hoekstra, A.Y. & Savenije, H.H.G. (2006a). Water saving through international trade of agricultural products. *Hydrology and Earth System Sciences*, 10: 455–468.

Falkenmark, M. & Rockström, J. (in press). Integrating agricultural water use with the global water budget. In: A. Garrido & H. Ingram (eds.), *Water for Food in a Changing World*. Routledge.

FAO (2009). *Aquastat*. [http://www.fao.org/nr/water/aquastat/main/index.stm].

Formas (ed.) (2008). *Water for Food*. Sweedish Research Council (Formas). Stockholm, Sweden.

Fraiture, C.; Cai, X.; Amarasinghe, U.; Rosegrant, M. & Molden, D. (2004). Does international cereal trade save water? The impact of virtual water trade on global water use. Comprehensive Assessment Research Report 4. Comprehensive Assessment Secretariat, Colombo, Sri Lanka [online]. Available from: [http://www.iwmi.cgiar.org/Assessment/FILES/pdf/publications].

Garrido, A. & Dinar, A. (eds.) (2009). *Managing Water Resources in a Time of Global Change: Mountains, Valleys and Flood Plains*. Routledge Publishers.

Garrido, A.; Llamas, M.R.; Varela-Ortega, C.; Novo, P.; Rodríguez-Casado, R. & Aldaya, M.M. (2010). *Water footprint and virtual water trade of Spain: policy implications*. Springer, New York, USA.

Gleick, P. (2009). *The World's Water: 2008–2009. The biennial report on freshwater resources*. Island Press, Washington, D.C., USA.

Hoekstra, A.Y. & Chapagain, A.K. (2008). *Globalization of water: Sharing the planet's freshwater resources*. Blackwell Publishing. Oxford, UK.

Hoff, H.; Falkenmark, M.; Gerten, D.; Gordon, L.; Karlberg, L. & Rockström, J. (in press). Synthesis article: Greening the global water system. *Journal of Hydrology*.

Kuylenstierna, J.; Destouni, G. & Lundqvist, J. (2008). Feeding the future world – securing enough food for 10 billion people. In: Swedish Research Council Formas (ed.), *Water for Food*. Formas, Stockholm, Sweden: 9–22.

MARM (2008). *Agricultural and statistics yearbook*. Spanish Ministry of the Environment, and Rural and Marine Affairs [online]. Available from: [http://www.mapa.es/es/estadistica/pags/anuario/introduccion.htm].

Novo, P.; Garrido, A. & Varela-Ortega, C. (2009). Are virtual water "flows" in Spanish grain trade consistent with relative water scarcity? *Ecological Economics*, 68: 1454–1464.

Oki, T. & Kanae, S. (2004). Virtual water trade and world water resources. *Water Science and Technology*, 49 (7): 203–209.

Rosegrant, M.; Cai, X. & Cline, S. (2002). *World water and food to 2025*. International Food Policy Research Institute (IFPRI). Washington, D.C., USA.

Roth, D. & Warner, J. (2008). Virtual water: Virtuous impact? The unsteady state of virtual water. *Agricultural and Human Values*, 25: 257–270.

The Economist (2009). *Outsourcing's third wave*, May 21st.

UNESCO (2008). *El agua, una responsabilidad compartida*. 2º Informe de las Naciones Unidas sobre el Desarrollo de los Recursos Hídricos el Mundo. Estatal Expoagua, Zaragoza, Spain.

Verma, S.; Kampman, D.A.; van der Zaag, P. & Hoekstra, A.Y. (2008). Going against the flow: A critical analysis of inter-state virtual water trade in the context of India's National River Linking Program. *Physics and Chemistry of the Earth*, 34(4–5): 261–269.

von Braun, J. & Meinzen-Dick, R. (2009). "Land Grabbing" by Foreign Investors in Developing Countries: Risks and Opportunities. IFPRI Policy Brief, 13. April.

Wichelns, D. (2004). The policy relevance of virtual water can be enhanced by considering comparative advantages. *Agricultural Water Management*, 66: 49–63.

Yang, H. & Zehnder, A.J.B. (2007). Virtual water: An unfolding concept in integrated water resources management. *Water Resources Research*, 43, W12301.

CHAPTER 9

Water commoditization: An ethical perspective for a sustainable use and management of water resources, with special reference to the Arab Region

Magdy A. Hefny
Regional Center for Studies and Research of Water Uses Ethics, Ministry of Water Resources and Irrigation, Shubra El Kheima, Egypt

ABSTRACT: The theme of the chapter is controversial since there is no clear cut argument to advocate whether water could be treated as a marketable commodity, or as a common good and a fundamental human right "Re-thinking water and food security" presents a good opportunity to review and reflect on this question and what could be a balanced integrated approach. The chapter has chosen to investigate and analyze the current trend of water commoditization in the Arab Region from an ethical perspective. The main objective is to call for a *water strategy*, that emphasizes basic ethical principles, and help improve on the design of the water and food security system. Water shortage and scarcity is becoming a major problem the world will face in the coming decade, and especially in relation to the Arab Region. The problem has been exacerbated by natural and manmade factors such as aridity, high population growth, degradation in water quality, institutional and policy aspects, public involvement and awareness, and other socio-economic complicating factors. There are many negative and positive effects due to the shift to *neo-liberal market-led economies*, which have surfaced in the past decade and a half throughout the Arab Region and the developing world at large. The question is turning towards how it will possible to mitigate the negative effects of increased levels of poverty and worsening environmental degradation.

Keywords: commoditization, pricing, ethical framework, sustainability, social equity, efficiency, virtual trade

1 INTRODUCTION

Among Earth's natural resources, water has always been perceived as a special natural resource essential for the whole existence of life. Globally, most of freshwater is used for agriculture (69%), while industrial use accounts for 23% and domestic human use accounts for just 8% of the total withdrawals. However, the Earth's hydrologic cycle has been profoundly altered and various components of this cycle are being torn apart and exploited making freshwater ecosystems severely fragile and threatened. The consequences are devastating to the environment, with serious and irreversible threats to aquatic ecosystems. Moreover, these activities have had severe social impacts in several regions around the globe, such as the loss of livelihoods and human displacement.

The Arab Region is no exception. The situation is becoming no longer sustainable, due to the high costs and negative environmental consequences associated with the deterioration of water resources and the resulting deteriorating livelihoods in the region.

The chapter – apart from the Introduction– addresses the following parts: Section 2 presents the various *aspects of water commoditization*, the debate on whether water is a commodity or a common good to be fairly distributed, our stand in this debate, and the premises of commoditization and marketization of water, as shown in the emerging processes of: a) coining water rights; b) having water pricing schemes, cost recovery approaches; and c) water privatization

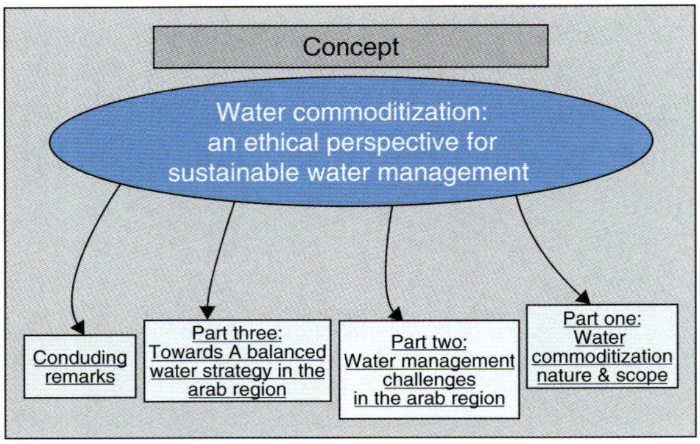

Figure 1. Concept of the Chapter.

and/or involving private sector investments in its development; as well as d) water in relation to trade which is called virtual water. Section 3 illustrates *water management challenges* in the Arab Region, giving due consideration to the water complexity and uncertainty question, major water challenges or *the gaps* in the development of the Arab water and agricultural sector, and reforms needed. Section 4 is calling for a *balanced water strategy* for the Arab Region. The strategy would be based on a comprehensive, integrated system approach that embraces the interface between major elements of the system: environment-technology-human, with emphasis on the basic ethical principles. This may improve on the design of the water and food security system, in relation to the Arab Region. Section 5 gives concluding remarks. Figure 1 shows the concept applied in the chapter.

2 WATER COMMODITIZATION: NATURE AND SCOPE

2.1 *Introduction*

It is usual to view water by nature as a common good of mankind. Water resources – both surface and groundwater – occur naturally and provide an open access to all. Thus water is considered *as a common good*, especially since it is an essential requirement for life. A common good is defined by three characteristics: 1) Non-rivalrous – one person does not deprive others from using it; 2) non-excludable – impossible to restrict others from using it; and 3) users are non-rejectable to use it – individuals cannot abstain from its consumption even if they decide to (Bannock *et al.*, 1987).

At the United Nations, the *Universal Declaration on Human Rights* was declared in 1948, where water was considered as a fundamental right, which is guaranteed an international legislation. The right to water was only expressed in the *Convention of the Rights of the Child* (Gleick, 1999). However, in 2002, the United Nations Committee on Economic, Cultural and Social Rights issued a General Comment, which openly expressed that "the concept of water management is not to be limited to its economic dimension only, and that access to water should be considered as a human right". The *Comment* clearly states that everyone is entitled to sufficient, affordable, accessible, safe and acceptable water for the purpose of personal and domestic usage (UNESCO, 2003).

However, the debate around water as a commodity started with the issuing of *The Dublin Statement* in 1992[1].

[1] A statement was issued in Dublin in January 1992 to emphasize that water should be recognized as an economic good. This was in preparation to the Rio Conference in 1992, and out of this preparatory meeting came *the Dublin Principles*.

- Freshwater is a finite and vulnerable resource, essential to sustain life, development and the environment.
- Water development and management should be participatory –involving users, planners, and policy makers at all levels.
- Women are central to providing, managing, and safeguarding water.
- Water has an economic value in all its competing uses and should be recognized as an economic good.

Meanwhile, the trend has evolved towards increased commoditization and marketing of water, and water rights are presently being examined and implemented in the UK and Australia. Policy and regulatory frameworks of these countries and others have taken example of *the American West* for more leadership in marketing of water rights (Turner, 2004)[2].

The question that raises itself here: will water become a commodity like oil one day in this 21st century? In my view, water is essential for life, whereas oil is not. Oil is fungible and can be replaced, but water cannot. Oil is the main focus in the globalization of the world economy, free trade, and market forces. Oil is favouring mechanisms of privatization and pricing in the debate, but water is life and it is a social and economic resource that needs to have special considerations for the poor and the marginalized people in society.

According to Hardin (1968), over-exploitation of a common good occurs due to the phenomenon called the *tragedy of the commons* where users ignore the impacts of using the resource on its current and future availability for other uses and users. Economists' prescription to solve the *tragedy of the commons* is to clearly define private water rights and establish water markets.

This can be difficult since basic water needs should be universally provided for all, and consequently water is not a typical private *economic good*. Private institutions working only under market rules cannot provide such goods because, once produced, they benefit the whole public and not specific individuals (Johansson *et al*., 2002). The only complete and viable solution to prevent the *tragedy of the commons* is by full recognition of the above-explained concepts of human rights, equality, and solidarity by water decision makers, managers, and users through Integrated Water Resources Management (IWRM) approaches within a societal ethical framework for good and rational water governance.

2.2 *The trend and its development*

The trend now is increasingly to deal more and more for dealing with water as a tradable good, something for sale. The privatisation of water services, that is taking place on a global scale, has opened the door for the concept of water as a service, and as a good. In this context, water is no longer perceived as a gift from God (or the State), but a commodity for which one has to pay. The following development emphasizes this trend:

2.2.1 *Bilateral water deals are concluded*

On March 25, 2004, Israel and Turkey concluded an agreement for the sale of 50 Mm^3/yr of water for 20 years. The water will be taken from the Manavgat River in Turkey, purified in Turkey and then transported in converted oil-tankers to Ashkelon in Israel. This shows that the idea of importing fresh water is not an academic one. In the future, this trend will, no doubt, increase due to increasing water scarcity worldwide. Jonathan Peled, spokesman for the Israeli Foreign Ministry said that "this landmark agreement turns water into an internationally accepted *commodity*".

Water is sold as a *good* by Lesotho to South Africa at the rate of 88 m^3/sec. Iran has finalized a US$ 2,000 million deal with Kuwait (June 2001) to pipe water from Northern Iran to Kuwait. Israel negotiated the price of water to be tankered from Manavgat in Turkey to Ashkelon. International tankering of water takes place in the Caribbean and the Philippines.

---

[2] Dr. M. Turner operates the Water Bank website [www.waterbank.com].

Intense controversy exists in Canada over the exportation of bulk water, but as a member of NAFTA and the WTO, Canada has to abide by the same rules, since both Organizations consider water to be a tradable good like any other good, and cannot claim that water is not a good. To cite here Maude Barlow (2001), has made it clear that "the cat has been out of the bag for years"[3]. And *Global Water Corporation* of Canada has contracted to ship 58 Mm^3/yr of Alaskan glacier water by tanker to be bottled in a free trade zone in China. They openly concluded that the venture "will substantially undercut all other imported products" because of China's cheap labor.

2.2.2 Staggering debate at the WWF 5[4]

The advent of Fifth World Water Forum at Istanbul in Turkey, March 2009, represented a kind of conferences diplomacy that brought together intergovernmental groups, ministers level, and parliamentarians, as well as non-governmental organizations on a joint summit. This time, the set-up of the crowded agenda and program consisted of six themes. The main theme of WWF5 was *Bridging Divides for Water* and under this theme, there were six sub-themes[5].

The Forum ended-up with a *Public Declaration*, which called for water conservation, especially for agriculture, as well as water harvesting and recycling of agricultural drainage and for clean water for health. The *Declaration* is not binding and therefore, is regarded as recommendations, which could be considered or neglected by decision-makers. France, Spain and many countries of Latin America tried to introduce an amendment to the Declaration, and to add a sentence: "access to safe drinking water was a human right rather than a need", but without success, to the degree that led some twenty countries to sign a statement of protest, including France, Spain, Switzerland, South Africa and Bangladesh.

The Declaration was also criticized by French Ecology Secretary, Chantal Jouann, who has announced and described this as a lack of political will, which is necessary for ensuring adequate water for drinking, at a time when about 80% of all diseases in developing countries are due to contaminated water.

There was another parallel Forum represented by *The Alternative People's Forum* that includes organizations of the rural poor, as well as organized labour movements. They have pronounced and aired their views on radio and television, as follow[6]:

a) Demonstrating and shouting "water for life, not for profit". Two activists from the non-governmental organization *International Rivers* were deported after holding up a banner just before the conference began that read "No Risky Dams?" The security was tight, with Turkish police was firing tear gas and detaining protesters.
b) Criticized the official World Water Day, during the Forum, as an non-inclusive, corporate-driven, event, a fraud pushing for water privatization and called for a more open, democratic and transparent forum.
c) The Forum is a big trade show, where decisions are made about who gets water and who doesn't get water. Strategies are developed and that's what it is all about. "They'll put on sessions on gender and water, but they don't mean any of it. This is really about one development model for water, and that is the privatization model. And that is what they are promoting, and that is what their consensus is, and they refuse to include the notion of the right to water. They set the agenda for the world.

[3] Maude Barlow is the National Chairperson, Council of Canadians.

[4] 25,000 of experts have participated in the forum in Istanbul (Turkey), which is considered to be the largest number of the forums witnessed for years. This is representing various sectors dealing with water, other than the governmental and non-governmental organizations, including politicians, parliamentarians and officials from drinking water in rural and urban areas, journalists and various media outlets.

[5] The six sub-themes are: 1) Global Changes & Risk Management; 2) Advancing Human Development and the MDGs; 3) Managing and protecting water resources to meet human and environmental needs; 4) Governance and management; 5) Finance; 6) Education, Knowledge and Capacity Building.

[6] Democracy Now! [democracynow.org], "The War and Peace Report".

d) Privatization is not more efficient. It's more expensive. It causes more environmental problems. And the incentive is not conserve water, but to use as much water as possible.

Maude Barlow[7] stated during the wrapping-up of the Forum, that "the World Water Forum is bankrupt of ideas and have no new ways to address the growing water crisis in the world, because they have maintained an adherence to an ideology that is not working, that has dramatically failed. What is happening here is no longer about the World Water Forum. So we will be less concerned – I mean, if they want to go to Marseilles, let them go to Marseilles next time. It won't matter. It really won't matter. The change has been here. It's been a transfer of power".

2.2.3 *At Stockholm water symposium*

At the 10th Stockholm Water Symposium (August 2000), participants came to a conclusion and agreed in their statement, that "a growing movement of people believe that the imperatives of economic globalization –unlimited growth, a seamless global consumer market, corporate rule, deregulation, privatization, and free trade– are the driving forces behind the destruction of our water systems. These must be challenged and rejected if the world's water is to be saved".

The human race has taken water for granted and massively misjudged the capacity of the earth's water systems to sustain the demands made upon it. Our supply of available fresh water is finite and represents less than 0.5% of the world's total water stock. 31 countries are facing water stress and scarcity and over 1,000 million people lack adequate access to clean drinking water.

In their consensus statement, participants recognized the terrible reality that "by the year 2025, as much as two-thirds of the world's population will be living with water shortages or absolute water scarcity". They have also acknowledged that "instead of taking great care with the limited water we have, we are diverting, polluting, and depleting it at an astonishing rate, as if there were no reckoning to come. But there is profound disagreement among those in the *water world*, around the nature of the threat and the solution to it".

3 PREMISES OF WATER COMMODITIZATION: GLOBAL AND IN THE ARAB REGION

3.1 *Economic globalization*

Economic globalization integrates the economies of nation-states into a single unified market and carries industrial production to new levels. It intensifies natural resource exploitation and exacerbates every existing environmental problem. The imperative of globalization is unlimited growth, making it impossible for participating countries to make preservation a priority (Maude Barlow, Council of Canadians). Economic globalization has also resulted in the exponential increase in the use of fossil fuels, dams and diversions, massive transportation systems needed to carry out global trade, and roads carved out of wilderness.

In the new economy, everything is for sale, even those areas of life once considered sacred, like seeds and genes, culture and heritage, food, air, and water (Maude Barlow). As never before in history, the public space, the vital commons of knowledge and our natural heritage, have been hijacked by the forces of private greed. It was often stated, on different occasions, that "given the current corporate practices, not one wildlife reserve, wilderness, or indigenous culture will survive the global economy. We know that every natural system on the planet is disintegrating. The land, water, air, and sea have been functionally transformed from life-supporting systems into repositories for waste. There is no polite way to say that business is destroying the world" (*ibid.*).

[7] Barlow is the senior adviser on water issues to the United Nations General Assembly and chair of the Council of Canadians, speaking out against the World Water Forum in Istanbul, Turkey.

3.2 *Water rights*

In general water right is frequently considered to be *a legal right* to abstract and use a quantity of water from national source (river, stream, and aquifer). Water rights may confer a legal right to impound or store special quantity of water in natural source behind hydraulic structure for abstraction or hydropower generation. Those and other activities are generally regulated by *water rights regime* and specific legal and administrative rules.

Beside water rights based on abstracting and using of water from natural sources (surface or groundwater), there are other categories of *Water Rights*, such as *contractual water rights*, such as giving legal rights to receive a quantity of water at a specified time from e.g. an irrigation canal, generally in return for payment. There are few advocates of tradable water rights, water sales and leasing, and other market based concepts, which are not common and are rarely used (except cases like e.g. Chile and New Zealand).

In areas with plentiful water and few users, such systems are generally not complicated or contentious. In other areas, especially arid areas where irrigation is practiced, such systems are often the *source of conflict*, both legal and physical. Some systems treat surface water and groundwater in the same manner, while others use different principles for each.

3.3 *Water pricing mechanisms*

Policymakers have been compelled to try new approaches to improve the management of water resources. Among these approaches are measures of *water pricing reform* aiming at encouraging water conservation and improving the efficiency of water use (Dinar, 2000). This is due to the increasing growth of world population, and the unequal spatial and temporal global distribution of water. By 2025 nearly 3,000 million people may be living in *water stressed* countries. Furthermore, nearly 1,000 million of the world population will be living in the Arab Region with less than 650 m^3/yr of water available per person, which is considered a severe water stress situation by all standards (Johansson *et al.*, 2002; Postel, 1996). Yet, the notion of an optimal water-pricing policy does not command consensus among economists and policymakers, as there is still disagreement regarding the appropriate means by which to derive the *right* water price.

3.3.1 *Basic assumptions for water pricing*

It has been argued that the *optimization of water allocative efficiency*, which is an effective tool to control water wastage, would maximize benefits to society from the available water supply, using water pricing mechanisms to enhance the efficiency of water use. However, ethical concerns for *human equality* and *fairness* of water allocation across economically disparate groups in society, could necessitate the acceptance of sub-optimal allocation efficiency for the sake of sustaining livelihoods and expanding access of the poor to water services (Dinar & Subrumanian, 1997). Water-pricing mechanisms do not aim at *redistributing income*, but in general it is in the governments' natural interest to increase water availability for all sectors of their citizenry. Consequently, based on ethical considerations of *solidarity*, it could be necessary to provide differing pricing mechanisms to account for disparate income levels so that affluent sectors of society would in effect subsidize the water supply for poor sectors.

An important factor also in pricing water is that, in principle, *long term sustainability* of water supply services is achieved only by fully recovering all the incurred costs of infrastructure, operation, maintenance, administration, and development of the needed facilities (Budds & McGranahan, 2003). Consequently, proponents of *water pricing cost-recovery mechanisms* claim that efficiency gains benefit all service users, including the poor who will gain by becoming connected to a reliable and sustainable system. The ability of governments or other service providers would be improved, with cost-recovery gains, to provide water and sanitation services for the poor. According to Abu-Zeid (2001), huge investments have been made globally in terms of physical water supply infrastructure, as well as institutional management organisations, to ensure access to water and sanitation. It is therefore very important to sustain these facilities through *water pricing cost-recovery mechanisms* in order to continue the provision of water and sanitation services, particularly for the poor and marginalized populations that lack any political clout.

3.3.2 *Water pricing in the Arab region*

The value of water in the Arab Region is given in Islam and relevant Islamic traditions and customary laws and should be duly respected. As water is vital for life, it is considered as a basic human right for all, its value should not be in economic terms only, but should also reflect the social, environmental, cultural, and religious morals placed on it. With respect to cost-recovery, different mechanisms can be adopted that are suited to the administrative, social, and economic conditions of various Arab societies into which water pricing is applied.

According to ESCWA (2003b), the implementation of an efficient water pricing policy requires the modification of existing water laws and regulations, taking into consideration water market requirements, privatization schemes, and enforcement mechanisms. However, in order to ensure equitable cost sharing, important considerations should be implemented regarding socio-economic standards and income levels. Moreover, ethical measures should be taken into consideration, such as respecting the human right to water, especially when it comes to lower income groups, as well as inclusiveness in water and sanitation services and transparency in water pricing mechanisms. Currently, water-pricing schemes in some Arab countries are not leading to water conservation and improved water use efficiency. Setting water prices at lower than production costs is still common in some of these countries. In Lebanon for example, the government still subsidizes water supply for agricultural, domestic, as well as industrial sectors (Abdurazzak & Kobeissi, 2002). In fact, in the irrigation sector, water is still highly subsidized and provided at a minimal cost in several Arab countries, in spite of the widespread water over-consumption in the agricultural sector throughout the region (ESCWA, 2003b).

Nevertheless, other Arab countries are already implementing efficient water pricing schemes, which take *ethical issues* of societal solidarity and equitable access to basic needs of water into consideration. In Jordan, for example, a water pricing scheme has been implemented using the concept of the *lifeline rate schedule* (ESCWA, 2003b). The scheme is designed to recover the service costs while keeping *lifeline* basic needs affordable for the poor. While cost recovery was included in the water *tariffs* to cover the operation and maintenance costs, these tariffs were structured in a way that guarantees the needed household minimum consumption at a subsidized fixed price. Subsidies are recovered from households with higher water consumption assuming that such consumers are wealthier. These tariffs have achieved basic financial objectives for the water authority and reduced the general demand for water. However, one main shortcoming in the Jordanian system is that it does not differentiate between urban and rural users, and no effort for raising awareness have been undertaken to address issues of social resistance (ESCWA, 2003b; Taha & Bataineh, 2002).

Other Arab countries also apply a *progressive block rate for water pricing*. In Tunisia, a progressive and selective pricing for drinking water has proved efficient in cost recovery, as well as a means to enhance equitable water use. Nevertheless, the system is not without limitations. The Tunisian tariff system has brought pressure to bear only on major consumers, while sparing –to a great extent– small and average consumers. As a result, the small and average consumers have been paying a subsidized price for water with only a weak water conservation incentive. It is important to not totally sacrifice water demand management considerations for the sake of equitable access, in order to induce commitment to water conservation among all users (Linam, 2002).

3.4 *Water privatization in the Arab region*

Concerns over transparency, reduction of costs, and productivity improvement have been addressed, albeit most of the time unsuccessfully through privatization plans in the Arab Region. On the other hand, as mentioned above, the commoditization of water in general is fraught with social and ethical problems, since water pricing mechanisms based on full cost recovery and total removal of subsidies lead mostly to unfair distribution of service costs. As such, lower-income people suffer the most from privatization schemes that do not account for social considerations and, consequently, do not provide for full access to water and sanitation.

In recent years, policy makers in most Arab countries have chosen privatization as a strategic decision that involves major reforms in line with overall structural adjustment programmes aimed at reducing budget deficits and meeting increasing demand for water and sanitation services. As such, privatization contracts are being implemented in Gaza, Jordan, Lebanon, Qatar, and Yemen; while in Bahrain, Egypt, Kuwait and Saudi Arabia, privatization agreements are being seriously considered (ESCWA, 2003b).

In Jordan, Lebanon, and Yemen, privatization has been initiated in the past few years to address the pertinent water supply problem. In Jordan and Yemen, increasing water demand has led to mining of groundwater resources; whereas in Lebanon, increasing water demand is expected to render the water balance negative, if present consumption and management practices continue. In all these countries, enabling legislative and institutional measures have been taken to promote the implementation of privatization schemes for water and sanitation (ESCWA, 1999b).

However, the region still suffers from many problems related to water privatization. Most of the countries have relatively little know-how in privatization planning and implementation. This means that significant *capacity building* of local institutions and expertise is needed at both the national and regional levels, if positive results are sought through privatization. Moreover, knowledge and information regarding the expected results of the chosen privatization scheme(s) should be made clear and available for local communities and societies throughout the region in order to avoid any marginalization or distrust among the public.

Moreover, while in some countries of the region there is strong political will and acceptable public adaptation to implementing water privatization, other countries are facing severe resistance to this process. Opponents to privatization are accusing the governments of selling off public assets cheaply and putting responsibility for a vital scarce resource in the hands of the private sector. Therefore, *dialogue with stakeholders is necessary* to create a better understanding of the issues at stake and to mobilize local support, which is necessary for the success and sustainability for the privatization of water service delivery (ESCWA, 2003b).

In general, countries opting for the privatization of water service delivery in the Arab Region should stress *transparency* and *effectiveness* in setting up the needed institutional frameworks. It is necessary also *to recognize social and ethical considerations* while developing the needed legal and regulatory frameworks to guide the implementation of privatization schemes. In addition, it is very important to achieve full coordination and consultation among various ministries and water institutions and authorities, as well as civil society structures. This allows for greater transparency and justice both during the planning and implementation processes, whether in utility selection, contract negotiations, monitoring of the bidding process, or the performance of private investors (Budds & McGranahan, 2003; ESCWA, 2003b).

Coordination and cooperation amongst various public and private sector institutions and other stakeholders is necessary to make these preparatory steps. These are certainly worthwhile in order to bring about a successful water and sanitation delivery operation that respects the human right for water, takes into consideration the socio-economic needs of various communities at stake, and protects the wise use of water, for it is certainly a scarce resource in the Arab Region (ESCWA, 2003b).

3.5 *Water in relation to trade (virtual water)*

The concept of *virtual water* has been introduced by Tony Allan in the early 1990s (Allan, 1993; 1994). It took nearly a decade to get global recognition of the importance of the concept for achieving regional and global water security. The first international meeting on the subject was held in December 2002 in Delft, the Netherlands. A special session was devoted to the issue of virtual water trade at both the Third and Fourth World Water Forums (Hoekstra, 2003). The *virtual water* content of a product, as often defined, is the volume of water used to produce the product, measured at the place where the product was actually produced (production site specific definition).

The virtual water content of a product can also be defined as the volume of water that would have been required to produce the product in the place where the product is consumed (consumption site specific definition) (Hoekstra & Chapagain, 2004).

It is not intended, here, to discuss the concept of virtual water nor the method used to calculate virtual water content of products. The concern is with the possibility of using the concept as a *planning tool* and to present issues that need to be tackled before considering virtual water trade as a policy option. This would be explained next in articulating *a water strategy* for the Arab Region that emphasizes basic ethical principles and help improve on the design of the water and food security system.

4 WATER MANAGEMENT CHALLENGES IN THE ARAB REGION

4.1 *Introduction*

Renowned water sciences institutes and eminent scholars (Abu-Zeid & Hamdy, 2004; El-Kady, 2003; ESCWA, 2003a) in their recent studies concur on certain factors that shape the Arab water crisis, especially in its management dimension. On the governance scale, Arab countries have been suffering from serious capacity gaps at various levels that hinder their ability to face the social, economic, and political challenges needed in water management. The general lack of familiarity with participatory and integrated management approaches; fragmented institutional structures; inadequate pricing schemes; imbalanced sectorial water allocation; persistence in resorting to expensive water supply augmentation projects; and centralized administrative structures are some of the problems facing effective water governance in the Arab Region.

In fact, some Arab countries have already embarked on the implementation of IWRM approaches in response to this current situation. It should be noted, nevertheless, that most of the national efforts in the region for IWRM implementation have been dominated by neo-liberal economic policies and, as such, interventions mostly constitute decentralization schemes; privatization of water services; insuring service cost recovery through water pricing; sectorial water (re)allocation; and expanding access to water and sanitation, as well as water quality management. Table 1 shows an identification and explanation of IWRM tools addressing governance failures.

Reviewing recent literature, one finds resources that can assist mapping out the key factors that make the Arab water crisis, and explain its complexity. These factors in their interaction are illustrating the complexity of the management dimension and its magnitude (Hefny, 2007). Reflecting on these sources can draw the attention and bring together the big picture, as shown in Figure 2, which reflects the complexity and interconnectedness among various factors that shape the crisis (Figure 2 shows mind mapping of the Arab water crisis in its complexity).

4.2 *Gaps in water and food production*

4.2.1 *Severest water scarcity in the region*

The Arab Region represents 10% of the world's surface; its population represents 5% of world population. However, it possesses only 0.5% of world renewable fresh water resources. This is due to the fact that arid and semi-arid weather dominates 82.2% of the whole region. Rainfall precipitation estimated to be 2,228 km^3. The losses amount to 90.4% due to evaporation. The Arab Region is home to 5% of the world's people. Meanwhile, water demand in the region is growing fast. The population has more than doubled in the past 30 years to about 280 million, and could double again in the next 30 years. So, the *population nexus water* makes the Arab Region to suffer from more and more water scarcity.

4.2.2 *Water availability is falling to crisis levels*

The Arab world draws its water resources from rainwater; rivers and underground water sheets as well as sources that are non-conventional (desalinated water and treated waste water). The quantities

Table 1. Soft Path of IWRM. IWRM Tools Addressing Governance failures.

IWRM Tools	Governance Failures
Policies Economic instruments Financing and incentive structures and polluters	– Failure to correct market distortions – Inappropriate price regulation – Perverse subsidies to resource users – Inappropriate tax incentives and credits – The existence of upstream and downstream externalities (environmental, economic and social)
Regulatory instruments Institutional capacity building	– Over-regulation or under-regulation – Conflicting regulatory regimes – No independence and impartiality of regulatory organisations – Provision of water services as natural monopolies
Information management Water campaigns and awareness raising	– Imprecise reflection of consumer preference systems – Short-sightedness – Voter ignorance and imperfect information – Special interest effects, including political weaknesses and vested interests
Role of the private sector	– Little entrepreneurial incentives for internal efficiency
Institutional roles Social change instruments	– The inability of the government to control and regulate the sustainable water use – The non-payment of services linked to water – Bureaucratic obstacles or inertia – Lack of an overall responsible authority
Water resource assessment Plans for IWRM	– The lack of effective knowledge of the resource, the demands imposed on water and the current uses that are made of it
Legislation Water rights	– Ill defined property rights, unclear ownership – Absence of or inappropriate legislation – Unclear ownership of property rights
Water resource assessment, risk assessment and management	– Ignorance and uncertainty about water markets, droughts, floods, etc., leading to inability to set prices correctly

Source: Rogers & Hall (2003).

that can be obtained from these sources differ widely from one country to another and from one location to another within a single country. Rainfall is inexistent in some countries and heavy in others, for example (over 1,000 km^3/yr in Sudan). Similarly, countries such as Sudan, Egypt, Syria, Iraq, Lebanon, Jordan and Morocco have large and small rivers but other Arab countries have neither rivers nor lakes.

Where water resources are rare and financial resources permit, the abundant water found underground is exploited (the East Al Jazira underground sheet covering 1.6 million km^2, and the Nubian underground sheet covering 2 million km^2). Other non-conventional sources are being promoted, with the result that Arab countries are at the forefront in the development of water desalination techniques, producing over 5 Mm^3/day, that is, over 70% of the world production (see Figure 3).

Meanwhile, population growth in Arab countries is the major factor for the increase in food consumption; such growth has almost doubled between 1970 and 1990 (Figure 4), and the projections for future growth are even more alarming. The rate of population growth is among the highest in the world. In addition, the economic development and the change in aspirations and standard of living due to oil revenues in the last few decades leads to a very high percentage of immigration to

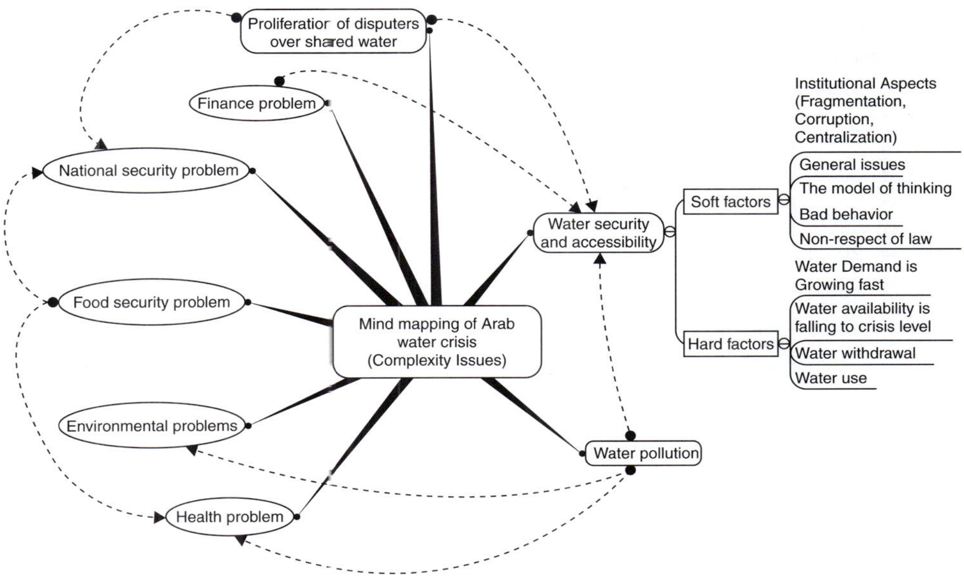

Figure 2. Mind Mapping of the Arab Water Crisis.

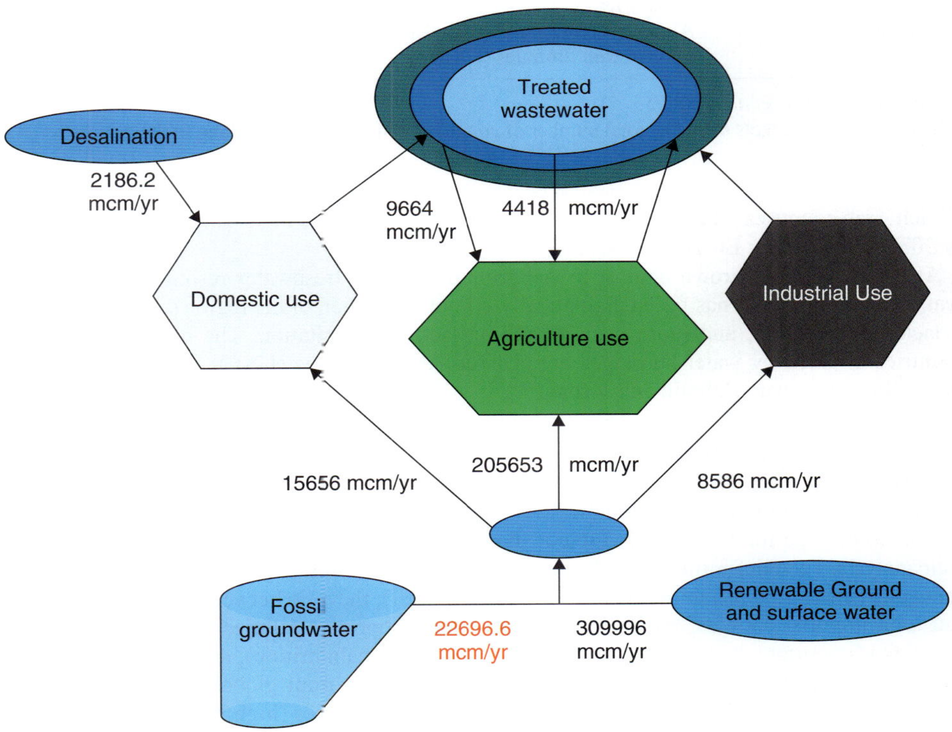

Figure 3. Water Resources Balance in the Arab Region.

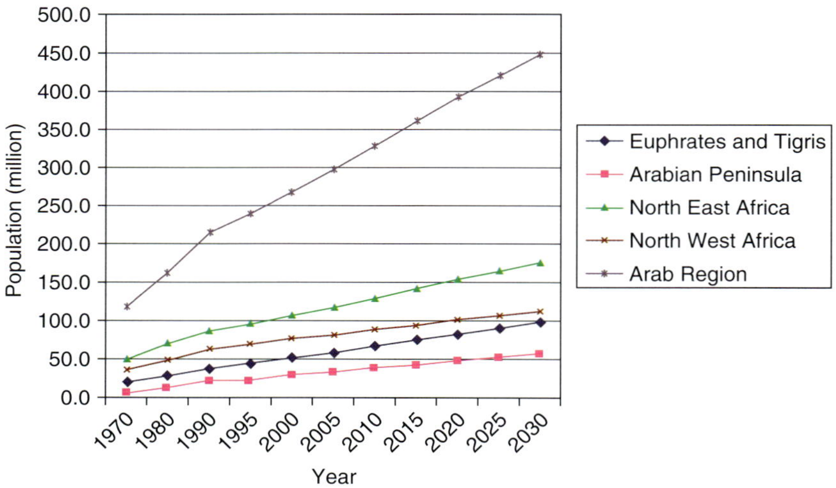

Figure 4. Population growth in the Arab Region.

Table 2. Food Security in Arab Countries, 1993–1997.

Food security	Group countries
Insecure (74)*	Sudan, Yemen
Neutral (51)*	Algeria, Egypt, Iran, Jordan, Kuwait, Lebanon, Libya, Morocco, Syria, Tunisia
Secure (37)*	United Arab Emirates (UAE)

Source: Diaz-Bonilla *et al.* (2000: 55–57).
*The numbers in brackets show the total number of countries in the world that belong to the group.

oil rich Arab countries. The total population is expected to continue increasing to reach 450 million by 2030 (Hamouda & El-Sadek, 2007).

As population has grown against a background of finite freshwater resources, so the water available to individuals has fallen dramatically. About 45 million of the region's population (16%) is lacking safe water, and more than 80 million lack safe sanitation. The most seriously Arab countries affected by water shortages are: United Arab Emirates (UAE), Saudi Arabia, Libya, Qatar, Kuwait, Oman, Bahrain and Jordan.

4.2.3 *Food security in Arab countries*

Diaz-Bonilla *et al.* (2000) have made an analysis to classify the countries of the world into three different groups: food insecure, food neutral, and food secure. For most countries, their analysis was based on data for 1993–1997. Table 2 shows the classification of Arab countries covered by their analysis. Their definition of food security is based on the following indicators: food production per capita (measuring the ability of a country to feed itself); the ratio between total export earnings and food imports (showing its ability to finance food imports); calories per capita and protein per capita (important explanatory variables for changes in malnutrition); and the non-agricultural population share (aimed at showing the extent of immunity from global changes in trade and agricultural policies) (Diaz-Bonilla *et al.*, 2000: 6–9). Trade-stress (high food imports relative to export earnings) tends to contribute to a lack of food security in the Arab Region more than in other regions. In another paper, Lofgren & Richards (2003) showed that in the second half of the 1990s, available indicators suggested that food insecurity at the national and household levels was

Table 3. Situation and future Projection of water resources and requirements in the Arab world.

Year	1990	2000	2025
Resources (10^3 Mm^3/yr)	257	274	278
Individual's share (m^3/yr)	1,431	1,142	801
Requirements (10^3 Mm^3/yr)	154	190	281
	+130 (surplus)	+84 (surplus)	−3 (deficit)

a serious problem in Iraq, Sudan, and Yemen, countries that also are likely to have the highest poverty rates in the region. At the household level, the performance for food security indicators was more positive up to the mid 1980s than in more recent years.

For national-level indicators, the picture is mixed. For the region as a whole, some of these indicators (per-capita food production and food import stress) show stronger gains since the mid 1980s, whereas others (including per-capita calorie consumption) improved less strongly in more recent years. Finally, the influence of political instability and conflicts (internal and external) on food security is obvious. Continued strife makes the challenge of improving food security much greater for the cases of Iraq and Sudan.

4.2.4 *Agricultural policies*

In the past 20 years Arab countries were aiming to achieve greater food self-sufficiency and accordingly the majority of Arab countries provided generous *subsidies* to the agricultural sector. Given the arid climate of these countries, production can only be increased by escalating the area of land under irrigation, by improving irrigation efficiency, or a combination of the two. Agricultural subsidies are given in several *forms*, for wells, canals, fuel, and other inputs, price support programs, trade protection in some countries, and lack of controls on groundwater extraction or charges.

These subsidies drastically increased the irrigated areas and are contributing to the depletion of aquifers (Hamouda & El-Sadek, 2007). Besides, these subsidies distort costs and revenues, and many of the agricultural activities in the Arab countries are financially profitable only because of Government subsidies and incentives. Although agriculture consumes about 80–90% of the available water in most Arab countries, it contributes to less than 10% of Gross Domestic Product (GDP) in most of these countries. The employment in the agricultural sector, as a proportion of total employment in 2003, ranged from 1% in Kuwait, Bahrain and Qatar to 58% in Sudan.

4.3 *Water resource balance in the Arab region*

The United Nations Environment Programme and a number of water and development experts have assessed the *individual's water requirements* at around 1,000 m^3/yr. This includes drinking water, agriculture, industry and other development sectors. Water is seen as being one of the main factors that determine the growth rate in most Arab countries. It is used domestically (drinking water), agriculturally (food production) and industrially, as well as for energy production, transport of people and goods, leisure, and environment conservation.

Regarding existing water potential in the Arab world, it appears that four countries are at present suffering from a lack of water and that, in the year 2025, eight countries will be in this situation (Table 3). This is because of the anticipated rise in water requirements resulting from demographic growth, from the extension of irrigated areas to feed those people, and from the consequent growth of industrial and commercial activity. It also springs from the rising rates of individual consumption due to an increase in living standards. Thus the problem of a water deficit and the need to guarantee permanent water resources will constitute the main challenge to be addressed to assure the future of coming generations.

4.4 *The need for sustainability measures*

The progressing developments and water policies in Arab countries resulted in the current difficult water shortage situation. Figure 3 shows a general picture for the current water balance or *cycle* in the whole of the Arab countries. It is clear that the huge amounts of water used for irrigated agriculture are leading to abstracting large amounts of non-renewable *fossil* groundwater. This creates real threats for the sustainability in the whole region. In order to restore and/or improve this imbalance, focus should be given mainly to the reduction of demands from the agricultural sector, and also to create additional sources of water to reduce groundwater abstractions. Achieving these objectives needs considerable efforts to reform both water and agricultural policies, and also significant cost to enhance the capacity of wastewater treatment as a safe and reliable source of water for irrigation (Hamouda & El-Sadek, 2007).

4.5 *The need for water and agricultural policy reform*

Agricultural policies in Arab countries also need to be reformulated to serve the purpose of increasing efficiency and conserve the already depleted water resources. As previously elaborated, subsidies are one of the major drawbacks of current agricultural policies. In addition to their burden on the general budget, agricultural subsidies lead to poor resource allocation and the depletion of aquifers which disregard the high opportunity cost of water and the contribution of groundwater aquifers to the sustainability of ecosystems (Hamouda & El-Sadek, 2007).

Reforming agricultural policy aims at: a) Conservation of water resources; b) Reduction of direct and indirect subsidies attributed to the agriculture sector; c) Water quality preservation; and d) Improved crop water productivity. It is vital that a number of steps be taken to economize on water consumption and protect water resources from pollution. These steps are, inter alia:

- Exploiting rainwater by building dams and reservoirs.
- Provision only of water needs, with guaranteed drinking water as a priority.
- Promoting water distribution and use techniques to monitor the modes of operation and upkeep.
- Setting water rates in a way that will avoid waste.
- Recycling water after it has been treated in the appropriate way suitable for reuse in urban use and irrigation.
- Developing water desalination techniques, especially since the Arab world is surrounded by seas and oceans and solar energy is available throughout the year.

These policy reforms include the following:

i. Limiting subsidies to technologies which promote water conservation. The strategic objective should be to optimize water productivity, since water is the limiting factor, rather than to maximize crop yields.
ii. Supporting the current policy of waiving charges for the use of treated wastewater in irrigation.
iii. Apply demand management principles to both municipal and irrigation water.
iv. Create efficiency indicators to monitor and evaluate water and agricultural policies.
v. Continue research efforts for the development of salt-tolerant and drought-resistant crops, and investigating additional water sources.
vi. Upgrade the assessment of existing water resources and demand projections.
vii. The promotion of a market-oriented policy through the cultivation of crops with high water productivity and low-water content.
viii. Progressively move from a policy of food security to a policy of water security and thus import crops with high water content (*virtual water*), and increased integration with global markets.
ix. Set up a legal/technical framework for surface and underground water allocation –Water Rights Adjudication Process– based on sound hydrological criteria. This should be used as a water management mechanism to approach resource sustainability (demand management),

providing a legal safeguard at the same time to farming investments (assurance of water availability).

x. Expanding the use of treated wastewater and drainage water which offer the opportunity for substituting part of the conventional water used in irrigation. However, three important bottlenecks still demand a solution: 1) building infrastructure for wastewater collection and treatment; 2) building infrastructure to store and transport treated wastewater to areas where it can be used; 3) upgrading national standards for treated wastewater reuse to encourage crop diversification.
xi. Adopting integrated policy between different elements influencing water resources management.
xii. Encourage private sector investment in developing water resources within the framework of national plans.
xiii. Enhancing the role of participatory water management in operation and maintenance.
xiv. Creating social and economic incentives for water conservation.
xv. Activating the role of education, training and information management in the field of water resources and management (for water users and officials).
xvi. Raising awareness regarding relevant water conservation and environmental issues.
xvii. Create regional initiatives to improve desalination technology, use of brackish water in agriculture, and investigate opportunities for import water.

4.6 *Virtual water as a policy option*

4.6.1 *Introduction*

It is now well recognized that Arab states are facing surmounting water/food challenges as a water-scarce region. This is evident by accounting the volume of virtual water that is embedded in the food imports of the countries concerned. There is a close relationship between water endowment and food import dependence. Studies have shown that although virtual water trade is going on in the region, it is yet to be considered as a policy option in planning and allocating water resources (Hamouda & El-Sadek, 2007).

However, virtual water as a policy option is often faced with skepticism and fear of economic or political control. The culture of the Arab Region constitutes an important element when discussing the need for change in current planning methods to accommodate virtual water trade. In this context, there are three important points to be considered: a) food imports are imperative for compensating water resource deficiency; b) cultural and behavioral changes are necessary for adapting to the current water scarce situation; and c) it is imperative that planners tackle the sources of skepticism first before introducing virtual water trade as a policy option (Hamouda & El-Sadek, 2007).

Having established the need for policy reform in all water consuming activities, it is important to investigate policy options provided by the virtual water trade concept and the suitability of such options to the Arab Region.

4.6.2 *Virtual water policy dimensions*

More often than not, scholars in the Arab Region contradict the political argument that has been put forward by Tony Allan from the beginning of the virtual water debate, that virtual water trade can be an instrument to solve geopolitical problems and even prevent wars over water (Allan, 1997). The contradiction is simply based on the perception that food exporting countries are mostly western countries, and the relationship between the Arab states and western countries is dominated by skepticism and fear of domination. Thus, looking at the bigger picture it is perceived by the Arab states that by depending on food imports they are giving in to foreign domination (Hamouda & El-Sadek, 2007).

Next to the political dimension, there is the economic dimension, equally stressed by Allan (1997; 1999; 2001). The economic argument behind virtual water trade is that, according to international trade theory, nations should export products in which they possess a relative or comparative advantage in production, while they should import products in which they possess a comparative disadvantage (Wichelns, 2001). In addition, growing food (i.e. feeding oneself) has other important aspects aside that of the economic or political ones; they are of major influence on the decision whether to grow a certain crop or not.

Hoekstra & Hung (2002; 2003) argue that –while pricing and technology can be a means to increase local water use efficiency and reallocating water at basin scale to its higher-value, alternative uses a means to increase water allocation efficiency– virtual water trade between nations can be an instrument to increase *global water use efficiency*. From an economic point of view it makes sense to produce the water-intensive products demanded in this world in those places where water is most abundantly available. In those places water is cheaper, there are smaller negative externalities to water use, and often less water is needed per unit of product. Virtual water trade from a nation where water productivity is relatively high to a nation where water productivity is relatively low implies that globally real water savings are made.

As mentioned before, past food security policies were based on area expansion to support the objectives of food self-sufficiency and to enhance exports. These expansions proved to be unfeasible with respect to available water resource and lead to real threats to the sustainability of current developments. In fact, the future increase in agricultural production must come from increased land and water productivity both in terms of higher yields and cropping intensities for which scope still exists. This will lead to greater water savings by reducing wasteful water losses to low economic value crops and achieving more efficient water use and better agronomic practices.

Before adopting the virtual water policy option, Arab countries need to be assured that they can have fair and secure trade with water-abundant nations. An advantage for the Arab countries is that they export enough to earn the foreign exchange required to purchase the food imports they need.

In short, the concept of virtual water is well founded, provided countries have a more transparent picture of its comparative advantage and accordingly they can translate this into a competitive advantage. The second issue pertains to the level of the economic base, i.e. whether the economy of the country is well developed and diversified to take the decision of reallocating water from cereals, which provide subsistence living to large sections of rural population.

Regional cooperation on this subject is of paramount importance as it would allow countries in the region to assess and analyze the situation on a broader basis, taking into consideration common strategic issues.

Beside the direct financial cost, other costs to be considered are related to imports by water deficit countries to solve food deficiency like (World Water Council, 2004): 1) increased dependency on main exporting countries; 2) local agriculture may be damaged because of importing food if it is not able to compete or adapt; 3) the exporting country may start interfering in the internal affairs of importing country; and 4) imports may result in foreign reserve depletion if there is no export compensation of less water intensive or higher value commodities.

5 TOWARDS A MORE BALANCED STRATEGY IN RESPONSE TO WATER MANAGEMENT CHALLENGES IN THE ARAB REGION

5.1 *Introduction*

In articulating water strategies, it is important to strike a balance that advocates water as a common good and a fundamental human right, and at the same time, recognizing that water has an intrinsic economic value, and must be used sustainably. In this case, it is important to recall the main principles of IWRM and its tools, and recommend some general guidelines for implementing such tools within an ethical framework for water management. These guidelines entail, inter alia: a) the implications of declining water resources and how to maximize the benefits of water under condition of increasing scarcity; b) the necessity for starting a process of re-thinking water and food security paradigms for allocating water to the most beneficial use as one of the basic elements of any water strategy; c) the increasing cost of water development, compounded by increasing cross-sectorial demand on the limited water resources, is forcing policy makers in the region; and

d) to focus on the economics of water and its efficient use and allocation among competing users. The economics of water is now considered one important aspect of water management.

Most notably, here, it is important to stress that "the cultural and socio-economic values of water are still a very elusive subject. Several learned meetings stressed the economic value of water, while others stressed its social and cultural values. The importance of one or the other will vary from one society to another and from time to time, depending on the specific historical background, cultural heritage, extent of fresh water availability and the socio-economic conditions of the concerned region. *Developing a unified approach is required*, with clearly defined associated conditions and limitations for its applicability, which should accommodate the diversity of the world's regions" (Mahmoud Abu-Zeid, 1998)[8].

I fully agree, with such recognition on the importance of the socio-economic and environmental factors. The soft factors and soft approaches are no less important than the hard hydrological engineering factors, in achieving a well balanced water strategy, as well as in following a more integrated and comprehensive approach (Hefny, 2006a; 2007).

There is simply no way to overstate the water crisis of the planet today. No piecemeal solution is going to prevent the collapse of whole societies and ecosystems. A radical rethinking of our values, priorities, and political systems is urgent and still possible. The answers lie in emphasizing a whole new water strategy that is based on ethical guidelines, and to declare clearly that, inter alia:

- Water belongs to the earth and all species, and is sacred to all life on the planet. All decisions about water must be based on ecosystem and watershed-based management. We need strong national and international laws to promote conservation, reclaim polluted water systems, develop water supply restrictions, ban toxic dumping and pesticides, control or ban corporate farming, and bring the rule of law to transnational corporations who pollute water systems anywhere.
- Water is a basic human right. This might sound elemental, but at the World Water Forum in The Hague, it was the subject of heated debate, with the World Bank and the water companies seeking to have it declared a human need. This is not semantic. If water is a human need, it can be serviced by the private sector. You cannot sell a human right.
- Water is a public trust to be guarded at all levels of government. No one has the right to appropriate it at another's expense for profit. Water must not be privatized, commoditized, traded, or exported for commercial gain.

Above all, we, as human beings, must change our behavior. We must emphasize identifying the capacity of our watersheds and, as communities, identify the limits we can place upon them. The world must accept conservation as the only model for survival, and we must all teach ourselves to live within our environment's capacity. The insidious problem with pricing and conservation by commoditisation is that actually undermines environmental science and activism, as well as governments' responsibility to protect their citizens and the environment by buying into the argument that the market will fix everything.

The question that matters here is to reflect on how it is best possible to get a balanced view of a water strategy, which would be designed on a sound basis for a more sustainable use and management of the water sector. It is more widely agreed that *in decision making processes*, there is a necessity to implement an integrative and interdisciplinary approach that takes care of three basic elements as follows:

- The necessity of attaining economic efficiency[9], which is an issue of considerable importance in economic theory, and is even more important today.

---

[8] The honorary president of the World Water Council and the President of the Arab Water Council.

[9] Economic efficiency is understood as the maximum production that can be obtained given the resources available.

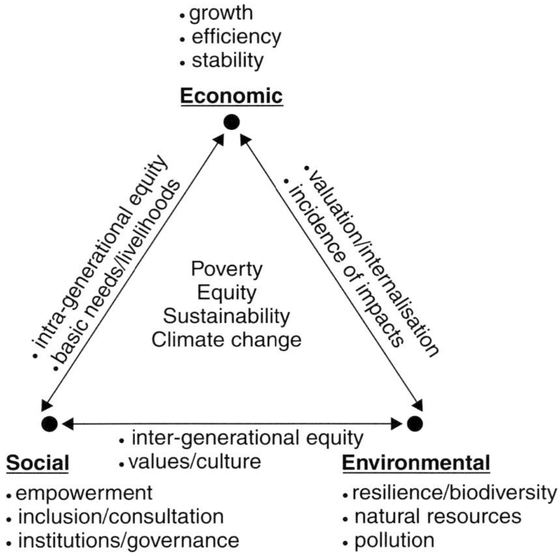

Figure 5. Sustainable development triangle –key elements and links (corners, sides, center).
Source: Adapted from Munasinghe (1992; 1994).

- Social equity[10] as well as human equality for all persons ought to be provided with what is needed on an equitable basis, having due regard to human rights and dignity, as there is no life without water, and those to whom it is denied are denied life.
- Environmental sustainability, which is centered on water conservation.

A major work has been done by Muhan Munasinghe entitled "Making Development More Sustainable". He has presented it in an *e-book of the Encyclopedia of Earth*, the sustainable development triangle with a balanced treatment of economic, social and environmental aspects. In chapter 12, he explored the question of how to make water resource management more sustainable. *Sustainomics* is introduced as a new language to explain the transdisciplinary integrative comprehensive balanced heuristic and practical framework for making development more sustainable. Figure 5 is adapted from Munasinghe (1992; 1994), to show the sustainable development triangle –key elements and links (corners, sides, center).

In this context of decision making and governance, one recognizes the trade-offs amongst the three *E's*: water efficiency, social equity and environmental sustainability, and even conflictive issues, in water use and management, could be reconciled.

In addition, the following illustration depicts major elements of a water strategy that is based on clear identification of an Arab Water Vision for 2025 that is emanated from the Millennium Development Goals and the World Summit on Sustainable Development (WSSD, 2002). The identification of challenges with its complexities is highlighted earlier for the food water nexus. However the mind mapping of ideas below is bringing the big picture, which is necessary in articulating and formulating *a water strategy* for the Arab Region that emphasizes basic ethical principles and helps improve on the design of the water and food security system (Figure 6).

[10] Social equity is understood as the redistribution that aims to reduce inequalities of income or wealth.

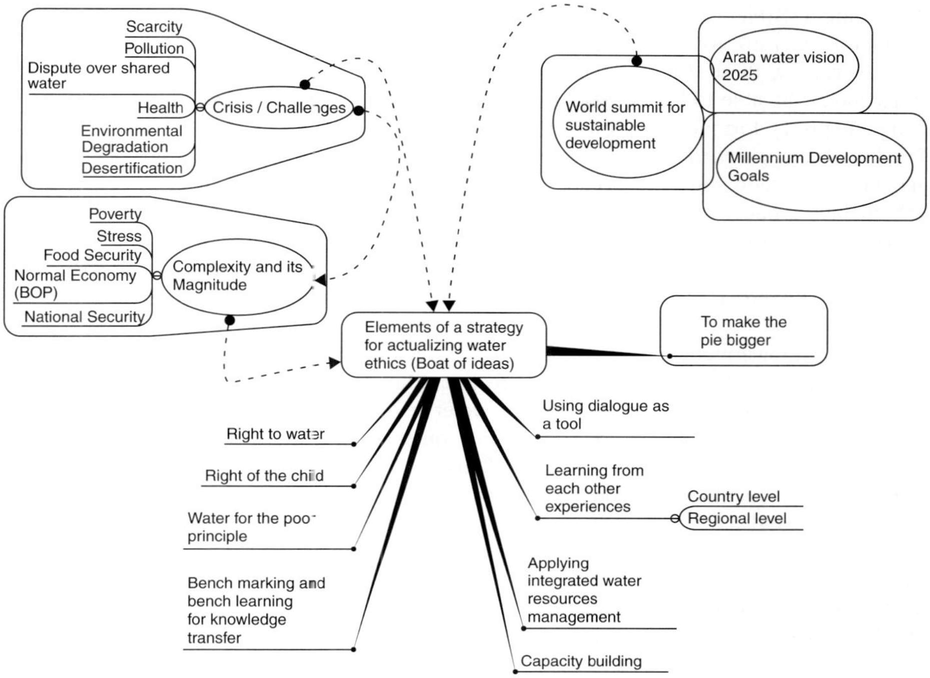

Figure 6. Mind mapping of ideas of an Arab Water Strategy.

6 CONCLUDING REMARKS: ETHICS IN DECISION MAKING AND GOVERNANCE

In decision making, water management is fundamentally a question of social and environmental justice based on three essential concepts: equity, fairness and access between and across generations. Its ethical dimensions may be perceived in the way answers are found to the following questions:

- Who participates in the decision-making process?
- Are these participants involved in formulating options or are they expected only to react to proposals that are already well-developed?
- How and what type of opportunity costs are considered?
- What kind of information is open to the public?
- How do professionals interact with non-professionals?
- Is there respect for cultural diversity and our heritage?
- How is a balance determined between the needs of human development and the need to preserve our natural resources?

Governance has to be based on shared values, and governments have to ensure that there are socially accepted moral standards on what can and cannot be done. These standards must determine what consequences of water management are or are not acceptable. For example, to what extent is damage to ecosystems acceptable? What loss to our heritage is tolerable? What impact on downstream water users is permissible? Successful civilizations have usually ensured that their water governance is rigidly enforced. When there is a breakdown of water regulation, conflict and economic failure often follow. Governments have a responsibility to ensure that an appropriate infrastructure is in place to allow these shared moral values to be debated and implemented.

According to Brelet (2004), *good water governance* is a prerequisite for ethical water management. Good water governance consists of a set of basic principles that guide water management. Most importantly, these principles include participation, transparency and equity. Accordingly, citizens should be able to have a say, directly or through civil society organizations, in decision making and policy formulation processes (UNDP, 2003).

The importance of *participation* does not just lie in being among the principles of ethical water management or good water governance. Experience shows that, with an open social structure which enables broader participation by civil society, water governance is more effective due to civil society's ability to influence government. Nevertheless, government regulations that facilitate local governance and participation are necessary for a clearer and more effective role of civil society and non-governmental organizations in enhancing water management (Rogers & Hall, 2003).

Therefore empowerment of local communities, especially remote and marginalized ones, through creating an enabling environment for more effective action by civil society is an important prerequisite to achieving a meaningful participation in water resources management. Moreover, the quality and effectiveness of government policies depends on ensuring participation throughout the policy formulation and decision making processes. This would create more confidence in the policies developed and the institutions through which such policies are formulated.

On the other hand, participation crucially depends on all levels of government following an inclusive approach when developing and implementing water policies. Inclusiveness can be achieved through social mobilization and freedom of association and speech. As such, institutions should communicate with all stakeholders and actors involved in the issues of water resources management. This entails conducting accurate and well-informed stakeholder analyses at all levels of decision-making, whether at the policy level (as mentioned above) or at the local project level that requires the inclusion, empowerment, and participation of local people.

Finally, water management should be done *in an open manner*. All decision-making processes and policy formulation should be transparent so that all involved stakeholders could follow the details of steps taken or required in policies developed in order to avoid mismanagement or misuse of water resources. For that, various roles in the legislative and executive branches need to be clear, and decision-makers, private sector, and civil society organizations should be accountable to the public. Accordingly, administrative procedures should be clear, as should the consequences for violation of policy provisions. Appropriate conflict resolution systems should also be attached for reconciliation among stakeholders through proper solutions (Rogers & Hall, 2003).

Moreover, governance institutions and systems need to communicate effectively among each other and with the public at large, because crucial to transparency and accountability is the free flow of information and direct communication. Language and communications between decision-makers and the public should be clear, accessible, and understandable for all in the interest of increasing public confidence in the relevant institutions (Rogers & Hall, 2003; UNDP, 2003).

The concept of *decentralization*, in theory, should improve efficiency, accountability, and equity because it correlates the benefits of local public services to the costs entailed. Nevertheless, in order to be considered ethical, decentralization has to be appropriately implemented because it has its own shortcomings and disadvantages, such as the possibility of elite capture that promotes clientelism instead of democratic participation.

In the Arab Region, different decentralization schemes for irrigation and domestic water supply have been adopted. Different experiences have led to different impacts, but in general they have raised to a certain extent the sense of responsibility among farmers and resulted in achieving higher water use efficiencies. It should be noted, however, that it is still too early to evaluate the overall impacts of decentralization on water management in the Arab Region because the whole idea is relatively very recent. Even in the countries that have had a good start with the process, it is still premature to claim that they already enjoy a good enabling environment based on a societal ethical framework for water management.

REFERENCES

Abdurazzak, M. & Kobeissi, L. (2002). *UNDP-ESCWA Initiative on National Framework for Water Resources Management in Lebanon*. Paper presented in the 2nd Water Demand Management Forum – IDRC, Beirut, Lebanon.

Abu-Zeid, M. (1998). Water and sustainable development: the vision for world water, life and the environment. *Water Policy*, 1(1): 9–19.

Abu-Zeid, M. (2001). Water Pricing in Irrigated Agriculture. *Water Resources Development*, 17(4): 527–538.

Abu-Zeid, M. & Hamdy, A. (2004). *Water Crisis and Food Security in the Arab World: Where we are and where do we go?* The Mediterranean Agronomic Institute of Bari, Italy.

Allan, J.A. (1993). Fortunately there are substitutes for water otherwise our hydro-political futures would be impossible. In: ODA, *Priorities for water resources allocation and management*. ODA, London, UK: 13–26.

Allan, J.A. (1994). Overall perspectives on countries and regions. In: P. Rogers & P. Lydon (eds.), *Water in the Arab World: perspectives and prognoses*. Harvard University Press, Cambridge, Massachusetts, USA: 65–100.

Allan, J.A. (1997). *Virtual water: A long term solution for water short Middle Eastern economies?* Paper presented at the 1997 British Association Festival of Science, University of Leeds, UK. 9 September 1997.

Allan, J.A. (1999). Water stress and global mitigation: Water, food and trade. *Arid Lands Newsletter*, 45.

Allan, J.A. (2001). *The Middle East water question: Hydropolitics and the global economy*. I.B. Tauris, London, UK.

Bannock, G.; Baxter, R.E. & Davis, E. (1987). *The Penguin Dictionary of Economics*. Penguin, New York, USA.

Barlow, M. (2001). Water as Commodity – The Wrong Prescription. Institute for Food & Development Policy, Oakland, USA. Backgrounder, *Food First Journal*, 7(3): 1–8.

Brelet, C. (2004). *Some Examples of Best Ethical Practice in Water Use*. UNESCO, COMEST, Paris, France.

Budds, J. & McGranahan, G. (2003). Are the debates on water privatization missing the point? Experiences from Africa, Asia and Latin America. *Environment & Urbanization*, 15(2): 87–113.

Diaz-Bonilla, E.; Thomas, M.; Robinson, S. & Cattaneo, A. (2000). Food Security and Trade Negotiations in the World Trade Organization: A Cluster Analysis of Country Groups. Discussion Paper 59, Trade and Macroeconomics Division. IFPRI, Washington, D.C., USA.

Dinar, A. (2000). Political Economy of Water Pricing Reforms. In: A. Dinar (ed.), *The Political Economy of Water Pricing Reforms*. Oxford University Press, New York, USA.

Dinar, A. & Subramanian, A. (1997). Water Pricing Experience. World Bank Technical Paper, 386. World Bank, Washington D.C., USA.

El-Kady, M. (2003). Water in the Arab World. The National Water Research Center of Cairo, Egypt. Introduced this theme at the 3rd World Water Forum, Kyoto, Japan.

ESCWA (1999a). Management of groundwater resources in ESCWA member states. Expert Group Meeting on Updating of Assessment of Water Resources in the ESCWA Member States. Beirut, Lebanon, 20–30 April 1999.

ESCWA (1999b). Current Water Policies and Practices in Selected ESCWA Member Countries. Economic and Social Commission for Western Asia. United Nations, New York, USA.

ESCWA (2003a). Updating the Assessment of Water Resources in ESCWA Member Countries. United Nations Publication, New York, USA.

ESCWA (2003b). Assessment of the Role of the Private Sector in the Development and Management of Water Supply in Selected ESCWA Member Countries. Economic and Social Commission for Western Asia, United Nations, New York, USA.

Gleick, P. (1999). The Human Right to Water. *Water Policy*, 1(5): 487–503.

Hamouda, A. & El-Sadek, A. (2007). Virtual Water Trade as a Policy Option for the Arab States. Paper presented in the 1st Arab Water Forum. Proceedings of the Forum, Cairo, Egypt.

Hardin, G. (1968). The Tragedy of the Commons. *Science*, 1628: 1243–1248.

Hefny, M. (2006a). A Cultural Approach to Actualize Water Use Ethics in the Arab Region. Proceedings of the 2nd Arab Water Forum. Cairo, Egypt.

Hefny, M. (2006b). Actualizing Water Ethics in the Regional Context of the Arab World. In: *Management of Intentional and Accidental Water Pollution*. Springer: 277–311.

Hefny, M. (2007). Water Ethics as a tool for Water Sustainable Development in the Arab Region. Proceedings of the 3rd Arab Water Forum. Cairo, Egypt.

Hoekstra, A.Y. (2003). Virtual water: An introduction. In: Proceedings of the International Expert Meeting on Virtual Water Trade. Value of Water Research Report Series, 12. IHE, Delft, the Netherlands.

Hoekstra, A.Y. & Chapagain, A.K. (2004). Water Footprints of Nations. Value of Water Research Report Series, 16. UNESCO-IHE Institute for Water Education.

Hoekstra, A.Y. & Hung, P.Q. (2002). Virtual water trade: A quantification of virtual water flows between nations in relation to international crop trade. Value of Water Research Report Series, 11. IHE, Delft, the Netherlands.

Hoekstra, A.Y. & Hung, P.Q. (2003). Virtual water trade: A quantification of virtual water flows between nations in relation to international crop trade. In: Proceedings of the International Expert Meeting on Virtual Water Trade. Value of Water Research Report Series, 12. IHE, Delft, the Netherlands.

Johansson, R.C.; Tsur, Y.; Roe, T.L.; Doukkali, R. & Dinar, A. (2002). Pricing Irrigation Water: A Review of Theory and Practice. *Water Policy*, 4: 173–199.

Linam, A.M. (2002). Water Valuation: The Case of Tunis, valuation of drinking water. Paper presented in the 2nd Water Demand Management Forum – IDRC, Beirut, Lebanon.

Lofgren, H. & Richards, A. (2003). Food Security, Poverty, and Economic Policy in the Middle East and North Africa. Trade and Macroeconomics Division, International Food Policy Research Institute, Washington, USA. TMD Discussion Paper, 111.

Munasinghe, M. (1992; 1994). The Encyclopedia of Earth [http:www.eoearth.org/article]. Overview and outline of Making Development More Sustainable: *Sustainomics* Framework and Applications (e-Book).

Postel, S. (1996). Dividing the Waters: Food Security, Ecosystem Health, and the New Politics of Scarcity. WorldWatch Paper 132, WorldWatch Institute.

Rogers, P. & Hall, A. (2003). Effective Water Governance. Technical Advisory Committee (TEC), TEC Background Papers, 7, Global Water Partnership, Stockholm, Sweden.

Taha, S. & Bataineh, F. (2002). Water Valuation: The Case of Jordan, valuation of water for domestic use. Paper presented in the 2nd Water Demand Management Forum – IDRC, Beirut, Lebanon.

Turner, M. (2004), The Commoditization and Marketing of Water. Water Bank website [waterbank.com], New Mexico, USA.

UNDP (2003). Water Governance and Poverty Reduction. Draft Practice Note for the Bureau for Development Policy. Energy and Environment Group. UNDP, New York, USA.

UNESCO (2003). *Water for People, Water for Life*. UN World Water Development Report. World Water Assessment Programme [www.unesco.org/water/wwap].

Wichelns, D. (2001). The role of *virtual water* in efforts to achieve food security and other national goals, with an example from Egypt. *Agricultural Water Management*, 49: 131–151.

World Water Council (2004). *E-Conference Synthesis: Virtual Water Trade – Conscious Choices*. March 2004. Synthesis, WWC Publication No. 2. Marseille, France: 31 pp.

WSSD (2002). World Summit on Sustainable Development. Johannesburg, 26 August – 4 September 2002. Available at: [http://www.un.org/events/wssd/].

CHAPTER 10

Water scarcity risks: Experience of the private sector

Stuart Orr
WWF-International, Gland, Switzerland

Anton Cartwright
University of Cape Town, South Africa

ABSTRACT: While almost everyone recognizes water scarcity to be a *public bad*, the manner in which water scarcity impacts government and business through complex social and ecological systems is less well understood. Risks associated with water scarcity can be classified as those which arise from insufficient water resources to meet the basic needs of people, the environment and business, risk from the consequences of insufficient water resources, and risk from poor water management decisions taken in reaction to water scarcity, often with negative consequences for some or all users. Yet risk is a concept that has multiple meanings and interpretations, and accordingly can motivate actors in different ways. Yet from the private sector (corporations and representative bodies/forums) there is an emerging recognition around their vulnerability to water stress throughout the supply chain. This chapter explores notions of risk, how risks manifest to companies and briefly discuss how aspects of the business sector have responded to their *water footprint* including interest in water public policy.

Keywords: water scarcity, risk, business, public policy

1 INTRODUCTION

An aversion to risk is one thing that rational people have in common. This chapter emphasizes the risks associated with insufficient water quantity, addressing the social and ecological problems that arise when water is scarce, as well as the problems created by disingenuous decisions in the face of water scarcity. The collation of risks that affect different people in different ways is highly specific, and involves the loss of certain *fine-grained* and often crucial information about the risk to specific stakeholders (Vatn & Bromley, 1994).

Aggregated water scarcity maps are useful for raising awareness, but reveal little about the implications of water scarcity for people and ecosystems. How does one go about identifying the regions and stakeholders that are most likely to be affected by shortages? How do we begin to understand the problems that water scarce regions and their people are likely to confront? The problem can only be understood, and action can only be effectively tailored, by focusing at the local level. Global and national scale macro assessments are useful for raising awareness, but do not reveal the dynamics, components and the biases within the water scarcity phenomenon.

Whilst exposure to water-related risk primarily manifests at the river basin or local scale, the origins of water scarcity and its impacts derive from the interaction of natural biophysical cycles, and the actions and decisions of people in a variety of sectors at local, national and international levels. While almost everyone recognizes water scarcity to be a *public bad*, the manner in which water scarcity impacts government and business through complex social and ecological systems is less well understood. What is more certain are that threats to people and all they value, necessitates a determined approach to exploring risks associated with water management failures (Kasperson & Kasperson, 2001).

Highlighting and explaining how securing adequate water supplies leads to major risk reduction, means engaging the diverse set of stakeholders who benefit from water use, and are often, unwittingly vulnerable, to water scarcity. An emerging consensus on the importance of water to society, as reflected through initiatives by NGOs, global business fora and in the increasing coverage of water issues in the media, provides impetus for challenging and redefining traditional paradigms of water management and public policy. Yet, there is a specific challenge in facilitating this paradigm shift, due to the generally differing languages and expectations of these groups around needs, time-frames and modes of communication. This chapter uses a risk lens to look at the rapid increase in private sector activity around water issues and makes a distinction between private sector companies who face water related risks, *versus* private equity investment or private water service delivery.

2 WATER SCARCITY

Ensuring a sustained supply of water to people and their economies became a matter of global concern at the United Nation's 1977 *Mar del Plata Conference*. Initiatives arising from the Conference had little impact upon water security, but did usher in a better understanding of the problem. There is an estimated 37×10^6 km^3 of fresh water on the planet, 8×10^6 km^3 of which is stored as groundwater. The World Health Organisation (WHO) believes that 25 L/day per person is the minimum required for domestic purposes[1]. The available resources are more than adequate for domestic consumption and washing, but this constitutes an insignificant fraction of total water use. When agricultural and industrial requirements are considered, between 1.4 m^3/day and 4.5 m^3/day per capita seems to be a critical current requirement. Yet water scarcity is caused more by the nature of demand and the allocation of water than the general availability of water, and is an issue best addressed through better water management and governance. Water scarcity is a "governance crisis, not a resource crisis" (Rogers *et al.*, 2006), and: "In the developing world, the problem is more often caused by policy and institutional failure, rather than by technical failure" (Castro, 2004).

Availability of sufficient water at the *global* level should not detract from the fact that at specific local levels, water is often acutely scarce and responsible for a wide range of adversities. What is more, the problem appears to be increasingly widespread and there are a number of reasons to suspect that the effective management of water will become more difficult:

a) Water consumption has grown at more than twice the rate of population expansion in the past century. The increasing demand is attributed to growing industrial demand, and more affluent lifestyles that consume more water. Increasing animal protein diets are more energy intensive. Also urbanisation which requires mass storage and concentrated supply of water must be factored into any view of shortages. On aggregate, a 13% increase in water consumption is anticipated in the first 25 years of this century (Rosegrant *et al.*, 2002a; 2002b).
b) Water use and distribution is uneven and many of the areas expected to experience the greatest *shortage* of water in the future are also the areas in which population *growth* is expected to be greatest.
c) The influence of anthropogenic warming is anticipated to cause increasing aridity over much of the mid-latitudes. In the short term, over-abstraction of water is often a problem distinct from climate change induced shortages, but in the longer-term, anthropogenic climate change and associated aridity will exacerbate problems caused by water mismanagement. It is that mismanagement which will undermine crucial environmental capacity and resilience that would have aided adaptation to climate change.

---

[1] Peter Gleick (1996) claims that 5 L/day were required for drinking, 20 L/day for sanitation services, 15 L/day for bathing and 10 L/day for food preparation.

3 USING RISK TO DESCRIBE WATER SCARCITY THREATS

Identifying the nature and location of future water scarcity risks requires imagination as well as knowledge. As the global insurance firm Swiss Re points out: "Emerging risks are not even being called risks. They are more like uncertainties that you have to search for". Nevertheless, Swiss Re has identified one big [emerging] risk: "the global unavailability of water" (Zanetti *et al.*, 2005). Risks associated with water scarcity can be classified as follows:

a) Risk from insufficient water resources to meet the basic needs of people, the environment and business, which in turn leads to . . .
b) Risk from the consequences of insufficient water resources, such as higher energy prices, loss of competitive advantage, political and economic instability, population migration, or lost economic opportunities to name a few; and as a result . . .
c) Risk from poor water management decisions taken in reaction to water scarcity, with negative consequences for some or all users. Such decisions may be a result of political or economic expediency, short-term thinking, lack of knowledge or capacity or simply desperation and lack of choice.

Discussions of risks imposed by water scarcity must involve those who are responsible for, and those who are affected by, the problems of water scarcity. Water is a public[2], a private and a social good, and a water scarcity event will have both private impacts and public repercussions on stakeholders. Accordingly, it is necessary in any risk analysis involving water to establish: *risk to whom?* With the understanding that the risk to an individual differs from societal or business risks and certain groups will be more vulnerable than others. It is also necessary to ask: *risk of what?* With the understanding that water scarcity is a subjective concept. For a farmer, the danger may be back-to-back years of below average rainfall. For the owner of a processing plant, the risk might be a temporary, sudden cessation of stream-flow during peak operation time. For a government, risks might include the increasing costs of accessing water for utilities and the implications of higher energy costs, or failing to deliver on economic growth and development pathways because of poor water management: failures that are passed on to business communities entrenched in various ways.

Most stakeholders have some means – formal or informal – of coping with water scarcity, and whilst water scarcity tends to impose costs on those who can least afford them, in some instances it can provide the catalyst for effective adaptation to a less risky state. In reality, judicious and the adoption of water-saving technologies might reduce risk even as people become more affluent, better able to adapt, and adopt water saving technologies, but thresholds exist beyond which hazards overwhelm societies or ecosystems, and water scarcity risk increases suddenly and unpredictably. These thresholds depend on location, value and activity (Parry *et al.*, 1996) and are difficult to predict. Incorrect threshold projections (including those of Malthus to food security, the Ehrlichs' to biodiversity in the 1980s, and UNEP to desertification in the 1970s) underpin many of the incorrect assessments of environmental risk.

Critically, water scarcity imposes risks on markets and social stability. In addition, the manner in which people respond to water scarcity (using groundwater more intensely, the opportunistic breaching of legislation, violation of environmental flow requirements, pursuing unilateral strategies of self-protection, and becoming embroiled in conflict) involve additional risks, many of which are not attributed directly to water scarcity. Analyses of such issues tend to under-represent the problem by ignoring the feedback loops that often compound water scarcity risks. It is also important to note that risks arise from water scarcities which are not directly related to the human

[2] Water supply is a public good because in most instances the supplying of water to one person does not prevent delivery of water to another. Similarly the ecological goods and services made possible by water can be consumed jointly. In cases of water scarcity such allocation issues do become critical. The opportunity cost of water increases imposes direct costs and benefits on individuals and companies. In these instances water is a private good.

use of water. For instance, scarcity and consequent reduction in river flows can increase the risk to subsistence and commercial fisherman; and reduction in groundwater levels can cause forest dieback, putting the livelihoods and businesses of another set of stakeholders at risk. A failure to consider indirect risks further distorts the time-scale of water scarcity events. Inadequate access to water (for people and business) is the obvious, but by no means the only, risk arising from water scarcity.

A more general problem with risk analyses relates to understanding of risk itself, and the fact that the influences of risk are dependent on perceptions. Levitt & Dubner (2005) compare the deaths caused by kitchen germs and BSE (Bovine spongiform encephalopathy, commonly known as *mad-cow disease*), and domestic swimming pools and domestic weapons respectively, and quote Peter Sandman[3] in noting that the "risks that scare people and the risks that kill people are very different" (kitchen germs and swimming pools being the less scary but more dangerous, in the respective cases). The authors note that risk could as easily be depicted as hazard + outrage, and that effective risk communication involves increasing or attenuating outrage depending on the issue. Applying this notion to this study it should be noted that there is very little natural outrage over water scarcity (much like climate change). As such, water scarcity events tend to be un-dramatic and silent. The threat of water scarcity is often perceived as being in the future or manageable by those people that have access to the public discourse, unlike for example, a terrorist attack that is largely uncontrollable, unpredictable and sudden and accordingly is associated with levels of outrage and perceived in popular mindsets to be a greater risk. When the hazard is high, but outrage is low, people tend to under-react. This situation could be redressed; that is, the risk of water scarcity could be increased to represent the actual hazard, through effective communication of how water scarcity impacts upon societies, people, businesses and the environment.

4 WATER-RELATED RISKS TO BUSINESS

There is increasing scrutiny of the manner in which companies, most notably multi-nationals, exploit natural resources. Public discourse highlighting the amount of water used by Coca-Cola, newspaper coverage of the impact of Kenya's cut-flower industry on that country's water resources, or the negative publicity given to the Spanish strawberry industry has given rise to greater business concern over reputations and actions. PricewaterhouseCoopers now emphasizes the need to consider environmental risk as a *portfolio issue* in the light of media vigilance; that is, fund managers should minimize their investment exposure to sectors and companies perceived to be at a high risk of an *exposé* or of litigation.

Today, there is an emerging recognition by the private sector (corporations and representative bodies/forums) around their vulnerability to water stress throughout the supply chain. Importantly, businesses from a range of industries and sectors are simultaneously engaging on water issues, from the food and beverage manufacturers concerned about upstream agricultural water requirements, through household chemical manufacturers concerned about negative water impacts through their products' use, to financial institutions concerned about the risk to their investments. Likewise, there are general categories of risk that will impact companies differently, depending on the purpose of the water use, amount, timing, location, price of water supply and discharge requirements. These factors differ greatly among sectors and companies *yet all* of these risks can hit companies' direct operations as well as their supply chain, ultimately affecting their operational costs, profits, and future growth.

Physical risk is directly related to too little water (scarcity), too much water (flooding) or water that is unfit for use (pollution), each of which is associated with the management of a water resource. Risks can be associated with water resources at the river basin level, or at the supply level; namely sanitation and other infrastructure systems. Even where water is readily available, physical

[3] See [http://psandman.com/index.htm].

risk can emerge from poor management of the resource from actors outside the direct control of companies.

In 2002, Swiss Re reported an increase in claims against *business interruption cover* as a result of periodic water shortages, suggesting that the problem had become more severe (Zanetti *et al.*, 2005). While this may not be the most prevalent or even the most immediate concern for most business operations today, there is every reason to believe that this risk will only increase in the future as demand for water from other users increases. Progressive companies have developed means of buffering themselves against water shortages, but always at an additional cost, and typically without guarantees. The purchase of water licenses has been a feature of asset management plans in Southern England since the 1980 droughts (SEI, 2006). The declaration of a drought triggers a series of costly interventions for water companies, including intensive monitoring, restrictions and public relations programmes. In 2005, for example, Vittel, the bottled water company, was forced to purchase US$ 9 million worth of land, and had to pay land owners an additional US$ 24.5 million in subsidies, simply to protect the supply of clean water to its French bottling plant (Perrot-Maitre, 2006).

Agriculture is commonly perceived as the most vulnerable sector with relation to absolute water shortages. Flower growers on the shores of Lake Naivasha (Kenya) have highly risky futures due, in part, to their own exploitation of the water resource on which their businesses depend, but to an even greater extent, to the cumulative impacts of the industry in a poor management framework. Although 70% of fresh water withdrawals are used in agriculture, industrial uses are also high. In California, for example, the electronics manufacturing industry used 24% of the available water in 1994/1995 (Faruqui, 2003), with every 30 cm of silicon computer chip requiring 8,622 L of de-ionised freshwater (Figuères *et al.*, 2003). In South Africa, the beer producer SABMiller was forced to halt production at one of its plants in 2007 due to water shortages, while Chile's flagship copper industry is being threatened by insufficient water to maintain operations (WWF, 2009).

Reputational risk with regard to water scarcity refers to the exposure of companies to censure and a resulting loss of customers due to perceptions around company decisions (WWF, 2009). Reputation is one of the most important corporate assets, and also one of the most difficult to protect. Reputational risk is harder to manage than other types of risk, largely because of a lack of established tools and techniques, and confusion about who is responsible for it.

The reputational risk to large water-using companies is greater where a catchment is in danger of habitat collapse; an exotic species is in danger of becoming extinct, or where water governance breaks down leading to a *tragedy of the commons* type race to the bottom. Where this scrutiny translates into public *outrage*, companies face dramatically amplified risks, especially when they are judged to be profligate or irresponsible (JPMorgan, 2008; CERES, 2010). Where such crises unfold, there is a tendency for governments and the media to apportion blame, sometimes fairly and sometimes opportunistically. High profile, multinational companies are easy targets for such blame regardless of their relative contribution to the problem.

Regulatory risks: both physical and reputational pressure for water use and discharge can result in more regulation, price increases, and even loss of operating licenses. Most businesses thrive in a stable regulatory regime, and change, particularly when unpredictable, can be a serious problem. Regulatory risks arise when a change in law or regulation increases the costs of operating a business, reduces the attractiveness of investment and/or changes the competitive landscape. Change to the regulatory regime around water can be one such risk. With increased recognition that water and environmental resources are threatened, many companies accept the need for reasonable regulation, as long as it is coherent, predictable and consistently applied. In some cases, business engagement is shifting to cooperative advocacy for regulation of water allocation and licensing from water resources and for regulation of water supply, sanitation access, and pricing in urban settings (WWF, 2009). Regulatory risks also arise when those charged with water management are incompetent, or where that particular water sector is open to corruption. In both cases, the lack of transparency and consistency undermines legislation and its reinforcement, and raises the level of uncertainty as to the long-term viability of business activities, spreading disincentives for future investment.

Financial risks: water shortages translate into higher energy prices, higher insurance and credit costs, and lower investor confidence, all of which further undermine business profitability. Water shortages translate into higher energy prices, higher insurance and credit costs, and lower investor confidence, all of which further undermine business profitability. More common than the risk of not having enough water is the risk that businesses find their comparative or competitive advantage undermined by cost water scarcity-driven cost inflation. As water becomes scarcer, water tariffs and other pricing mechanisms tend to increase, due to greater competition for water between sectors, higher water search costs, the need to drill deeper boreholes, higher pumping costs and the need to recoup the cost of expensive water transport schemes. Water scarcity also adds to energy costs. For example, cold water is essential for the cooling of coal-fired and inland nuclear power plants, and water shortages leading to higher water prices increase the cost of power generation from these plants. Switzerland is forecast to experience a 25% decrease in nuclear power generation by 2020 due to declining water supplies from glaciers which ordinarily power those facilities (OcCC, 2007). This problem became acutely clear in Italy in May 2007, when power plants in the Po Basin were forced into outages due to a lack of water. Ongoing water shortages in Australia, caused by below average rainfall with excessive water use could disrupt power supplies from the Snowy Hydro plant, responsible for 3.5% of Australia's grid energy. Shortages would raise energy costs and threaten almost half of the energy supplied to Canberra (WWF, 2009).

5 COMPANY ACTION ON RISK

Other risk categories have been described related to market requirements, geopolitics and investment criteria (Pegram, 2010) and have a bearing on how companies respond to risk incidents and water scarcity. Increasingly, companies are recognizing these risks and the importance of water in their production and supply chains. As the interest and research in risk has increased, a number of corporations have adopted water footprint studies as a first step (see SAB Miller & WWF, 2009). While business uptakes of water footprint measures has led to water requirements and discharges being better accounted for in terms of direct operations and value chains, the complex characteristics of the water cycle necessitate best practice in areas such as policy engagement, as well as measures and assessments if meaningful and lasting outcomes are to be obtained.

JPMorgan is applying the water risk concept to key sectors and studying the implications for financing; and arguing for increased disclosure of water dependencies in supply chains. At the same time, the mainstreaming of water scarcity and climate change risk in the popular media has increased public awareness of water issues. It is apparent that certain companies may feel the water squeeze at an operational level and throughout their supply chains, from investors who are wary of risk. Governments that are managing stressed water resources will do so in the increasing glare of media coverage and customer awareness.

At a particular moment in time, the risk exposure of a company to water may be acceptable, but it can change rapidly. Historically change has generally not impacted much on business operations, because they are generally involved in higher value or strategic use of water than agriculture, as the dominant use and therefore storage has been developed and/or water has continued to be allocated, even in stressed watersheds. This buffering is not guaranteed to continue, because an increasing number of river basins internationally are already stressed or *closed* to further development and growth. Consider also:

a) Further population and economic growth, together with increasing climate variability will exponentially increase stress.
b) Globalization and communications technology has increased communities' power to exert political and reputational pressure
c) Awareness, understanding and application of environmental, social and economic regulation by governments is improving (albeit marginally).

The consequence is that acceptable risk today may become unacceptable under changing contexts. With physical water stress, communities, interest groups and politicians become more vociferous (particularly where poor water management arrangements exist). The associated dispute and conflict often leads to poor (knee-jerk) water management decisions, inappropriate regulatory responses and/or unfair targeting of large water users, with corresponding physical, regulatory and reputational risks, respectively.

6 BEYOND FOOTPRINT: ENGAGING PUBLIC POLICY

Public policy attempts to define the rules, the intent and the instruments for government to implement water management. Public policy functions that have direct bearing on companies' interface with water include: development of policy and legislation around water, planning and implementation of water resource allocation and management, water infrastructure development and operation, management and delivery of water supply and sanitation services, and protection of water resources and natural systems.

While it is clear that many firms have sought to drive down risk through activities such as community engagement, or efficiency measures, for most some level of engagement *beyond footprint* is inevitable. That is, certain activities outside the policy arena serve to maintain a social license to operate and buffer against sudden shocks in water use, pollution or regulation. However, where uncertainty remains, the consideration of public policy engagement for stability and consistency is strong. Deciding how much uncertainty a company may be willing to live with, *versus* spreading risk to search for potential new areas of sourcing or manufacture are key considerations. While regulatory compliance (or even exceeding requirements) in a company's operations is a necessity to manage these risks, it is typically not sufficient. Engaging in a public policy process allows for opportunities to articulate the common interests of stability and cooperation, as opposed to merely competing over a resource that is becoming scarcer and therefore more socially, ecologically, and economically valuable.

Moreover, there is a growing recognition by businesses that they can and should play a larger role in achieving water-related policy goals, as well as increasing expectations by society for businesses to transparently participate in regional and international water governance efforts[4]. Especially in the regions under high water stress, or where substantial populations lack safe and affordable water for basic needs, work with local stakeholders including water agencies, community groups and other industry water users to share and manage limited resources more equitably and efficiently may be both desirable and expected.

There are various drivers for corporate engagement in public policy such as:

a) To manage short-term (physical) water risks: when accident or natural disasters cause disruption of local water supply, government and business have shared interest and need to quickly address impacts, requiring collaboration such as information sharing/dissemination, management support, financial contribution.
b) To reduce mid- to long-term (physical and regulatory) water risks: systemic water-related risks and uncertainty in water supply and quality may be reduced by helping government establish and implement stable and effective water resource policy. Activities include ongoing policy engagement with local/state government and formulation of water policy through multi-stakeholder fora.
c) To reduce reputational risks: alignment of corporate water management strategies with public policy/public interests in reliable and accessible clean water supply will reduce risk.

---

[4] See: UN CEO Water Mandate, part of the UN Global Compact and the Guide for Responsible engagement in public policy, as well as World Economic Forum debates.

The engagement can take various forms, including advocacy or lobbying, self/voluntary regulation, partnership with government and local authorities, financial support to build water infrastructure and/or to advance police objectives, etc. These can be done at different levels, ranging from local, to catchment/regional, to state/national level.

7 CONCLUSION

Risk is a language that the private sector understands well. The complex nature of water has become real to many companies and the recent upsurge in sectoral attention to the issue, while one of opportunities for investment is also one of great uncertainty. Risk is transferred when water is poorly managed and there are varying degrees to which a stakeholder can cope. Companies are generally well equipped to handle such events, but in responding to crisis might be accused of capturing gains through access to power and debate. Water is ideally managed in the public interest which places many sectors into cooperative as opposed to competitive environments with poorer users, other sectors and the environment.

It must be noted that global trade and investment in agricultural commodities is increasing from private sector interests. Their risk exposure differs greatly form those companies who face consumer boycotts, media attention or have large sunk capital in water scarce environments.

Outstanding questions remain over their activities and the potential to exacerbate water policy, water trade, water management or policy capture, and to what degree these activities are interpreted in the media and regulatory environments, potentially increasing and transferring risk on to those least able to cope, including more risk exposed private sector entities.

REFERENCES

Castro, J.E. (2004). Urban water and the politics of citizenship: the case of the Mexico City Metropolitan Area during the 1980s and 1990s. *Environment and Planning A* 36(2): 327–346.

CERES (2010). *Murky Waters: Corporate Reporting on Water Risk. A Benchmarking Study of 100 Companies*. Available at: [www.ceres.org].

Faruqui, N. (2003). Balancing between the eternal yesterday and the eternal tomorrow: economic globalization, water and equity. In: C.M. Figuères, C. Tortajada & J. Rockström (eds.), *Rethinking Water Management: innovative approaches to contemporary issues*. Earthscan, London, UK: 41–68.

Figuères, C.M.; Tortajada, C. & Rockström, J. (2003). *Rethinking Water Management: innovative approaches to contemporary issues*. Earthscan, London, UK. 224 pp.

Gleick, P. (1996). Basic water requirements for Human Activities: Meeting Basic Needs. *Water International*, 21: 83–92.

JPMorgan (2008). *Watching water: A guide to evaluating corporate risks in a thirsty World*. Global Equity Research. 60 pp. Available at: [http://pdf.wri.org/jpmorgan_watching_water.pdf].

Kasperson, J.X. & Kasperson, R.E. (eds.) (2001). *Global Environmental Risk*. Earthscan and United Nations University Press. 574 pp.

Levitt, S.D. & Dubner, S.J. (2005). *Freakonomics: a rogue economist explores the hidden side of everything*. Harper Collins, USA. 245 pp.

OcCC (2007). *Climate change and Switzerland 2050. Expected impacts on environment, society and economy*. Organe consultatif sur les changements climatiques [Advisory Body on Climate Change] (OcCC). Bern, Switzerland. 168 pp. Available at: [http://www.occc.ch/reports_e.html].

Parry, I.; Schimmelpfennig, D.; Lewandrowski, J.; Reilly, J. & Tsigas, M. (1996). *Agricultural Adaptation to Climate Change: Issues of Long Run Sustainability*. U.S. Department of Agriculture (USDA), USA.

Pegram, G. (2010). *Global water scarcity: risks and challenges for business*. Lloyd's 360° Risk Insight, London, UK; WWF-UK, Godalming, UK. 30 pp. Available at: [http://assets.wwf.org.uk/downloads/water_scarcity_aw.pdf].

Perrot-Maître, D. (2006). *The Vittel payments for environmental services: a "perfect" PES case?* International Institute for Environment and Development, London, UK. 24 pp. Available at: [http://www.iied.org/pubs/pdfs/G00388.pdf].

Rogers, P.P.; Llamas, M.R. & Martínez-Cortina, L. (2006). *Water crisis: myth or reality*? Marcelino Botin Water Forum 2004. Taylor & Francis, London, UK. 333 pp.

Rosegrant, M.W.; Cai, X. & Cline, S.A. (2002a). *World Water and Food to 2025: Dealing with Scarcity*. International Food Policy Research Institute (IFPRI), Washington, D.C., USA. 338 pp. Available at: [http://www.ifpri.org/publication/world-water-and-food-2025].

Rosegrant, M.W.; Cai, X. & Cline, S.A. (2002b). *Global Water Outlook to 2025. Averting an Impending Crisis*. International Food Policy Research Institute (IFPRI), Washington, D.C., USA. 36 pp. Available at: [http://www.ifpri.org/publication/global-water-outlook-2025].

SAB Miller & WWF (2009). *Water Footprinting. Identifying and addressing water risks in the value chain*. SAB Miller, Woking, UK; WWF, Godalming, UK. 28 pp. Available at: [http://www.sabmiller.com/files/reports/water_footprinting_report.pdf].

SEI (2006). *Asset management Plans as Risk Reducing Instruments for UK Water Utilities*. Stockholm Environment Institute (SEI), Sweden.

Vatn, A. & Bromley, D.W. (1994). Choices without prices, without apologies. *Journal of Environmental Economics and Management*, 26(2): 129–148.

WWF (2009). *Investigating Shared Risk in Water: Corporate Engagement with the Public Policy Process*. WWF-UK (By G. Pegram, S. Orr & C. Williams), Godalming, UK. 48 pp.

Zanetti, A.; Schwartz, S. & Enz, R. (2005). *Natural catastrophes and man-made disasters in 2004: more than 300,000 fatalities, record insured losses*. Swiss Reinsurance Company. Vol. Sigma 1/2005.

IV
Environmental conservation *vs.* food production

CHAPTER 11

Incorporating the water footprint and environmental water requirements into policy: Reflections from the Doñana Region (Spain)

Maite M. Aldaya
Twente Water Centre, University of Twente, the Netherlands

Francisco García Novo
University of Sevilla, Spain

M. Ramón Llamas
Water Observatory of the Marcelino Botin Foundation, Spain; Complutense University of Madrid, Spain; Royal Academy of Sciences, Spain

ABSTRACT: This chapter is a preliminary hydrologic and economic analysis of the water footprint and environmental water requirements in the Doñana region, which includes the Doñana National and Natural Parks and its surroundings. This region is located at the coastal end of the Guadalquivir valley and comprises the marshes and estuary of the Guadalquivir river in south-western Spain. It preserves one of the largest and most important remaining wetlands in Europe. This initial analysis shows that the environment is the main water user amounting to about 59% of total water use, followed by agriculture with around 40%, and urban and industrial water supply 1%, for an average rainfall year. The green water use by forests represents about 44% of the total water use. Groundwater is also a key factor. It is well known since decades that the small but meaningful depletion of the water table due to groundwater abstraction for irrigation, for urban water supply, and by eucalyptus plantations, has an impact on the natural vegetation. Following the ELOHA approach (Ecological Limits of Hydrologic Alteration) the theoretical environmental flows in the Doñana region, needed to recover the wetlands, amount to about 200 Mm^3/yr including surface and groundwater bodies in average years. The blue water available for human use (total surface and groundwater available minus environmental water requirements) is lower than the agricultural blue water footprint alone (243 and 282 Mm^3/yr, respectively). Within agriculture however the largest amount of blue water is used to produce water-intensive low-economic value crops such as rice. These preliminary results seem to indicate that the application of an integrated approach using the water footprint and environmental water requirement analysis, could improve the practice of water resources planning and management, the preservation of ecosystems and the associated livelihoods.

Keywords: Doñana, environmental water requirements, water footprint, green water, groundwater

1 INTRODUCTION

Assessment of water availability, water use and water scarcity has been the subject of increasingly intensive research over the past years. However, the requirements of ecosystems for water have rarely been considered explicitly in such assessments. The present chapter analyses the environmental water requirements for the Doñana region within this framework.

This is a very important issue from a global perspective. Many authors state that the global annual precipitation over continents amounts to about 110,000 km^3 whereas the global water footprint (consumptive water use) is around 8,000 km^3 (Falkenmark & Rockström, 2004; Comprehensive

Assessment of Water Management in Agriculture, 2007; Hoekstra & Chapagain, 2008; WWAP, 2009). Therefore the human direct use of water is less than 8%. However, more accurate figures are needed in relation to the water quantity that is accessible and reliable for human use considering the water requirements for the ecosystems, which is in principle a smaller quantity than the absolute raw water available in nature. Estimations of this type are rare and generally without enough factual evidence. The present study is a preliminary attempt to contribute to fill this gap in the water resources literature. Furthermore, it is the first time that the environmental water requirements are analysed together with the water footprint.

The water footprint is an indicator of water use that looks at both direct and indirect water use of a consumer or producer (Hoekstra, 2003). The water footprint of an individual, community or business is defined as the total volume of freshwater that is used to produce the goods and services consumed by the individual or community or produced by the business (Hoekstra *et al.*, 2009). Water use is measured in terms of water volumes consumed (evaporated) and/or lost polluted per unit of time. A water footprint can be calculated for any well-defined group of consumers (e.g. an individual, family, village, city, province, state or nation) or producers (e.g. a public organization, private enterprise or economic sector). The water footprint is a geographically explicit indicator, not only showing volumes of water use and pollution, but also the locations. The total water footprint breaks down into three components: the blue, green and grey water footprint (Hoekstra *et al.*, 2009). The blue water footprint is the volume of freshwater that evaporates from the global blue water resources (surface water and groundwater) or is lost as a result of the production process. The green water footprint is the volume of water evaporated from the global green water resources (rainwater stored in the soil as soil moisture). The grey water footprint refers to the volume of polluted water as a result of the production process. The grey water footprint method and data however are currently being refined.

The water footprint focuses on human water use. A significant innovation of this work is to emphasize the imperative challenge of considering the environmental water requirements as well. Environmental water requirements or environmental flow requirements refer to the water considered sufficient for protecting the structure and function of a surface or groundwater ecosystem and its dependent species. These requirements are defined by both the long-term availability of water and its variability, and are established through environmental, social, and economic assessment (King *et al.*, 2000; IUCN, 2003). Obtaining clear data on environmental water requirements would permit comparisons between the actual water footprint with the theoretical amount of water available for human use (total water available minus environmental water requirements). This may enable building a transparent, multidisciplinary framework for informing water allocation decisions. It is critically important that a certain volume of water is planned for the maintenance of ecosystem functions and the services they provide to humans. Planning water allocation taking into account the environmental water needs would be helpful to achieve the right balance between allocating water for direct human use (e.g. agriculture, power generation, domestic purposes and industry) and indirect human use (maintenance of ecosystem goods and services) (Acreman, 1998). This is one way to approach a win-win situation balancing and harmonising human and environmental water requirements. This could also contribute to the implementation of the EU Water Framework Directive (2000/60/EC) (WFD), which sets the clear objective of achieving the *good status*, both ecological and chemical, of all water bodies in the EU (surface as well as groundwater) by 2015. This is particularly relevant since Spain is the first country that has included the water footprint analysis into governmental policy making in the context of the WFD (Official State Gazette, 2008).

This report analyses the water footprint of the Doñana region and presents a first and preliminary attempt to quantify the environmental water requirements from the ecosystem perspective. The Doñana region is located at the coastal end of the Guadalquivir valley and comprises the delta and the estuary of the Guadalquivir river in south-western Spain. It preserves one of the largest and most important remaining wetlands in Europe. The final objective is to improve the practice of water resources planning and management, and the condition of ecosystems and associated livelihoods through the application of the water footprint analysis. This report also aims to illustrate the role of the environment as a legitimate *water user* in water resources assessments. For this

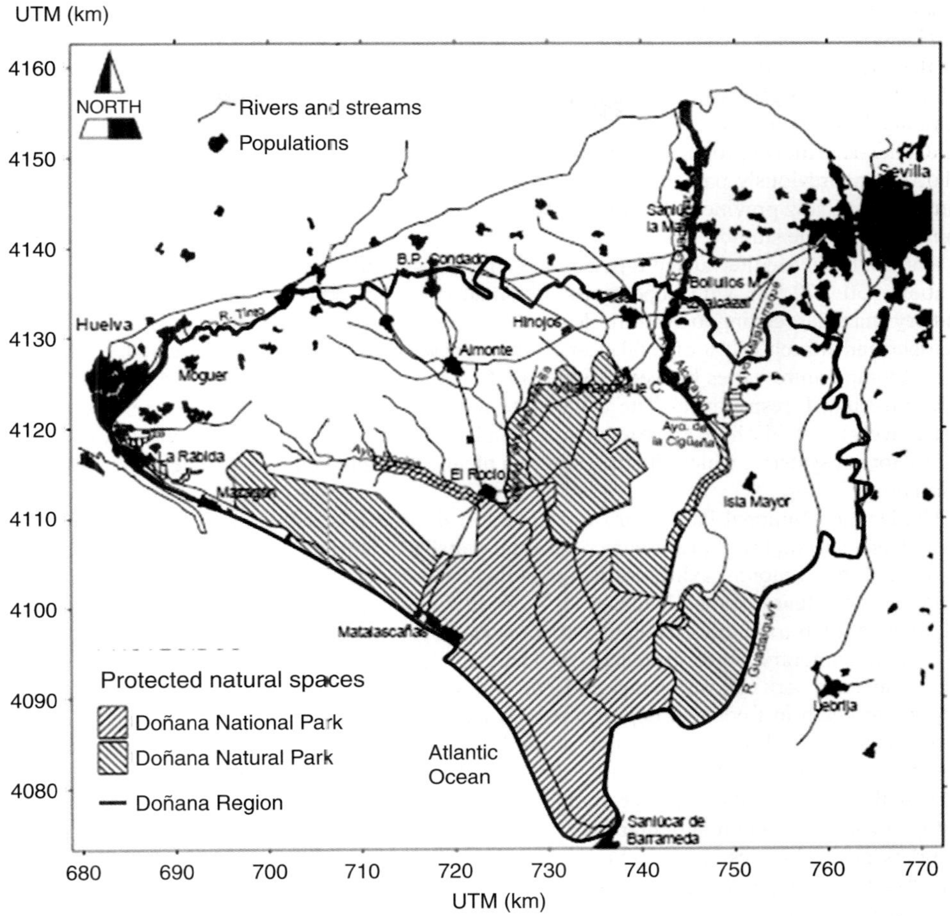

Figure 1. Location of the Doñana National and Natural Parks within the Doñana region.
Source: Modified from Custodio *et al*. (2006).

purpose the present study analyses the current environmental green and blue (surface and ground) water use and the theoretical environmental water requirements following the ELOHA approach (Ecological Limits of Hydrologic Alteration). Due to the limited information on environmental water requirements, rough figures are provided. The water footprint for agriculture, household and industry is provided from the production perspective. In the case of agriculture, notably the main water consumer (representing about 98% of the total human water consumption), a hydrologic and economic analysis is presented. Finally, the actual blue water consumption is compared with the theoretical blue water available for human use (total blue water available minus environmental flow requirements), analysing the challenges and opportunities for water resources management in the Doñana region. Steps for future research in this field are also suggested.

2 STUDY AREA

The Doñana region (2,900 km^2), in south-western Spain, extends along the coastal plain of the Gulf of Cádiz from the right bank of the estuary of the Guadalquivir river to the estuary of the Tinto river (Custodio *et al*., 2009). It includes several territories with a different degree of environmental protection: Doñana National Park, Doñana Natural Park and its area of influence (Figure 1). The

Doñana National Park (*Parque Nacional de Doñana*, traditionally referred as *Coto de Doñana*) is a UNESCO World Heritage and a Ramsar site, and a prised wildlife refuge. It covers an area of 542 km^2, while the peripheral protected buffer zone or Natural Park has 537 km^2 (Andalusian Regional Government, 2009). The Doñana region includes the following 14 municipalities: Almonte, Bonares, Bollullos Par del Condado, Hinojos, Lucena del Puerto, Moguer, Palos, Rociana del Condado (Huelva province), Aznalcázar, Puebla del Río, Pilas, Villamanrique de la Condesa, Isla Mayor, previously named Villafranco del Guadalquivir (Sevilla province) and Sanlúcar de Barrameda (Cádiz province). Data from Sanlúcar, on the left bank of Guadalquivir river, were not considered for the study.

Doñana has a Mediterranean type climate with oceanic influences. The average annual rainfall is about 550–570 mm, with important interannual variations; rainy years approaching 1,000 mm and dry years descending to 300 mm (Muñoz-Reinoso, 2001). This rainfall mainly occurs between October and March (80% of total yearly amount), whereas the summer is very dry. The mean annual temperature varies between 16 and 17°C, with mean temperatures of 24.7 and 24.4°C in July and August, respectively. The winters are mild (10.9 and 10.0°C in December and January, respectively). Both the high mean temperature and the variation of rainfall determine the water availability for ecosystems, which fluctuates from a plentiful supply in winter to a persistent drought in summer.

The Doñana National Park is one of the largest European breeding areas for migrating birds. It shelters a vast and well preserved continental marsh (25,000 ha), which remains flooded during winter and spring months exhibiting a high biological productivity. Although it originated as a tidal marsh in the estuary of the Guadalquivir river, it is now separated from the river only receiving the flow of small tributaries. Doñana Parks present a great diversity of biotopes: marsh, swamps, with large temporary ponds, beaches, fixed and wandering dunes, with temporary and permanent ponds, pine and cork oak woodlands and a diverse shrubbery bearing Mediterranean scrub in dune ridges and heath in floodable depressions (García Novo & Marín Cabrera, 2006). The marshes host four threatened bird species, one of the biggest heronries in the Mediterranean, and host a figure of 500,000–600,000 wintering waterfowl. An estimated figure of 6 millions of migratory birds will pass by the Park on a yearly basis, either commuting from N to S Europe or else from Europe to Africa through Gibraltar Strait. 78% of European birds have been recorded in Doñana. The Park also preserves an almost complete mammal fauna with fallow deer and red deer, wild boar, European badger, Egyptian mongoose, and endangered species such as the Iberian Lynx and the Spanish Imperial Eagle. The flora includes some 900 vascular plants. A summary of Doñana species biodiversity including vertebrates, plants and plankton species (2,200 species in all) has been compiled by García Novo & Marín Cabrera (2006).

In 1963, the World Wildlife Fund for Nature (WWF), together with the Spanish government, purchased a section of 6,974 ha of Guadalquivir marshes and surrounding sandy areas building up the *Reserva de la Estación Biológica de Doñana*. A few years later, in 1969, a wider area including the Reserva was preserved as Doñana National Park (34,625 ha). On a later date (1979) the protected area was further enlarged to 77,260 ha including the National Park and surrounding areas with limited resource exploitation for environmental protection. The Natural Park of Doñana was created in 1982 incorporating the protection areas and after some later additions, the National Park extends for 54,720 ha and the Natural Park for 53,709 ha making a vast protected area of 1,084 km^2. On 1992 an International Commission of Experts presented a Sustainability Plan for the development of Doñana region (Castell *et al.*, 1992). It was the earliest example of sustainability plans ever published and its implementation has secured the preservation of the National Park to the present day.

The most serious incident occurred in 1998 when the retention walls of a mine tailing deposit from a pyrite mine (in Aznalcollar, some 60 km N of the Park) collapsed (Grimalt *et al.*, 1999). The discharge of about 2 Mm^3 of minerals and 4 Mm^3 of mineralised water flowed into the Guadiamar River, reaching the Natural Park and the Guadalquivir Estuary in the following days. The spill covered 4,286 ha of land surface, with 2,500 ha of cropland, contaminating soils with arsenopyrites and pyrolusite, releasing Ag, As, Bi, Cd, Co, Cu, Fe, Hg, Pb, S, Sb, Se, Tl and Zn. The accident had significant ecological and economic consequences, severely affecting organisms of Guadiamar river and destroying the agriculture of the valley. Considerable efforts by the Spanish Government succeeded in an early control of the spill, and the longstanding effort to remediate the contamination and restoring of the area (Manzano *et al.*, 1999). Along with clean up and restoration, the Spanish Ministry of Environment formulated in 1998 the *Doñana 2005* project, which encompasses a series of strategic actions to restore the traditional hydraulic dynamics of Doñana Parks.

More than twenty years ago concern was expressed over the impact of mass tourism and intensive irrigated agriculture in the region outside but around the National Park (Llamas, 1988; 1989; Suso & Llamas, 1993). This concern continues and has also been expressed by international institutions (UNEP, 2004; WWF, 2009b). There have been fears that these activities were causing the depletion of regional aquifers, leading to a fall in groundwater levels and a gradual reduction in the extent and duration of seasonal flooding in the marshes (Zunzunegui *et al.*, 1998; Serrano & Zunzunegui, 2008).

3 METHOD

The present study estimates the water footprint within the Doñana region considering the green, blue and grey water components for the most representative crops and the blue water component for industrial products and urban water use for an average rainfall year (2001). Within the blue water component, the volumes of surface and groundwater consumption are differentiated.

Water footprint of primary crops

The green, blue and grey water footprints of primary crops were calculated following the methodology described in Hoekstra & Chagapain (2008) and Hoekstra *et al.* (2009). The total crop water requirement, effective rainfall (fraction of rainfall that is stored in the soil and available for the growth of plants) and irrigation requirements per region have been estimated using the CROPWAT model (Allen *et al.*, 1998; FAO, 2009a). The calculations have been done using climate data from the nearest and most representative meteorological station located in the major crop-producing regions (Sevilla-Tablada) and a specific cropping pattern for each crop according to the type of climate.

The green water footprint of the crop (m^3/t) [t = tonne = 1,000 kg] has been estimated as the ratio of the green water use (m^3/ha) to the crop yield (t/ha), where total green water use is obtained by summing up green water evapotranspiration over the growing period. Green water evapotranspiration is calculated based on the CROPWAT model outputs, as the minimum of effective rainfall and crop water requirement with a time step of ten days.

The blue water footprint of the crop has been taken equal to the ratio of the volume of irrigation water consumed to the crop yield. The irrigation water consumed is based on the CROPWAT model output and estimated as the difference between the crop water requirement and effective rainfall on a ten-day basis. When the effective rainfall is greater than the crop water requirement, the irrigation requirement is equal to zero. The total evapotranspiration of irrigation water is obtained by summing up the blue water evapotranspiration over the crop growing period.

Even with some limitations, the grey water footprint was used to estimate the nitrogen pollution from agriculture. The grey water footprint of a primary crop (m^3/t) is calculated as the load of pollutants that enters the water system (kg/yr) divided by the maximum acceptable concentration for the pollutant considered (kg/m^3) and the crop production (t/yr) (Hoekstra & Chapagain, 2008). In this study, nitrogen was chosen as an indicator of the impact of fertiliser use in the production systems. The total volume of water required per tonne of N is calculated considering the volume of nitrogen leached (tonne/tonne) and the maximum allowable concentration in the surface or groundwater bodies. The quantity of nitrogen that reaches free flowing water bodies has been assumed to be 10% of the applied fertilization rate (in kg/ha/yr) (following Hoekstra & Chapagain, 2008). In line with the European Nitrates, Groundwater and Drinking Water Directives, the standard for nitrate is 50 mg/L (measured as NO_3^-). This is very similar to the drinking water standard recommendation by the US Environmental Protection Agency (EPA, 2005), which is 10 mg N/L, equivalent to about 45 mg NO_3^-/L. The standard of 10 mg N/L was used to estimate the volume of water necessary to dilute polluted leaching flows to permissible limits. This is a conservative approach, since the natural background concentration of N in the water used for dilution has been assumed negligible.

Finally, in this study we have included the concept of economic water productivity (€/m^3) to assess the production value, expressed in market price (€/t) per m^3 of water consumed when producing the commodity (m^3/t).

Environmental water requirements

Concerning the environmental water requirements, the blue water requirement values (including surface and groundwater) were based on already existing calculations for the Doñana region, including the National Park, Natural Park and its region of influence, by WWF (2009a). These estimations were based on the ELOHA approach (Poff *et al.*, 2009), which many consider a scientifically robust and flexible framework for assessing and managing environmental flows, where knowledge is systematically organized within a context of decision making. This method, considering the flow-ecology relationships and covering a wide range of issues, can optimally support comprehensive regional flow management. However, it can only be applied where hydrology is understood. It reflects the published research to date and not all ecosystem responses. Furthermore, one has to be very careful when making predictions based on this method.

Similar to the water footprint of primary crops, the green water use by natural forests was estimated using the CROPWAT model (FAO, 2009a). The eucalyptus water consumption was taken from CSIC.

4 DATA SOURCES

Data have been compiled from different sources.

- *Agricultural data*
  Data related to area (crop area both rainfed and irrigated) were taken from Custodio *et al.* (2006). Data on average rainfed and irrigated crop yield (kg/ha) at provincial level (Sevilla) and crop market prices were taken from the Agro-alimentary Statistics Yearbook of the Spanish Ministry of Agriculture, Fisheries and Food (MARM, 2009) for the average rainfall year 2001. As a first approximation an average rainfall year has been analysed. In further studies, however, it would be interesting to account for temporal variability within and between years. Climate data for the water footprint calculations have been taken from the CLIMWAT database (FAO, 2009b), using climate data from Sevilla-Tablada meteorological station. Crop coefficients for the different crops were obtained from Chapagain & Hoekstra (2004), Allen *et al.* (1998) and FAO (2009b). Data on the application of nitrogen fertilisers for Andalucía have been obtained from the Spanish Ministry for the Environment (MIMAM, 2007).
- *Hydrologic data*
  Data related to water origin (surface and groundwater) by agricultural region were taken from the 1999 Agrarian Census of the National Statistics Institute (INE, 1999). Data concerning urban water supply were obtained from Custodio *et al.* (2006).
- *Environmental data*
  Limited data are available on environmental water requirements. Data on environmental blue water requirements were taken from WWF (2009a). As previously mentioned, data on environmental green water use by forests were estimated using the CROPWAT model (FAO, 2009a). A constant crop coefficient of 1 –as recommended for rubber trees, tea and conifer trees (Allen *et al.*, 1998)– was assumed representative for natural forests (Roost *et al.*, 2008).

Table 1. Land uses in the Doñana region.

Area	km^2
Doñana region	2,900
Aquifer Almonte-Marismas (AAM)	2,400
Permeable area of AAM	1,840
Wetland	1,500
National Park	542
Natural Park	537
Cultivated area	877
Forested area	1,210
Santa Olalla lake	0.48

Source: Modified from Custodio *et al.* (2006).

5 LAND AND WATER USE IN THE DOÑANA REGION

As mentioned in Section 3, the analysis has been carried out for the Doñana region (2,900 km^2), which includes the Doñana National Park (542 km^2), the Natural Park (537 km^2), and its area of influence (see Figure 1).

5.1 *Land uses in the Doñana Region*

The Doñana region is characterised by an extreme polarization between natural, urban, and agricultural uses (García Novo & Marín Cabrera, 2006) (Table 1 and Figure 2). On the one hand, natural uses consist of marshes, river banks and forest of great ecological and environmental value and limited economic use (Andalusian Regional Government, 1999). On the other hand, urban and agricultural uses are intensive, highly productive farming (rice fields and strawberries in greenhouses), which coexist with more extensive and traditional farming (vineyards and olive groves) (Andalusian Regional Government, 1999).

Urban uses are concentrated in the Sevilla-Huelva corridor, where all the main municipal towns are located. Apart from this corridor, just two coastal touristic resorts (Matalascañas and Mazagón) are present.

Concerning the agricultural land use, during the last 30 years, new crops have been introduced in the region, such as strawberries and raspberries. Even if they are not spatially extensive, they are very significant, not only in economic but also in environmental terms since the agricultural intensification and the widespread use of plastic greenhouses brings along environmental consequences (Andalusian Regional Government, 2002).

Ecosystems
According to the vegetation composition and ecological processes, Doñana comprises the following ecosystems (García Novo, 1997; Custodio *et al.*, 2009):

- *The beach*. The littoral of the National Park extends for 25 km, from the Guadalquivir estuary in the East to Matalascañas, a touristic resort in the West. A second leg about 30 km long, primarily belonging to the Natural Park, extends from Matalascañas to Mazagón resort in the Ría de Huelva. This long beach, one of the few unspoiled littoral landscapes of the Iberian Peninsula, describes a wide arch. The light coloured and fine sands of mixed eolian and estuarine origin indicate a long distance transport downcurrent in the littoral West-East stream that from South Portugal reaches the Guadalquivir estuary. The beach is rich in marine birds (seagulls, oystercatchers) and plovers.
- *Stabilized sands*. Within the stabilized sands, it is the hydrological behaviour of the soil surface as a discharge or a recharge area of the aquifer that makes the main environmental difference.

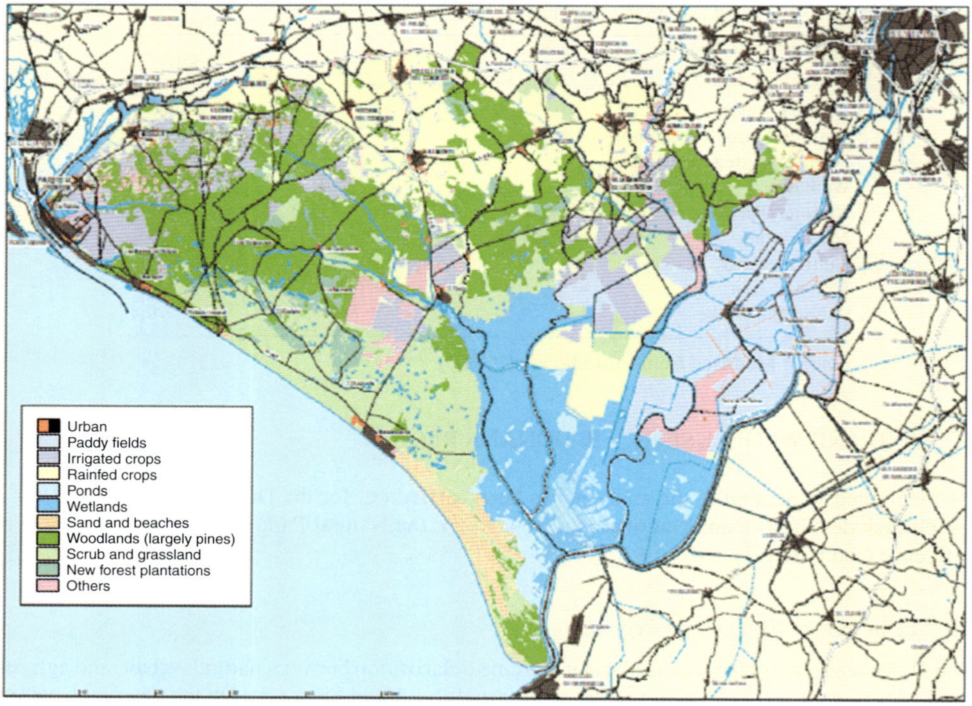

Figure 2. Land uses in the Doñana region.
Source: Modified from Andalusian Regional Government (2003).

A modest topographic gradient of about 6 m in height difference, opposes dry *elevations* to wet *depressions* where soil water is available during the dry summer. Aquifer discharge of low conductivity waters (0.5–4 mS/cm) in depressions feeds shallow temporary ponds or deeper permanent ponds. The inherited dune morphology of dune crests or sand bars (naves) further differentiates elevations.

- *Woodlands*. Woodlands represent the mature terrestrial ecosystem with the largest structural development and more intense environmental control. In the mobile dunes there are two types: planted or seminatural *Pinus pinea* woodlands with remnants of junipers, *Juniperus oxycedrus ssp. macrocarpa*. In the stabilized sands woodlands are represented by the modest sized *Juniperus phoenicea ssp. turbinata*, in the dry sand bars, cork-oaks *Quercus suber* and ash tree, *Fraxinus angustifolia* in the moister sands and poplars, *Populus alba* and tamarisks, in some foodable areas: arranged according to water conductivity from low (*Tamarix africana*) to medium (*T. canariensis*) and high (*T. gallica*) values. Eucalypts *Eucalyptus globulus* and *E. camaldulensis* were extensively planted in the early 1960s and 1970s for timber production. Most of the eucalyptus plantations have been largely eradicated, few of them surviving as abandoned plantations with a poor understory. The predominant woodland type is dominated by cork-oaks.
- *The Mediterranean scrublands*. Forest degradation resulted in sandy soil impoverishment and the spread of shrubbery vegetation. Prescribed fire regimes for range management induced pyrophitic scrub vegetation that largely dominates Doñana vegetation. Sharp boundaries between scrub types rarely occur. Transitions or clines are the rule both in space (horizontal or vertical structures) and time (succession). Down slope, *monte blanco* (white scrub) ecosystem develops with the thorny *Ulex australis* and *Ulex argenteus ssp. subsericeus*, *Halimium halimifolium*, whose light colored leaves dominate vegetation shade. *Monte blanco* type occurs on the 1.5 to 3 m range above the piezometric surface, receiving some summer water supply. In the wetter

areas, at the bottom of the slopes, the *monte negro* ecosystem (black scrub), a heath, dominates the scrubbery with a composition closely resembling that in the wetter slacks: *Calluna vulgaris*, *Erica umbellata*, *E. scoparia ssp. scoparia*, *Genista triacanthos ssp. triacanthos*, *Cistus psilosepalus*, *C. salvifolius* and their hybrids are widespread. Sites suffering temporary flooding let other species be present in the *hygrophytic monte negro* ecosystem with *Erica ciliaris*, *Molinia caerulea ssp. arundinacea*, *Imperata cylindrica*, *Ulex minor*, *Scirpus holoschoenus* and *Saccharum ravennae*, that can attain 4 m of height. *Monte negro* type grows to a short distance above the water table on the permanent discharge surfaces of the aquifer, benefiting from permanent soil water supply. Soil water logging occurs when water table surfaces, inducing the *hygrophytic scrub* at the bottom of the slopes or in upwelling surfaces at any height (Muñoz-Reinoso & García Novo, 2004).

- *The grasslands*. In the stabilized sands some permanent grasslands are recognized. In the small patches among shrubs a seasonal therophytic vegetation develops. In the wide ecotone (locally known as Vera) making the transition from sands to the marsh, a series of grassland types have been described. Water availability, flooding duration, water conductivity and grazing pressure control plant species composition.
- *Hygrophytic vegetation*. In large depressions and around the ponds, fringes of hygrophytic vegetation develop, often forming linear forests closely resembling the river bank forests with ashes, *Fraxinus angustifolia ssp. angustifolia*, poplars, *Populus alba*, and willows, *Salix atrocinerea* and the dense hygrophytic scrubbery described above, that withstands temporary flooding. Closer to the water surface, grassland appears which is rich in sedges, rushes, buttercups and flooding resistant species.
- *The ponds*. Locally known as lagunas, the ponds of the stabilized sands are variable ecosystems both in time and space. Shallow (20–40 cm deep) ponds are small (0.5–1 ha), usually presenting terrestrial vegetation which will be killed during heavy rainfall periods when ponds are filled. A strong successional cycle starts after pond desiccation. Deeper ponds with a stable aquifer discharge are lined along dune fronts in both active and stabilized dunes. These large pond attain 4 m of depth and 100 ha (Lagunas de Santa Olalla-Dulce) under high water conditions, showing no terrestrial vegetation invasions.
- *The marshes*. Doñana marshes, extending for around 28,000 ha, constitute a vast seasonal freshwater wetland of international importance. They are Europe's largest sanctuary for migrating birds. They were declared National Park in 1969, Biosphere Reserve in 1980, Important Wetland Site under the Ramsar Convention in 1982 and Natural World Heritage Site in 1984 (García Novo & Marín Cabrera, 2006). The variable winter rains flood Doñana marshes forming a wide shallow lake that dries up during spring and summer. Flooded areas are very variable in depth and turbidity, and they change depending on the amount and pattern of rainfall and time of the year.

5.2 *Water in the Doñana region*

Water availability is the controlling factor for the maintenance of both aquatic and terrestrial ecosystems, the composition of plant communities, their biomass, productivity and succession (Muñoz-Reinoso, 2001). Under natural conditions, most of the contributions of water were coming from precipitation, several rivers and streams (Guadalquivir, Guadiamar –now diverted, La Rocina, El Partido, Las Cañadas, etc.) including regular entries through the Guadalquivir estuary and aquifer discharge (García Novo & Marín Cabrera, 2006) (Figure 3). The upstream and midstream of the Guadiamar riverflow (there is a dam upstream and diversion at the end) and the Guadalquivir river (densely regulated watershed), however, are to some extent anthropically controlled by means of hydraulic engineering.

Groundwater is crucial for the maintenance of rivers and wetlands of Doñana. The aquifer system of Almonte-Marismas feeds upwelling surfaces at various points on the periphery of the marshes, allowing the formation of temporary ponds characteristics of these areas, as well as springs that drain into the marsh. Groundwater recharge occurs mainly by infiltration of rainfall through

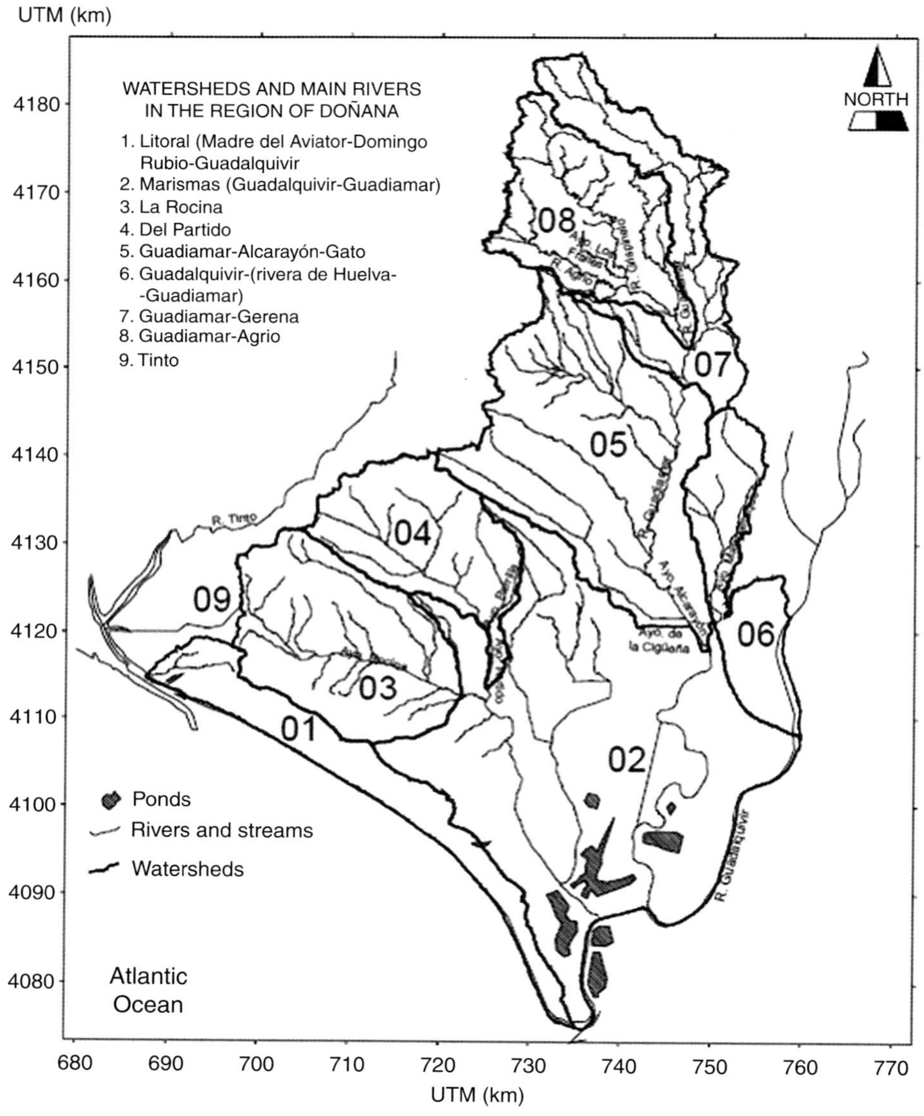

Figure 3. Rivers and their watersheds in the region of Doñana.
Source: Custodio *et al.* (2006).

permeable saturated sands and to a lower extent by irrigation return flows, by infiltration of used water and by small contributions of the Guadiamar river (Custodio *et al.*, 2006). The aquifer on the other hand continuously discharges to the sea along the coast, to the main streams (La Rocina), to the Tinto river and its affluents and to the ecotones (contact areas between the sands and the marsh), which at the same time maintain phreatophitic vegetation, both natural (*monte negro* and the remaining gallery forests) and those planted trees capable to reach the aquifer level with deep roots (as eucalypts). Temporary and permanent ponds (Santa Olalla-Dulce system) are maintained by aquifer discharges the water level closely following aquifer fluctuations (Serrano & Zunzunegui, 2008). These natural discharges of groundwater and the small streams (or *caños*) and associated streams are an essential trait to explain the rich ecological diversity of Doñana region (García Novo *et al.*, 1996; 2007; Custodio *et al.*, 2006).

6 ENVIRONMENTAL WATER USE AND REQUIREMENTS WITHIN THE DOÑANA REGION

In this section, the current environmental green and blue water use is estimated, followed by the analysis of the theoretical environmental blue water requirements.

6.1 *Environmental blue water use*

The current environmental blue water use in the Doñana region amounts to about 154 Mm^3/yr, 116 related to surface and 38 to groundwater (Andalusian Regional Government, 2002; 2003). Within the Doñana region, in spite of the complexity and interrelation of the hydrologic cycle, the following blue water units can be differentiated with certain degree of independence, based on their hydrodynamic behaviour (Andalusian Regional Government, 2002):

- The Tinto river basin is an independent hydrodynamic unit composed on the one hand by small streams draining into the Tinto River, and on the other hand by the Domingo Rubio and Madre del Aviator streams, that drains into the ocean.
- The coastal strip of the Almonte-Marismas aquifer. To the south of the hydrogeological watershed defined by the Almonte-Marismas aquifer, rainfall infiltrates in large amounts into the aquifer, subsequently draining into the Atlantic Ocean. In total this flow adds up to about 38 Mm^3/yr (Andalusian Regional Government, 2002).
- Strip edge of the natural marsh and streams of the central area of the Doñana Natural Park (La Rocina, El Partido, Cañada Mayor, Juncosilla and Portachuelo). In this area the underground flows of the aquifer towards La Vera and La Retuerta (approximately 30 Mm^3/yr), together with the irregular surface water contributions (of the order of 86 Mm^3/yr), feed the marshes, or to a lesser extent other wetlands and endoreic systems (Andalusian Regional Government, 2002).
- Guadiamar river and its affluent streams. The non-regulated flows of the Guadiamar river, along with the urban and industrial waste water discharges, are partially used for irrigation. The surpluses are dumped into the Guadalquivir river. Just in extraordinary avenues it overflows and incorporates its contributions to the National Park marshes through the Lucio del Cangrejo.
- Guadalquivir river no longer flows into the Park marshes but a large network of channels and ditches provided with floodgates let estuary waters from Brazo de la Torre and Guadalquivir river enter marshes and paddy fields for irrigation and aquaculture. About 400 Mm^3/yr are pumped from the Guadalquivir river to irrigate the rice fields, most of which returns directly to the river or through the Brazo de la Torre. Just in extraordinary avenues some defence systems are overflowed and marshes are flooded from the estuary. This has occurred for the last time in 1973. The large precipitations during 2009/10 winter made marshes flood from rainfall and small tributaries to unusually high water levels (1.5–2 m). Large water volumes were finally discharged to estuary.

6.2 *Environmental green water use*

Another question that should also be incorporated into water allocation decisions and environmental water requirement analyses is the green water used by natural vegetation. Hitherto, the environmental flow studies (e.g. Smakhtin *et al.*, 2004) have mainly focused on blue water use without considering the green water evapotranspiration from the natural vegetation.

Due to the limited data availability, the current environmental green water use was estimated only for woodlands, which are mainly located in the western part of the region. This rough estimation is probably underestimating the environmental green water use in the region since neither grasslands,

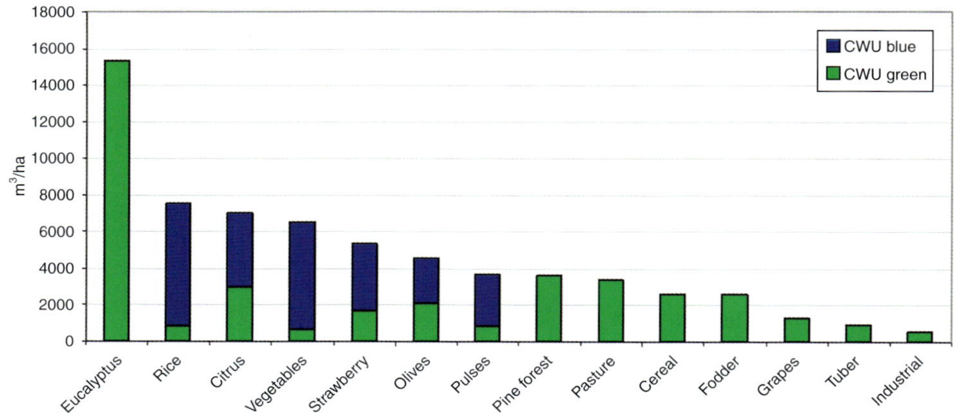

Figure 4. Green and blue water consumption (CWU) of agricultural crops and natural forests in the Doñana region for an average rainfall year (m^3/ha).
Source: Own elaboration.

marsh vegetation nor the scrubland have been considered. Comparing the water consumed by the different crops and natural forests in terms of m^3/ha suggests that the pine forests use less water than all the irrigated crops in the region (3,650 m^3/ha versus 3,700–7,600 respectively), whilst using more water than rainfed crops (550–2,600 m^3/ha) (Figure 4). However, these figures vary depending on the context; for instance, pine trees in the western Doñana region (area of El Abalario) are stunted probably with low evapotranspiration, whereas in Hinojos trees are huge possibly having a high evapotranspiration. In total, in the Doñana region, the environmental green water used by forests adds up to about 442 Mm^3/yr. This calculation is based on pine forest (*Pinus pinea*) water requirements since these are currently the main forest species in the region. The non-agricultural evapotranspiration or environmental green water used by forests contributes significantly to the use of water resources, amounting to about 44% of total water use. These relevant figures, even if generally included in the water balance studies, have not been accounted in the traditional national water accounting systems. In the case of environmental flow studies (Smakhtin *et al.*, 2004), they have mainly focused on blue water use without considering the green water evapotranspiration from natural vegetation. Further research is needed on this topic.

Figure 4 shows the high water consumption by eucalyptus trees. The introduction of eucalyptus trees in most of the western water table area (e.g. Moguer, El Abalario, La Mediana, La Rocina) some 50 years ago also impacted local phreatic wetlands. The increase of evapotranspirated volumes led to a water table drawdown large enough to reduce or eliminate local phreatic discharges to many ponds between El Abalario height, La Rocina stream (Manzano *et al.*, 2005). This seems to have been the case of the formerly permanent pond-complexes of Ribetehilos and La Mediana. Nowadays, however, most eucalypt trees have been removed.

6.3 *Environmental blue water requirements*

The environmental blue water requirements were taken from a detailed estimation by WWF (2009a) using the ELOHA approach. This approach is considered by many a scientifically robust and flexible framework for assessing and managing environmental flows, where knowledge is systematically organized within a context of decision making. According to this study the theoretical environmental surface and groundwater requirements of the Doñana region, in order to recover the wetlands in the National Park, amount to about 200 Mm^3/yr in average years.

In the mentioned study the relationship between hydrology and ecology was considered to link the dynamics of ecosystems, habitats and species. Vegetation is a good performance indicator for its

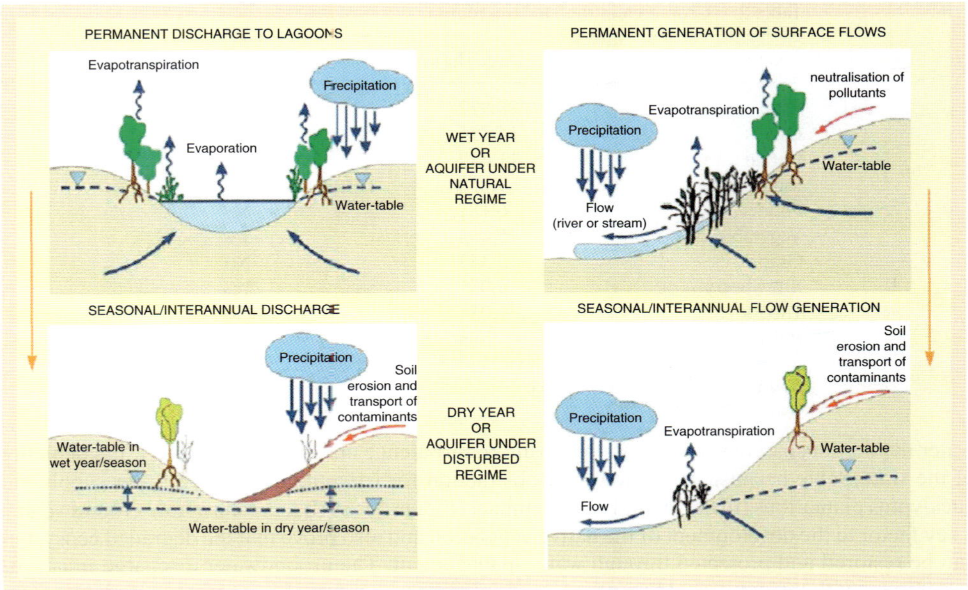

Figure 5. Most frequent types of wetlands depending on groundwater in Doñana.
Source: Manzano *et al.* (2002; 2005).

intrinsic importance and the role played on the various animal groups. For this reason the vegetation response to hydrological changes in Doñana through modelling and analysis of thematic mapping was analysed. For the period 1990–2004 some of the species that need more water have reduced their surface area more than 60%, in some cases reaching 80% reduction. This knowledge should be incorporated in the environmental flow proposals as explained below.

Environmental blue groundwater requirements

As previously mentioned, the Doñana area overlies a large sedimentary aquifer that is of outstanding environmental and human importance, as groundwater plays an essential role both in the generation and maintenance of a wide variety of ecologically important natural habitats, and in the subsistence and development of a human population that depends almost entirely on this groundwater for their domestic water supply and for crop irrigation. This aquifer is officially known as the Almonte-Marismas Hydrogeological Unit, and in practice simply as the Doñana Aquifer (García Novo & Marín Cabrera, 2006).

It is well known since decades that in various ecosystems in the Doñana area there is a clear relation between the depth of the groundwater table and the type of vegetation (Allier *et al.*, 1974; Custodio, 2005) (Figure 5). For instance, as explained in a previous paragraph, the depressions of fixed dunes with winter flooding and where summer water table is shallower than 2 m exhibit *Monte Negro* vegetation with heather *Erica scoparia*, *E. ciliaris* and heath, *Calluna vulgaris* and remnants of former cork oak *Quercus suber* woodlands with strawberry trees *Arbutus unedo*. When water table in summer remains below 2 m of soil surface but not surfacing in winter, *Monte Blanco* vegetation prevails where *Halimium halimifolium* dominates). Should water table remain below 2 m and therefore inaccessible to plant roots all year around, a Mediterranean vegetation resembling garrigue or maquis types, rich in rockroses and aromatic plants grow in an open cover. When water table intersects the land surface forming small ponds in winter, permanent meadows often dominate the scenery (Custodio *et al.*, 2009). These four types of vegetation have decisive influence in the animal community. In a parallel way, aquatic ecosystems are controlled at every level by

Table 2. Crop area in the Doñana region.

Crop type	Area (ha)	Irrigation
Rice	28,922	Yes
Olives	14,755	Yes
Cereal	11,477	No
Industrial	9,034	No
Fodder	6,181	No
Vegetables	5,463	Yes
Grapes	5,343	No
Strawberry	5,081	Yes
Tubers	1,031	No
Pulses	396	Yes

Source: Custodio *et al.* (2006).

water depth, transparency, chemical composition, turbulence and others. Obviously, the depletion of the water table because of groundwater abstraction for irrigation, for urban water supply and eucalyptus plantations (in the 1950s) had an impact (Figure 5). Such facts show how groundwater is a key factor to the development of the native vegetation and also prove that the degraded ecosystem can be restored and protected through water management. The groundwater discharge in *natural* conditions (without significant human intervention) ranges between 25–32 Mm^3/yr to the Vera and 10–15 Mm^3/yr to the Northern ecotone (Custodio *et al*., 2006). Groundwater allocation processes have traditionally not considered environmental values, although this is being included in some Australian states. The long-term viability of groundwater dependent ecosystems requires that they be identified, their water requirements understood and this understanding be built into groundwater allocation processes. This knowledge should also be incorporated in the ecological flow proposals which must find the ultimate conformity and consistency among the legal set.

7 WATER FOOTPRINT WITHIN THE DOÑANA REGION

7.1 *Agricultural water footprint*

The Doñana region crop area is about 87,683 ha according to Custodio *et al*. (2006). Table 2 shows the area dedicated to each type of crop in the year 2001. When looking at rainfed agriculture, cereals, industrial crops, fodder and grapes are the main crops in the Doñana region. Concerning irrigated agriculture, rice, olives, vegetables and strawberries are the major crops in the region. In the case of olive trees complementary irrigation is used. The rice in Doñana amounts to about 33% of total cropped area, completely grown in the province of Sevilla. Rice cultivation is located in the north and east of the Guadalquivir marshes, replacing the natural vegetation. The hydrologic regime of the marshes is almost completely artificial, with the inflows and tidal flats substituted by dikes, channels and gates. The water for the rice is in a way pumped from the Guadalquivir river.

Apart from the environmental water use, agriculture is the second main water user in the region, being the main source of employment and income in the area as well. This sector thus plays a highly important role in its conservation.

Figure 6 shows the water footprint figures in terms of m^3/t for the different crops. The traditional rainfed vineyard and tubers (e.g. potatoes) are two of the most water efficient crops: apart from having a low water footprint are entirely based on green water resources. Vegetables and strawberries, grown under plastic also present a low water footprint, though they are almost entirely based on blue water. On the other hand pulses, cereals, olives and rice show higher water footprints, being the rice almost based on blue water resources.

Within the agricultural sector, rice seems to be the major water user in the Doñana region (Figure 7). The total water used (partly evaporated, partly returned to the catchment) to grow rice is

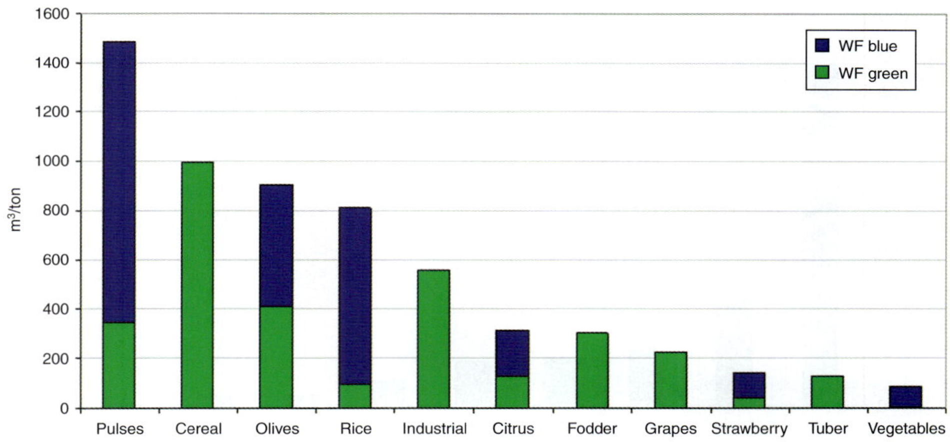

Figure 6. Green and blue water footprint (WF) of agricultural crops in the Doñana region for an average rainfall year (m^3/t).
Source: Own elaboration (see Table 3).

Table 3. Evapotranspiration (ET), crop water use (CWU), yield (Y), production (Prod) and water footprint (WF) for primary crops in the Doñana region.

	ET_g	ET_b	ET	CWU_g	CWU_b	CWU	Y	WF_g	WF_b	WF	Prod	WF_g	WF_b	WF
	mm			m^3/ha			t/ha	m^3/t			t/yr	Mm^3/yr		
Cereal	258	0	258	2,579	0	2,579	3	995	0	995	29,748	30	0	30
Fodder	258	0	258	2,579	0	2,579	9	303	0	303	52,539	16	0	16
Grapes	128	0	128	1,282	0	1,282	6	229	0	229	29,921	7	0	7
Industrial	55	0	55	548	0	548	1	559	0	559	8,853	5	0	5
Olives	209	430	638	2,086	2,500	4,586	5	412	493	905	74,778	31	37	68
Pulses	87	285	372	868	2,852	3,720	3	347	1,141	1,488	990	0	1	1
Rice	89	668	758	893	6,682	7,575	9	95	713	808	271,144	26	193	219
Strawberry	166	375	541	1,662	3,745	5,407	39	43	97	140	195,619	8	19	27
Tuber	97	0	97	966	0	966	7	130	0	130	7,661	1	0	1
Vegetables	62	591	653	618	5,909	6,527	72	9	82	91	391,675	3	32	36
Total												**127**	**283**	**410**
Pine forest	365	0	365	3,652	0	3,652								
Pasture	343	0	343	3,427	0	3,427								

Source: Own elaboration based on area data from Custodio *et al.* (2006) and yield and production from MARM (2009).
ET_g = green water evapotranspiration; ET_b = blue water evapotranspiration; ET = total evapotranspiration; CWU_g = green crop water use; CWU_b = blue crop water use; CWU = total crop water use; Y = yield; WF_g = green water footprint; WF_b = blue water footprint; WF = total water footprint; Prod = production

about 13,000 m^3/ha (CHG, 2009), however, the total water consumed (evaporated or incorporated into the product, not returned to the catchment) is about half this figure (7,600 m^3/ha) according to our estimations (Figure 4). Rice is grown on a very well defined monoculture basis, cross-linked by channels, drainages and tracks. Out of the transformed marsh, rice has also expanded, mainly in an illegal way, entering the public domain in the channeling of Guadiamar and Guadalquivir beaches, in the surroundings of the National Park (Andalusian Regional Government, 2002). Rice is the oldest and most stable crop in the transformed marsh. The paddy fields in the lower Guadalquivir

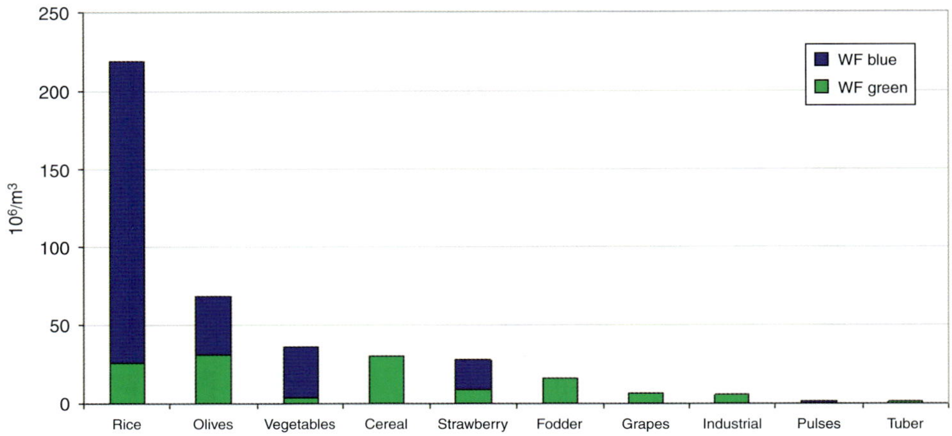

Figure 7. Total green and blue water consumption of agricultural crops in the Doñana region for an average rainfall year (Mm^3).
Source: Own elaboration (see Table 3).

produce about 44% of the national rice production, with about 80% of it located in the former marshes.

Strawberry cultivation has had large environmental consequences such as the lost of forested areas, erosion increase, habitat fragmentation, groundwater depletion, plastic residues from greenhouses and agrochemical pollution (Andalusian Regional Government, 2002; WWF, 2009b). Apart from these environmental problems, strawberry cultivation has a vital economical significance for the region. It is the crop with the highest water apparent productivity adding up to about 6 €/m^3 (Figure 8). The sector generates 55,000 jobs (WWF, 2009b) for a total of 4.5 million day's stints. Most of the workers though are immigrants from Eastern Europe and Northern Africa, with a salary of about 36€/day. This sector receives 24 million €/yr in form of direct public subsidies (average of 3,636 €/ha/yr) (WWF, 2009b), which represents 10–12% of the revenue. Water only accounts for 3.42% of the total costs (WWF, 2009b).

It seems widely known that illegal or alegal water use affects the area neighbouring the Doñana National Park, especially in the rice-growing area of Los Hatos, north of the marsh, and at the west of the National Park, where the strawberry fields are concentrated (WWF, 2006). In Los Hatos area, 12 Mm^3/yr of groundwater are abstracted without formal rights, mainly to irrigate rice, which causes important depletion in the Doñana aquifer.

All in all, the total blue agricultural water footprint amounts to 282 Mm^3/yr (71% surface and 29% groundwater) (Table 4) versus the 200 Mm^3/yr of blue water needed by the environment and 443 Mm^3/yr total blue water available in the region. Along these lines, the Chanza-Doñana project plans to transfer 5 Mm^3 from the Chanza-Piedra-Andévalo system to the Doñana region in order to replace groundwater use in the Doñana region. This volume of water however does not seem enough to satisfy the agricultural demand. Perhaps an option, part of the solution, would be to limit the number of (a)legal water abstractions from the Almonte-Marismas aquifer.

Grey water footprint

Agriculture also implies a threat of eventual damage to the ecosystems from the use of fertilizers and other chemicals (Custodio & Palancar, 1995). To analyse this type of pollution the grey water footprint of nitrogen fertilizer was estimated within the Doñana region. The grey water footprint shows the volume of water required to assimilate the fertilisers that reached the water system based on the average N fertiliser application rate, an assumed leaching percentage of 10% and a nitrogen water quality standard of 10 mg/L (around 50 mg/L of NO_3) (Table 4).

Table 4. Nitrogen fertilizer application and the grey water footprint (volume of water required to assimilate the fertilizers leached to the water bodies) (m^3/t) in the Doñana region.

	Average N fertilizer application rate**	Area*	Total N fertilizer applied	Nitrogen leached to the water bodies	EPA (2005) standard g/m^3	Volume of dilution water required	Production*	Grey water footprint
	kg/ha	ha	t/yr	t/yr	mg/L	Mm^3/yr	t	m^3/t
Cereal	91	11,477	1,044	104	10	10	29,748	351
Fodder	28	6,181	173	17	10	2	52,539	33
Industrial	30	9,034	271	27	10	3	8,853	306
Olives	85	14,755	1,254	125	10	13	74,778	168
Potatoes	51	1,031	53	5	10	1	7,661	69
Pulses	15	396	6	1	10	0	990	60
Vegetables	150	5,463	819	82	10	8	391,675	21

* *Source*: Custodio *et al.* (2006).
** *Source*: MIMAM (2007).

Contrary to what one might expect, the grey water footprint, in terms of m^3/t, is noticeably higher for cereals and industrial crops than for vegetables (see Table 4 and Figure 8). For wheat, fertiliser application rates are on average about 40% lower than for vegetables, but cereal yields per hectare are on average more than 90 times less than vegetable yields (see Table 3). It is widely known, however, that vegetable production is a very intensive form of agriculture in terms of water use and chemical inputs. This becomes clear when one considers the nitrogen load per hectare: the average fertiliser application rate in terms of kg/ha is higher for vegetables than for cereals (150 versus 91 kg N/ha/yr, respectively). One can thus see that the grey water footprint of vegetables compared to cereals is low when expressed per tonne but high when expressed per hectare, which is in line with previous studies (Aldaya & Hoekstra, 2009; Chapagain & Orr, 2009). The same can be observed for the blue water footprint: relatively low for vegetables when expressed per tonne, but relatively high when expressed per hectare. The total grey water footprint related to nitrogen amounts to 27 Mm^3, which is about 6% of the total available blue water in the region (443 Mm^3). This means that 6% of the assimilative capacity of the blue water is being used by nitrogen fertilizers.

7.2 *Industrial and urban water footprint*

Industrial and urban water use refers to total withdrawals of blue water. In the case of Doñana, and according to Custodio *et al.* (2006), all the water abstracted for industrial and urban water supply comes entirely from local groundwater resources and amounts to about 6.7 Mm^3/yr. The main towns within the Doñana region are Matalascañas, El Rocío, Villamanrique de la Condesa and Hinojos. In its boundaries one can find Moguer (and Mazagón), Palos, Lucena, Rociana, Pilas, Bollullos Par del Condado, and Aznalcázar.

Groundwater is mainly abstracted to supply the towns and the large touristic resorts of El Rocío, Matalascañas and Mazagón, which also affect wetlands (Manzano *et al.*, 2005). Matalascañas tourist resort is located on the coastline within the municipality of Almonte, and has a set of wells for urban water supply along its northern fringe, just bordering the National Park. Daily cycle of water abstraction in wells is followed by water level fluctuations in ponds within the National Park with a few hours delay (Serrano & Zunzunegui, 2008). Matalascañas has less than 2,000 permanent inhabitants, but the population increases to more than 100,000 in summer and many

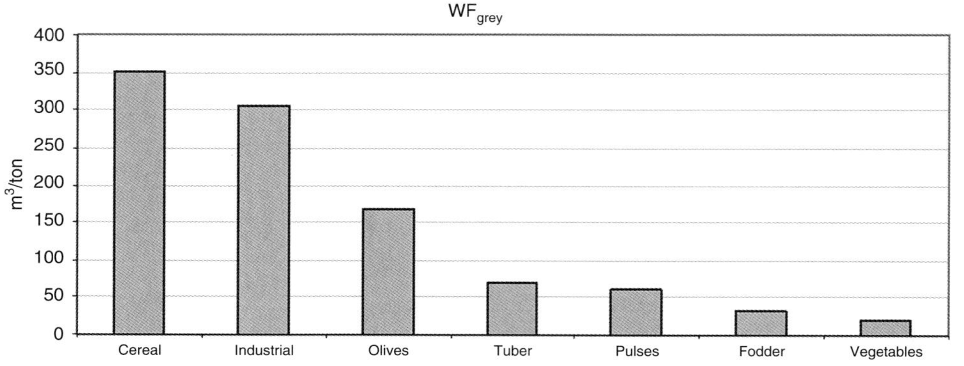

Figure 8. Grey water footprint related to nitrogen fertilizer of agricultural crops (WF_{grey}) in the Doñana region (m^3/t).
Source: Own elaboration.

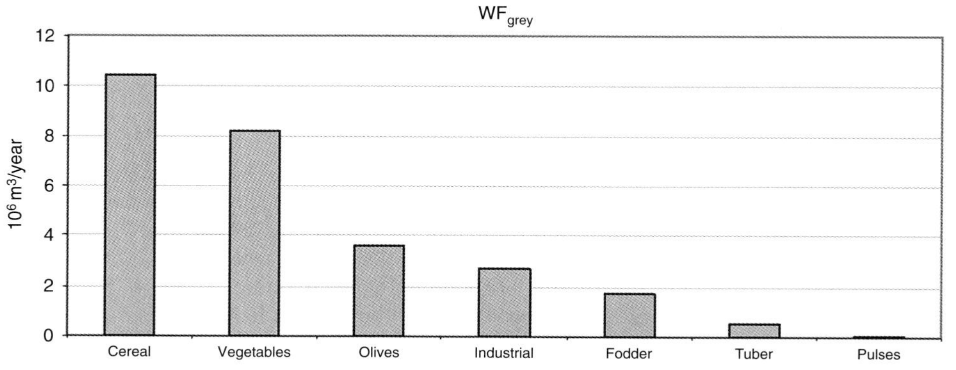

Figure 9. Total grey water footprint related to nitrogen fertilizer of agricultural crops (WF_{grey}) in the Doñana region (Mm^3/yr).
Source: Own elaboration.

weekends, showing a strong seasonal occupation (Muñoz-Reinoso, 2001). The urban water supply to Matalascañas is about 2.5–3 Mm^3/yr (Castell *et al.*, 1992), mainly for human use and garden watering (DGOT, 1992). There is also a religious festival at El Rocio Shrine, which brings large crowds of about one million people every spring and large groups on holidays for all year long.

8 WATER FOOTPRINT AND ENVIRONMENTAL WATER REQUIREMENTS WITHIN THE DOÑANA REGION

Water is the strategic basis for the maintenance and development of Doñana from the ecological perspective: Doñana contains wetlands, marshes, ponds and terrestrial vegetation, which is largely dependent on water level depth. From the economic perspective, there is again a strong dependence for domestic supply, intensive agriculture and tourism. A first approximation of the water footprint and environmental water requirements is presented below.

Currently the main green and blue water user in the Doñana region is the environment, using about 59% of total water consumption in an average rainfall year (Table 5). Agriculture is the second main water user amounting to about 40% of total water consumption. Finally, urban water supply and industry use around 1% of the total water used. However, the latter two use exclusively blue water resources, which generally have a higher opportunity cost than green water use

Table 5. Water footprint (WF) and environmental water requirements within the Doñana region for an average rainfall year (Mm^3/yr).

		blue				
	green	surface	ground	total	grey[5]	Total[6]
Agricultural WF[1]	127	199	83	282	27	409
Urban and industrial WF[2]			7	7		7
Current environmental water use[3]	442	116	38	154		596
Current total water use	**569**	**315**	**128**	**443**		**1,012**
Environmental water requirements[4]	**200**					**200**

[1]Water footprint related to agricultural production. *Source*: Own estimation. Surface and groundwater distinction for Marismas and Condado Litoral regions according to INE (1999).
[2]Water footprint related to urban and industrial water supply. *Source*: Custodio *et al.* (2006).
[3]Current environmental water use. *Source*: blue water use is based on data from the Andalusian Regional Government (2002) and green water use refers to own estimations for forests.
[4]Environmental water requirements. *Source*: blue water requirements for a dry (80 Mm^3) and average year (200 Mm^3) from WWF (2009a).
All of them are consumptive water uses except for the case of urban water supply and industrial water footprint. In the case of the coastal urbanizations of Matalascañas and Mazagón however their water use is consumptive since they discharge into the sea.
[5]Grey water footprint referring to nitrate pollution.
[6]Consumptive water use, without including grey water.

(Hoekstra & Chapagain, 2008). Green water can be productively used only for crop production (not in the sands) and natural biomass production (support of ecosystem functioning), while blue water can be used not only for irrigating crops and the ecosystems but also for various other types of domestic, agricultural and industrial water use. On the other hand, all the figures provided refer to consumptive water uses except some urban and industrial water withdrawal (non-consumptive use of water). In the case of the coastal touristic resorts of Matalascañas and Mazagón, their water use is consumptive since they discharge into the sea; they could however avoid this through water reuse. In fact Matalascañas golf course is irrigated with treated urban sewage. If consumptive uses were presented for the whole urban and industrial water uses, these numbers would be somewhat lower but the picture would not change noticeably.

Agriculture and tourism, the two main economical activities, compete for water with environmental preservation (Manzano *et al.*, 2005). Part of this area is under the two highest protection figures for natural spaces existing in Spain: the Doñana National Park and Doñana Natural Park both merged in a wider figure as the Parks of Doñana under the administration of the Regional Government of Andalusia (Custodio *et al.*, 2009). Within the National Park boundaries, human activity is greatly restricted, while across the Natural Park some traditional activities like pinecone collection, logging, beekeeping, fish farming and old-style charcoal preparation are allowed (UNEP, 2005). Cattle breeding (caws, horses and sheep) are maintained.

Over the years pristine Doñana wetlands have been profoundly modified from the hydrodynamic point of view because of a series of activities, mainly the drainage of the wetlands into agricultural land (mainly rice), but also by channelling the Guadiamar river. According to a report by WWF (2009a) currently Doñana receives less than 20% of the contributions in natural regime. In line with the same report, the environmental flows from the rivers needed to recover the wetlands in the National Park amount to about 200 Mm^3/yr in average years and 80 Mm^3/yr in dry years. The mentioned report follows the ELOHA approach for its estimation (WWF, 2009a). In our analysis we have focused on an average rainfall year (Figure 10). We have adopted in a preliminary way these estimations but this is obviously a simplification and first approximation since environmental flows not only refer to quantity but also to quality and timing of water flows along the year required to sustain healthy freshwater ecosystems and the benefits they provide to human communities.

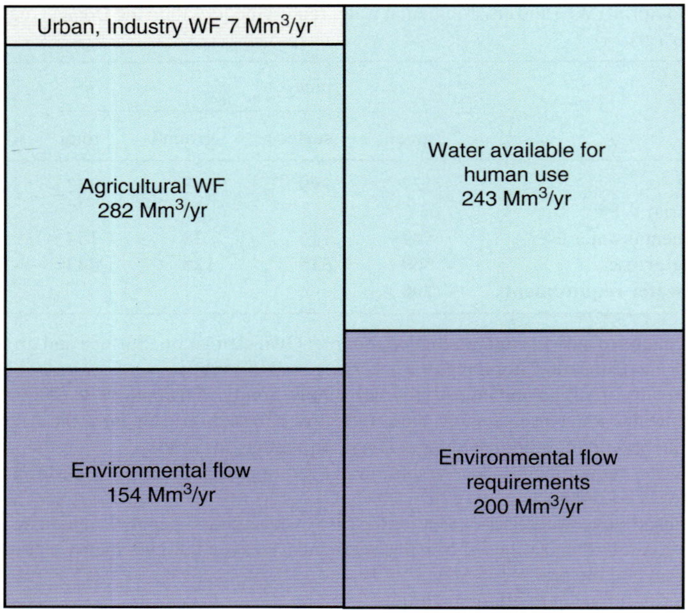

Figure 10. Theoretical water requirements, actual environmental water use and the total water footprint for an average rainfall year.
Source: Own estimation, based on data from Custodio *et al.* (2006) and WWF (2009a).

Currently, the total water available in the Doñana region is roughly about 443 Mm^3/yr (Figure 10). However, according to the environmental water requirements theoretically the water needed in an average rainfall year for the environment is around 200 Mm^3/yr (WWF, 2009a). This suggests that at present the environmental water requirements are slightly constrained in average rainfall years due to irrigation.

9 ECONOMIC ANALYSIS

The water apparent productivity analysis can be very useful in order to identify possible water uses not justified in economic efficiency terms and achieve an efficient allocation of water resources. This has been carried out for the main water using sectors, namely, agriculture and the environment.

Within the agricultural sector, in line with Figure 11, there are crops that are very profitable such as strawberries or vegetables. There are other crops that, even if less lucrative (e.g. grapes), are not irrigated or little irrigated and use the soil water coming from rainfall (green water). Besides having a lower opportunity cost, the use of green water for the production of crops has generally less negative environmental externalities than the use of blue water (irrigation with water abstracted from ground or surface water systems) (Aldaya *et al.*, 2009).

Rice is by far the most freshwater consuming crop (Figure 11). The water used for rice cultivation is imported into the region; it is pumped from the Guadalquivir river to irrigate the rice fields, most of which returns to the river. Rice in Sevilla, Spain's leading producer, is located on the right bank of the Guadalquivir river, in particular in the municipalities of Isla Mayor, Puebla del Río, Coria del Río and Villamanrique de la Condesa. The crop area in this region is about 28,000 ha, amounting to about 310,000 t harvested, which represents around 40% of the Spanish production. In recent years an effort has been made to reconcile the rice farming with biodiversity conservation, enhancing water use efficiency, upgrading production techniques and adopting integrated farming practices. More recently, during 2008–2009, the pilot project "Organic rice cultivation in the surroundings

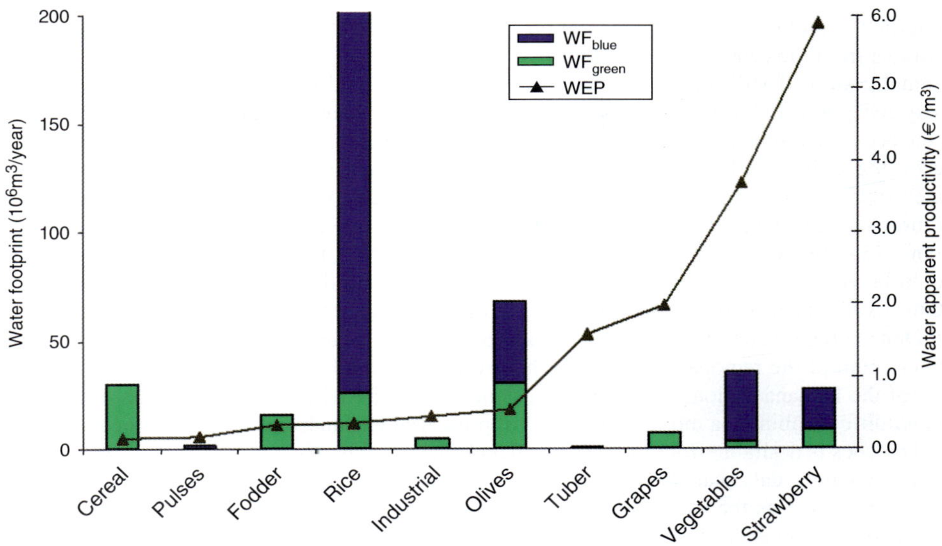

Figure 11. Total green and blue water use of agricultural crops (WF_{green}, WF_{blue}) and water apparent productivity (WEP) in the Doñana region for an average rainfall year.
Source: Own elaboration (see Table 6).

Table 6. Water footprint hydrologic and economic analysis.

	WF_{green} Mm³/yr	WF_{blue} Mm³/yr	WF_{grey} Mm³/yr	BWAP €/m³	WAP €/m³
Cereal	30	0	10	0	0.1
Fodder	16	0	2	0	0.3
Grapes	7	0	No available	0	2.0
Industrial	5	0	3	0	0.5
Olives	31	37	4	1.0	0.5
Pulses	0	1	0	0.2	0.2
Rice	26	193	No available	0.4	0.3
Strawberry	8	19	No available	8.5	5.9
Tuber	1	0	1	0	1.6
Vegetables	3	32	8	4.1	3.7
Total	**127**	**283**	**27**		
Wetland*					**10.0**

Source: Own elaboration.
*Source: Own elaboration based on Costanza *et al.* (1997).
WF_{green} = green water footprint; WF_{blue} = blue water footprint; WF_{grey} = grey water footprint; BWAP = blue water apparent productivity; WAP = total (green and blue) water apparent productivity.

of the Guadalquivir protected coastal wetlands" was launched, which aims to promote the organic rice farming that fosters harmonious coexistence with the natural values of this area.

However, the apparent productivity of rice is about 0.30 €/m³, one of the lowest in the region (Figure 11). That is probably one of the reasons why, during the 2007–2008 drought, the Central Board of Users of the Lower Almanzora, Almeria, began negotiations with irrigating farmers from the Lower Guadalquivir to arrange water rights transfers (also called Negratín-Almanzora transfer). They even bought 1,600 ha of land devoted to rice cultivation in order to use these water rights in

Almería (Corominas, 2008). An integration of rice fields and natural wetlands patches seems to be desirable from the ecosystem viewpoint.

Finally, even if still difficult and controversial, a first and rough attempt to value Doñana's ecosystem goods and services has been made. In line with Costanza *et al.* (1997), the economic value of the ecosystem services provided by a wetland is about 14,785 US$/ha/yr, that is, 10,665 €/ha/yr. According to those authors these goods and services include gas, disturbance and water regulation, water supply, waste treatment, habitat/refugia, food production, raw materials, recreation and cultural services. Considering that the Doñana region has a wetland area of 1,500 km^2 and uses around 154 Mm^3/yr (Andalusian Regional Government, 2002), the wetland economic productivity would be very roughly 10 €/m^3, notably higher than crop productivities, which amount to 0.10–6 €/m^3. Although a simplification, these figures seems to point that investments in conservation in the Doñana region can provide benefits both to communities and their environment. An absolute preservation of the regional environment however would not be beneficial for the society. In the case of the Doñana region, wetlands for instance, an agri-environment integration of rice fields and wildlife combining a mosaic composed of preserved wetland, aquaculture surfaces and paddy field patches is desirable from the ecological and social perspectives alike.

Unfortunately, data scarcity prevented enlarging the scope of the present chapter to cover some other topics such as the water footprint of floodable areas, such as natural marshes, scrubbery, grasslands, and other ecosystem types. Out of the scope of this study are the ecosystem services of the Parks such as the habitats provided by the rice fields. The evaluation of blue water flow which is maintained in the Guadalquivir river during summer drought to keep water salinity below a threshold favourable to irrigation, is again a specific footprint to be incorporated in future analysis.

10 CONCLUSIONS

Having achieved *more crops and jobs per drop*, the present analysis of the water footprint and environmental water requirements in the Doñana region suggests that achieving *more cash and care of nature per drop* simultaneously could be feasible in this region if policy makers take action.

In the Doñana region, the environment is the main water user amounting to about 59% (596 Mm^3/yr) of total water use, followed by agriculture with about 40% (409 Mm^3/yr) and urban water supply and industry: 1% (7 Mm^3/yr).

The total blue water available in the region amounts to 443 Mm^3 for an average rainfall year. If the environmental water requirements were met, 200 Mm^3/yr according to WWF (2009a) including marshes, streams and aquifer, the total water available for human use in the region would be 243 Mm^3/yr. However, the current agricultural blue water consumption alone adds up to 282 Mm^3/yr. The environmental blue water requirements thus seem to be mildly constrained in the Doñana region because of irrigation. The largest amount of irrigated water is for producing a low value crop (rice) (0.30 €/m^3).

According to these preliminary figures, and in spite of the positive measures for the protection of Doñana taken by the Spanish authorities (including hydrological restoration, the enlargement of the protected area and the substantial reduction of the irrigated area), there is still a certain risk of change in the ecological character of Doñana National Park because of surface and groundwater extraction mainly for agriculture.

Groundwater plays a significant hydrological and ecological role in the natural functioning of Doñana. It also supports a very diverse flora and fauna and is one of the most important National Parks in Spain. The groundwater regime is threatened by abstraction of water for irrigation and the fertilizer leaching from agriculture. In this context, it is relevant to clarify the ownership and legality of wells and water withdrawals through them, as well as to establish ways of measurement. Additionally there is a need to further reduce fertilizer use and other agrochemicals shifting to organic agriculture for instance. Even if in the short term the effects will go unnoticed (the water-table changes in Doñana progress at a slow pace), in the long term the reduction of pollution and water consumption is crucial to ensure the conservation of Doñana.

Finally, the green water used by forests contributes significantly to the use of water resources (442 Mm^3/yr, which is about 44% of total water use). This analysis is a first approximation since data on the environmental water uses are limited. If the evapotranspiration by scrublands had been taken into account, these figures would have probably been higher. Further research is needed on this topic.

In conclusion, it seems clear that integrated water allocation, planning and management is needed in the Doñana region, considering the environmental water requirements together with the blue (surface and ground), green and grey water footprints, to achieve a more compatible agricultural production with the protection of ecosystems and this is possible without impairing the livelihoods of the farmers because the main water consumer is a low economic value and water intensive crop –the rice.

REFERENCES

Acreman, M.C. (1998). Principles of water management for people and the environment. In: A. de Shirbinin & V. Dompka (eds.), *Water and Population Dynamics*. American Association for the Advancement of Science, 321 pp.

Aldaya, M.M. & Hoekstra, A.Y. (2009). The water needed to have Italians eat pasta and pizza. Value of Water Research Report Series, 36 (online). UNESCO–IHE, Delft, the Netherlands. Available from: [http://www.waterfootprint.org/Reports/Report36-WaterFootprint-Pasta-Pizza.pdf].

Aldaya, M.M.; Allan, J.A. & Hoekstra, A.Y. (2009). Strategic importance of green water in international crop trade. *Ecological Economics*, 69: 887–894.

Allen, R.G.; Pereira, L.S.; Raes, D. & Smith, M. (1998). *Crop evapotranspiration – Guidelines for computing crop water requirements*. FAO Irrigation and drainage, paper 56. Food and Agriculture Organization, Rome, Italy.

Allier, C.; González Bernáldez, F. & Ramírez Díaz, L. (1974). Mapa Ecológico de la Reserva Biológica de Doñana 1:10.000. CSIC. Seville, Spain.

Andalusian Regional Government (1999). Mapa de Usos y Coberturas Vegetales del Suelo de Andalucía 1999. Andalusian Regional Government.

Andalusian Regional Government (2002). Plan de ordenación del territorio. Ámbito de Doñana. Andalusian Regional Government. [http://www.labclima.ua.es/complegis/ot/andalucia/plan_subreg/planes/do%F1ana.pdf]. Accessed on January 2010.

Andalusian Regional Government (2003). Plan de ordenación del territorio. Ámbito de Doñana. Andalusian Regional Government. [http://www.donana.es/arch_var/POTAD.pdf]. Accessed on January 2010.

Andalusian Regional Government (2009). Parque Nacional Doñana. Junta de Andalucía. Consejería de Medio Ambiente. [http://www.juntadeandalucia.es/medioambiente/servtc5/ventana/informeFichaEspacioNatural.do]. Accessed on January 2010.

Castell, M.; Cruz, J.; Custodio, E.; García Novo, F.; Gaudemar, J.P.; González Vallvé, J.L.; Granados, V.; Magraner, A.; Román, C.; Smart, M. & van der Maarel, E. (1992). Dictamen sobre estrategias para el desarrollo socioeconómico sostenible del Entorno de Doñana. Junta de Andalucía, Sevilla, Spain, 125 pp.

Chapagain, A.K. & Hoekstra, A.Y. (2004). Water footprints of nations. Value of Water Research Report Series, 16. UNESCO–IHE, Delft, the Netherlands. [http://www.waterfootprint.org/?page=files/Research%20data]. Accessed on January 2010.

Chapagain, A.K. & Orr, S. (2009). An improved water footprint methodology linking global consumption to local water resources: A case of Spanish tomatoes. *Journal of Environmental Management*, 90(2): 1219–1228.

CHG (2009). Superficie de los Cultivos de Regadío y sus Necesidades de Riego en la Demarcación de la CHG. Guadalquivir River Basin Authority.

Comprehensive Assessment of Water Management in Agriculture (2007). Water for Food, Water for Life: A Comprehensive Assessment of Water Management in Agriculture. D. Molden (ed.). Earthscan, London, UK.

Corominas, J. (2008). ¿Modernización o reconversión de regadíos? Dimensiones socio-económicas, ambientales y territoriales. 6° Congreso Ibérico sobre planificación y gestión del agua, 4–7 December 2008, Vitoria, Spain.

Costanza, R.; d'Arge, R.; de Groot, R.; Farber, S.; Grasso, M.; Hannon, B.; Limburg, K.; Naeem, S.; O'Neill, R.V.; Paruelo, J.; Raskin, R.G.; Sutton, P. & van den Belt, M. (1997). The value of the world's ecosystem services and natural capital. *Nature*, 387, 15 May.

Custodio, E. (2005). *Aspectos hidrológicos de los humedales que dependen del agua subterránea* [Hydrological aspects of wetlands relying on groundwater]. 1ª Reunión Internacional de Expertos sobre la Regeneración Hídrica de Doñana [1st International Expert Meeting on Water Regeneration of Doñana].

Custodio, E. & Palancar, M. (1995). Las aguas subterráneas en Doñana. Ciencia y Técnica de la Ingeniería Civil. *Revista de Obras Públicas*, February: 3340.

Custodio, E.; Manzano, M. & Dolz, J. (2006). *Recursos de agua en la Comarca de Doñana* [Water resources in the Region of Doñana]. II Plan de Desarrollo Sostenible de Doñana. Informe Técnico [II Sustainable Development Plan of Doñana. Technical Report].

Custodio, E.; Manzano, M. & Montes, C. (2009). Las aguas subterráneas en Doñana. Aspectos ecológicos y sociales. Junta de Andalucía, Seville, Spain. 244 pp.

DGOT (1992). *Información-diagnóstico de los aspectos urbano-turísticos de Matalascañas (Almonte)* [Information-diagnosis of the urban-tourist aspects of Matalascañas (Almonte)]. Dirección General de Ordenación del Territorio, Cons. Obras Publ. y Transp. Junta de Andalucía [Directorate General of Spatial Planning, Ministry of Public Works and Transportation. Andalucian Regional Government].

EPA (2005). *List of drinking water contaminants: ground water and drinking water*. US Environmental Protection Agency.

Falkenmark, M. & Rockström, J. (2004). Balancing water for humans and nature: The new approach in ecohydrology. Earthscan, London, UK.

FAO (2009a). CROPWAT 8.0 model, Food and Agriculture Organization, Rome, Italy. [www.fao.org/nr/water/infores_databases_cropwat.html]. Accessed on January 2010.

FAO (2009b). CLIMWAT 2.0 database, Food and Agriculture Organization, Rome, Italy. [www.fao.org/nr/water/infores_databases_climwat.html]. Accessed on January 2010.

García Novo, F. (1997). The ecosystems of Doñana National Park. In: F. García Novo; R.M.M. Crawford & M.C. Díaz Barradas (eds.), *The ecology and conservation of European Dunes*. EUDC-Universidad de Sevilla: 97–116.

García Novo, F. & Marín Cabrera, C. (2006). Doñana: Water and Biosphere. Doñana 2005 Project. Guadalquivir River Basin Authority, Spanish Ministry of the Environment and UNESCO. Madrid, Spain.

García Novo, F.; Zunzunegui, M.; Muñoz Reinoso, J.C.; Gallego Fernández, J.B. & Díaz Barradas, M.C. (1996). Surface and groundwater control on ecosystem development: the case of Doñana National Park (SW Spain). In: J. Cruz Sanjulián & J. Benavente (eds.), *Wetlands: a Multiapproach Perspective*. University of Granada, Spain: 81–101.

García Novo, F.; Martín Vicente, A. & Toja Santillana, J. (2007). La Frontera de Doñana. Universidad de Sevilla. Seville, Spain.

Grimalt, J.O.; Ferrer, M.; & MacPherson, E. (1999). Special issue: the mine tailing accident in Aznalcollar. *The Science of the Total Environment*, 242: 3–11.

Hoekstra, A.Y. (ed.) (2003). Virtual water trade. Proceedings of the International Expert Meeting on Virtual Water Trade, Delft, the Netherlands, 12–13 December 2002. Value of Water Research Report Series, 12. UNESCO-IHE, Delft, the Netherlands.

Hoekstra, A.Y. & Chapagain, A.K. (2008). *Globalization of water: Sharing the planet's freshwater resources*. Blackwell Publishing, Oxford, UK.

Hoekstra, A.Y.; Chapagain, A.K.; Aldaya, M.M. & Mekonnen, M.M. (2009). Water footprint manual: State of the art 2009. Water Footprint Network, Enschede, the Netherlands. [http://www.waterfootprint.org/downloads/WaterFootprintManual2009.pdf]. Accessed on January 2010.

INE (1999). 1999 Agrarian Census. Instituto Nacional de Estadística [National statistics institute]. [http://www.ine.es/]. Accessed on January 2010.

IUCN (2003). Flow: the essentials of environmental flows. IUCN, Water and Nature Initiative. M. Dyson, G. Bergkamp & J. Scanlon (eds.). IUCN, Gland, Switzerland.

King, J.M.; Tharme, R.E. & DeVilliers, M.S. (eds.) (2000). Environmental Flow assessments for Rivers: Manual for the Building Block Methodology. Water Research Commission, Pretoria, South Africa.

Llamas, M.R. (1988). Conflicts between wetland conservation and groundwater exploitation: two case histories in Spain. *Environmental Geology and Water Sciences*, 11(3): 241–251.

Llamas, M.R. (1989). Consideraciones en relación con el impacto negativo de la extracción de aguas subterráneas en dos importantes ecosistemas españoles. Tecnología del Agua, 9(61): 23–34.

Manzano, M.; Ayora, C.; Domenech, C.; Navarrete, P.; Garralón, A. & Turrero, M.J. (1999). The impact of the Aznalcollar mine tailing spill on groundwater. Sci. Total Environ. 242: 189–209.

Manzano, M.; Borja, F. & Montes, C. (2002). Metodología de tipificación hidrológica de los humedales españoles con vistas a su valoración funcional y a su gestión. Aplicación a los humedales de Doñana. *Boletín Geológico y Minero*, 113 (3): 313–330.

Manzano, M.; Custodio, E.; Mediavilla, C. & Montes, C. (2005). Effects of localised intensive aquifer exploitation on the Doñana wetlands (SW Spain). In: A. Sahuquillo, J. Capilla, L. Martínez-Cortina & X. Sánchez-Vila (eds.), *Groundwater Intensive Use*. Taylor & Francis.

MARM (2009). *Agro-alimentary Statistics Yearbook*. Spanish Ministry of the Environment and Rural and Marine Affairs. [http://www.mapa.es/es/estadistica/pags/anuario/introduccion.htm]. Accessed on January 2010.

MIMAM (Ministerio de Medio Ambiente [Spanish Ministry of Environment]) (2007). El agua en la economía española: Situación y perspectivas [Water in the Spanish Economy: current affairs and perspectives]. Spanish Ministry of the Environment. [http://circa.europa.eu/Public/irc/env/wfd/library?l=/ framework_directive/implementation_documents_1/wfd_report s/member_states/spain/article_5/completo_nivel 1pdf/_EN_1. 0_&a=d]. Accessed on January 2010.

Muñoz-Reinoso, J.C. (2001). Vegetation changes and groundwater abstraction in SW Doñana, Spain. *Journal of Hydrology*, 242(3–4): 197–209

Muñoz-Reinoso, J.C. & García Novo F. (2004). Multiscale control of vegetation patterns: the case of Doñana (SW Spain). *Landscape Ecology*, 20: 51–61.

Official State Gazette (2008). Approval of the water planning instruction. Ministry of the Environment and Rural and Marine Affairs. Boletín Oficial del Estado [Official State Gazette] 229. 22 September 2008. [http://www.boe.es/boe/dias/2008/09/22/pdfs/A38472-38582.pdf]. Accessed on January 2010.

Poff, N.L.; Richter, B.D.; Aarthington, A.H.; Bunn, S.E.; Naiman, R.J.; Kendy, E.; Acreman, M.; Apse, C.; Bledsoe, B.P.; Freeman, M.C.; Henriksen, J.; Jacobson, R.B.; Kennen, J.G.; Merritt, D.M.; O'Keeffe, J.H.; Olden, J.D.; Rogers, K.; Tharme, R.E. & Warner, A. (2009). The ecological limits of hydrologic alteration (ELOHA): A new framework for developing regional environmental flow standards. Freshwater Biology, doi:10.1111/j.1365-2427.2009.02204.x.

Roost, N.; Cai, X.L.; Turral, H.; Molden, D. & Cui, Y.L. (2008). Adapting to intersectoral transfers in the Zhanghe Irrigation System, China, Part II: Impacts of in-system storage on water balance and productivity. *Agricultural Water Management*, 95 (6): 685-697.

Serrano, L. & Zunzunegui, M. (2008). The relevance of preserving temporary ponds during drought: hydrological and vegetation changes over a 16-year period in the Doñana National Park (South-West Spain). *Aquatic Conservation: Marine Freshwater Ecosystems*, 18: 261–279.

Smakhtin, V.; Revenga, C. & Döll, P. (2004). Taking into account environmental water requirements in global-scale water resources assessments. Comprehensive Assessment Research Report 2, IWMI, Colombo, Sri Lanka.

Suso, J. & Llamas, M.R. (1993). Influence of groundwater development of the Doñana National Park ecosystems (Spain). *Journal of Hydrology*, 141(1–4): 239–270.

UNEP (2004). Freshwater in Europe, Major European Water sheds, Guadalquivir. United Nations Environment Programme. [http://www.grid.unep.ch/product/publication/freshwater_europe/guadalquivir.php]. Accessed on January 2010.

UNEP (2005). Doñana National Park, Spain. United Nations Environment Programme, World Conservation Monitoring Centre. [http://www.unep-wcmc.org/sites/wh/pdf/Donana.pdf]. Accessed on January 2010.

WWAP (2009). Water in a changing world: The 3rd United Nations world water development report. UNESCO Publishing, Paris, France; and Berghahn Books, Oxford, UK.

WWF (2006). Illegal water use in Spain: Causes, effects and solutions. World Wide Fund for Nature (WWF). [http://assets.panda.org/downloads/illegal_water_use_in_spain_may06.pdf]. Accessed on January 2010.

WWF (2009a). Caudales ecológicos de la marisma del Parque Nacional de Doñana y su área de influencia. World Wide Fund for Nature-Spain (WWF).

WWF (2009b). Strawberry farms in Doñana. World Wide Fund for Nature (WWF). [http://assets.wwf.es/downloads/factsheetstrawberry.pdf]. Accessed on January 2010.

Zunzunegui, M.; Díaz Barradas, M.C. & García Novo, F. (1998). Vegetation fluctuation in Mediterranean dune ponds in relation to rainfall variation and water abstraction. *Applied Vegetation Science*, 1: 151–160.

CHAPTER 12

The blue, green and grey water footprint of rice from both a production and consumption perspective

Ashok K. Chapagain
WWF–UK, Godalming, UK

Arjen Y. Hoekstra
University of Twente, the Netherlands

ABSTRACT: The aim of this chapter is to make a global assessment of the green, blue and grey water footprint of rice, using a higher spatial resolution than earlier studies and using local data on actual irrigation. Evapotranspiration from rice fields is calculated with the CROPWAT model; the distinction between green and blue water evapotranspiration is based on data on precipitation and irrigation. Water pollution from N-fertilizers is estimated based on application rates. The calculated green, blue and grey water footprints of paddy rice are converted into estimations of the green, blue and grey water footprints of derived rice products on the basis of product and value fractions. International virtual water flows related to trade in rice products are estimated by multiplying export volumes by their respective water footprints in the exporting countries. Per nation, the water footprint of rice production is estimated by aggregating the water footprints per production region. The water footprint of rice consumption is estimated for each nation by aggregating the water footprints in the regions where the rice consumed in a nation is grown. For rice importing countries, the water footprint related to rice consumption is thus partly (or fully) outside the country itself.

In the period 2000–04, the global average water footprint of paddy rice is 1,325 m^3/t (48% green, 44% blue, and 8% grey), which is much lower than previous estimates. There is about 1,025 m^3/t of percolation in rice production. The global water footprint of rice production is 784,000 Mm^3/yr. The ratio of green to blue water varies greatly, both over time and space. In countries like India, Indonesia, Vietnam and Thailand, Myanmar and the Philippines, the green water fraction is substantially larger than the blue water fraction. In the USA, however, the blue water fraction is 3.7 times the green water fraction and in Pakistan it is 5.6 times.

During the period 2000–04, the global virtual water flows related to international rice trade was 31,000 Mm^3/yr (45% green, 47% blue, and 8% grey). The share of blue water component in the average rice export is a bit higher than in the average rice production.

The consumption of rice products in the EU27 alone is responsible for the annual evaporation of 2,279 Mm^3 of water and polluted return flows of 178 Mm^3 around the globe, mainly in India, Thailand, the USA and Pakistan. The water footprint of global rice consumption creates relatively low stress on the water resources in India compared to that in the USA and Pakistan, as in the latter cases rice is extensively irrigated with scarce blue water resources.

Keywords: rice, trade, water footprint, green water, blue water, fertilizer, pollution

1 INTRODUCTION

Rice is one of the major crops feeding the world population and is most important in South Asia and Africa. Large irrigation projects are often constructed to meet the water demand in rice production. As a result, rice is one of the largest water consumers in the world. This chapter quantifies how much fresh water is being used to produce rice globally, distinguishing between two different sources: irrigation water withdrawn from gound- or surface water (blue water) and rainwater (green water). It also quantifies the volume of polluted water related to the use of nitrogen fertilizers in rice production (grey water).

Rainwater and irrigation water are necessary for rice growth in two ways: to maintain soil moisture and –in wet irrigation– to maintain the standing layer of water over the paddy field. In the major rice producing regions of the world, the crop is grown during the wet (monsoon) season, which reduces the irrigation demand by effectively using rainwater.

As much of the standing water in paddy fields percolates and recharges groundwater and surface water, there is a substantial contribution to the local blue water availability. Percolation can be seen as a loss to the paddy field, but for the catchment area it is not considered as a loss, because the water can be captured and reused downstream (Bouman *et al.*, 2007b). In some irrigation systems in flood plains with impeded drainage or systems in low lying deltas a continuous percolation can even create shallow groundwater tables closer to the surface (Belder *et al.*, 2004). Although the document focuses on the estimation of evapotranspiration from rice fields, it also estimates percolation flows, because evapotranspiration and percolation are both part of the soil water balance.

2 METHOD AND DATA

There are mainly two systems of rice production: wetland systems and upland systems. About 85% of the rice harvest area in the world is derived from wetland systems. About 75% of rice production is obtained from irrigated wetland rice (Bouman *et al.*, 2007b). In Asia, rice fields are prepared by tillage followed by puddling. The soil layer is saturated and there is standing water during the entire growth period of the crop. In the USA, Australia, parts of Europe and some Asian countries, rice land is prepared dry and flooded later.

In the production database of the FAOSTAT data (2009), 115 countries are reported as rice producers. During the period of 2000–04, the average annual global production of rice was 592 million tonnes (Mt) [1 t = 1,000 kg] with an average yield of 4.49 t/ha. The yield ranges from nearly 1 t/ha (Jamaica, Micronesia, Congo, Brunei Darussalam, Comoros, Chad, Liberia, Mozambique, Congo D.R., Sierra Leone, etc.) to 8.7 t/ha (Australia). In India, the rainfed yield ranges between 0.5–1.6 t/ha, whereas that of irrigated rice is 2.3–3.5 t/ha (Gujja *et al.*, 2007).

Table 1 presents production data for the thirteen countries with the largest average annual production during the period 2000–04. These countries account for more than 90% of the global production during this period. These thirteen countries together account for more than 82% of the total export of rice-equivalent globally. About 6–7% of the world rice production is traded internationally. A complete list of rice producing countries with production statistics is presented in Appendix B.

The average fertilizer application rates for the top-13 rice producing countries have been taken from IFA *et al.* (2002) and are presented in Table 1. The use of animal manure reduces the need for chemical fertilizer use in crop fields. This is reflected in lower fertilization application rates in the database, mainly in developing countries. There is no readily available global dataset on use of animal manure in rice fields. Moreover, the spatial distribution of the fertilizer within a country is also not well known, therefore results on water pollution should be treated cautiously.

The reference crop evapotranspiration (ETo) and monthly average rainfall data for the concerned climate stations are taken from the CLIMWAT database (FAO, 1993) for all countries, but from FAOCLIM (FAO, 2001) for the USA. The ETo data in these databases are derived using the Penman-Monteith equation as described in Allen *et al.* (1998). Using the CROPWAT model (FAO, 1992), the crop evapotranspiration (ETc) and the available effective rainfall are calculated for the given set of data on ETo, monthly rainfall, Kc and the crop calendar. Rice crop coefficients are taken from Allen *et al.* (1998). Monthly data on rainfall and ETo are distributed within the month to obtain data per five days. As CROPWAT 4 (FAO, 1992) is not suitable to calculate the crop water requirement for rice (Clarke *et al.*, 1998), we have used it only to get the values of ETc and the available effective rainfall for a time step of 5 days. For each of the thirteen countries, the crop

Table 1. Statistics for the thirteen largest rice producing countries during the period 2000–04.

Countries	Average production (Mt/yr)[1]	Global share (%)[1]	Average area harvested (Mha/yr)[1]	Average yield (t/ha)[1]	N (kg/ha)[2]	P_2O_5 (kg/ha)[2]	K_2O (kg/ha)[2]
China	177.66	30.0	28.67	6.19	145	60	40
India	126.50	21.4	43.06	2.93	67.7	24.2	9.4
Indonesia	52.01	8.8	11.64	4.47	105	22	14
Bangladesh	37.22	6.3	10.64	3.50	72	15	10
Vietnam	33.96	5.7	7.51	4.52	115	45	42
Thailand	26.80	4.5	10.04	2.67	62	33	17
Myanmar	22.58	3.8	6.43	3.51	35	12	4
Philippines	13.32	2.3	4.06	3.28	51	15	11
Brazil	11.07	1.9	3.37	3.28	40	50	30
Japan	10.99	1.9	1.71	6.44	78	92	72
USA	9.52	1.6	1.29	7.40	150	60	60
Pakistan	6.91	1.2	2.34	2.95	52.2	6.9	0.2
Korea, Rep.	6.81	1.2	1.05	6.51	110	70	80
Sub total	535.35	90.5	131.80	–	–	–	–
Global total	591.75	100.0	150.67	4.49	–	–	–

[1]*Source*: FAOSTAT data (2009). [Mt = 10^6 t]; [1 t = 1,000 kg]; [Mha = 10^6 ha]
[2]Average fertilizer use in rice cultivation. *Source*: IFA *et al.* (2002).

evaporative demand (ETc) is calculated for each season of rice production in all the regions using the climate data (FAO, 1993; 2001) for the concerned regions (Appendix A).

The CROPWAT model suggests a number of methods to estimate the effective rainfall, and the method of the USDA SCS (United States Department of Agriculture, Soil Conservation Service) is one of them. As this method does not take into account the soil type and the net depth of irrigation, it gives a lower estimation of effective rainfall compared to the water balance approach (Mohan *et al.*, 1996) and is not very accurate. However, as the water balance approach is highly data demanding, the USDA SCS method is widely used in estimating the effective rainfall in agriculture water management (Cuenca, 1989; Jensen *et al.*, 1990). We have also chosen the USDA SCS method for the present study. The USDA SCS equations to estimate effective rainfall are:

$$P_{eff} = \frac{P_{total}}{125} \times (125 - 0.2 \times P_{total}) \text{ for } P_{total} \leq 250\,\text{mm}$$

$$P_{eff} = (125 + 0.1 \times P_{total}) \text{ for } P_{total} \geq 250\,\text{mm}$$

where *Peff* is the effective rainfall and *Ptotal* is the total rainfall in the concerned period.

For rice cultivation in wetland systems, paddy fields are prepared and the soil is kept saturated. The common practice is to first prepare land by puddling. This is done by saturating the soil layer for one month prior to sowing. The volume of water (SAT) necessary for this stage is assumed to be 200 mm as suggested by Brouwer & Heibloem (1986). As lowland rice is grown in a standing layer of water, there is a constant percolation and seepage loss during this period. Percolation loss (PERC) is primarily a function of soil texture. It varies from 2 mm/day (heavy clay) to 6 mm/day for sandy soil. As rice is mostly grown in soil with more clayey texture, for the present study we have taken 2.5 mm/day as an average (Brouwer & Heibloem, 1986) for the entire period of rice cultivation except for the last 15 days when the field is left to dry out for easy harvesting. A water layer is established during transplanting or sowing and maintained throughout the growing season. Although the volume of water needed for maintaining the water layer (WL) is available for percolation losses and to meet the evaporative demand of the crop during the last phase of paddy growth, it is necessary to get this volume of water at the beginning of the crop period (Figure 1).

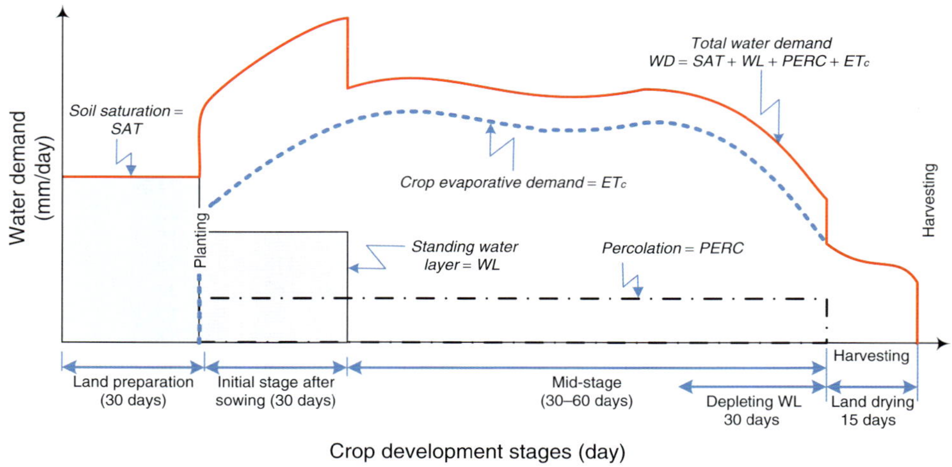

Figure 1. The schema used to estimate the water demand at different stages of crop growth.

In this study, it is assumed that a water layer of 100 mm is established in the month of sowing. A time step of five days is chosen for the calculation. The total water demand (WD) is calculated by adding ETc, WL, SAT and PERC for each time step.

For the last 15 days prior to the harvesting when the land is left to dry out, the volume of water required for evaporation is supplied by the effective rainfall in the period and any residual soil moisture maintained from the previous stages. Approximately 30 days before the land is left to dry out, the standing layer of water is slowly left to deplete without any augmenting water supply to maintain the water layer. This practice makes the best use of water supplied to maintain the WL in the previous stages. The method, thus, accounts the storage of water in time either as soil moisture or as water layer over the rice field.

Any residual soil moisture after the harvest is not included in the water footprint estimation. It is assumed that the initial soil moisture before the land preparation is negligible. It is also assumed that the contribution of capillary rise from the shallow groundwater table in the rice fields is negligible. The net inflow and outflow of the overland runoff from the bunded rice fields are assumed to be zero as well. The schema to measure the depth of water available (WA) for use in different stages of crop development is presented in Figure 2.

The water use in the rice fields is calculated for each 5-day cumulative period using the schema as presented in Figure 3. If the total water demand (WD) is less than total water available (WA), green water use is equal to the demand WD. In cases where the WD outstrips WA, the deficit is met by irrigation water supply. This deficit is called irrigation water demand. If a paddy field is 100% irrigated, it is assumed that the *blue water* use in crop production is equal to the deficit. For areas equipped with partial irrigation coverage, the blue water use is estimated on a pro-rata basis.

In order to show the sort of detail we have applied, we give an example here for India. There are two major rice production seasons in India, known as *Kharif* (monsoon season) and *Rabi* (dry season). For the period of 2000–04, the share of *Kharif* production to the gross national production (GNP) is 86% and the remaining 14% is from *Rabi*. The data for harvested area, crop period, irrigated share, crop yield and total production are taken from the Directorate of Rice Development (2001). Crop water use depends on the crop calendar adopted and it is difficult to analyse multiple crop calendars that possibly exist in a region. The study assumes a single representative calendar is valid per region in India. The planting and harvesting time for the crop are assumed to be at the average of these dates gathered from various sources such as the Directorate of Rice Development (2001), IRRI (2006), and Maclean *et al.* (2002). The major *Kharif* rice producing regions in India

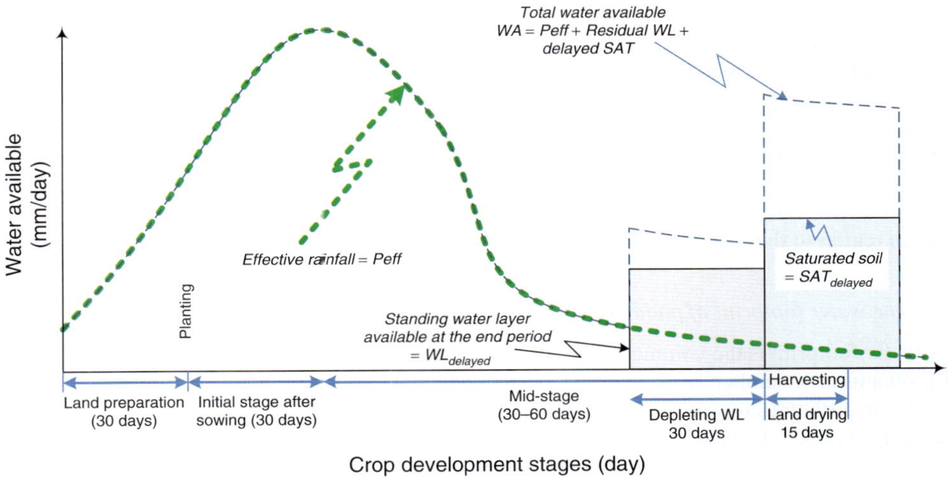

Figure 2. The schema used to estimate the total water available at different stages of crop growth.

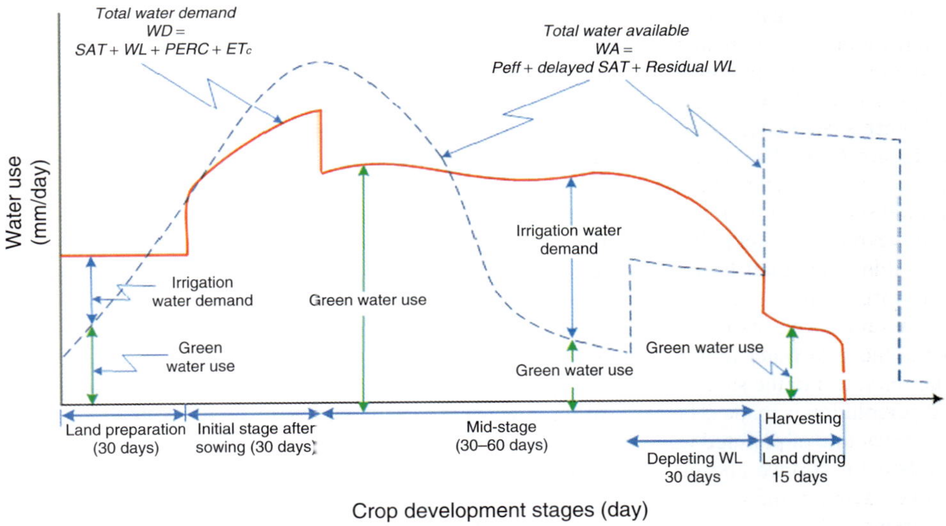

Figure 3. Distinguishing the green water use and irrigation water demand.

are Uttar Pradesh, West Bengal, Punjab, Bihar, Andhra Pradesh, Tamil Nadu, Madhya Pradesh, Orissa and Assam, producing 85% of the national *Kharif* rice production (Appendix A). The major *Rabi* rice producing regions are Andhra Pradesh, West Bengal, Tamil Nadu, Karnataka and Orissa, producing 92% of the national *Rabi* rice production. The state-wise data for irrigated area are taken from the Directorate of Rice Development (2001). The rice production in *Rabi* is assumed to be fully irrigated and the remainder of the total irrigated area is attributed to the *Kharif* rice. The irrigation water requirement (m^3/ha) and the green water use (m^3/ha) are calculated per state for the major rice producing regions. For the remaining regions, the average irrigation water requirement and green water use are calculated based on the data for the major regions. Blue water use is calculated by multiplying the irrigation requirement with the irrigated area in each season per state. The green water use in irrigated areas is calculated by multiplying the green water use (m^3/ha) by the total area in each season.

The example of India is followed for each of the other twelve countries. The planting and harvesting dates for all of the crop producing regions in these countries are chosen based on the major crop season in these regions (USDA, 1994; Directorate of Rice Development, 2001). The climate stations representing the production regions, regional share of production (%) to the total national production and irrigation coverage per region are presented for all countries in Appendix A. For each production region, we have estimated the green water use, irrigation demand and blue water use based on whether it is a *wetland system* or an *upland system*. The national averages of green and blue water use are calculated based on the data per region and the share of production of each region to the total national production.

2.1 *The water footprint of paddy rice*

The water footprint is the volume of water used to produce a particular good, measured at the point of production. A number of studies have been conducted to quantify the water footprint of a large variety of different crop products (Chapagain & Hoekstra, 2003; Chapagain & Hoekstra, 2004; Oki & Kanae, 2004 ; Hoekstra & Hung, 2005; Chapagain, 2006; Chapagain & Hoekstra, 2007; Hoekstra & Chapagain, 2008). These studies provided a broad-brush to the global picture since the primary focus of these studies was to establish a first estimate of global virtual water flows and/or national water footprints. More detailed crop-specific studies have been produced for cotton (Chapagain *et al.*, 2006), tea and coffee (Chapagain & Hoekstra, 2007), tomato (Chapagain & Orr, 2009) and sugar beet, sugar cane and maize (Gerbens-Leenes & Hoekstra, 2009). This is the first detailed global assessment of rice.

The calculation framework to quantify the water footprint of rice is based on Hoekstra & Chapagain (2008), Chapagain (2006), and Hoekstra *et al.* (2009). The water footprint of a product (m^3/unit) is calculated as the ratio of the total volume of water used (m^3/yr) to the quantity of the production (t/yr). The water footprint has three components: the green water footprint (evaporation of water supplied from the rain in crop production), blue water footprint (evaporation of the irrigation water supplied from surface and renewable groundwater sources) and the grey water footprint (volume of fresh water polluted in the production process). Most studies on the calculation of water footprints of products have taken the two evaporative components only (i.e. green and blue water footprint), excluding the grey water footprint. In an earlier study, Chapagain & Hoekstra (2004) have assumed a constant percolation loss of 300 mm of water per year from the rice field and added that to the total water footprint of rice. This is inconsistent, however, with the approach taken for other products in the same study. In the present study, a clear distinction between the evaporation and percolation is made. The percolation flow is not included in the water footprint.

The volume of polluted water depends both on the pollutant load and the adopted concentration standard for surface and ground water bodies for individual categories of pollutants. To avoid double counting, the grey water use in crop production should take the maximum of any of these requirements for individual pollutant categories. Due to data limitations, this study looks at nitrogen (N) as a representative element for estimations of the grey water footprint.

Nitrogen recovery rarely exceeds 30–40% in wet-land rice production systems (De Datta, 1995). In these systems, rice is primarily grown in clay soils thus restricting the nitrogen loss by leaching. Loss of nitrogen by runoff is also controlled in most rice fields. Ammonia (NH_3) volatisation and denitrification are recognized as major nitrogen loss mechanisms that affect the efficiency of urea and other N fertilizers in irrigated wetland rice (De Datta, 1995). In general, irrigated systems have higher fertilizer application rates than rainfed systems. For example, in India during the period of 2003–04, the fertilizer application in irrigated crop land amounted to 22% of the total national fertilizer application, whereas for rainfed crops was only 9.6% (Table 2). In Indonesia 52% of the fertilizers used are applied to rice (FAO, 2005b).

In wetland rice cultivation, the global NH_3 loss to the atmosphere from the annual use of 12 Mt of mineral fertilizer (N) amounts to 2.3 Mt N/yr, or 20% of the N application, and 97% of this occurs in developing countries (FAO & IFA, 2001). For a continuous flooding rice system, the denitrification is never more than 10%, where for an intermittent fallow system it is up to 40%

Table 2. Fertilizer used for rice production in India during 2003–04.

	Gross harvested area (10^6 ha)	Share in national fertilizer consumption (%)	Fertilizer consumption (kg/ha)			
			N	P_2O5	K_2O	Total
Irrigated	24.0	22.2	103.4	32.8	18.8	155.0
Rainfed	20.7	9.6	56.6	14.5	6.5	77.6
National	44.7	31.8	81.7	24.3	13.1	119.1

Source: FAO 2005a.

(Fillery & Vlek, 1982). As reported in Xing & Zhu (2000), there is about 0–5% of leached nitrogen from upland rice fields, though this varies from 10 to 30% if the surface runoff is taken into account. Zhu *et al.* (2000) have suggested the leaching losses to be 2% of the application rate. The magnitude of nitrogen leaching depends on soil conditions (irrigation frequencies, rainfall pattern, soil texture, percolation rate, etc.) and methods of fertilization application (application rate, time, agronomical practices, etc.). However, as the focus of this chapter is global in nature, a first-order estimation of the volume of water polluted is made following the method presented in Chapagain *et al.* (2006). In this document, we have taken a flat rate of nitrogen leaching equal to 5% of the nitrogen application rate.

Since 1991, the European Union (EU) member states have had to comply with the Nitrates Directive which aims to protect ground and surface waters from pollution by nitrogen (nitrates) originating from agriculture. The permissible limit of nitrates in surface and ground water bodies as set by the EU is 50 mg/L nitrate-NO_3. The standards recommendation by the EPA (2005) is 10 mg/L (measured as nitrogen). We have taken the number from the EU Nitrates Directive to estimate the volume of water necessary to dilute leached nitrogen to the permissible limit.

2.2 *The water footprint of processed rice*

Paddy is the most primary form of rice. The actual rice kernels are encased in an inedible and protective hull. Brown rice or husked rice has the outer hull removed, but still retains the bran layers that give it a characteristic tan color and nut-like flavor. Brown rice is edible but has a chewier texture than white rice. Milled rice is also called white rice. Milled rice is the product after milling which includes removing all or part of the bran and germ from the paddy.

On average, rice varieties are composed of roughly 20% rice hull, 11% bran, and 69% starchy endosperm. The endosperm is also known as the total milled rice which contains whole grains or head rice, and broken grains. Rice milling can be a one step, two steps or multi-step process. In a one step milling process, husk and bran removal are done in one pass and milled or white rice is produced directly out of paddy. In a two-step process, husk and bran are removed separately, and brown rice is produced as an intermediate product. In multi-stage milling, rice goes through a complex set of different processing steps. The maximum milling recovery (total milled rice obtained out of paddy, expressed as a weight percentage) is 69–70% depending on the rice variety. The global average of milling recovery is only 67%.

The water footprint of the primary rice crop is attributed to the processed products on the basis of so-called product fractions and value fractions (Chapagain & Hoekstra, 2004; Hoekstra *et al.*, 2009). The product fraction is defined as the weight of a derived product obtained per tonne of root product. For example, if one tonne of paddy rice (the root product) produces 0.95 t of husked rice (the derived product), the product fraction of husked rice is 0.95. If there are more than two products obtained while processing a root product, we need to distribute the water footprint of the root product over its derived products based on value fractions and product fractions. The value fraction of a derived product is the ratio of the market value of the derived product to the aggregated market value of all the derived products obtained from the root product. To estimate

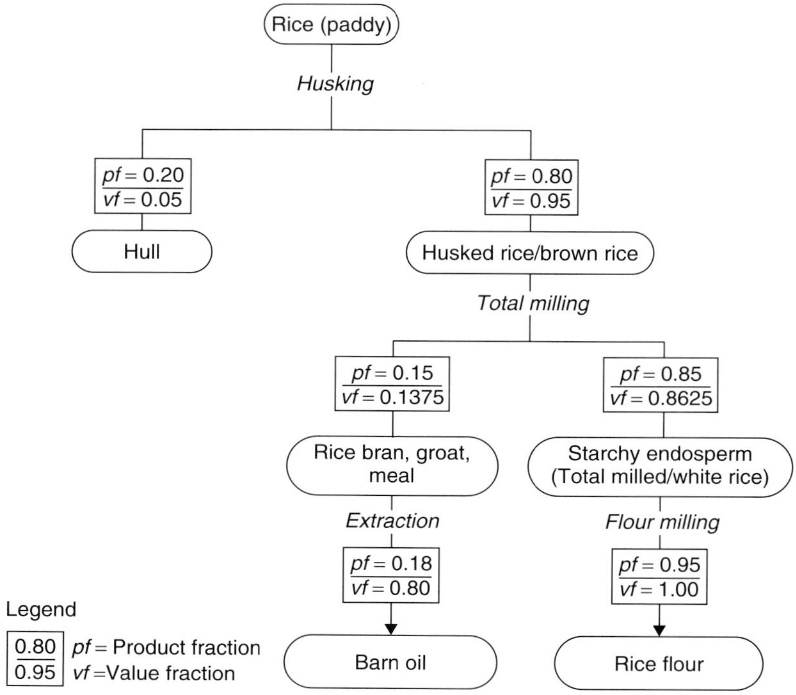

Figure 4. Product tree of rice showing value fraction and product fraction per rice processing stage.

the water footprint of various rice products originating from paddy, a product tree (Figure 4) is constructed showing the various products at various levels (primary, secondary and tertiary) along with their product fraction and value fraction. Based on these, the water footprints of the various derived rice products are calculated.

2.3 *Calculation of international virtual water flows*

The virtual water flow between two nations is the volume of water that is being transferred in virtual form from one place to another as a result of product trade. The virtual water flows between nations related to trade in rice products have been calculated by multiplying commodity trade flows (t/yr) by their associated water footprint (m^3/t) in the exporting country (Chapagain & Hoekstra, 2008). The virtual water export of a country is the volume of water used to make export goods or services. Similarly, the virtual water import of a country is the volume of virtual water imported through goods or services. Data on international trade of rice products are taken from PC–TAS (Personal Computer – Trade Analysis System) (ITC, 2006) for the period 2000-04[1].

2.4 *Calculation of the water footprint related to rice consumption in a country*

The water footprint of the consumption of rice in a nation can be divided into two parts, an internal and an external component. The internal water footprint of rice consumption refers to the consumption and pollution of national water resources for the part of rice produced and consumed

[1] The trade data on rice imports by Papua N. Guinea is erroneous in PC–TAS and thus discarded in estimating the international virtual water flows with all of its trading partner countries.

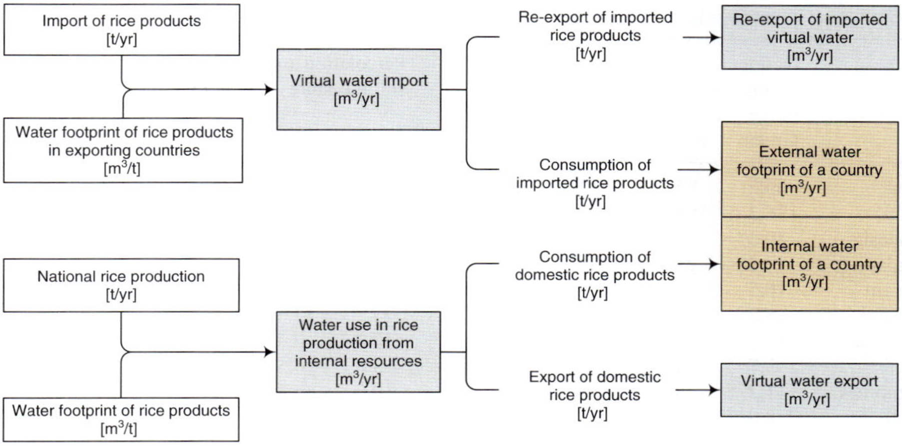

Figure 5. The calculation scheme for assessing the water footprint of national consumption of rice products.

Table 3. Depth of water used in rice production (mm/yr) for the thirteen major rice-producing countries. Period 2000–04.

	Evaporation		Pollution	Percolation	
	Green	Blue	Grey	Rain water	Irrigation water
China	228	302	73	209	277
India	314	241	34	231	178
Indonesia	260	217	53	226	188
Bangladesh	192	202	36	192	202
Vietnam	139	92	58	190	125
Thailand	252	149	31	210	125
Myanmar	297	133	18	268	120
Japan	219	258	39	224	264
Philippines	277	139	26	254	127
Brazil	260	220	20	227	192
USA	168	618	75	104	383
Korea, Rep.	232	253	55	198	216
Pakistan	124	699	26	73	412

internally. Any consumption of the part of imported rice would create equivalent size of the external water footprint of the country in locations where the rice is imported from. The internal and external water footprints are assessed following the scheme shown in Figure 5.

3 WATER FOOTPRINT OF RICE PRODUCTION

The calculated average water depth used in rice production in each of the thirteen major rice producing countries is presented in Table 3. In the USA, the evaporation is relatively high, at the same time the effective rainfall is much lower, making the irrigation volume one of the highest. Rice fields in both the USA and Pakistan are 100% irrigated, making the blue water footprint high in these countries.

Table 4. Total national water footprint of rice production and percolation of water in the thirteen major rice-producing countries (10^3 Mm^3/yr). Period 2000–04.

	National water footprint of rice production (evaporation + pollution)				Percolation and residual soil moisture			Total water use (WF + percolation)
	Green	Blue	Grey	Total	Green	Blue	Total	
China	65.2	86.5	20.8	172.5	60.0	79.5	139.5	312.0
India	136.3	104.5	14.7	255.5	100.4	77.0	177.4	432.9
Indonesia	30.3	25.3	6.1	61.7	26.3	21.9	48.2	110.0
Bangladesh	20.4	21.5	3.8	45.7	20.5	21.5	42.0	87.7
Vietnam	10.5	6.9	4.3	21.7	14.3	9.4	23.7	45.3
Thailand	25.2	15.0	3.1	43.3	21.1	12.5	33.6	76.9
Myanmar	19.1	8.5	1.1	28.8	17.2	7.7	24.9	53.7
Japan	3.7	4.4	0.7	8.8	3.8	4.5	8.3	17.1
Philippines	11.2	5.6	1.0	17.9	10.3	5.2	15.5	33.4
Brazil	8.8	7.4	0.7	16.8	7.6	6.5	14.1	31.0
USA	2.2	8.0	1.0	11.1	1.3	4.9	6.3	17.3
Korea, Rep.	2.4	2.6	0.6	5.6	2.1	2.3	4.3	10.0
Pakistan	2.9	16.3	0.6	19.9	1.7	9.6	11.3	31.2

The total water use (m^3/yr) for rice production in each country is calculated by multiplying the national harvested area of rice crops (ha/yr) with the corresponding depth of water (mm/yr) used in paddy fields. The water footprint of rice production is the sum of water evaporated from the rice fields and the volume of water polluted in the process. The results are presented in Table 4. It also presents the volume of water percolated or left over as residual soil moisture after the crop harvest in the fields. Total water use is the sum of the water footprint and percolation. The water footprint refers to a real loss to the catchment, while the percolation is actually not a loss to the catchment.

Table 5 shows the water footprint and percolation per unit of paddy rice produced (m^3/t). The figures follow from dividing total national water footprint and percolation related to rice production (m^3/yr) by the gross national paddy production per year (t/yr). The volume of water evaporated per tonne of rice is quite similar to the evaporation per tonne of wheat, as also noted in Bouman & Tuong (2001). The higher evaporation rates per hectare as a result of the standing water layer in rice fields are apparently compensated by the relatively higher yields of rice (Bouman *et al.*, 2007b).

Table 5 also shows the global average water footprint of rice, calculated based on the share of national production of the top-13 rice producing countries to the total global production. Since the export share of these 13 countries to the total export volume during the period 2000–04 differs widely, the global average water footprint of rice paddy is also calculated weighing the export share of these countries. As the top-13 largest rice producing countries contribute 82% to the global share of rice export, the difference between these two averages is not big. Global average results presented in the following sections are based on the global average water footprint based on production. Table 6 shows the global average water footprints of rice products.

Using the global average water footprint of paddy calculated and the production data for the rest of the countries, the global water footprint of rice production is estimated to be 784,000 Mm^3/yr (48% green, 44% blue and 8% grey) (Figure 6). The volume of water percolated in the rice fields plus any residual soil moisture left in the field after rice harvest is equal to 607,000 Mm^3/yr, about half of which (52%) is sustained by rainfall in the rice field. Including percolation, the total blue water use in the rice field becomes 636,000 Mm^3/yr, which is the number often quoted in the literature while referring to the total water used in rice production. If we add the total water footprint

Table 5. Water footprint and percolation per unit of paddy rice produced (m^3/t) in the thirteen major rice-producing countries. Period 2000–04.

	Water footprint				Percolation		
	Green	Blue	Grey	Total	Rain water	Irrigation water	Total
China	367	487	117	971	338	448	785
India	1,077	826	116	2,020	794	609	1,403
Indonesia	583	487	118	1,187	505	422	927
Bangladesh	549	577	103	1,228	550	578	1,128
Vietnam	308	203	127	638	420	277	697
Thailand	942	559	116	1,617	787	467	1,253
Myanmar	846	378	50	1,274	763	341	1,103
Japan	341	401	61	802	348	409	757
Philippines	844	423	78	1,345	775	388	1,163
Brazil	791	670	61	1,521	691	585	1,276
USA	227	835	101	1,163	141	517	658
Korea, Rep.	356	388	84	829	303	331	634
Pakistan	421	2,364	88	2,874	248	1,394	1,642
Average based on weighted production data	632	584	109	1,325	535	490	1,025
Average based on weighted export data	618	720	112	1,450	522	538	1,060

Table 6. The global average water footprint of rice products (m^3/t). Period 2000–04.

PC–TAS code	Product description	Green	Blue	Grey
100610	Rice in the husk (paddy or rough)	632	584	109
100620	Rice, husked (brown)	750	693	130
110314	Rice groats and meal	688	636	119
100630	Rice, semi-milled, milled, whether or not polished or glazed	761	704	132
100640	Rice, broken	761	704	132
110230	Rice flour	801	741	139

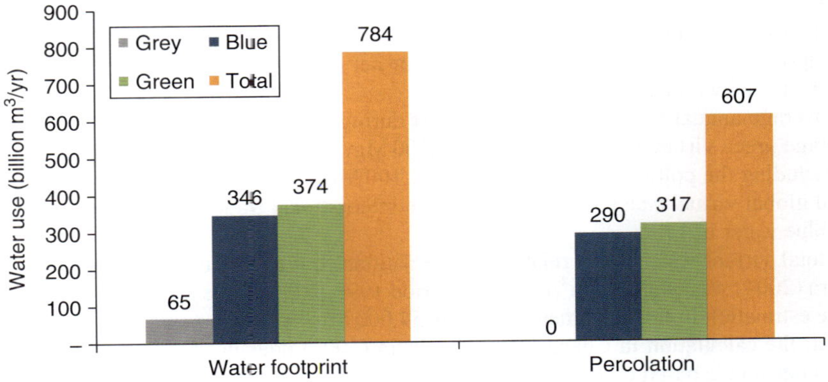

Figure 6. The global water footprint of rice production and the total volume of water percolated in rice fields (km^3/yr). Period 2000–04.

Table 7. Largest net-exporters and net-importers of virtual water related to the international trade of rice products.

Largest net exporters (Mm^3/yr)					Largest net importers (Mm^3/yr)				
	Green	Blue	Grey	Total		Green	Blue	Grey	Total
Thailand	5,607	3,327	691	9,625	Nigeria	1,528	1,204	211	2,943
India	2,764	2,119	298	5,181	Indonesia	788	682	149	1,620
Pakistan	428	2,405	90	2,923	Iran	670	721	97	1,489
USA	237	2,172	245	2,654	Saudi Arabia	650	694	82	1,426
Vietnam	595	392	246	1,233	Senegal	756	482	107	1,344
Uruguay	428	395	74	897	South Africa	701	509	88	1,298
Italy	417	370	74	861	Philippines	490	386	103	979
Egypt	307	284	53	644	Brazil	433	459	83	974
China	87	410	106	602	Japan	340	514	83	937
Australia	215	196	40	451	Malaysia	399	349	66	814

and the percolation water volume, it is equal to 1,391,000 Mm^3/yr, which is nearly the same as the global water use in rice fields (1,359,000 Mm^3/yr) as reported in Chapagain & Hoekstra (2004). Water footprints of rice production for all countries are presented in Appendix B.

4 INTERNATIONAL VIRTUAL WATER FLOWS

International trade in rice during the period 2000–04 resulted in a total international virtual water transfer of 31,100 Mm^3/yr (45% green water, 47% blue water, 8% grey water). This means that international rice trade is linked to the evaporation of 28,700 Mm^3/yr of water with an additional 2,400 Mm^3 of fresh water being polluted each year in the exporting countries.

The top ten largest gross virtual water exporters are Thailand (9,627 Mm^3/yr), India (5,185 Mm^3/yr), USA (3,474 Mm^3/yr), Pakistan (2,923 Mm^3/yr), China (1,296 Mm^3/yr), Vietnam (1,233 Mm^3/yr), Italy (1,048 Mm^3/yr), Uruguay (899 Mm^3/yr), Egypt (644 Mm^3/yr) and Australia (599 Mm^3/yr) covering nearly 87% of the total virtual water export international trade in rice products globally. The largest gross importers are Nigeria (2,944 Mm^3/yr), Indonesia (1,637 Mm^3/yr), Iran (1,506 Mm^3/yr), Saudi Arabia (1,429 Mm^3/yr), South Africa (1,348 Mm^3/yr), Senegal (1,346 Mm^3/yr), Brazil (1,010 Mm^3/yr), Japan (988 Mm^3/yr) and Philippines (979 Mm^3/yr) covering about 42% of the total import. Appendix D shows gross virtual water export and import for all countries.

Net imports of water are calculated by subtracting the gross export volume of water from the gross import volume of water, and vice versa for net exports. The largest net exporters and net importers are shown in Table 7.

The average annual blue virtual water import during the study period was 14,600 Mm^3/yr and the average green virtual water import was 14,100 Mm^3/yr. The total average annual virtual water flows including the pollution component was 31,100 Mm^3/yr. The share of green virtual water to the total global virtual water flows related to the international trade of rice products is 45%, and that of blue water is 47%.

The total virtual water flows related to international trade of rice according to Chapagain & Hoekstra (2004) was 64,000 Mm^3/yr for the period 1997–2001 (Table 8). This is quite comparable with the estimation in this document, which is 54,000 Mm^3/yr when percolation is also included. However, the calculation in Chapagain & Hoekstra (2004) does not separate the green and blue components, and is based on national average climate dataset. The earlier study does not separate percolation from the total virtual water flows. The former study gave an overestimation, as it was assumed that the total crop water requirements in the rice fields are always met either by rainfall or

Table 8. Global international virtual water flows by rice product (Mm^3/yr).

Product description	Current study *					Chapagain & Hoekstra **
	Green	Blue	Grey	Percolation	Total	Total virtual water flows
Rice flour	108	89	17	162	375	511
Rice groats and meal	6	5	1	9	21	24
Rice in the husk (paddy or rough)	662	1,392	192	1,430	3,675	2,776
Rice, broken	2,121	1,800	351	3,311	7,583	10,853
Rice, husked (brown)	1,417	1,715	258	2,423	5,813	5,302
Rice, semi-milled or wholly milled	9,768	9,591	1,561	15,447	36,367	44,741
Total	14,081	14,592	2,379	22,782	53,834	64,207

* Period 2000–2004. The assessment includes grey water.
** Period 1997–2001. The assessment does not separate different components of virtual water flows. It excludes grey water, but includes percolation in rice fields.

Table 9. Top-15 of countries with the largest water footprints related to rice consumption (Mm^3/yr). Period 2000–04.

Country	Water footprint (Mm^3/yr)				Per capita water footprint (m^3/yr)
	Green	Blue	Grey	Total	
India	133,494	102,425	14,385	250,305	239
China	65,154	86,050	20,680	171,884	134
Indonesia	31,097	26,005	6,262	63,364	299
Bangladesh	20,560	21,574	3,846	45,980	317
Thailand	19,640	11,654	2,421	33,714	547
Myanmar	18,989	8,483	1,118	28,591	612
Vietnam	9,860	6,496	4,074	20,430	256
Philippines	11,736	6,020	1,137	18,893	238
Brazil	9,186	7,869	757	17,812	99
Pakistan	2,480	13,935	521	16,936	117
Japan	4,084	4,923	748	9,755	77
USA	1,924	5,779	719	8,422	29
Egypt	3,467	3,203	599	7,269	105
Nigeria	3,478	3,005	548	7,031	54
Korea, Rep.	2,491	2,732	592	5,814	122

by irrigation water supply, which is not the case in general. On the other hand, the earlier estimate does not include the volume of water polluted in the process.

5 WATER FOOTPRINT OF RICE CONSUMPTION

The largest consumer of rice in terms of water is India, followed by China, Indonesia, Bangladesh, Thailand, Myanmar, Vietnam, the Philippines and Brazil. The composition of the water footprint related to rice consumption for the 15 largest countries is presented in Table 9. The per-capita water footprint of rice consumption is quite high in Thailand (547 m^3/yr) compared to India (239 m^3/yr), Indonesia (299 m^3/yr), China (134 m^3/yr) and the USA (29 m^3/yr). This variation is also because

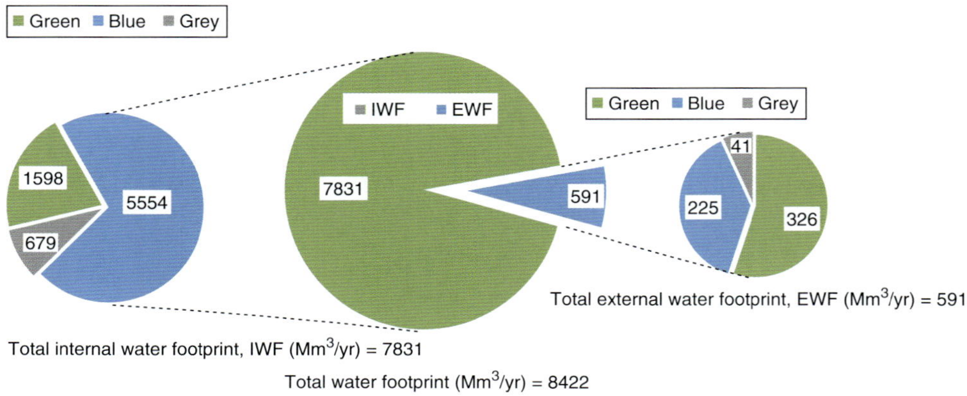

Figure 7. Water footprint of rice consumption in the USA (Mm^3/yr). Period 2000–04.

Table 10. External water footprint (EWF) of the USA by locations (Mm^3/yr). Period 2000–04.

	Green	Blue	Grey	Total	Share to the total EWF
Thailand	245	137	29	411	69.5%
India	47	34	5	86	14.5%
Pakistan	5	25	1	30	5%
China	9	12	3	24	4%
Australia	11	10	2	23	4%
Others	8	7	1	17	3%
Total	326	225	41	591	100%

the diet contains more rice in some countries compared to others. The complete list of countries with their water footprints related to rice consumption is presented in Appendix C.

From the perspective of food security as well as from the viewpoint of sustainable consumption it is interesting to know where water footprints related to national consumption actually *land*. We give here two examples, one for the USA and one for Europe. The total water footprint of the USA is 8,422 Mm^3/yr. The internal water footprint is relatively large (93% of the total water footprint) (Figure 7). The external water footprint of the USA is 591 Mm^3/yr and largely refers to water use in Thailand, India, Pakistan, China and Australia (Table 10).

In contrast to the USA, the sizes of the rice-consumption related internal and external water footprints of the EU27 are fairly comparable. Out of 5,335 Mm^3/yr, the internal component is 2,878 Mm^3/yr and the external one is 2,457 Mm^3/yr (Figure 8). More than 70% of the total external water footprint of the EU27 rests on eight countries, namely India, Thailand, USA, Pakistan, Egypt, Guyana, China and Vietnam. Figure 9 shows the external water footprint of the EU27 in each of these countries, distinguishing between the green, blue and grey water footprint. The largest share of the blue water footprint is for rice imported from the USA and Pakistan. Although the total footprint on India is the largest, a large fraction of it is made up of green water. Though the total footprint on Egypt, Guyana and Vietnam is much lower than in Pakistan, the grey component on these countries is relatively higher than in Pakistan.

6 DISCUSSION AND CONCLUSION

Rice is a staple food for 3,000 million people (Maclean *et al.*, 2002), especially in Southeast Asia, the Middle East, Latin America, and the West Indies. In terms of human nutrition and caloric

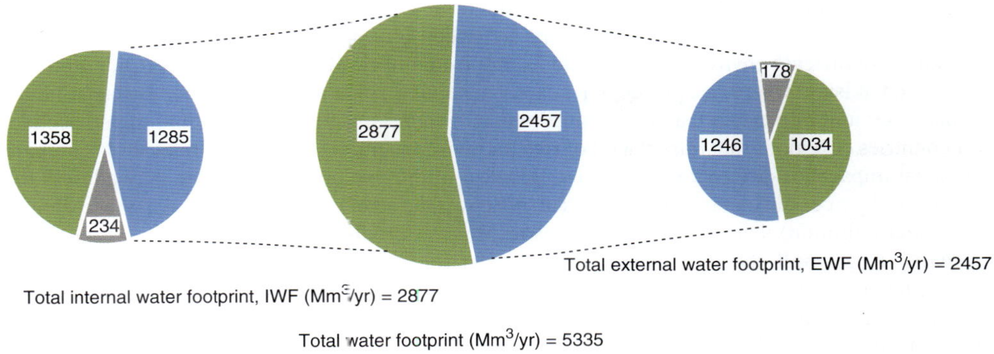

Figure 8. Water footprint of rice consumption in EU27 countries (Mm^3/yr). Period 2000–04.

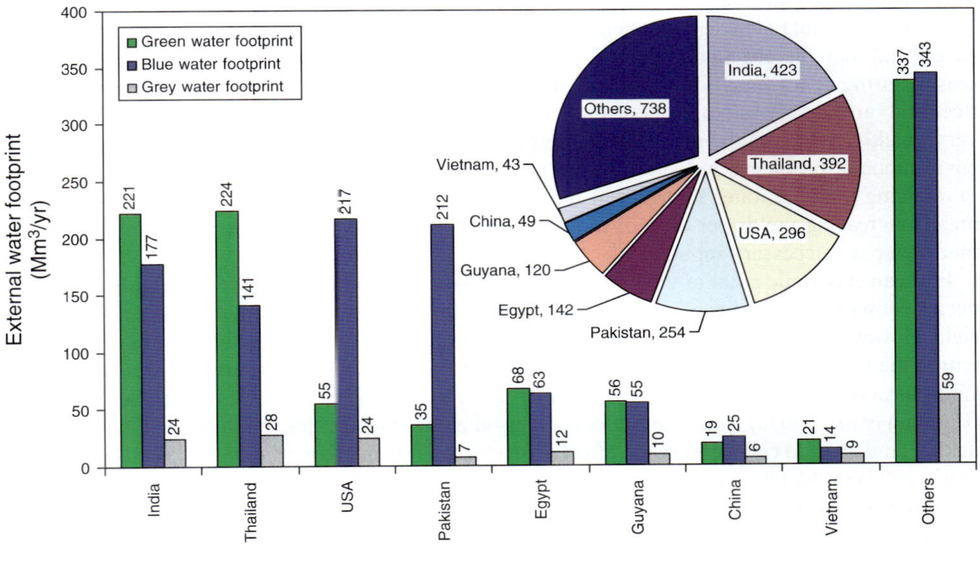

Figure 9. The external water footprint of rice consumption in the EU27. Period 2000–04.

intake, it provides nearly one fifth of the direct human calorie intake worldwide, making it the most important food crop (Smith, 1998; Zeigler & Barclay, 2008). Rice consumption exceeds 100 kg per capita annually in many Asian countries (compare for example with the USA average of 10 kg) and is the principal food for most of the world's poorest people, particularly in Asia, which is home to 70% of those who earn less than US$ 1 a day (Zeigler & Barclay, 2008). Rice production is deeply rooted in the socio-political culture in Asia which nearly produces nearly 90% of the global rice (Bouman *et al.*, 2007a).

The water footprint of rice production and consumption is quite significant in South Asian countries. However, in these countries most of the water footprint is rooted in the wet season, so that the contribution to water scarcity is relatively low in contrast to our general perception. Globally, there is nearly an equal share of green and blue water use in the total water footprint of rice. The green water footprint (rain) has a relatively low opportunity cost compared to the blue water footprint (irrigation water evaporated from the field). The environmental impact of the blue water footprint in rice production depends on the timing and location of the water use. It would need a dedicated analysis to estimate where and when blue water footprints in rice production constitute significant environmental problems, but from our results it is obvious that rice from the USA and Pakistan, where rice production heavily depends on blue water, will generally cause larger impacts per unit of product than rice from Vietnam. From a sustainable-consumption perspective, for countries or regions that import a lot of rice for own consumption, it may be relevant to compare the local impacts of different rice sources. Besides, in international context one may address the question why rice consumers like in the EU do not cover the actual water cost (costs of water scarcity and water pollution) that occurs in the countries from where the rice is obtained. Since irrigation systems are generally heavily subsidized and water scarcity is never translated into a price, the economic or environmental costs of water are not contained in the price of rice. The water cost may actually largely vary from place to place, depending on whether the rice comes from e.g. India, Thailand, the USA, Pakistan or Egypt, and depending on whether the rice is produced in the dry or the wet period.

In probably a majority of cases, the green water footprint of rice production does not constitute significant negative environmental or economic impacts. Rainwater allocated for rice production generally has no opportunity cost, which means that alternative uses of the rain (natural vegetation, other crops) would not give higher benefits. Storing rainwater in the fields reduces or delays surface runoff and may thereby flatten peak flows in downstream rivers, which may be useful in the wet season during heavy rains. On the other hand, this mechanism may be absent or even reversed when rice fields are already full of water up to the point of overflow, in which case rain will become runoff very quickly. Although the green water footprint in rice production may not constitute significant environmental problems, reduction of the green water footprint at a global level is probably key in reducing the blue water footprint in rice production. Better use of rain wherever possible, that means increasing yields per drop of rainwater, will reduce the demand for rice from areas where blue water is a necessary input.

From an economic point of view, reducing percolation of blue water in the rice fields is relevant, because it will reduce costs of water supply by reducing the absolute volume of water needed in the field. However, the environmental benefit may not be quite big as percolated blue water will remain within the same catchment as from where it was abstracted. As a lot of water is percolating in the first phase of the land preparation, a number of water saving technologies have been suggested (Bouman *et al.*, 2007a), which are effectively used in the Phillippines, India and China. The direct dry seeding method can increase the effective use of rainfall and reduce irrigation needs (Cabangon *et al.*, 2002) in the phase of land preparation. Another way to reduce percolation from fields is to use System of Rice Intensification (SRI). SRI suggests ways to improve rice yields with less water, the main advance is that it uses water just enough to keep the roots moist all the time without any standing water at any time. The argument behind SRI is that the main benefit of flooding the rice plant is to check the proliferation of weeds, thereby saving labour (Gujja *et al.*, 2007), which can be a favourable option where the supply is limited or scarce.

Rice production is a so-called diffuse source of pollution and hence difficult to mitigate. The option to have optimal application of fertilizer so that the application matches exactly the plant uptake, as in the case of dry crops, is not suitable in rice production. There is inevitably percolation leaching a part of the fertilizer. The grey component of the water footprint can only be reduced with a reduction in the leaching of fertilizers and pesticides from the field, e.g. by increasing water use efficiency, using slow-release fertilizers and nitrification inhibitors, puddling the rice fields, planting catch and cover crops and using crop residues in situ (Choudhury & Kennedy, 2005). The loss of nitrogen may cause environmental and health problems. Although these problems cannot be alleviated completely, there are enough research findings that indicate that these problems can be minimized through a number of management practices (Choudhury & Kennedy, 2005). The fate of nitrogen in soil is mainly governed by different processes: plant-uptake, ammonia volatilization, de-nitrification and losses to surface (runoff) or ground water bodies (leaching). All these three processes are intertwined and it is hard to study them in isolation. A systematic analysis of fate of nitrogen should be carried out at field level to reveal any specific impacts on the system.

REFERENCES

Allen, R.G.; Pereira, L.S.; Raes, D. & Smith, M. (1998). Crop evapotranspiration – Guidelines for computing crop water requirements. In: *FAO Irrigation and Drainage*, Paper 56. FAO, Rome, Italy.

Belder, P.; Bouman, B.A.M.; Cabangon, R.; Guoan, L.; Quilang, E.J.P.; Yuanhua, L.; Spiertz, J.H.J. & Tuong, T.P. (2004). Effect of water-saving irrigation on rice yield and water use in typical lowland conditions in Asia. *Agricultural Water Management*, 65(3): 193.

Bouman, B.A.M. & Tuong, T.P. (2001). Field water management to save water and increase its productivity in irrigated lowland rice. *Agricultural Water Management*, 49(1): 11.

Bouman, B.; Barker, R.; Humphreys, E. & Tuong, T.P. (2007a). Rice: feeding the billions. In: D. Molden (ed.), *Water for Food, Water for Life*. Earthscan, London, UK & International Water Management Institute, Colombo, Sri Lanka.

Bouman, B.A.M.; Humphreys, E.; Tuong, T.P.; Barker, R. & Donald, L.S. (2007b). Rice and Water. In: *Advances in Agronomy*: 187–237. Academic Press.

Brouwer, C. & Heibloem, M. (1986). Irrigation water management, Training manuals, 3. Food and Agriculture Organization of the United Nations (FAO). Rome, Italy.

Cabangon, R.J.; Tuong, T.P. & Abdullah, N.B. (2002). Comparing Water Input and Water Productivity of Transplanted and Direct-Seeded Rice Production Systems. *Agricultural Water Management*, 57(1): 11–13.

Chapagain, A.K. (2006). *Globalisation of water: Opportunities and threats of virtual water trade*. A.A. Balkema Publishers, Taylor & Francis Group, Leiden, the Netherlands.

Chapagain, A.K. & Hoekstra, A.Y. (2003). Virtual water flows between nations in relation to trade in livestock and livestock products. Value of Water Research Report Series, No. 13. UNESCO-IHE.

Chapagain, A.K. & Hoekstra, A.Y. (2004). Water footprints of nations. Value of Water Research Report Series, No. 16. UNESCO-IHE. Delft, the Netherlands.

Chapagain, A.K. & Hoekstra, A.Y. (2007). The water footprint of coffee and tea consumption in the Netherlands. *Ecological Economics*, 64(1): 109–118.

Chapagain, A.K. & Hoekstra, A.Y. (2008). The global component of freshwater demand and supply: An assessment of virtual water flows between nations as a result of trade in agricultural and industrial products. *Water International*, 33(1): 19–32.

Chapagain, A.K. & Orr, S. (2009). An improved water footprint methodology linking global consumption to local water resources: A case of Spanish tomatoes. *Journal of Environmental Management*, 90(2): 1219–1228.

Chapagain, A.K.; Hoekstra, A.Y.; Savenije, H.H.G. & Gautam, R. (2006). The water footprint of cotton consumption: An assessment of the impact of worldwide consumption of cotton products on the water resources in the cotton producing countries. *Ecological Economics*, 60(1): 186–203.

Choudhury, A.T.M.A. & Kennedy, I.R. (2005). Nitrogen fertilizer losses from rice soil and control of environmental pollution problems. *Communications in Soil Sciences and Plant Analysis*, 36: 1625–1639.

Clarke, D.; Smith, M. & El-Askari, K. (1998). *CropWat for Windows: User guide*. Food and Agriculture Organization of the United Nations (FAO), Rome, Italy.

Cuenca, R.H. (1989). *Irrigation system design: An engineering approach*. Prentice-Hall, Englewood Cliffs, New Jersey, USA.

De Datta, S.K. (1995). Nitrogen transformations in wetland rice ecosystems. *Nutrient Cycling in Agroecosystems*, 42(1-3): 193–203.

Directorate of Rice Development (2001). Rice in India – A Handbook of Statistics. Available at: [http://drdpat.bih.nic.in].

EPA (2005) List of drinking water contaminants: Ground water and drinking water. Available at: [http://www.epa.gov/safewater/mcl.html#1].

FAO (1992). CROPWAT: A computer program for irrigation planning and management. *Irrigation and Drainage*, Paper 46. Developed by Martin Smith. Food and Agriculture Organization of the United Nations (FAO), Rome, Italy.

FAO (1993). CLIMWAT for CROPWAT: A climatic database for irrigation planning and management. *Irrigation and Drainage*, Paper 49. Developed by Martin Smith. Food and Agriculture Organization of the United Nations (FAO), Rome, Italy.

FAO (2001). FAOCLIM2: World-Wide Agroclimatic database. CD ROM. Food and Agriculture Organization of the United Nations (FAO), Rome, Italy.

FAO (2005a). *Fertiliser use by crop in India*. Food and Agriculture Organization of the United Nations (FAO), Rome, Italy.

FAO (2005b). *Fertiliser use by crop in Indonesia*. Food and Agriculture Organization of the United Nations (FAO), Rome, Italy.

FAO & IFA (2001). Global estimates of gaseous emission of NH_3, NO and N_2O from agricultural land. Food and Agriculture Organization of the United Nations (FAO), Rome, Italy & International Fertilizer Industry Association (IFA).

FAOSTAT data (2009). FAO Statistical Databases [http://faostat.fao.org/].

Fillery, I. & Vlek, P. (1982). The significance of denitrification of applied nitrogen in fallow and cropped rice soils under different flooding regimes. Greenhouse experiments. *Plant and Soil*, 65(2): 153.

Gerbens-Leenes, P.W. & Hoekstra, A.Y. (2009). The water footprint of sweeteners and bio-ethanol from sugar cane, sugar beet and maize. *Value of Water Research Report Series*, No. 38, Delft and Enschede, The Netherlands; UNESCO-IHE and University of Twente.

Gujja, B.; Riddell, P.; Goud, V.V.; Dalai, S.; Murty, M.; Holland, R.; Rupela, O.P.; Rao, P.; Kumar, M.; Rao, K. & New Concept team (2007). More rice with less water: SRI – System of Rice Intensification. Dialogue on Water, Food and Environment. WWF.

Hoekstra, A.Y. & Chapagain, A.K. (2008). *Globalization of Water: Sharing the Planet's Freshwater Resources*. Blackwell Publishing Ltd., Oxford, UK.

Hoekstra, A.Y. & Hung, P.Q. (2005). Globalisation of water resources: International virtual water flows in relation to crop trade. *Global Environmental Change*, 15(1): 45–56.

Hoekstra, A.Y.; Chapagain, A.K.; Martínez-Aldaya, M. & Mekonnen, M.M. (2009). Water footprint manual: State of the Art 2009. Water Footprint Network, Enschede, the Netherlands.

IFA, IFDC, IPI, PPI & FAO (2002). *Fertilizer use by crop*. Available at: [http://www.fertilizer.org/ifa/statistics/crops/fubc5ed.pdf].

IRRI (2006). *World Rice Statistics*. Available at: [http://www.irri.org/science/ricestat/].

ITC (2006). PC–TAS version 2000-2004 in HS or SITC. CD-ROM. International Trade Centre, Geneva, Switzerland.

Jensen, M.E.; Burman, R.D. & Allen, R.G. (eds.) (1990). *Evapotranspiration and irrigation water requirements*. American Society of Civil Engineers, New York, USA.

Maclean, J.L.; Dawe, D.C.; Hardy, B. & Hettel, G.P. (eds.) (2002). Rice Almanac. CABI Publishing, UK.

Mohan, S.; Simhadrirao, B. & Arumugam, N. (1996). Comparative study of effective rainfall estimation methods for lowland rice. *Water Resources Management*, 10(1): 35.

Oki, T. & Kanae, S. (2004). Virtual water trade and world water resources. *Water Science & Technology*, 49(7): 203–209.

Smith, B.D. (1998). *The Emergence of Agriculture*. Scientific American Library, a Division of HPHLP, New York, USA.

USDA (1994). *Major world crop areas and climatic profiles*. World Agricultural Outlook Board, U.S. Department of Agriculture, Washington D.C., USA.

Xing, G.X. & Zhu, Z.L. (2000). An assessment of N loss from agricultural fields to the environment in China. *Nutrient Cycling in Agroecosystems*, 57(1): 67.

Zeigler, R.S. & Barclay, A. (2008). The relevance of rice. *Rice*, 1(1): 3–10.

Zhu, J.G.; Han, Y.; Liu, G.; Zhang, Y.L. & Shao, X.H. (2000). Nitrogen in percolation water in paddy fields with a rice/wheat rotation. *Nutrient Cycling in Agroecosystems*, 57(1): 75.

Appendix A. Data on main regions of rice production within major rice producing countries.

Country	Crop season, major rice harvesting regions, share to the national production, irrigated area in % or ha, crop planting date, crop length in days and relevant climate stations
Bangladesh	Aus (14%, 100%, 15-Apr, 130d, Guwahati), T.Aman (40%, 100%, 01-Aug, 130d, Guwahati), Aman broadcast (6%, 100%, 15-Apr, 115d, Guwahati), Boro (40%, 100%,01-Dec, 170d, Guwahati)
Brazil	Rio grande (50%, 100%, 15-Nov, 120d, Passo Fundo & Bage), Minas Geralas (8%, 40%, 15-Nov, 120d,Pocos de Caldas), Mato Grosso (8%, 40%, 15-Nov, 120d, Cuiaba), Santa Caatarina and Parana (9%, 100%,15-Nov, 120d, Londrina & Puerto Stroessner), Goias (5%, 40%,15-Nov, 120d, Goiania), Maranhao and others (10%, 40%, 01-Jan, 120d, Quixeramobim), Tocantins (3%, 40%, 01-Jan, 120d, Tocantin), Sao Paulo (3%, 40%,15-Nov, 120d,Pocos de Caldas), Mato Grosso do sul (2%, 40%, 15-Nov, 120d, Campo Grande), Para (2%, 40%, 01-Jan, 120d, Quixeramobim)
China	**Single crop:** Hunan (1.44%, 90%, 1-May, 135d, Changsha), Sichuan (12%, 90%, 1-May, 135d, Chungking), Jiangsu (9.12%, 90%,1-May, 135d, Hangzhou), Hubei (4.32%, 90%, 1-May, 135d, Changsha), Anhui (3.84%, 90%,1-May, 135d, Hangzhou), Fujian (0.96%, 90%, 1-May, 135d, Hangzhou),Yunnan (2.4%, 90%, 1-May, 135d, Kunming), Liaoning 1.92%, 90%,1-May, 135d, Shenyang), Guizhou (1.92%, 90%,1-May, 35d, Chungking), Heilongjiang (1.92%, 90%,1-May, 135d, Harbin), Jilin (1.44%, 90%,1-May, 35d, Shenyang), Henan (1.44%, 90%,1-May, 135d, Heze), Shanghai (0.96%, 90%,1-May, 135d, Hangzhou), Others (4.32%, 90%,1-May, 135d) **Early double:** Hunan (5.46%, 90%, 1-Mar, 120d, Changsha), Hubei (2.34%, 90%, Changsha), Guangdong (4.42%, 90%, Guangzhou), Jiangxi (3.64%, 90%, Changsha), Anhui (1.3%, 90%, Hangzhou), Zhejiang (3.12%, 90%, Hangzhou), Guangxi (3.38%, 90%, Kunming), Fujian (1.56%, 90%, Hangzhou), Others (0.78%, 90%,) **Late double:** Hunan (6.24%, 90%, 1-Aug, 120d, Changsha), Hubei (2.6%,90%, Changsha), Guangdong (4.16%, 90%, 1-Aug, 120d, Guangzhou), Jiangxi (3.64%,90%,1-Aug, 120d, Changsha), Anhui (1.3%, 90%, 1-Aug, 120d, Hangzhou), Zhejiang (3.38%, 90%, 1-Aug, 120d, Hangzhou), Guangxi (2.34%, 90%, 1-Aug, 120d, Kunming), Fujian (1.56%, 90%, 1-Aug, 120d, Hangzhou), Others (0.78%, 90%,1-Aug, 120d)
India	**Khariff:** West Bengal (12.34%, 195000ha, 01-Jun, 150d, Chandbali), Uttar Pradesh (16.43%, 3716000ha, 15-Jun, 120d, Bareilly), Andhra Pradesh (8.70%, 2503000ha, 01-Apr, 180d, Begampet), Punjab (11.08%, 2447000ha, 01-Jul, 120d, Amritsar), Tamil Nadu (8.55%, 1764000ha, 01-May, 150d, Banglore), Bihar (9.40%, 1942000ha, 15-Jun, 120d, Bareilly), Orissa (6.73%, 1375000ha, 01-Jun, 180d, Chandbali), Madhya Pradesh (7.20%, 1282000ha, 15-Jul, 150d, Pendra). Assam (4.10%, 296000ha, 15-Mar, 150d, Guahati), Karnataka (3.38%, 615000ha, 150d, Banglore), Haryana (3.42 %, 1024000ha, 150days,-), Maharashtra (3.25%, 385000ha, 150d,-), Gujarat (1.37%, 371000ha, 150d,-), Kerala (0.83%, 139000ha, 150d,-), Jammu & Kashmir (0.69%, 239000ha, 150d,-), Tripura (0.55%, 150d,-), Manipur (0.50%, 73000ha, 150d,-), Rajasthan (0.29%, 63000ha, 150d,-), Nagaland (0.28%, 65000ha,150d,-), Meghalaya (0.21%, 45000ha, 150d,-), Goa (0.20%, 14000ha, 150d,-), Arunachal Pradesh (0.17%, 34000ha, 150d,-), Himachal Pradesh (0.16%, 51000ha, 150d,-), Mizoram (0.14%, 4000ha, 150d,-), Sikkim (0.03%, 16000ha, 150d,-). *Note: Rainfed area in Khariff season in '000ha: West Bengal 4413, Uttar Pradesh 2104, Andhra Pradesh 172, Punjab 22, Tamil Nadu 164, Bihar 3005, Orissa 2845, Madhya Pradesh 4139, Assam 1980, Karnataka 452, Haryana 4, Maharashtra 1071, Gujarat 282, Kerala 165, Jammu & Kashmir 26, Tripura 202, Manipur 88, Rajasthan 115, Nagaland 81, Meghalaya 60, Goa 43, Arunachal Pradesh 85, Himachal, Pradesh 32 Mizoram 57.* *Total rainfed area = 21606000ha* **Rabi:** West Bengal (36.43%, 1386000ha, 01-Dec, 150d, Chandbali), Uttar Pradesh (0.12%, 6000ha, 01-Dec, 150d, Bareilly), Andhra Pradesh (32.15%, 1232000ha, 01-Jan, 150d, Begampet), Tamil Nadu (9.20%, 318000ha, 01-Nov, 150d, Banglore), Bihar (2.24%, 128000ha, 01-Nov, 150d, Bareilly), Orissa (5.19%, 295000ha, 01-Jan, 150d, Chandbali), Assam (3.92%, 231000ha, 01-Jan, 150d, Guahati), Karnataka (8.21%, 341000ha, 15-Jan,

(Continued)

Appendix A. (Continued)

	150d, Banglore), Maharashtra (0.55%, 33000ha,-), Kerala (1.18%, 59000ha,-), Tripura (0.81%, 55000ha,-), Mizoram (0.01%, 1000 ha,-). *Note: Rabi crop is 100% irrigated in India. Total area under Rabi crop = 4085000ha*
Indonesia	**Main crop:** Java (41.3%,60%,, 15-Nov, 120d, Jakarta and Semarang), S.Sulawesi (8.1%, 60%, 01-Jun, 120d, Manado), N.Sumatra (10.5%, 60%, 01-Aug, 120d, Medan), S.Sumatra (8.4%, 60%,15-Nov, 20d, Palemban), Kalimanthan (5.0%, 60%,15-Nov, 120d, Banjamarsin), Bali and Nusa (6.0%, 60%,15-Nov, 120d, Bali (Denpasa) and Nusa (Kupang)) **Second crop:** Java (17.7%, 100%, 15-May, 120d, Jakarta and Semarang), S.Sulawesi (0.9%, 100%, 01-Nov, 120d, Manado), N.Sumatra (1.1%,100%,01-Apr, 120d,Medan), S.Sumatra (1.1%,100%,15-Jun, 120d, Palemban)
Japan	Tohoku (27%, 100%, 01-Jun, 120d,Akita and Ishinomaki), Kanto Tosan (17%, 100%, 01-Jun, 120d, Niigata), Hokuriku (13%,100%,01-Jun, 120d, Kumagaya), Kyushu (11%, 100%, 01-Jun, 120d, Saga), Chuguko (7%, 100%, 01-Jun, 120d, Fukuyama), Hokkaido (7%, 100%, 01-Jun, 120d, Iwamizawa), Kinki (7%,100%, 01-Jun, 120d, Kyoto),Tokai (6%, 100%, 01-Jun, 120d, Gifu), Shikoku and others (5%, 100%, 01-Jun, 120d, Kochi)
Korea, R.	Inchon (33.33%, 75%, 01-Jun, 120d, Inchon), Taegu (33.33%, 75%, 01-Jun, 120d, Taegu), Mok-poh (33.33%, 75%, 01-Jun, 120d, Mok-poh)
Myanmar	Irrawaddy (23.15%, 15.00%, 01-Jun, 150d, Bassein),Pegu (15.23%, 15.00%, 01-Jun, 150d, Pyinmana), Rangoon (8.60%, 15.00%,01-Jun, 150d, Rangoon), Sagaing (7.88%,15.00%, 1-Jun, 150d, Monywa), Arakan (5.74%, 15.00%, 01-Jun, 150d, Sittwe),Shan (6.12%, 15.00%, 01-Jun, 150d, Mandalay), Mon (4.61%, 15.00%, 01-Jun, 150d, Moulmein),Mandalay (4.29%, 15.00%,01-Jun, 150d, Mandalay), Karen & others (9.38%, 15.00%, 01-Jun, 150d, Molumein), all other regions growing 2nd crop (15.00%,100.00%, 01-Jun, 150d, Rangoon)
Pakistan	Punjab (43%, 100%, 01-Jun, 120d, Lahore), Sind (46%, 100%, 01-Jun, 120d, Hyderabad/ Karachi), Baluchistan (8%, 100%, 01-Jun, 120d, Hyderabad/Dadu), North West Frontier Province (3%, 100%, 01-Jun, 120d, Peshawar)
Philippines	Central Luzon, Southern Tagalog and Ilocos (35%, 77%, 01-Jun, 120d, Manila Airport), Cagyan Valley and Cordillero AR (15%, -%, 01-Jun, 120d, Aparri),Western Vsyas and Central Visyas (9%, -%, 15-Nov, 120d, Ilolo), Western Mindano and Central Mindano (12%, 100%, 15-Nov, 120d, Dipolog), Northern Mindano and Southern Mindano (23%, 100%, 15-Nov, 120d, Hinatuan), Bicol (4%, -%, 01-Jun, 120d, LegaspiEastern Visyas (2%, -%, 15-Nov, 120d, Massin)
Thailand	**Main crop** Northern (23.2%,23%,01-Jun, 120d,ChiangM ai and NakhonS awan),Central (19.2%,23%,01-Jul, 120d,KrungT hep (Bangkok)),North East (33.6%,23%,15-Jun, 120d, Udon Thani and Ubon Ratchathani),South (4%,23%,15-Oct, 120d,Ba nD on) **2nd crop:** Northern (5%, 100%, 15-Feb, 120d, Chiang Mai and Nakhon Sawan), Central (12.8%, 100%, 15-Feb, 120d, KrungT hep (Bangkok)), North East (1.8%, 100%, 15-Feb, 120d,Udon Thani and Ubon Ratchathani), South (0.4%, 100%, 15-Feb, 120d, BanD on)
USA	Arkansas (43.3%, 100%, 01-May, 120d, Memphis), California (13.7%, 100%, 01-Jun, 120d, Sacramento and Frenso), Louisiana (18.8%, 100%, 01-May, 120d, Lafayette),Texas (12.4%, 100%, 01-May, 120d,Victoria), Mississippi (8.6%, 100%, 01-May, 120d, Memphis), Missiouri (3.1%,100%, 01-May, 120d, Memphis)
Viet Nam	**Winter crop** North (16.4%, 85%, 01-Dec, 105d, Hanoi), Central (6.8%, 0%, 01-Dec, 120d, QuiNhon), South (16.8%, 60%, 01-Jan, 120d, Ho Chi Minh (Saigon) Main crop North (11.47%, 85%,15-Jun, 105d, Hanoi), Central (8.14%, 0%, 15-Jun, 120d, QuiNhon), South (17.39%, 60%, 01-Jul, 105d, Ho Chi Minh (Saigon)) **Autumn crop** North (7.13%, 85%, 01-May, 105d, Hanoi), Central (5.06%, 0%, 01-May, 120d, QuiNhon), South (10.81%, 60%, 01-May, 105d, Ho Chi Minh (Saigon))

The K_c values for the initial, mid and end crop development stages are taken as 1.05, 1.2 and 0.6 respectively. The K_c for the end period in China is taken equal to 0.90. For India, the crop water requirement is calculated only for the top 10-states contributing to 85% of national production during Kharif season and 91% during Rabi season. For the remaining regions the national average values for each season are taken. Source: irrigation percentage of Khariff rice is from (Directorate of Rice Development, 2001). All other data were compiled from (USDA, 1994) and various other online national statistical data sources.

Appendix B. Water footprint of national rice production. Period 2000–04.

	Area* ha	Yield* ton/ha	Production* ton/yr	Water footprint of production (Mm^3/yr) Green	Blue	Grey	Total	Percolation (Mm^3/yr)
China	28,670,030	6.2	177,657,605	65,241	86,460	20,786	172,486	139,518
India	43,057,460	2.9	126,503,280	136,258	104,544	14,683	255,486	177,427
Indonesia	11,642,899	4.5	52,014,913	30,309	25,323	6,113	61,744	48,213
Bangladesh	10,641,271	3.5	37,217,379	20,415	21,463	3,831	45,708	41,985
Viet Nam	7,512,160	4.5	33,960,560	10,455	6,888	4,319	21,663	23,661
Thailand	10,038,180	2.7	26,800,046	25,247	14,980	3,112	43,339	33,591
Myanmar	6,431,364	3.5	22,581,828	19,111	8,538	1,125	28,774	24,918
Philippines	4,056,577	3.3	13,322,327	11,246	5,633	1,034	17,914	15,491
Brazil	3,371,562	3.3	11,068,502	8,753	7,411	674	16,838	14,120
Japan	1,706,000	6.4	10,989,200	3,744	4,408	665	8,818	8,317
USA	1,285,671	7.4	9,520,015	2,161	7,951	964	11,076	6,262
Pakistan	2,339,200	3.0	6,910,650	2,909	16,340	611	19,859	11,345
Korea, R	1,045,173	6.5	6,808,450	2,423	2,644	575	5,641	4,320
Egypt	630,353	9.5	5,972,257	3,774	3,487	653	7,913	6,126
Nepal	1,545,156	2.7	4,220,395	2,667	2,464	461	5,592	4,329
Cambodia	2,045,837	2.0	4,165,772	2,632	2,432	455	5,520	4,273
Nigeria	2,211,800	1.4	3,085,600	1,950	1,802	337	4,088	3,165
Sri Lanka	809,552	3.5	2,822,732	1,784	1,648	308	3,740	2,896
Madagascar	1,219,074	2.2	2,715,380	1,716	1,585	297	3,598	2,785
Colombia	499,532	5.1	2,579,150	1,630	1,506	282	3,417	2,646
Iran	577,372	4.2	2,464,653	1,557	1,439	269	3,266	2,528
Laos	746,177	3.2	2,371,400	1,498	1,385	259	3,142	2,433
Malaysia	680,660	3.2	2,190,829	1,384	1,279	239	2,903	2,247
Korea, DPR	571,371	3.7	2,110,040	1,333	1,232	231	2,796	2,164
Peru	301,409	6.6	2,003,010	1,266	1,170	219	2,654	2,055
Ecuador	367,290	3.8	1,419,705	897	829	155	1,881	1,456
Italy	221,009	6.1	1,359,921	859	794	149	1,802	1,395
Guinea	649,437	1.7	1,123,543	710	656	123	1,489	1,153
Uruguay	168,635	6.3	1,069,425	676	624	117	1,417	1,097
Australia	113,307	8.7	985,385	623	575	108	1,306	1,011
Tanzania	498,186	1.7	861,572	544	503	94	1,142	884
Argentina	153,400	5.6	852,764	539	498	93	1,130	875
Spain	117,248	7.3	852,050	538	497	93	1,129	874
Mali	407,607	2.0	808,799	511	472	88	1,072	830
Venezuela	152,577	4.9	751,797	475	439	82	996	771
Côte d'Ivoire	340,713	1.9	648,855	410	379	71	860	666
Dominican Republic	136,683	4.7	643,747	407	376	70	853	660
Cuba	188,867	3.2	610,100	386	356	67	808	626
Russia	141,600	3.5	498,952	315	291	55	661	512
Guyana	118,627	4.1	487,027	308	284	53	645	500
Turkey	62,400	6.2	386,400	244	226	42	512	396
Sierra Leone	376,643	1.0	381,767	241	223	42	506	392
Afghanistan	145,200	2.4	357,400	226	209	39	474	367
Bolivia	145,134	2.3	329,117	208	192	36	436	338
Congo, DR	426,004	0.8	321,633	203	188	35	426	330
Nicaragua	87,129	3.2	277,059	175	162	30	367	284
Mexico	62,032	4.4	271,416	171	158	30	360	278
Panama	119,921	2.2	264,672	167	155	29	351	272
Ghana	122,088	2.1	256,783	162	150	28	340	263
Kazakhstan	73,367	3.2	232,256	147	136	25	308	238

(Continued)

Appendix B. (Continued)

	Area*	Yield*	Production*	Water footprint of production (Mm^3/yr)				Percolation
	ha	ton/ha	ton/yr	Green	Blue	Grey	Total	(Mm^3/yr)
Costa Rica	57,875	3.6	210,747	133	123	23	279	216
Senegal	83,761	2.4	203,045	128	119	22	269	208
Uzbekistan	84,562	2.2	182,526	115	107	20	242	187
Suriname	46,854	3.8	176,061	111	103	19	233	181
Greece	22,119	7.3	161,522	102	94	18	214	166
Iraq	83,500	2.1	155,600	98	91	17	206	160
Mozambique	173,892	0.9	153,901	97	90	17	204	158
Portugal	25,051	5.8	146,301	92	85	16	194	150
Chile	27,086	5.0	136,072	86	79	15	180	140
Liberia	126,700	1.0	129,680	82	76	14	172	133
Uganda	81,400	1.5	119,200	75	70	13	158	122
Haiti	53,340	2.1	114,400	72	67	12	152	117
Chad	93,877	1.2	111,356	70	65	12	148	114
Paraguay	28,252	3.9	109,490	69	64	12	145	112
France	19,348	5.6	109,166	69	64	12	145	112
Burkina Faso	48,549	2.0	94,411	60	55	10	125	97
Guinea-Bissau	68,314	1.3	91,315	58	53	10	121	94
Ukraine	21,330	3.7	79,680	50	47	9	106	82
Malawi	49,332	1.6	78,937	50	46	9	105	81
Mauritania	17,450	4.4	75,390	48	44	8	100	77
Turkmenistan	47,800	1.6	73,160	46	43	8	97	75
Niger	23,132	3.0	70,376	44	41	8	93	72
Togo	31,482	2.1	64,832	41	38	7	86	67
Burundi	18,840	3.2	60,207	38	35	7	80	62
Benin	25,550	2.3	57,838	37	34	6	77	59
Timor-Leste	34,710	1.8	57,087	36	33	6	76	59
Tajikistan	17,154	3.3	56,563	36	33	6	75	58
Cameroon	32,168	1.9	52,993	33	31	6	70	54
Kenya	12,817	3.6	46,429	29	27	5	62	48
Bhutan	21,310	2.1	44,485	28	26	5	59	46
Guatemala	14,956	2.6	39,557	25	23	4	52	41
El Salvador	5,325	6.2	32,611	21	19	4	43	33
Gambia	13,080	2.1	28,050	18	16	3	37	29
Morocco	5,460	5.1	27,542	17	16	3	36	28
Central Afr. Rep.	14,500	1.9	27,480	17	16	3	36	28
French Guiana	8,042	3.1	24,511	15	14	3	32	25
Rwanda	7,111	3.3	24,464	15	14	3	32	25
Bulgaria	4,587	4.6	21,044	13	12	2	28	22
Kyrgyzstan	6,126	3.0	18,609	12	11	2	25	19
Comoros	14,000	1.2	17,000	11	10	2	23	17
Azerbaijan	3,545	4.6	16,664	11	10	2	22	17
Sudan	6,388	2.3	15,750	10	9	2	21	16
Honduras	4,620	3.2	15,272	10	9	2	20	16
Zambia	11,113	1.3	14,354	9	8	2	19	15
Fiji	6,055	2.3	14,099	9	8	2	19	14
Ethiopia	7,593	1.8	13,882	9	8	2	18	14
Macedonia	2,627	4.8	12,582	8	7	1	17	13
Belize	4,018	2.9	11,367	7	7	1	15	12
Hungary	2,608	3.8	9,904	6	6	1	13	10
Somalia	1,604	5.9	9,600	6	6	1	13	10
Angola	6,196	1.3	7,378	5	4	1	10	8

Appendix B. (Continued)

	Area* ha	Yield* ton/ha	Production* ton/yr	Water footprint of production (Mm^3/yr) Green	Blue	Grey	Total	Percolation (Mm^3/yr)
Solomon Islands	1,300	3.9	5,100	3	3	1	7	5
Trinidad & Tbg.	1,154	2.9	3,368	2.13	1.97	0.37	4.46	3.45
South Africa	1,380	2.3	3,160	2.00	1.85	0.35	4.19	3.24
Romania	889	2.3	2,183	1.38	1.27	0.24	2.89	2.24
Congo	1,959	0.7	1,299	0.82	0.76	0.14	1.72	1.33
Gabon	500	2.0	1,000	0.63	0.58	0.11	1.33	1.03
Papua New Guinea	390	2.0	780	0.49	0.46	0.09	1.03	0.80
Zimbabwe	266	2.3	620	0.39	0.36	0.07	0.82	0.64
Brunei Darussalam	579	0.7	438	0.28	0.26	0.05	0.58	0.45
Algeria	200	1.5	300	0.19	0.18	0.03	0.40	0.31
Swaziland	50	3.4	170	0.11	0.10	0.02	0.23	0.17
Micronesia	80	1.1	90	0.06	0.05	0.01	0.12	0.09
Réunion	40	2.0	80	0.05	0.05	0.01	0.11	0.08
Jamaica	14	1.1	16	0.01	0.01	0.00	0.02	0.02
Total	150,666,851	4.49**	591,751,209	373,907	345,512	64,655	784,073	607,019

* *Source: FAO (2009)*

** *Weighted average, calculated based on production per country.*

Appendix C. Water footprint of national rice consumption. Period 2000–04.

	Internal water footprint (Mm^3/yr)				External water footprint (Mm^3/yr)				Total water footprint (Mm^3/yr)*			
	Green	Blue	Grey	Total	Green	Blue	Grey	Total	Green	Blue	Grey	Total
India	133,493	102,423	14,385	250,301	1	3	0	4	133,494	102,425	14,385	250,305
China	64,754	85,812	20,630	171,195	400	238	50	688	65,154	86,050	20,680	171,884
Indonesia	30,301	25,316	6,111	61,727	797	689	151	1,637	31,097	26,005	6,262	63,364
Bangladesh	20,414	21,462	3,831	45,707	146	112	16	273	20,560	21,574	3,846	45,980
Thailand	19,639	11,653	2,421	33,713	1	1	0	2	19,640	11,654	2,421	33,714
Myanmar	18,989	8,483	1,118	28,591	–	–	–	–	18,989	8,483	1,118	28,591
Viet nam	9,860	6,496	4,074	20,430	–	–	–	–	9,860	6,496	4,074	20,430
Philippines	11,246	5,633	1,034	17,914	490	386	103	979	11,736	6,020	1,137	18,893
Brazil	8,735	7,396	673	16,804	451	474	84	1,008	9,186	7,869	757	17,812
Pakistan	2,480	13,935	521	16,936	–	–	–	–	2,480	13,935	521	16,936
Japan	3,724	4,386	662	8,772	360	537	86	983	4,084	4,923	748	9,755
USA	1,598	5,554	679	7,831	326	225	41	591	1,924	5,779	719	8,422
Egypt	3,467	3,203	599	7,269	–	–	–	–	3,467	3,203	599	7,269
Nigeria	1,949	1,801	337	4,088	1,528	1,204	211	2,943	3,478	3,005	548	7,031
Korea R.	2,409	2,628	572	5,609	82	103	21	205	2,491	2,732	592	5,814
Nepal	2,667	2,464	461	5,592	15	14	2	31	2,682	2,478	463	5,623
Cambodia	2,628	2,428	454	5,511	46	31	6	83	2,674	2,459	461	5,594
Iran	1,552	1,434	268	3,254	676	726	98	1,500	2,227	2,160	367	4,754
Madagascar	1,715	1,585	297	3,597	114	280	21	414	1,829	1,865	318	4,012
Sri lanka	1,782	1,647	308	3,737	80	124	10	214	1,862	1,771	318	3,951
Malaysia	1,366	1,262	236	2,865	417	366	70	852	1,783	1,628	306	3,717
Colombia	1,629	1,506	282	3,417	73	65	12	150	1,703	1,570	294	3,567
Laos	1,498	1,385	259	3,142	–	–	–	–	1,498	1,385	259	3,142
Peru	1,266	1,170	219	2,654	38	42	7	87	1,304	1,212	226	2,741
Ecuador	863	797	149	1,809	1	1	0	3	864	798	149	1,812
Guinea	710	656	123	1,489	90	112	20	223	800	768	143	1,711
Senegal	128	118	22	269	756	482	107	1,344	884	600	129	1,613
Saudi arabia	–	–	–	–	650	694	82	1,426	650	694	82	1,426
Tanzania	541	500	94	1,135	90	95	21	207	631	595	115	1,341
South africa	2	2	0	4	701	509	88	1,298	703	511	89	1,302
Mexico	171	158	30	359	160	579	70	809	331	737	100	1,168
Russian fed	304	282	53	639	231	224	58	513	535	505	111	1,152
Mali	511	472	88	1,072	19	15	4	38	530	488	92	1,110

Turkey	244	225	42	511	172	263	42	477	416	488	84	988
Venezuela	456	421	79	956	4	15	2	21	460	437	81	977
Cuba	385	356	67	808	72	71	26	169	457	428	92	977
Italy	403	380	69	853	39	44	6	89	442	424	75	941
Cote divoire	408	377	71	856	–	–	–	–	408	377	71	856
Australia	366	339	62	767	41	40	5	87	407	379	68	854
Dominican R	403	372	70	845	3	3	1	6	406	375	70	851
UK	–	–	–	–	331	423	55	808	331	423	55	808
Spain	321	301	56	678	52	57	9	118	373	358	65	796
France	57	53	10	120	306	302	49	658	364	356	59	778
Ghana	162	149	28	339	213	185	37	435	375	334	65	774
Argentina	358	331	62	750	8	7	1	16	365	338	63	766
Hong kong	–	–	–	–	367	235	48	651	367	235	48	651
Singapore	–	–	–	–	346	219	49	614	346	219	49	614
Kenya	29	27	5	62	98	349	20	467	127	376	25	529
Uruguay	248	229	43	519	0	0	0	1	248	229	43	520
Afghanistan	224	207	39	470	–	–	–	–	224	207	39	470
Bolivia	207	191	36	434	5	5	1	11	212	196	37	445
Nicaragua	174	161	30	366	15	42	5	62	189	203	36	428
Congo DR	203	188	35	426	–	–	–	–	203	188	35	426
Guyana	193	178	33	405	0	0	0	0	193	178	33	405
Costa rica	130	121	23	273	26	94	11	131	156	215	34	405
Mozambique	95	89	16	200	56	127	9	192	151	215	26	392
Canada	–	–	–	–	118	219	28	365	118	219	28	365
Panama	160	148	28	335	3	10	1	14	163	158	29	349
Germany	–	–	–	–	140	178	26	344	140	178	26	344
Portugal	89	82	15	186	75	69	13	156	163	150	28	342
Oman	–	–	–	–	83	241	14	339	83	241	14	339
Kazakstan	130	120	22	272	1	2	0	4	131	122	23	276
Niger	44	41	8	93	69	94	13	176	113	135	21	269
Burkina faso	60	55	10	125	67	56	10	133	127	111	20	258
Belgium	–	–	–	–	108	117	18	243	108	117	18	243
Uzbekistan	115	106	20	241	–	–	–	–	115	106	20	241
Uganda	75	69	13	157	24	53	7	83	99	122	20	241
Chile	85	79	15	179	30	24	4	58	115	103	19	237
Iraq	92	85	16	192	–	–	–	–	92	85	16	192

(Continued)

Appendix C. (Continued)

	Internal water footprint (Mm^3/yr)				External water footprint (Mm^3/yr)				Total water footprint (Mm^3/yr)*			
	Green	Blue	Grey	Total	Green	Blue	Grey	Total	Green	Blue	Grey	Total
Jordan	–	–	–	–	79	91	15	184	79	91	15	184
Mauritius	–	–	–	–	51	119	8	179	51	119	8	179
Suriname	85	78	15	178	–	–	–	–	85	78	15	178
Liberia	82	75	14	171	–	–	–	–	82	75	14	171
Poland	–	–	–	–	69	86	14	170	69	86	14	170
Benin	36	33	6	76	47	39	7	93	83	73	13	169
Togo	39	36	7	82	37	38	7	81	76	74	13	164
Haiti	72	67	12	151	–	–	–	–	72	67	12	151
Ukraine	50	46	9	105	24	19	4	46	74	65	13	151
Yemen	–	–	–	–	71	62	9	142	71	62	9	142
Greece	60	56	10	126	7	7	1	15	67	63	12	141
Qatar	–	–	–	–	31	103	5	139	31	103	5	139
Kuwait	–	–	–	–	57	74	7	138	57	74	7	138
Gambia	18	16	3	37	50	42	8	99	68	58	11	137
Romania	1	1	0	3	61	60	12	133	62	62	12	136
Paraguay	63	58	11	131	2	1	0	3	64	59	11	134
El salvador	20	19	4	43	19	62	8	89	39	81	11	131
Guatemala	25	23	4	52	16	52	6	75	41	75	11	127
Switz.liecht	–	–	–	–	57	51	8	116	57	51	8	116
Malawi	50	46	9	104	3	4	0	7	53	50	9	112
Gabon	1	1	0	1	63	38	8	109	64	38	9	111
Taiwan (poc)	–	–	–	–	40	63	8	110	40	63	8	110
Turkmenistan	46	43	8	97	3	3	1	7	50	46	9	104
Hungary	6	6	1	13	36	34	6	76	42	40	7	89
Algeria	0	0	0	0	41	37	10	87	41	37	10	87
Netherlands	–	–	–	–	20	63	4	87	20	63	4	87
Sweden	–	–	–	–	40	38	6	85	40	38	6	85
Czech R.	–	–	–	–	37	38	7	83	37	38	7	83
Burundi	38	35	7	80	0	0	0	0	38	35	7	80
Bahrain	–	–	–	–	19	57	3	80	19	57	3	80
Israel	–	–	–	–	37	28	6	70	37	28	6	70
Cameroon	33	31	6	70	–	–	–	–	33	31	6	70
Jamaica	0	0	0	0	27	35	6	68	27	35	6	68
Tajikistan	31	29	5	66	–	–	–	–	31	29	5	66

Lebanon	–	–	–	–	30	29	5	65	30	29	5	65
New zealand	–	–	–	–	29	29	4	62	29	29	4	62
Bhutan	28	26	5	59	–	–	–	–	28	26	5	59
Bulgaria	13	12	2	27	13	15	3	31	26	26	5	58
Honduras	9	9	2	19	7	26	3	37	16	35	5	56
Belarus	–	–	–	–	24	26	4	54	24	26	4	54
Azerbaijan	10	10	2	22	14	15	2	32	25	25	4	54
Fiji	9	8	2	19	19	13	3	35	28	21	4	54
Slovakia	–	–	–	–	24	25	4	53	24	25	4	53
Rwanda	15	14	3	32	6	11	1	18	21	25	4	50
Macau	–	–	–	–	27	17	4	47	27	17	4	47
Denmark	–	–	–	–	22	20	4	46	22	20	4	46
Austria	–	–	–	–	22	20	4	46	22	20	4	46
Trinidad tbg	2	2	0	4	16	19	3	37	18	20	3	41
Cent.Afr. R.	17	16	3	36	1	2	0	3	19	18	3	40
Finland	–	–	–	–	19	17	3	40	19	17	3	40
Morocco	17	16	3	36	1	1	0	3	19	17	3	39
Ethiopia	9	8	2	18	8	9	1	18	17	17	3	36
Norway	–	–	–	–	17	15	2	34	17	15	2	34
Sudan	10	9	2	21	6	5	1	12	16	15	3	33
Zambia	9	8	2	19	7	6	1	14	16	14	3	33
Maldives	–	–	–	–	16	15	2	33	16	15	2	33
Albania	–	–	–	–	14	14	3	31	14	14	3	31
Kyrgyzstan	12	11	2	25	1	1	0	3	13	12	2	27
Tunisia	–	–	–	–	15	9	2	26	15	9	2	26
Brunei dar.	0	0	0	1	11	7	1	20	12	7	1	20
Lithuania	–	–	–	–	9	9	1	20	9	9	1	20
Ireland	–	–	–	–	8	9	2	18	8	9	2	18
Armenia	–	–	–	–	7	8	1	17	7	8	1	17
Macedonia, tfyr	7	7	1	16	0	0	0	1	8	7	1	16
Fr.polynesia	–	–	–	–	8	7	1	16	8	7	1	16
Croatia	–	–	–	–	8	7	1	16	8	7	1	16
Moldova rep.	–	–	–	–	6	8	1	15	6	8	1	15
Belize	7	7	1	15	0	0	0	0	7	7	1	15
Zimbabwe	0	0	0	1	7	6	1	14	8	6	1	15
Others	5	5	1	11	57	74	13	143	62	79	13	154
Grand total	360,336	331,511	62,360	754,208	13,570	14,000	2,295	29,865	373,907	345,512	64,655	784,073

* Note: It is the total water footprint of a nation related to rice consumption. It does not include water losses as a result of percolation and left over soil moisture in the rice fields.

Appendix D. Virtual water fluxes related to the international trade in rice products. Period 2000–04.

	Gross virtual water import (Mm^3/yr)				Gross virtual water export (Mm^3/yr)				Net virtual water import (Mm^3/yr)			
Countries	Green	Blue	Grey	Total	Green	Blue	Grey	Total	Green	Blue	Grey	Total
Afghanistan					1.8	1.6	0.3	3.7	−1.8	−1.6	−0.3	−3.7
Albania	14.2	14.0	2.8	30.9					14.2	14.0	2.8	30.9
Algeria	40.9	36.6	9.6	87.1	0.0	0.0	0.0	0.0	40.9	36.6	9.6	87.1
Andorra	0.3	0.3	0.0	0.6	0.2	0.1	0.0	0.3	0.1	0.1	0.0	0.3
Anguilla	0.1	0.3	0.0	0.4					0.1	0.3	0.0	0.4
Antigua Barbados	0.2	0.2	0.0	0.4					0.2	0.2	0.0	0.4
Argentina	11.3	10.7	1.5	23.5	184.9	170.9	32.0	387.8	−173.6	−160.1	−30.5	−364.2
Armenia	7.3	8.2	1.3	16.9					7.3	8.2	1.3	16.9
Aruba					9.8	9.0	1.7	20.5	−9.8	−9.0	−1.7	−20.5
Australia	70.4	68.0	9.4	147.7	285.7	264.0	49.4	599.2	−215.3	−196.0	−40.0	−451.4
Austria	23.0	21.5	3.9	48.4	1.3	1.2	0.2	2.8	21.6	20.3	3.7	45.5
Azerbaijan	14.5	15.3	2.2	32.0	0.2	0.2	0.0	0.4	14.3	15.1	2.2	31.6
Bahamas	0.9	3.4	0.4	4.7	0.9	0.8	0.1	1.8	0.1	2.6	0.3	2.9
Bahrain	19.6	57.7	3.1	80.3	0.3	0.2	0.0	0.5	19.3	57.4	3.0	79.8
Bangladesh	145.6	111.8	15.7	273.1	0.5	0.6	0.1	1.2	145.0	111.2	15.6	271.8
Barbados	2.1	6.1	0.8	9.0	0.0	0.0	0.0	0.0	2.1	6.1	0.8	8.9
Belarus	24.2	25.8	4.2	54.2					24.2	25.8	4.2	54.2
Belgium	188.5	191.5	32.4	412.3	80.9	74.8	14.0	169.7	107.6	116.7	18.4	242.6
Belize	0.0	0.1	0.0	0.1	0.2	0.1	0.0	0.3	−0.1	0.0	0.0	−0.2
Benin	47.0	39.4	6.9	93.3	0.6	0.6	0.1	1.3	46.3	38.8	6.8	91.9
Bhutan					0.0	0.0	0.0	0.0	0.0	0.0	0.0	0.0
Bolivia	5.3	5.1	0.9	11.3	1.2	1.1	0.2	2.6	4.0	4.0	0.7	8.7
Bosnia Herzg	1.6	1.6	0.3	3.5					1.6	1.6	0.3	3.5
Botswana	4.4	4.1	0.8	9.3					4.4	4.1	0.8	9.3
Br.Virgin Island					1.0	0.9	0.2	2.1	−1.0	−0.9	−0.2	−2.1
Brazil	451.7	474.5	84.0	1010.2	18.8	15.9	1.4	36.1	432.9	458.6	82.6	974.1
Brunei	11.6	7.1	1.5	20.2	0.2	0.2	0.0	0.4	11.4	7.0	1.4	19.8
Bulgaria	13.9	15.2	3.1	32.2	1.2	1.1	0.2	2.5	12.7	14.1	2.9	29.7
Burkina Faso	67.5	55.7	10.2	133.4	0.3	0.2	0.0	0.5	67.2	55.4	10.2	132.8
Burundi	0.0	0.0	0.0	0.1					0.0	0.0	0.0	0.1
Cambodia	46.1	30.8	6.1	83.0	4.2	3.9	0.7	8.8	41.9	26.9	5.4	74.1
Cameroon					0.1	0.1	0.0	0.1	−0.1	−0.1	0.0	−0.1
Canada	122.3	223.6	29.0	374.9	4.5	4.2	0.8	9.5	117.8	219.4	28.2	365.4
Cape Verde	5.1	3.5	1.5	10.1					5.1	3.5	1.5	10.1
Cayman Islands					0.2	0.1	0.0	0.3	−0.2	−0.1	0.0	−0.3

Cent.Afr. R.	1.4	1.8	0.2	3.4					1.4	1.8	0.2	3.4
Chile	30.2	24.0	4.5	58.6	1.0	0.9	0.2	2.1	29.2	23.1	4.3	56.5
China	403.3	239.9	50.4	693.7	490.2	649.7	156.2	1296.1	−87.0	−409.8	−105.7	−602.5
Colombia	73.5	64.7	12.1	150.3	0.3	0.3	0.1	0.6	73.2	64.4	12.0	149.7
Costa Rica	26.4	95.6	11.6	133.6	3.9	3.6	0.7	8.2	22.5	91.9	10.9	125.3
Cote Divoire					1.7	1.5	0.3	3.5	−1.7	−1.5	−0.3	−3.5
Croatia	8.1	7.1	1.4	16.5	0.4	0.4	0.1	0.8	7.7	6.7	1.3	15.7
Cuba	71.6	71.4	25.5	168.5	0.0	0.0	0.0	0.1	71.5	71.4	25.5	168.5
Cyprus	3.9	3.8	0.6	8.3	11.4	10.5	2.0	23.8	−7.4	−6.7	−1.3	−15.5
Czech R.	40.9	41.8	7.5	90.3	3.6	3.3	0.6	7.6	37.3	38.4	6.9	82.7
Denmark	25.7	23.4	4.4	53.5	3.4	3.2	0.6	7.2	22.2	20.3	3.8	46.3
Djibouti					1.9	1.8	0.3	4.1	−1.9	−1.8	−0.3	−4.1
Dominica	0.8	0.7	0.1	1.6					0.8	0.7	0.1	1.6
Dominican R.	2.6	3.0	0.5	6.2	3.9	3.6	0.7	8.1	−1.2	−0.6	−0.2	−2.0
Ecuador	1.3	1.2	0.2	2.7	34.3	31.7	5.9	72.0	−33.0	−30.5	−5.7	−69.3
Egypt					307.0	283.7	53.1	643.9	−307.0	−283.7	−53.1	−643.9
El Salvador	19.7	62.3	7.8	89.7	0.8	0.7	0.1	1.6	18.9	61.6	7.6	88.1
Eritrea	0.5	0.5	0.1	1.1					0.5	0.5	0.1	1.1
Estonia	2.3	3.7	0.4	6.4					2.3	3.7	0.4	6.4
Ethiopia	8.0	9.0	1.2	18.2	0.1	0.1	0.0	0.2	7.9	8.9	1.1	18.0
Fiji	19.5	12.8	2.6	34.9					19.5	12.8	2.6	34.9
Finland	19.9	18.0	3.4	41.3	0.8	0.7	0.1	1.6	19.1	17.3	3.2	39.7
Fr.Polynesia	7.8	6.7	1.3	15.8					7.8	6.7	1.3	15.8
France	369.7	361.4	59.9	791.0	75.2	69.5	13.0	157.6	294.6	291.9	46.9	633.4
Gabon	63.1	37.9	8.4	109.4	0.0	0.0	0.0	0.1	63.0	37.8	8.4	109.3
Gambia	49.9	41.6	8.0	99.5	0.1	0.1	0.0	0.2	49.8	41.5	8.0	99.4
Georgia	2.5	1.7	0.4	4.6					2.5	1.7	0.4	4.6
Germany	205.0	238.4	37.7	481.0	65.2	60.2	11.3	136.7	139.8	178.1	26.4	344.4
Ghana	214.2	186.0	37.0	437.2	1.6	1.4	0.3	3.3	212.6	184.5	36.7	433.9
Greece	11.4	11.8	1.9	25.1	46.7	43.1	8.1	97.8	−35.3	−31.4	−6.1	−72.8
Grenada	0.9	1.1	0.2	2.2					0.9	1.1	0.2	2.2
Guatemala	16.2	52.7	6.5	75.5	0.6	0.5	0.1	1.2	15.6	52.2	6.4	74.2
Guinea	90.1	112.2	20.4	222.7					90.1	112.2	20.4	222.7
Guyana	0.1	0.1	0.0	0.1	114.6	105.9	19.8	240.4	−114.6	−105.9	−19.8	−240.3
Haiti					0.1	0.1	0.0	0.2	−0.1	−0.1	0.0	−0.2
Honduras	7.7	27.3	3.3	38.4	1.4	1.3	0.2	3.0	6.3	26.0	3.1	35.4
Hong Kong	368.5	236.9	48.8	654.2	1.3	1.8	0.4	3.5	367.1	235.2	48.4	650.7
Hungary	36.1	34.3	6.2	76.6	0.1	0.1	0.0	0.2	36.0	34.3	6.1	76.4
Iceland	0.7	0.8	0.1	1.6	0.0	0.0	0.0	0.1	0.6	0.8	0.1	1.5

(Continued)

Appendix D. (Continued)

Countries	Gross virtual water import (Mm^3/yr)				Gross virtual water export (Mm^3/yr)				Net virtual water import (Mm^3/yr)			
	Green	Blue	Grey	Total	Green	Blue	Grey	Total	Green	Blue	Grey	Total
India	1.3	2.7	0.4	4.4	2765.4	2121.7	298.0	5185.1	−2764.0	−2119.0	−297.6	−5180.7
Indonesia	796.9	689.2	151.2	1637.3	8.6	7.2	1.7	17.5	788.3	682.0	149.5	1619.8
Iran	678.0	729.0	98.8	1505.9	8.3	7.7	1.4	17.4	669.8	721.4	97.4	1488.5
Iraq					6.8	6.3	1.2	14.2	−6.8	−6.3	−1.2	−14.2
Ireland	10.4	10.8	1.9	23.1	2.3	2.1	0.4	4.7	8.1	8.7	1.5	18.4
Israel	37.0	27.7	5.6	70.3	0.0	0.0	0.0	0.0	37.0	27.7	5.6	70.3
Italy	82.5	91.7	12.8	187.1	499.6	461.7	86.4	1047.7	−417.1	−369.9	−73.6	−860.6
Jamaica	27.3	35.2	5.7	68.2	0.0	0.0	0.0	0.0	27.3	35.2	5.7	68.2
Japan	361.9	539.6	86.9	988.3	21.6	25.5	3.8	51.0	340.2	514.1	83.0	937.3
Jordan	80.3	91.9	15.0	187.1	1.3	1.2	0.2	2.8	78.9	90.7	14.7	184.3
Kazakstan	1.7	2.1	0.5	4.2	17.4	16.1	3.0	36.4	−15.7	−13.9	−2.5	−32.2
Kenya	98.1	349.0	20.1	467.2	0.0	0.0	0.0	0.1	98.0	349.0	20.1	467.1
Korea Rep.	82.1	103.7	20.8	206.6	14.5	15.8	3.4	33.7	67.7	87.9	17.3	173.0
Kuwait	57.1	74.6	7.0	138.6	0.2	0.2	0.0	0.5	56.8	74.4	6.9	138.1
Kyrgyzstan	1.0	1.4	0.3	2.7					1.0	1.4	0.3	2.7
Latvia	4.9	5.8	0.9	11.6	3.1	2.9	0.5	6.5	1.8	2.9	0.3	5.1
Lebanon	30.1	29.5	5.1	64.7	0.1	0.1	0.0	0.2	30.1	29.4	5.1	64.6
Liberia					0.3	0.3	0.1	0.6	−0.3	−0.3	−0.1	−0.6
Lithuania	9.2	9.6	1.4	20.3	0.3	0.3	0.1	0.6	8.9	9.4	1.4	19.6
Luxembourg	1.1	1.0	0.2	2.3	0.0	0.0	0.0	0.1	1.1	1.0	0.2	2.2
Macau	27.0	16.7	3.5	47.2					27.0	16.7	3.5	47.2
Macedonia	0.4	0.4	0.1	0.9	0.5	0.5	0.1	1.1	−0.1	−0.1	0.0	−0.2
Madagascar	113.6	279.7	21.3	414.5	0.3	0.3	0.0	0.6	113.3	279.4	21.2	413.9
Malawi	2.8	4.0	0.5	7.3	0.1	0.1	0.0	0.2	2.7	3.9	0.4	7.0
Malaysia	422.2	371.0	70.5	863.6	23.4	21.7	4.1	49.1	398.7	349.3	66.5	814.5
Maldives	16.3	14.5	1.9	32.7					16.3	14.5	1.9	32.7
Mali	19.3	15.5	3.6	38.4					19.3	15.5	3.6	38.4
Malta	1.2	1.3	0.2	2.6					1.2	1.3	0.2	2.6
Mauritius	51.3	119.3	8.3	178.8					51.3	119.3	8.3	178.8
Mexico	160.1	579.5	70.5	810.0	0.6	0.6	0.1	1.3	159.4	578.9	70.4	808.7
Moldova Rep.	5.8	7.7	1.4	14.9					5.8	7.7	1.4	14.9
Mongolia	3.3	4.2	1.0	8.5	0.3	0.2	0.0	0.5	3.1	4.0	0.9	7.9
Montserrat	0.0	0.1	0.0	0.2					0.0	0.1	0.0	0.2
Morocco	1.1	1.4	0.2	2.7					1.1	1.4	0.2	2.7
Mozambique	57.3	128.5	9.4	195.2	3.4	3.2	0.6	7.2	53.9	125.3	8.8	188.1
Myanmar					121.5	54.3	7.2	182.9	−121.5	−54.3	−7.2	−182.9

N.Caledonia	6.4	5.4	1.0	12.8					6.4	5.4	1.0	12.8
Namibia	4.0	4.5	0.7	9.3					4.0	4.5	0.7	9.3
Nepal	15.4	13.9	1.9	31.2	0.2	0.1	0.0	0.3	15.2	13.8	1.8	30.8
Neth.Antiles					7.6	7.0	1.3	15.9	−7.6	−7.0	−1.3	−15.9
Netherlands	154.2	187.0	27.0	368.3	133.9	123.8	23.2	280.9	20.3	63.2	3.9	87.4
New Zealand	29.1	28.6	4.5	62.2	0.1	0.1	0.0	0.1	29.1	28.5	4.5	62.1
Nicaragua	15.0	42.2	5.4	62.6	0.8	0.8	0.1	1.7	14.2	41.4	5.3	60.9
Niger	69.0	93.9	13.1	175.9					69.0	93.9	13.1	175.9
Nigeria	1528.5	1203.9	211.4	2943.8	0.4	0.3	0.1	0.8	1528.1	1203.6	211.3	2943.0
Norway	18.7	16.3	2.6	37.6	1.9	1.8	0.3	4.1	16.7	14.5	2.3	33.5
Oman	83.3	241.5	14.4	339.2	0.3	0.3	0.1	0.6	83.0	241.2	14.3	338.5
Pakistan					428.1	2405.2	89.9	2923.2	−428.1	−2405.2	−89.9	−2923.2
Panama	2.8	10.2	1.2	14.3	7.6	7.0	1.3	15.9	−4.7	3.2	−0.1	−1.6
Paraguay	1.7	1.5	0.2	3.3	6.8	6.3	1.2	14.3	−5.1	−4.8	−1.0	−11.0
Peru	38.0	42.2	7.3	87.5					38.0	42.2	7.3	87.5
Philippines	489.8	386.4	103.0	979.2	0.1	0.1	0.0	0.2	489.7	386.3	103.0	979.0
Poland	69.9	86.9	13.8	170.6	0.5	0.5	0.1	1.1	69.4	86.4	13.7	169.5
Portugal	77.9	71.7	13.4	163.0	7.3	6.7	1.3	15.2	70.7	65.0	12.1	147.8
Qatar	31.1	103.3	5.2	139.6	0.1	0.1	0.0	0.2	31.0	103.2	5.2	139.4
Romania	61.2	60.6	12.0	133.8	0.2	0.2	0.0	0.5	61.0	60.4	11.9	133.3
Russia	238.8	231.7	59.8	530.3	19.0	17.6	3.3	39.9	219.8	214.1	56.5	490.4
Rwanda	5.7	11.0	1.3	18.0					5.7	11.0	1.3	18.0
S.Vincent-Gr	2.8	7.5	1.0	11.2	2.5	2.3	0.4	5.3	0.3	5.2	0.5	5.9
Samoa	1.3	1.7	0.4	3.4					1.3	1.7	0.4	3.4
Sao Tome Prn	0.6	0.7	0.1	1.4					0.6	0.7	0.1	1.4
Saudi Arabia	651.3	695.4	82.0	1428.7	1.3	1.2	0.2	2.8	650.0	694.2	81.8	1425.9
Senegal	756.7	482.3	106.8	1345.7	0.8	0.8	0.1	1.7	755.8	481.6	106.6	1344.0
Serbia, Mtneg	4.0	3.6	0.7	8.3					4.0	3.6	0.7	8.3
Seychelles	5.0	4.0	0.6	9.7					5.0	4.0	0.6	9.7
Singapore	383.0	252.5	55.3	690.7	36.5	33.7	6.3	76.5	346.5	218.8	49.0	614.2
Slovakia	24.0	24.9	4.5	53.4	0.1	0.1	0.0	0.3	23.8	24.8	4.5	53.1
Slovenia	6.0	5.8	1.1	12.9	1.6	1.5	0.3	3.4	4.4	4.2	0.8	9.4
Somalia					1.0	0.9	0.2	2.0	−1.0	−0.9	−0.2	−2.0
South Africa	724.5	531.1	92.4	1347.9	23.7	21.9	4.1	49.8	700.8	509.1	88.3	1298.2
Spain	87.0	94.4	15.8	197.2	252.7	233.5	43.7	529.9	−165.7	−139.1	−27.9	−332.6

(*Continued*)

Appendix D. (Continued)

Countries	Gross virtual water import (Mm^3/yr)				Gross virtual water export (Mm^3/yr)				Net virtual water import (Mm^3/yr)			
	Green	Blue	Grey	Total	Green	Blue	Grey	Total	Green	Blue	Grey	Total
Sri Lanka	79.7	124.1	10.2	214.0	1.5	1.4	0.3	3.2	78.1	122.7	9.9	210.8
St.Kitts Nev	0.3	1.0	0.1	1.4					0.3	1.0	0.1	1.4
St.Lucia	2.0	2.0	0.4	4.4	0.0	0.0	0.0	0.0	2.0	2.0	0.4	4.3
Sudan	5.7	5.5	1.0	12.2					5.7	5.5	1.0	12.2
Suriname					26.5	24.5	4.6	55.6	−26.5	−24.5	−4.6	−55.6
Swaziland	6.1	5.7	1.1	12.9					6.1	5.7	1.1	12.9
Sweden	41.3	38.9	6.3	86.6	0.8	0.8	0.1	1.7	40.5	38.2	6.2	84.9
Switzerland-Liecht.	72.7	65.8	10.5	149.1	15.6	14.4	2.7	32.7	57.2	51.4	7.8	116.4
Syria	140.8	117.7	22.2	280.7	261.9	242.0	45.3	549.2	−121.1	−124.3	−23.1	−268.5
Taiwan, PoC	49.7	76.4	11.3	137.4	10.2	13.5	3.2	27.0	39.5	62.9	8.0	110.5
Tajikistan					4.4	4.1	0.8	9.3	−4.4	−4.1	−0.8	−9.3
Tanzania	90.9	96.0	21.0	207.9	3.9	3.6	0.7	8.1	87.0	92.4	20.4	199.8
Thailand	0.8	1.1	0.2	2.1	5608.3	3327.6	691.2	9627.1	−5607.5	−3326.5	−691.0	−9625.0
Togo	38.7	39.1	7.0	84.8	3.3	3.0	0.6	6.9	35.5	36.0	6.4	77.9
Tokelau					0.7	0.6	0.1	1.4	−0.7	−0.6	−0.1	−1.4
Tonga					0.1	0.1	0.0	0.2	−0.1	−0.1	0.0	−0.2
Trinidad Tbg	17.2	20.1	3.2	40.4	1.7	1.6	0.3	3.6	15.5	18.5	2.9	36.9
Tunisia	14.9	9.2	1.9	25.9	0.0	0.0	0.0	0.0	14.9	9.1	1.9	25.9
Turkey	172.4	263.5	41.6	477.5	0.8	0.8	0.1	1.7	171.6	262.8	41.4	475.8
Turkmenistan	3.4	3.2	0.6	7.2	0.1	0.0	0.0	0.1	3.3	3.2	0.6	7.1
Turks Ca. Islands	1.8	6.7	0.8	9.3	0.5	0.5	0.1	1.0	1.3	6.2	0.7	8.3
Uganda	24.3	52.8	6.7	83.8	0.5	0.4	0.1	1.0	23.8	52.3	6.6	82.7
Ukraine	23.6	19.0	4.0	46.6	0.4	0.4	0.1	0.9	23.2	18.6	3.9	45.7
UAE					139.2	128.6	24.1	291.9	−139.2	−128.6	−24.1	−291.9
UK	392.0	479.6	65.3	936.8	61.3	56.6	10.6	128.5	330.7	423.0	54.7	808.4
Uruguay	0.7	0.6	0.1	1.4	428.6	396.1	74.1	898.8	−428.0	−395.5	−74.0	−897.4
USA	440.7	321.9	57.6	820.2	677.7	2493.8	302.4	3473.9	−237.0	−2171.8	−244.8	−2653.6
Uzbekistan					0.3	0.3	0.0	0.6	−0.3	−0.3	0.0	−0.6
Venezuela	4.3	15.9	1.9	22.1	19.7	18.2	3.4	41.2	−15.3	−2.3	−1.5	−19.1
Viet Nam					595.2	392.1	245.9	1233.2	−595.2	−392.1	−245.9	−1233.2
Yemen	70.6	61.7	9.2	141.6					70.6	61.7	9.2	141.6
Zambia	6.9	6.1	1.1	14.1					6.9	6.1	1.1	14.1
Zimbabwe	7.3	5.6	1.1	14.1	0.1	0.1	0.0	0.2	7.2	5.5	1.1	13.9
Others	0.0	0.0	0.0	0.0	12.2	11.2	2.1	25.5	−12.1	−11.2	−2.1	−25.5
Total	14081.2	14591.7	2379.0	31051.8	14069.0	14580.5	2376.9	31026.4				

CHAPTER 13

Water quality and nonpoint pollution: Comparative global analysis

José Albiac
Agrifood Research and Technology Center (CITA), Zaragoza, Spain

Elena Calvo
Department of Economic Analysis, University of Zaragoza, Spain

Javier Tapia & Encarna Esteban
Agrifood Research and Technology Center (CITA), Zaragoza, Spain

ABSTRACT: The improvement of the management of water resources quantity and quality in countries around the world requires better information and biophysical knowledge, both on water resources and on their associated ecosystems. The lack of basic information and knowledge makes very difficult the whole process of designing, implementing and enforcing water policies, and additionally promotes detrimental strategic behavior by stakeholders, regions and economic sectors. Sound water policies entail also the cooperation of stakeholders leading to collective action, because of the public good dimension of water. Therefore, besides information and knowledge, another essential feature is the right institutional setting. One component of this institutional setting is to have strong basin authorities, and the other component equally important is the involvement of stakeholders in basin authorities, so that all decisions are taken by them. Economic instruments advanced by some policy makers and water experts, can not substitute for collective action based on strong water institutions. Without the attainment of collective action, water policies are quite likely to fail.

Keywords: water quality, nonpoint pollution, economic instruments, cooperation, collective action

1 INTRODUCTION

The pressure on water resources has been mounting worldwide during the last century, creating problems in basins of rich and poor countries alike linked to the ever-increasing growth in population and economic activities at global scale. This pressure on water and the resulting damages have built up rapidly during last decades. The current situation is that water degradation is pervasive in many basins around the world, driven by the impacts of the escalating anthropogenic effects of human activities.

The problems created by these growing demands on water resources are twofold: one is water scarcity in watersheds brought about by excessive surface and groundwater withdrawals. The other is water degradation from pollution loads leading to many tracts of rivers and whole aquifers being spoiled, and losing their capacity to sustain ecosystem functioning and human activities.

The total amount of precipitations is 110,000 km^3, of which 40,000 km^3 are runoff and aquifer recharge or blue water, and 70,000 km^3 are soil storage or green water returning to the atmosphere through evapotranspiration. Anthropogenic water extractions have climbed from 600 km^3 to 3,600 km^3 between 1900 and 2000, along with the growth of population from 1,700 to 6,000 million. Of these extractions, 2,300 km^3 are employed for irrigation, 900 for industrial use and 400 for urban use (FAO-IFAD, 2006). Although water returns to basins from these human activities amount to 2,500 km^3, these returns contain large pollution loads that degrade heavily water receiving media.

In last decades, water scarcity has become widespread in most arid and semiarid regions around the world. There is a severe scarcity problem in almost all the important rivers in these regions, such as in the Nile, Ganges, Indus, Yellow, Yangtze, Tigris, Euphrates, Amu and Syr Daria, Murray-Darling, Colorado and Rio Grande. Surface and subsurface resources in these river basins are being depleted and their quality degraded. The scarcity problems in basins of arid and semiarid regions were created at first by extractions of surface waters, but at present they are compounded by the huge development of groundwater by individual wells during last decades, brought about by the adoption of pumping technologies with falling costs worldwide.

The largest groundwater extractions by country take place in India, USA and China with extractions above 100 km^3, and in Bangladesh, Pakistan and Iran with extractions above 50 km^3 (Björklund *et al.*, 2009; Vrba & van der Gun, 2004). All these countries have large aquifer systems being depleted located in the Indus basin, the Ganges basin, the Northern China plain, and the North America high plains. The region of the Indus, Ganges and Brahmaputra basins is the larger irrigated area in the world, extending 2.7 million hectare (Mha) over northern India, Pakistan, Bangladesh, Nepal and eastern Afghanistan. Large-scale estimates of irrigation acreage are available from Siebert *et al.* (2007). Irrigated acreage covers 1.2 Mha in the Indus basin and 1.1 Mha in the Ganges basin, and the mean flow per year of the Indus is 175 km^3 and that of Ganges is 400 km^3. Groundwater overdraft in this region has been estimated at 50 km^3/yr from satellite data (Tiwari *et al.*, 2009), with overdraft estimates by basin close to 35 km^3 in the Ganges-Brahmaputra and 10 km^3 in the Indus basin. In the India part of these basins, groundwater extractions per year are estimated at 280 km^3 with an overdraft amounting to 30 km^3. The problems created by this huge depletion of aquifers result from the declining water tables, and from the degradation of water quality by pollution loads or saline intrusion in coastal aquifers. One important health problem is the arsenic pollution detected in Bangladesh, which is poisoning the impoverished population. The poorest farmers are the first group threatened by this massive depletion of resources, because they have limited access to infrastructures such as electricity and food distribution facilities, and to farm inputs, production technologies and capital financing.

The Ogallala aquifer in the North America high plains covers 450,000 km^2 and supplies water to irrigate 5 Mha. Withdrawals for irrigation are 26 km^3/yr which include an overdraft of around 10 km^3. The current storage amounts to 3,610 km^3 and the accumulated depletion is estimated at 310 km^3, with a water table decline that could be up to 30 meters (McGuire, 2007). The main worries at present among stakeholders and public administrations are the increase of pumping costs, the effects on dwindling surface flows, and the impact on near-stream habitats. The only measure taken so far by federal, state and local public agencies is the monitoring of water level changes, started in the 1990s. But no control measures to stabilize of reduce overdraft have been taken yet.

Groundwater depletion in the Indus-Ganges basins and the Ogallala aquifer, together with groundwater depletion in Southwestern USA, Australia, Spain and Mexico, demonstrate that aquifer mismanagement is the rule, and that sustainable management of groundwater is a complex task very difficult to achieve. The reason behind the pervasive aquifer mismanagement worldwide is that groundwater is a common pool resource with environmental externalities, and adequate management can only be brought about by cooperation of stakeholders through the right institutional setting.

The absence of regulation has been supported by the arguments put forward by Gisser & Sanchez (1980), and by the ensuing literature on groundwater management (Koundouri, 2004). The so-called *Gisser-Sanchez effect* claims that welfare gains from policy interventions are negligible in aquifer management, when comparing with non regulation or *free-market* outcomes. An essential element for the validity of this approach is the disregard for aquatic ecosystems linked and dependent on large aquifers. When environmental externalities from overdraft are taken into account, this approach becomes untenable (Esteban, 2010). The policy issue is important, because of the mentioned severe groundwater depletion problems in the Indus and Ganges basins, the Northern China plain, the North America high plains and other regions, causing large scale degradation of aquatic ecosystems. This *laissez faire* approach deserves revision, in order to demonstrate that policies and social interventions for sustainable aquifer management not only make sense but are also very much needed.

There are only very few real world cases of aquifers attempting good management through collective action. To the best of our knowledge, this collective action has been achieved in only two small aquifers in Santa Clara (California, USA) and Vall d'Uxo (Spain), and only in one large aquifer in Eastern La Mancha (Spain). The common feature in the three cases has been the organization by stakeholders of extractions control, but with distinctive features in each case: huge imports of water in Santa Clara, changes in irrigation technologies in Vall d'Uxo, and social agreement to reduce extractions and become rightful users in Eastern La Mancha. These cases are important lessons on how to design incentives for the management of water quantity and quality.

2 WATER QUALITY AND NONPOINT POLLUTION

Water quality is an essential condition for having living rivers with healthy aquatic ecosystems. At present, the pressure on water resources is growing rapidly both in terms of expanding water extractions and quality degradation from pollution loads. Water quality degradation is pervasive in most water courses around the world, driven by the escalating pollution loads from anthropogenic point and nonpoint sources.

In high income countries, there have been large investments in sewage networks and water treatment facilities during recent decades to control point pollution, which have stabilized or in some cases reduced the concentration of pollutants in rivers. Nonpoint pollution is much more difficult to tackle, because control measures are very difficult to design, implement and enforce. As a consequence of the abatement of point pollution, the relative importance of nonpoint pollution loads is increasing in high income countries. In medium and low income countries, rivers and aquatic ecosystems are being degraded by the surge in point pollution loads from urban and industrial sources, and large tracts of water courses become unsuitable for many water uses.

Policies to control nonpoint pollution are not so easy to design, and some authors such as Vitousek *et al.* (2009) mention the USA and the EU as examples of reductions in nutrient imbalances, in spite that pollution remains very high in their water media. In Europe, results appear disappointing after the considerable efforts to curb pollution. European regulations include the Urban Wastewater Directive (with investments above 100,000 million €), the Nitrates Directive, both of 1991, and the Water Framework Directive of 2000.

The huge investments of the Wastewater Directive should have reduced urban pollution, but the European data (EEA, 2009) for the last 15 years on nitrate concentration indicate a slight reduction in rivers and a 50% increase in aquifers. The data from OECD (2008) confirm this poor quality improvement, with most major European rivers showing no abatement of nutrients or even a worsening in some rivers (Table 1). These data show this poor behavior which hampers the recovery of water quality in the last thirty years. The Biochemical Oxygen Demand (BOD) has improved in most European countries except in Belgium (Escaut), UK (Thames) and Netherlands (Maas) which show no improvement. The improvement in BOD took place in Germany and Denmark in the beginning of the 1990s, and in France, Spain and Italy in the beginning of the 2000s.

The worst water quality results are for nitrates, with most countries showing no improvement in the last thirty years, and some rivers such as the Loire, Guadalquivir and Strimonas even increasing nitrate loads in the beginning of the 2000s. The only countries that reduce nitrate loads are Germany (Rhein, Elbe, Wesser) and Norway (Skienselva) during the late 1990s. Phosphorus pollution loads show no improvement in the majority of rivers, with pollution reductions taking place at end of the 1990s in the Rhein, Elbe and Wesser (Germany), Thames (UK), Gudena (Denmark), Maas (Netherlands) and Ebro (Spain).

The Nitrates Directive was based on voluntary compliance, and recently farmers have been required to keep a nitrogen balance book, with uncomplying farmers drawn by chance being penalized in their agricultural subsidies. The Nitrates Directive only applies to cultivation over aquifers declared officially polluted, but not to cultivation over whole basins or very polluting

Table 1. Water quality in selected European rivers (average 2002–2004).

Country	Watershed	BOD (mg O_2/L)	Nitrates (mg N/L)	Phosphorus (mg P/L)	Lead (μg/L)	Cadmium (μg/L)	Chromium (μg/L)	Copper (μg/L)
Norway	Skienselva	2.0*	0.2	0.01	0.2	0.02	0.11	0.62
Sweden	Dalalven		0.1	0.02	0.5*	0.02	0.37*	1.48
Denmark	Gudena	1.9	1.3	0.09				
UK	Thames	3.4	6.6	0.66	2.9	0.10	1.17	6.63*
Netherlands	Maas	2.5	3.6	0.21	2.8	0.15	1.77	3.77
Belgium	Escaut	3.6	4.7	0.66	12.0	0.67	9.93	10.10
Germany	Rhein	3.0	2.5	0.14	3.0	0.20	2.55	6.22
	Elbe	6.9	3.0	0.17	2.2	0.18	1.20	4.36
	Weser	2.8	3.7	0.14	4.5*	0.20	2.03*	3.56
France	Loire	3.2	3.1	0.21		0.40*		
	Seine	3.1*	5.6	0.63*	22.1*	2.18*	24.67*	15.03*
Spain	Guadalquivir	4.2*	6.1*	0.95*	10.2*	1.87*		5.73*
	Ebro	1.9	2.2	0.09	7.5	0.23*	0.92*	1.61*
	Guadiana	1.6	1.8	0.69*		3.39		
Portugal	Tejo	2.3	1.0	0.20	11.0	3.00	22.33*	2.10
Italy	Po	1.3	2.5	0.25				
Greece	Strimonas		1.8	0.14		0.64*		
Turkey	Porsuk	1.4	1.5	0.06	12.2	6.50	7.50	5.67

The symbol * indicates that the average is for years 1999–2001 or before. The Biochemical Oxygen Demand (BOD) measures pollution by organic matter, and water is considered drinkable for BOD between 0.75 y 1.50 mg O_2/L.
Source: OECD (2008).

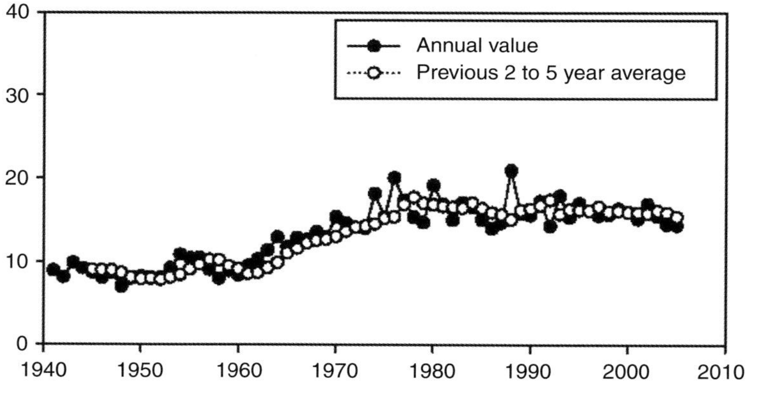

Figure 1. Net nitrogen inputs for the Mississippi Basin (kg N/ha/yr).
Source: EPA (2007).

crops not receiving subsidies (e.g. greenhouses). Water pricing advanced by the Water Framework Directive as the key policy measure does not seem either a good instrument to curb nitrate pollution, since the pollution driver is fertilizer not water.

In the USA, it seems that there is no improvement in nonpoint pollution loads over the last decade. The large study completed by NOAA (National Oceanic and Atmospheric Administration) in 2000 on hypoxia in the Northern Gulf of Mexico, has not spurred any significant reduction of nitrogen loads in the Mississippi basin as Figure 1 indicates (EPA, 2007). The major effort in the USA to curb nonpoint pollution has been made in the Chesapeake Bay, but results there show only moderate reductions. From 1985 to 2007, the nitrogen loads have fallen from 153,000 t to 119,000 t [t = tonne = 1,000 kg] and the phosphorus loads from 12,300 to 8,300 t, which are still far from the sought thresholds of 79,000 t of nitrogen and 5,800 t of phosphorus. The implication is that the current voluntary measures have to be supplemented with more strong regulatory measures (Linker *et al.*, 2009).

3 CONCLUSIONS

The achievements of developed countries to control nonpoint pollution are quite modest, and the nonpoint pollution policies of the USA and Europe are not good enough examples to inspire policies in developing countries (Albiac, 2009). A case in point is Chinese water policies that must deal with serious water scarcity problems in both the Yellow River and the Yangtze River basins, and also with severe water quality degradation problems in many tracts of rivers around the country. Chinese policy makers are worrying about nonpoint pollution from agriculture, when in fact they should worry first on urban and industrial point pollution. Ongley & Tao (2009) indicate that appropriate water policies in China require undertaking good studies and assessments on pollution loads from all sources in the whole country.

Regulation in both USA and EU has contributed to the abatement of point source pollution from urban and industrial sources, due to the construction of treatment facilities, and the decline in some emissions of dangerous substances from industrial processes. But the improvement of water quality in USA and European rivers is far from obvious for the majority of basins and pollutants, despite all legislation and investments.

There has been a certain improvement of some quality parameters in several surface and coastal water bodies, with the resulting reduction in pressure on their aquatic ecosystems. However, no substantial improvements are detected in the water quality of USA and EU rivers and aquifers. The problems of agricultural nonpoint source pollution remain, in particular those of nutrients and pesticides, and also the problems of water scarcity in arid and semiarid regions of the USA and European countries (Albiac *et al.*, 2009). One of the factors explaining these difficulties is that nonpoint pollution is a *common pool resource* (or public bad) where economic instruments such as taxes and subsidies fail.

The current policy practice in the protection of natural resources consists on actions to compensate private benefits of local agents causing damages through market instruments (Hardner & Rice, 2002; Pagiola *et al.*, 2004), or to promote conservation with large mitigation investments by governments, international agencies and private foundations in protected natural reserves (Pimm *et al.*, 2001; Balmford *et al.*, 2002).

However, these policy practices are unable to curtail the massive degradation of natural resources and ecosystems worldwide. What seems to be needed is the cooperation of the stakeholders managing and using the resources. Therefore, the policy effort has to be focused on nurturing cooperation. The economic argument supporting this collective action approach is that natural resources are mostly common pool resources, requiring cooperation rather than just economic instruments that are likely to fail with public goods. The recent collapse of climate change negotiations illustrates the consequences of cooperation failure among countries and stakeholders.

The key policy question in nonpoint pollution is that an appropriate institutional setting is required to induce farmers' cooperation, because pollution abatement is impossible without farmers' involvement and active support in order to spur the needed collective action.

REFERENCES

Albiac, J. (2009). Nutrient Imbalances: Pollution Remains. *Science*, 326 (5953): 665.

Albiac, J.; Mema, M. & Calvo, E. (2009). Sustainable Water Management and Nonpoint Source Pollution Control in Spain and the European Union. In: J. Albiac & A. Dinar (eds.), *The Management of Water Quality and Irrigation Technologies*. Earthscan, London, UK.

Balmford, A.; Bruner, A.; Cooper, P.; Costanza, R.; Farber, S.; Green, R.; Jenkins, M.; Jefferiss, P.; Jessamy, V.; Madden, J.; Munro, K.; Myers, N.; Naeem, S.; Paavola, J.; Rayment, M.; Rosendo, S.; Roughgarden, J.; Trumper, K. & Turner, R. (2002). Economic Reasons for Conserving Wild Nature. *Science*, 297 (5583): 950–953.

Björklund, G.; Burke, J.; Foster, S.; Rast, W.; Vallée, D. & van der Hoek, W. (2009). Impacts of Water Use on Water Systems and the Environment. Water in a Changing World, Chapter 8. *The United Nations World Water Development Report 3*. UNESCO and Earthscan, London, UK.

EEA (European Environment Agency) (2009). *Core Set of Indicators No. 020. Nutrients in Freshwater*. European Environment Agency, Copenhagen, Denmark.

EPA (Environmental Protection Agency) (2007). *Hypoxia in the Northern Gulf of Mexico*. An Update by the EPA Science Advisory Board. Environmental Protection Agency, Washington, D.C., USA.

Esteban, E. (2010). *Water as a Common Pool Resource: Collective Action in Groundwater Management and Nonpoint Pollution Abatement*. Ph.D. Thesis. Department of Applied Economics, University of Zaragoza, Spain.

FAO-IFAD (Food and Agricultural Organization – International Fund for Agricultural Development) (2006). Water for Food, Agriculture and Rural Livelihoods. Water, a Shared Responsibility, Chapter 7. *The United Nations World Water Development Report 2*. UNESCO and Berghahn Books, New York, USA.

Gisser, M. & Sanchez, D. (1980). Competition versus optimal control in groundwater pumping. Water Resources Research, 16(4): 638–642.

Hardner, J. & Rice, R. (2002). Rethinking Green Consumerism. *Scientific American*, 286(5): 89–95.

Koundouri, P. (2004). Potential for Groundwater Management: Gisser-Sanchez Effect Reconsidered. Water Resources Research, 40(6). doi:10.1029/2003WR00216.

Linker, L.; Shenk, G.; Wang, P. & Batiuk, R. (2009). Integration of Modelling, Research and Monitoring in the Chesapeake Bay Program. In: J. Albiac & A. Dinar (eds.), *The Management of Water Quality and Irrigation Technologies*. Earthscan, London, UK.

McGuire, V. (2007). *Water-Level Changes in the High Plains Aquifer, Predevelopment to 2005 and 2003 to 2005*. Scientific Investigations Report 2006–5324. U.S. Geological Survey. U.S. Department of the Interior. Reston, USA.

OECD (Organisation for Economic Co-operation and Development) (2008). *OECD Environmental Data. Compendium 2006-2008*. OECD, Paris, France.

Ongley, E. & Tao, Y. (2009). Problems in Assessing Nonpoint Source Pollution in China: Links to Policy and Regulation. In: J. Albiac & A. Dinar (eds.), *The Management of Water Quality and Irrigation Technologies*. Earthscan, London, UK: 13–39.

Pagiola, S.; von Ritter, K. & Bishop, J. (2004). *How much is an ecosystem worth? Assessing the economic value of conservation*. World Bank. Washington, D.C., USA.

Pimm, S.; Ayres, M.; Balmford, A.; Branch, G.; Brandon, K.; Brooks, T.; Bustamante, R.; Costanza, R.; Cowling, R.; Curran, L.M.; Dobson, A.; Farber, S.; Fonseca, G.A.; Gascon, C.; Kitching, R.; McNeely, J.; Lovejoy, T.; Mittermeier, R.A.; Myers, N.; Patz, J.A.; Raffle, B.; Rapport, D.; Raven, P.; Roberts, C.; Rodríguez, J.P.; Rylands, A.B.; Tucker, C.; Safina, C.; Samper, C.; Stiassny, M.L.; Supriatna, J.; Wall, D.H. & Wicove, D. (2001). Can We Defy Nature's End? *Science*, 293 (5538): 2207–2208.

Siebert, S.; Döll, P.; Feick, S.; Hoogeveen, J. & Frenken, K. (2007). *Global Map of Irrigation Areas, version 4.0.1*. Johann Wolfgang Goethe University and FAO, Rome, Italy.

Tiwari, V.M.; Wahr, J. & Swenson, S. (2009). Dwindling groundwater resources in northern India, from satellite gravity observations. *Geophysical Research Letters*. doi:10.1029/ 2009GL039401.

Vitousek, P.; Naylor, R.; Crews, T.; David, M.B.; Drinkwater, L.E.; Holland, E.; Johnes, P.J.; Katzenberger, J.; Martinelli, L.A.; Matson, P.A.; Nxiguheba, G.; Ojima, D.; Palm, C.A.; Robertson, G.P.; Sanchez, P.A.; Townsend, A.R. & Zhang, F.S. (2009). Nutrient Imbalances in Agricultural Development. *Science*, 324 (5934): 1519–1520.

Vrba, J. & van der Gun, J. (2004). The world's groundwater resources. *World Water Development Report 2*, contribution to Chapter 4. Report IP 2004-1. International Groundwater Resources Assessment Centre (IGRAC), Utrecht, the Netherlands.

V
The strategic role of groundwater in achieving food security

CHAPTER 14

Intensive groundwater development: A water cycle transformation, a social revolution, a management challenge

Emilio Custodio
Royal Academy of Sciences of Spain; Department of Geo-Engineering and Foundation International Centre for Groundwater Hydrology, Technical University of Catalonia (UPC, Barcelona, Spain)

ABSTRACT: Groundwater is the slow path of the water cycle. It is generally associated to a very large water storage relative to recharge/discharge annual flows, which means generally a long average water turnover time, quite delayed hydraulic responses and very retarded chemical water composition changes, especially for large aquifer systems. This behaviour is often beyond the normal human temporal experience. Groundwater development needs adequate drilling and pumping technology, as well as relatively cheap and easy-to-obtain energy. This has not been possible until mid the 20th century. Before that time most aquifers were close to natural conditions and groundwater use was mainly from winning natural outflows. Between the 1930s and the 1950s an exponential groundwater development took place in some countries, mostly in arid and semi-arid areas. In many developing countries this happened between the 1970s and 1980s, or later. This has produced and is producing large economic and social benefits with only very few exceptions. However, intensive groundwater development modifies groundwater flow in aquifer systems, as well as the relationships with surface water. The consequences are groundwater level drawdown, discharge decrease to rivers, springs and wetlands, chemical changes – which is an important issue and a future growing challenge–, and also economic, environmental and social costs. Also, aquifer development is generally associated to a rather long transient evolution that may involve progressive water level drawdown and increasing costs, up to a new steady situation. This should be compensated by benefits from socio-economic development, as is in most cases, with only a few exceptions, but attention is needed to assure this is in this way and will continue to be. Intensive aquifer development has been produced mostly by individual and small groups, especially farmers, often without public funding. This is like a social *silent revolution*. Since groundwater is a limited resource in order to avoid the problems associated to unrestricted access, management is needed to move towards sustainable development. The complexity of dealing with a very large number of stakeholders over an extended territory needs a combination of government and public management institutions, regulations and means, and stakeholders' involvement and co-responsibility. This is a current challenge due to short experience, and the need to create knowledge, data, and awareness on the need to manage a valuable and essential common asset. Collective stakeholders' institutions, tailored to the characteristics of each local situation, seem an effective component. Sustainable use is possible, although under different economic and social conditions as those existing when development started, and inside integrated water resources management plans, agreed by all interested parts.

Keywords: *groundwater, intensive development, management, collective institutions, silent revolution*

1 GROUNDWATER: THE SMOOTHER OF THE WATER CYCLE

Groundwater, which is an essential branch of the Earth's water cycle, is still poorly known by a large part of many water professionals, the population, and the media. It often suffers from knowledge errors and deviated principles, or *hydromyths* (Custodio, 2003; Custodio & Llamas,

Insert 1. Key groundwater characteristics.

- Essential part of the Earth's water cycle
- Large storage to flow ratio
- Long turnover times: 10s to 1000s of years
- Recharge and storage distributed over the territory
- Closely linked to surface water
- Delayed response: 10s to 1000s of years
- Important role in natural processes and environmental services to humans
- Key source for human water needs through:
  - Tapping natural outflows
  - Using groundwater-sustained vegetation and fauna
  - Pumping and drainage

1997). Groundwater is not directly perceived or recognized by lay people, although it shows up as springs and stream base flow and as contributions to permanent lakes and wetlands.

Groundwater is associated with a large water storage in the pores, fissures and voids of the terrain. It flows slowly from recharge (inflow) areas to discharge (outflow) areas. In recharge areas, often extensive ones, rain, snowmelt, runoff and losing streams – besides anthropic-generated irrigation return flows and leakages – infiltrate and increase water storage in the ground. Discharge areas can be both diffuse and concentrated, and they appear in a small part of the territory as springs, along streams, at the sea and lakes coasts, as more or less permanent wetland areas and lagoons, and as shallow water tables. Discharge means a depletion of storage (reserves) that is continuously or discontinuously compensated by recharge. The large water storage smoothes out recharge variability – which may be almost constant for deep water tables – and produces a discharge that is much less variable than recharge.

Average renewal time of water in aquifers is typically from several years in small, highly permeable formations, up to many millennia in large ones; especially in deep, confined aquifer systems (see Insert 1). This means a slow and delayed aquifer response to external actions (extractions, land use modifications, surface water alteration, climate and global changes) that is beyond the common experience of human beings, which is at the scale of daily to monthly, at most a few years scale.

What has been commented above refers to water flow and storage, but it is also true for water chemical composition, a key issue as well, whose changes may be even slower and more complex. Salinity and chemical composition is the result of the evaporative concentration of rainfall diffuse recharge, interaction with soil gas and reaction with ground materials. The event related and seasonal surface water chemical composition variability, to which people are used to, for groundwater is highly smoothed out and it can even disappear. However at the outflow point, independent of whether it is natural or artificial, seasonal and interannual, the mixing of variable groundwaters with different origins and characteristics is in fact possible, which may be indirectly related to seasonality.

Poor hydrodynamical and physico-chemical knowledge by humans has not been a deterrent to groundwater development; the main deterrents so far have been available technology and energy sources to extract the water. However, development without knowledge that is good-enough on groundwater systems has often led to serious interference problems with Nature and with other groundwater users, other water resources, and even with land surface conditions. There are serious possible water quality problems. The current and common poor management of aquifers, and the associated water quantity and quality problems, and increasing costs, that are being observed in many areas of the world can be explained by the multiple developers, who often have adequate technology for obtaining groundwater but poor knowledge of the resource, compounded by a water administration institution, that often is unprepared, unstaffed and uninterested. Groundwater problems are often highlighted by the media, without analysing the actual causes. However, groundwater is a key, cheap and reliable freshwater resource (Burke & Moench, 2000; Custodio, 2005b; Bocanegra *et al.*, 2005), which has been and will continue to be essential to the local economy and for social development. This is rarely mentioned. Serious aquifer development failures are rare and local, mostly reflecting very extreme situations of lack of water resources (e.g. Sana'a in Yemen) or serious contamination. In water scarce areas brackish groundwater is currently becoming a water resource through modern membrane technology desalinization, provided the residual brines produced can be safely disposed, and the cost can be afforded.

2 GROUNDWATER AND AQUIFERS

Groundwater is held in porous and fissured geological formations that are called aquifers when water can be abstracted at rates that allow supplying local needs, or when there are measurable discharge flows to rivers, lakes, coastal areas and wetlands. When water can be only abstracted with small rates, the geological formation is called an aquitard. An aquifer system consists of aquifers and aquitards that may exchange quite large groundwater quantities due to their large contact areas. Aquifers and aquifer systems exchange water with surface water: rivers, lakes, the sea and wetlands, with other groundwater systems, feeding springs, wetlands and river baseflows.

Groundwater is not subjected to significant evaporation to the atmosphere by solar radiation, except for shallow water tables in arid and semi-arid areas, or when transpired by phreatophyte plants.

In what follows surface areas will be expressed in km^2 ($10^6\,m^2$) and in ha (hectares, $10^4 m^2$), and water volumes in hm^3 (million cubic meters, $10^6\,m^3$) and in km^3 (cubic kilometers, $10^9\,m^3$). Time is generally given in years (yr).

Aquifer characteristics and circumstances are highly variable. Their size vary from very local, a few km^2, to very large aquifers, of continental size, often across regional and national boundaries. $1\,m^2$ of land may contain free-water depths from a few up to some tens and even a few hundreds of meters, depending on aquifer thickness and porosity, although only a fraction (0.1 to 0.3) can be extracted due to capillary retention, and at least some meters of saturated thickness has to be left to allow well functioning. Thus, the large figures of aquifer reserves given in many reports may be misleading and they should be changed to usable reserves under specified conditions (well depth, economic cost, water quality degradation with depth or in some areas of the aquifer system, . . .). Renewable resources may vary from almost zero (in very arid lands) up to more than 0.5 m/yr ($0.5\,m^3/m^2/yr$) in rainy areas with slightly retentive soils and poor vegetation cover.

World aquifers have been mapped in WHYMAP (2004) at the 1:50,000 scale, with an associated Geographical Information System (GIS) for downscaled details. The large transboundary aquifers have been mapped in WHYMAP (2006) and IGRAC (2009).

Table 1 shows a selection of the world's largest aquifer systems. Some have quite important renewable resources, at least on part of the territory (see Box 1 for the Guaraní aquifer system). Others are in arid areas and -under current climatic conditions- their renewable resources are very small, but contain huge developable water reserves that have accumulated for millennia.

Small aquifers are also important, since they may provide key local water resources. Besides they may be ecologically significant and provide a manageable flow and storage reserve that may be very important for dealing with seasonality and short-term inter-annual fluctuations, especially when integrated into the water resource system. An example are the small, highly permeable, largely river-recharged aquifers of the Lower Llobregat (about $120\,km^2$), that are key pieces for the urban and industrial water supply of the Barcelona area (Catalonia, Spain) (see Box 2). Examples like these are very numerous around the world. They have a special significance in densely populated coastal areas, often highly developed for tourism, and with important intensive agricultural developments around (Custodio, 2010; Bocanegra *et al.*, 2010). Many of them, after initial stages of deterioration, are moving towards sustainable development through management, as explained later on.

Table 1. Some of the largest aquifer systems of the world (data derived from WHYMAP, 2006; Foster & Loucks, 2006; Margat, 2008; IGRAC, 2009; UNESCO, 2007; other sources). Uncertain extent if bracketed.

Aquifer/Aquifer system	Type	Countries	Extension (km^2)
Northern Great Plains	C	Canada, USA (N)	[500,000]
Ogallala (High Plains)	B	USA (central)	450,000
Gulf Coast Plains	C	USA, Mexico	1,150,000
W Amazonia (Solimões – Alter do Chão)	A/B	Brazil, (Peru, Bolivia, Ecuador)	[950,000]
Yrendá – Toba – Tarijeño	B	Paraguay, Argentina, Bolivia	[500,000]
Guaraní	A	Brazil, Paraguay, Argentina, Uruguay	1,200,000
La Pampa	C	Argentina (central)	[200,000]
NW Sahara + Murzuk	A	Algeria, Libya, Tunisia, Niger	1,500,000
Senegal – Mauritanian Basin	A/B	Mauritania, Senegal, Gambia, Guinea-Bissau	300,000
Nubian Sandstones	A	Chad, Egypt, Libya, Sudan	2,200,000
Iullemeden – Irhaser	B	Mali, Niger, Nigeria, Algeria	635,000
Chad Basin	A/B	Niger, Nigeria, Chad, Cameroon	1,900,000
Northern Kalahari	B	Angola, Botswana, Namibia, Zambia, Zimbabwe	[700,000]
Karoo Basin	B	South Africa	600,000
Arabian Platform	C	Iraq, Jordan, Saudi Arabia, Syria, Yemen, Bahrain, Kuwait, Oman, Qatar	>2,000,000
Central Europe Lowland Plains	C	From the Netherlands to Russia	[3,000,000]
Central Asia	C	Kazakhstan, Kyrgyzstan, Uzbekistan, Tajikistan	660,000
Indus Plain	B	Pakistan	320,000
Ganges – Brahmaputra Plain	B	Bangladesh, India	600,000
New Guinea	B/C	Indonesia, Papua New Guinea	870,000
Great Artesian Basin	A	Australia	1,700,000

Types:
A: Largely confined, outcropping at the boundaries
B: More or less continuous, largely outcropping aquifer
C: Compound of a series of smaller aquifers

Box 1. The Guaraní aquifer, South America

The Guaraní aquifer is part of the 1,600,000 km^2 Paraná sedimentary basin. It covers an area recently evaluated to have 1,200,000 km^2 (Rebouças & Amore, 2002; Kemper *et al.*, 2003), mostly in Brazil (70%), and extending into Argentina (19%), Paraguay (6%), and Uruguay (5%). The aquifer crops out over 13% of its surface area, mostly in Brazil, and contains freshwater. It is recharged directly through rain infiltration, and indirectly through leakage from upper formations: $20 + 140 = 160$ km^3/yr of freshwater.

The much larger confined area (70%) contains most of the freshwater reserves, although their quality degrades progressively, and presents salinity, sulphate and fluoride excesses due to the very slow flow rate. The aquifer thickness goes from almost nil to more than 800 m toward the southern and southwestern areas, with an average value of 250 m. The confining beds include the huge Paraná basalt lava flows –an aquitard yielding moderate quantities of groundwater – and the Bauru aquifer. Depths to the Guaraní aquifer varies from a few metres to up to 1,800 m. Water reserves are about 45,000 km^3, but only a small fraction is sustainably developable. The area containing warm water, mostly related to their depth and very slow flow, is assumed to be close to 400,000 km^2, about 30,000 km^2 of them with an expected temperature of more than 60°C, up to 70°C. Warm water is currently mined in Argentina and Uruguay for spas, mostly as brackish to saline water (Tujchneider *et al.*, 2007).

Freshwater development from the Guaraní aquifer is currently intensive in some areas of Brazil (e.g. Saõ Paulo and Curitiba), with local cumulative drawdowns up to 120 m. A GEF (United Nations Global Environmental Facility) project has been carried out recently (2004–2008) by the World Bank, to advance towards the management of this huge transboundary groundwater reserve through improved, shared knowledge, and means for technical, administrative and political coordination, fostering international and national entities to be involved in the job.

This is an attempt to try to deal with possible future problems at the time that existing regional and local problems are addressed, including the important general economic and social components.

Box 2. Managed small aquifers: the Lower Llobregat aquifers (Barcelona, Catalonia, Spain).

The Lower Llobregat aquifer system, close to Barcelona (Catalonia, Spain), is a small, about 120 km^2, river valley filled with gravels that continues into a delta formation formed by an Upper aquifer and a Deep aquifer separated by an aquitard (Custodio, 2008). The Deep aquifer opens to the sea bottom at about 4 km offshore, at about 120 m depth. Groundwater dynamic storage, mostly in the valley, is up to about 200 km^3. Currently, recharge is largely from infiltration from the Llobregat river, canal irrigated areas, and inflow from the mountain sides. There is intensive agriculture, largely supplied through surface water canals, and partly by using groundwater. The large urban and industrial areas of the metropolitan area of Barcelona are partly on the aquifer. Groundwater use started late in the 19th century through wells pumped with steam engines for urban supply, and many flowing wells in the delta for town and rural supply, and later on for industrial use. Intensive industrial use started in the 1920s and especially in the 1940s, peaking in the 1970s, jointly with urban supply, up to 130 hm^3/yr. After this moment, extraction has been decreasing down to the current 60 hm^3/yr. This is due to management efforts (Box 16), as well as a changing industrial pattern, which has evolved from high water demanding factories to less water demanding ones. Extraction for supply to the Barcelona area in the 1960s changed from continuous to complementary, and during droughts and emergencies, when local and imported treated river water was made available. The small groundwater reserve relative to annual extraction was a concern in dry summers and drought periods, so the Barcelona's Water Supply Company started enhanced river infiltration in 1948, and occasional artificial recharge of treated river water through wells in 1969.

The intensive aquifer development produced low groundwater levels in the whole Deep aquifer, below sea level, starting a serious seawater intrusion, first detected in 1965. Many salinized wells were abandoned but others continued to be operated for industrial cooling, thus helping to slow the seawater intrusion process. The current extraction decrease and the increased artificial recharge are improving the situation. Starting in 2007 an injection well barrier of deeply treated (including reverse osmosis) municipal waste water is operating to successfully control and redress seawater intrusion (Niñerola *et al.*, 2009). Increased artificial water recharge through infiltration basins in the river valley is also under way. Currently, groundwater levels are rather high, available storage for droughts and emergencies has been increased, and groundwater quality is improving.

Upstream from this aquifer system there are two small, rather isolated, gravel-filled river tracts, the *Cubeta d'Abrera* and the *Cubeta de Sant Andreu de la Barca*. Currently they supply industrial settlements and provide water to the towns around. Groundwater resources depend heavily on river water infiltration and may suffer reserve exhaustion at the end of the summer season. Exhausted reserves recover when river flow increases. These aquifers are used as 15 to 20 hm^3 water reservoirs due to their high transmissivity. Management includes artificial recharge of river water through basins.

Small aquifers may be key pieces and their use can be made sustainable through management. There are noticeable costs involved in management but benefits compensate for these costs.

3 GROUNDWATER DEVELOPMENT: HISTORICAL EVOLUTION AND INTENSIVE USE

Groundwater has been traditionally developed from its natural discharge by tapping springs or deriving stream baseflow, or by excavating shallow wells in high water-table areas and using manual or simple mechanical pumping devices. At most this development had a small influence in natural groundwater flow and on its chemical composition. Only in arid areas have humans devised other more sophisticated groundwater winning methods such as water galleries –the well known *khanats* (or other denominations like *foggaras*, *viajes de agua*, galleries), in Persia, Yemen, North Africa, the eastern Iberian Peninsula, and the Spanish archipelagos– or deep excavated wells. The Bible mentions some in the eastern Mediterranean area. All of them were costly and slow to construct, expected to endure for generations. In some cases these were complemented with works to increase aquifer recharge by retaining storm flow in creeks by means of transversal works, or by deriving them to flat areas to favour infiltration. Examples can be found in the Middle East, the Atlas, eastern Canary Islands, and the Iberian Peninsula. Even in these cases, aquifer conditions remained close to natural.

Important changes did not appear until the mid 18th century, when steam-driven pumps were used to dewater mines, as done in England and Wales. In the 19th century this technology was applied for town and factory water supply in some urban and industrial areas by means of quite expensive, large diameter, hand or mechanically excavated wells, down to at least a few meters below the water table, or penetrating through low permeability confining beds. In these shaft-wells, water-proof chambers to hold the bulky piston pumping machinery were installed and operated by means of steam engines. Some are preserved as museums.

In the mid 19th century, small diameter, cased deep-bore technology was made available in Europe, partly inspired by old methods developed in China. This allowed the construction of flowing wells under favourable hydrogeological conditions. A first success was the Rue de Grenelle well in Paris, in 1841. This technology rapidly expanded to other places, with variable success, and helped to solve acute town water-supply problems. Prat de Llobregat, near Barcelona, is one of the examples; late in the 19th century this was a water-supply revolution for the village and farmhouses, and later on fostered an important industrial development.

In Madrid, the attempts to drill flowing wells in the mid second part of the 19th century failed due to an erroneous hydrological conceptual model that was assumed to be similar to that of Paris. After Llamas (1983), this failure discouraged the centralized public Spanish administration to consider groundwater as a reliable water resource. The consequence was that groundwater was not taken into account in government-financed projects for more than a century. This was in spite of the success in other areas of Spain, which were mostly promoted through private funding.

Many flowing wells decreased or soon ceased to flow due to what is now a well-known hydrodynamical evolution of confined aquifers. Flowing wells were substituted by pumping from drilled wells once the first mechanical means were available. Development was accelerated, following progress in hydrogeological science (de Vries, 2007). Groundwater abstraction was made possible through the submergible centrifugal pump technology. This went on in parallel with easy and relatively cheap energy availability by means of internal combustion engines coupled to electricity generators, and afterwards through the electricity network, as is currently the case, except in remote locations or in poorly developed areas.

The combination of easily available drilling technology, pumping machinery, and energy, led to a rapid expansion of groundwater development in many areas of the world, often exponentially at the early stages (Shah *et al.*, 2007). This is what happened in many aquifers and is what currently happens in many developing countries.

Intensive groundwater development means a water extraction that significantly modifies aquifer natural flow (Llamas & Custodio, 2003; Sahuquillo *et al.*, 2004; Custodio *et al.*, 2005; Llamas & Garrido, 2007). This situation comes from small early developments. Then, the benefits derived from groundwater development encourage accelerated development. Overexploitation, and other similar designations, are often used, but they are poorly defined concepts, corrupted by careless use – alike to what happened with the concept of sustainability–, often with poor understanding of hydrological processes and putting the accent on negative aspects (Custodio, 2002; Hernández-Mora *et al.*, 2001; Collin & Margat, 1993), thus downplaying the often quite important, associated benefits of aquifer development, and its role on the assumed world water crisis (Mukherji, 2006).

The starting time of intensive aquifer development vary according to the area. This may be as early as the 1910s in central Australia, the 1920s and 1930s in the arid southwestern USA (Texas, Arizona, California) and in Long Island (New York), and also in the volcanic Canary Islands by means of deep shaft-wells and long water galleries. In Europe this happened largely in the 1940s in industrial areas of northern France, Belgium, Germany, western England, and Catalonia and Valencia in eastern Spain, and later on in northern Italy, and in the 1960s and the 1970s in intensively irrigated agricultural and tourist areas along the Mediterranean Sea coast. In Mexico and Brazil intensive development started in the 1950s. Large areas are now under intensive aquifer use in developing countries. Some of the most important ones are in India, northern China, Pakistan and Indonesia (Shah *et al.*, 2000; 2003; Ragone *et al.*, 2007; Margat, 2008). Very large quantities of groundwater are currently extracted in some of these areas as shown in Table 2. The evolution in a few countries is shown in Figure 1.

4 CONSEQUENCES OF LOCAL WATER-CYCLE MODIFICATION THROUGH AQUIFER INTENSIVE DEVELOPMENT

The consequences of the modification of the local water-cycle through intensive aquifer development can be summarized by considering the different involved aspects (Custodio & Cardoso da Silva, 2008; Custodio, 2001a) (see Insert 2), aside from the economic and social implications, which are explained in the next section.

4.1 *Effects on hydrodynamics and water quantity*

Aquifer development has two main aspects: a) winning groundwater flow on its way from recharge to discharge areas; b) depleting groundwater storage. Both of them are closely related and simultaneous. The consequence is water level drawdown, first around the wells and later on over progressively larger areas. Initially groundwater reserves are preferentially used and later on the behaviour evolves toward decreasing aquifer natural discharges (Figure 2). This is a transient process that may last years to centuries, depending on aquifer size and hydraulic characteristics. In many cases the evolution is toward a new steady state if development does not change; then what is abstracted is detracted from natural discharge, discounting possible enhanced natural recharge or artificial recharge.

In extreme situations, when water extraction exceeds recharge, groundwater reserves are progressively depleted. This situation is known as groundwater mining. Examples are given in Box 3 (Ogallala aquifer), Box 4 (Columbia River Basalts), Box 5 (Hermosillo aquifer), and Box 6 (Canary Islands). Groundwater mining is possible due to the large water reserves of aquifers. This is often a common situation in arid and semiarid lands (Foster & Loucks, 2006), where the extracted water is mostly for irrigation. Groundwater mining may be the consequence of unplanned intensive aquifer development, but may be an accepted transient situation, as in the Ogallala aquifer and in south-eastern Spain (Box 7), or a wanted development as in Libya (Box 8) and Saudi Arabia. Some

Table 2. Some examples of aquifer depletion of large aquifers (data modified from Brown, 2002; Foster *et al.*, 2005; other diverse sources).

	Water-table depletion		Use km^3/yr					
Aquifer	m/yr	m (a)	Total	Exc. (b)	Main use	Evolution (c)	Results	Notes
USA (as a whole)		>12	>100		I, U	TM, M	Δ cost, LS	(1)
Ogallala (USA)	up to 3	80	36	10	I	TM	Δ cost, IA	Box 8
Columbia River Basalts (USA)	up to 0.5	>200	–	–	I, U	TM	Δ cost	Box 4
Guanajuato (Mexico)	2–3	–	–	–	I, U	TM	Δ cost, LS	Box 15
Hermosillo (Mexico)	–	>100	0.5	0.05	I	TM	Δ cost, S	Box 5
Sana'a (Yemen)	7	–	0.27	0.22	I, U	U	Δ cost, Ex	(2)
Qatar				–	I, U	TM	S	(3)
Chenaram Plain (NW Iran)	3 up to 8	–	> 50	–	I, U	U	Δ cost	(4)
Indus Plain (India, Pakistan)	–	–	> 50	10	I	U	Diverse	(5)
India (as a whole)	up to 1–3	–	280	30	I	SD, U	Δ cost	(6)
Ganges-Brahmaputra Plains (India, Banglad.)	–	–	> 100	50		SD, TM	Δ cost	
N Plain (China)	0.5–3	35	>100	–	I	SD	Δ cost, IA	(7)
Great Artesian Basin (Australia)	–	80	>100	50	cattle	TM	Less flow	(8)
SE Spain	up to 10	250	0.8	0.5	I	U, TM	Δ cost, S	(9) Box 7
Canary Islands (Spain)	up to 10	300	0.25	0.1	I, U	TM	Δ cost, IA, S	(10) Box 2

Notes:
(1) Country as a whole. Excess extraction in California, Arizona, New Mexico, Texas, Kansas, areas in the southeast, east coast, northwest.
(2) Small but important aquifer. Urban reserves estimated to last some years (NWRA, 2009).
(3) Small area; mostly freshwater lens depletion from 1984. Saline water below.
(4) Seasonal problems.
(5) Start in 1960. Some million wells.
(6) Many states. In some areas extraction rate may double recharge. About 30 million wells (including paddle pumps); they were 11 million in 1960.
(7) About 3.5 million tube-wells.
(8) Flowing wells starting in the 1880s; flows dwindling. About 60% of recharge.
(9) Small carbonate aquifers being depleted. In the whole Spain there are officially 0.5 million wells but other estimations are up to 2 million, including springs for supply (Custodio *et al.*, 2009a).
(10) The three main islands: Gran Canaria, Tenerife and La Palma (Custodio & Cabrera, 2002; 2008).
(a) Total water-table depletion in the more seriously affected areas
(b) Excess refers to storage depletion. This may be the transient hydrodynamic evolution towards a new steady state when extraction is less than recharge (both uncertain), or permanent depletion, or a mix of them. Often this is unclear when there are not enough monitoring and serious studies
(c) SD = starting development; TM = towards management; M = managed; U = uncertain

Main use: I = irrigation; U = urban

Results: Δcost = increasing water cost; IA = irrigation abandonment; LS = land subsidence; S = increasing salinity; Ex = exhaustion and serious local supply problems.

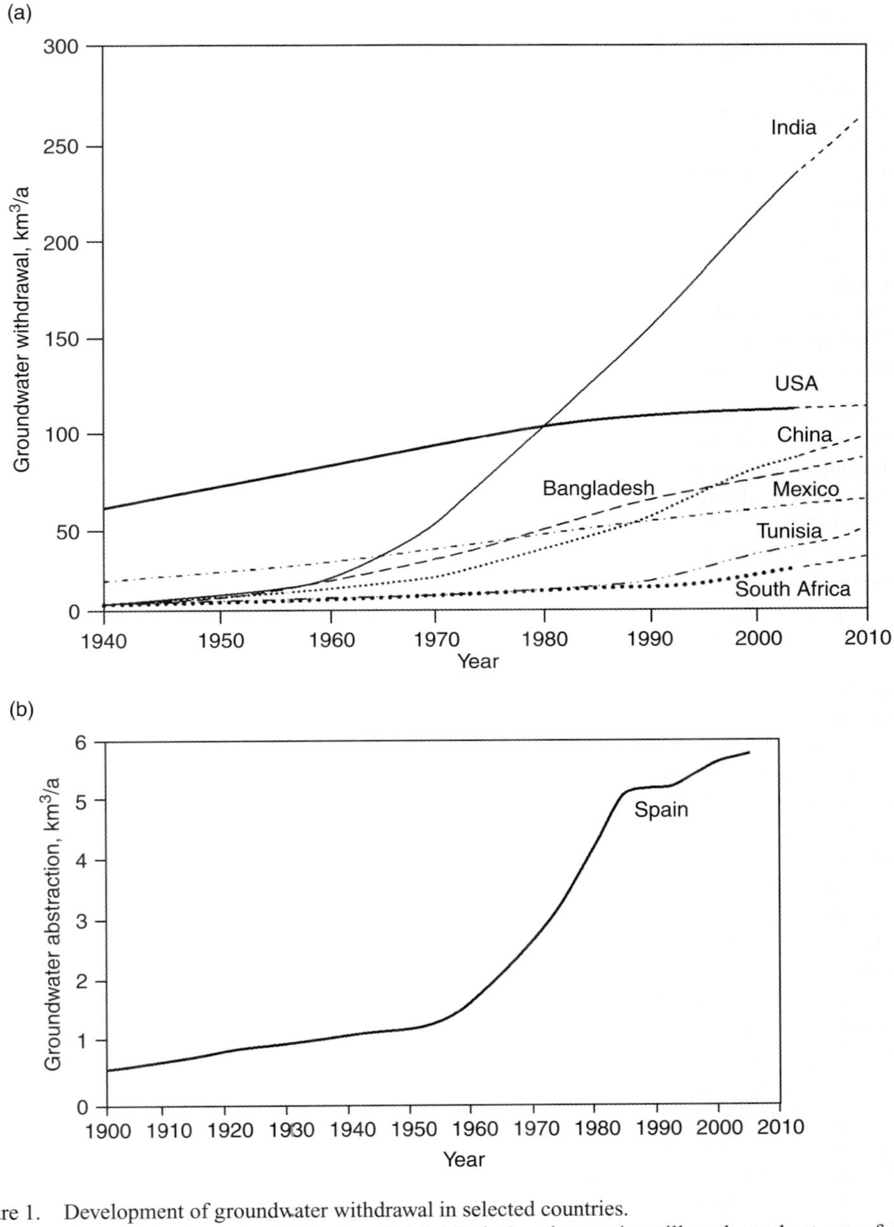

Figure 1. Development of groundwater withdrawal in selected countries.
A) In countries evolving towards stabilization (USA, Mexico) and countries still on the early stages of accelerated development (India, China, Tunisia) (modified from Shah, 2005).
B) Development in Spain since early the 20th century with the accelerated development initiated in the 1950s and the current trend towards stabilization from the 1980s (modified from MIMAM, 2000).

Insert 2. Main consequences of groundwater (intensive) development.

Drawbacks:
- On water quantity (progressive, slow appearance):
  - groundwater level drawdown and storage decrease
  - discharges to springs, rivers and wetlands decrease
  - reduction of phreatophyte and riparian vegetation areas
- On water quality (progressive, slow appearance):
  - penetration of poor quality water, including marine intrusion
  - solutes from the ground (direct, hydrolysis, redox changes)
- On ecological values: decrease of Nature services to humans
- On land surface: subsidence and collapses
- On economic aspects: increasing costs (extraction, water quality treatment)

Benefits:
- Increased freshwater availability
- Taming water resources variability
- Resilience to droughts and catastrophic events
- General low cost (even if progressively increasing)
- Availability over large areas (shorter water transport lines)

examples are included in Table 2. In the large Nubian Aquifer System there is also groundwater mining (Box 9).

Piezometric level drawdown results in a modification of aquifer-river relationships by decreasing and even stopping river baseflow, in which case the stream converts from draining the aquifer to a temporary or permanently recharging situation if there are ephemeral or allochtonous flows (Sophocleous, 2002; 2010). The same can be said for lakes and wetlands (Custodio, 2000a; 2001b). The effect may often be quite clear as spring-flow decreases and even dries out. Lake level may be lowered and groundwater contribution may be lessened with respect to surface water inflow. Groundwater-dependant wetlands tend to reduce their size. This may be conspicuous in semi-arid and arid areas, in which the vegetation cover evolves toward sparse, deep-rooted plants, or just disappear.

In coastal aquifers the marine water-freshwater relationships are modified due to the freshwater head changes relative to seawater level. The result is often a landward salt water wedge penetration into the ground, often a complex situation. Also relationships with already existing saline and brackish water in the aquifer may be changed, often slowly decreasing the depth of these water bodies. Below pumping wells and drainage structures (ditches, drains) the freshwater head decreases and this favours brackish and saline water upconing. This may also affect nearby extraction works, which may be accompanied by a serious saline water disposal problem (Custodio, 2005a).

Wells are often constructed to yield as much groundwater as possible for a given penetration into the aquifer, thus connecting different aquifer layers through the well bore. Consequently the well yields a mixture of water from different depths, with different recharge areas, transit time, and chemical characteristics. The mixing pattern varies with discharge, time, well construction and local aquifer properties.

4.2 *Effects on water chemical composition and quality*

These effects are often less conspicuous and may appear long delayed, but they are by no means less important. They may be the result of water mixing in the abstraction work when different aquifer layers are penetrated and interconnected, even allowing the inflow of brackish or saline water

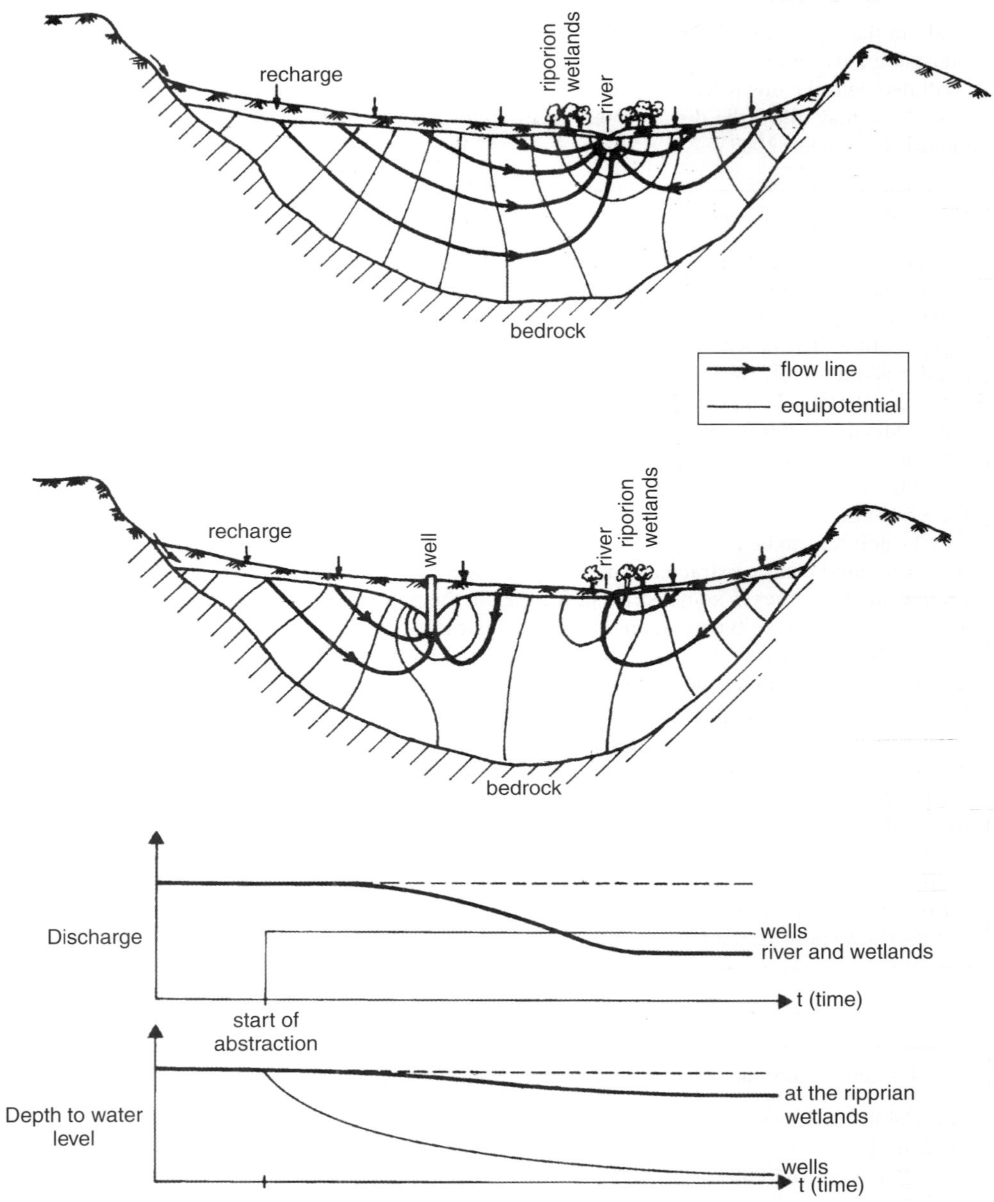

Figure 2. Schematic representation of the groundwater withdrawal effect.
a) Natural situation in a sedimentary basin recharged by rainfall infiltration and main discharge into a river and through the associated riparian vegetation.
b) Long-term effect of intensive groundwater development, in which a generalized water-table drawdown is produced in order to concentrate a large fraction of recharge in to the well area, thus reducing discharge into the river and riparian vegetation area.
c) Schematic evolution of river and riparian wetlands discharge, and of groundwater level at the wells and the riparian wetlands; the time scale may vary from months to centuries depending on aquifer hydraulic characteristics and size. In this case only a fraction of recharge is withdrawn and in the long-term a new steady situation is possible.

already in the aquifer, or the result of development-induced seawater intrusion, and/or upconing. Water-table lowering may induce surface water recharge having a different chemical composition or polluted, and the down-leaching of wastes, fertilizers, agrochemicals, leakage from pipes and tanks, … or may allow the displacement of groundwater in nearby or overimposed aquifers and aquitards (Custodio, 2007).

Box 3. Aquifer intensive development: The Ogallala aquifer, USA.

The Ogallala aquifer is a large unconfined aquifer (450,000 km^2) extending over large, semi-arid areas of the High Plains of Texas and Kansas, and also South Dakota, Wyoming, Nebraska, Colorado, Oklahoma and New Mexico, in Central USA. Saturated thickness is from less than 15 m up to 300 m in some areas in Nebraska. Depth to the water table is 30 to 200 m. The aquifer is intensively exploited, mainly for irrigation, since late in the 19th century, but mostly since the 1950s, when pumping reached up to the current level of about 26 km^3/yr (Dennehy, 2000; McGuire, 2007). The result is a continuous water-table drawdown in Texas and Kansas since average recharge is about 5 to 10 mm/yr and extraction in Kansas was about 110 mm/yr in 1990 and 90 mm/yr in 2000. A total depletion of about 320 km^3 has been produced, out of the 3,600 km^3 remaining storage, although such a large figure is misleading because it assumes that the aquifer can be completely desaturated as a whole, which is not practically feasible, and only a fraction can be extracted. The result is water cost increase and important depletion of groundwater reserves in some areas. In Kansas, in the period 1991–2001 drawdown in critical areas has been 6 m, or 0.6 m/yr, and the saturated thickness has been reduced from 21 to 15 m. Exploitable reserves are reckoned to last about 10 years in some areas, about 25 years or more in other areas, assuming no action is undertaken [http://www.kgs.ku.edu/HighPlains/atlas/]. Reaction is variable according to the State, from little action in Texas to decrease extractions in Kansas. The Kansas Water Plan goals are aimed at stopping or sensibly decreasing water-table drawdown by 2020, and restoring the flow of some streams to natural conditions. This means reducing extractions, even purchasing rights and lands by state agencies, and changing from irrigated to rain-fed agriculture. Problems are important but known in advance, and may be afforded through the richess created by previous and outgoing groundwater depletion, by means of innovative management approaches. The creation of an interstate groundwater commission for the High Plains aquifer is a possibility (Sophocleous, 2010).

This shows how after a long period of depletion there is action for sustainable use.

Box 4. Intensive groundwater development: The Columbia River Basalts aquifer, USA.

The Columbia River Basalt Group is the northern part of the 370,000 km^2 Columbia Lava Plateau of flood basalts with interbedded sediments. The Columbia Basalt Group underlies an area, more than 164,000 km^2, in the States of (western) Idaho, Oregon and Washington (USA), and contains more than 17.4 km^3 of mostly freshwater. The formation consists of more than 300 extensive continental tholeiitic flood-basalt flows, in total some hundred metres thick, up to 3,200 m (Eaton *et al.*, 2009). The N-S Cascade Range, parallel to the Pacific Ocean coast, divides the area in two: the western side is more humid and rather well recharged; the eastern side is semi-arid and poorly recharged, generally less than 50 mm/yr. Groundwater development has been intensive since the 1960s, especially in the eastern side, mostly for agricultural irrigation. In some areas there are up to 7 deep wells per km^2. In Willamette Valley it is reported that there are about 100,000 wells, growing at a rate of 3,000 to 4,000 per year. Groundwater level drawdown of up to 0.5 m/yr is accompanied not only by increased pumping costs (the depth to water level increased from 20 m in 1960 down to 40 m in 2000 in Willamette Valley area) but also a progressive well yield decrease.

Some Groundwater Management Areas have been created to deal with problems and the possible loss of jobs, and in some areas ASR (Aquifer Storage and Recovery) practices have started for urban area supply (Eaton *et al.*, 2009) and for some large agricultural establishments, through injection of treated-to-potable quality impounded surface water, when it is available. The economic analysis shows that ASR technology is affordable.

When effects of intensive use are detrimental, action starts to make aquifer use sustainable.

Box 5. Intensive groundwater use: The Hermosillo coastal aquifer, Mexico.

The Hermosillo coastal aquifer underlays a 3,200 km^2 area, Sonora State, northwestern Mexico. This and the close-by Guaymas area, hold important irrigated agricultural areas, the basket of Mexico. These Miocene coastal aquifers, with an open, long shore front, are 150 to 500 m, up to 800 m, thick, although the lower part may be confined by a coastal clay layer up to 400 m thick near the shore but thinning out at about 40 km inland. In Hermosillo, starting about 1947, the groundwater development was exponential. Around 1967 extraction was estimated at 1,200 hm^3/yr (Rangel-Medina *et al.*, 2005; Moreno Vázquez, 2006). Current groundwater level drawdown in the central areas is up to 135 m below the natural situation, which means up to 65 m below sea level. Studies showed growing salinization problems in some areas about 30 km inland from the coast (Steinich *et al.*, 1998), about 27 km^2 with up to 16 g/L salinity, and a progressive water cost increase. Farmers did not pay much attention to government agencies warnings and continued to be largely unorganized to confront the situation. Groundwater exploitation was predicted to fail in about 25 years according to models. Some agreement permitted reducing extractions and to move inland some of the close-to-the-coast salinized wells. Current extraction is about 550 hm^3/yr out of 350 hm^3/yr of recharge, including 70 hm^3/yr infiltrated from surface water (canals and storage reservoirs) and about 100 hm^3/yr of aquifer water substituted by seawater or saline water, mostly upflowing from the deep aquifer layers (Rangel Medina *et al.*, 2005; Szynkiewicz *et al.*, 2008). This means 200 hm^3 of freshwater storage depletion. This has to be compared with an average rainfall recharge of 100 hm^3/yr. However, the aquifer, as well as the smaller Guaymas aquifer, continues to supply water, complemented with surface water from the Hermosillo dam and water taken from close-by aquifers, thus permitting to move towards sustainable use of these important water resources, at a cost, while sustaining the important current economic development.

This shows the resilience of a large aquifer to be exhausted, thus allowing time for awareness on the common asset to be protected.

Box 6. Intensive groundwater development: The Canary Islands, Spain.

The Canary Islands is a volcanic archipelago in the eastern Atlantic Ocean, facing the Sahara region. It consists of seven main islands, 250 to 2,000 km^2 of surface area, with altitudes up to 3,700 m, one of the highest in Europe, although the easternmost islands, the oldest and most eroded ones, are only 650 to 850 m high. Gran Canaria, Tenerife and La Palma are the most populated islands, with important economic activities that include important tourist establishments and large irrigated areas since the 16th century although fully developed after the 1930s. The low and medium altitude areas, especially at the southern slopes, are arid to semi-arid, but the highlands receive up to 1,000 mm/yr. Groundwater is obtained by tapping springs (if still flowing), drilling deep wells and excavating long water galleries (Custodio & Cabrera, 2002), to penetrate into the low permeability, although with vertical fissures, volcanic core formations (Custodio & Cabrera, 2008). Exploitation of this type of volcanic islands, with high water table gradients, is subject to hydrodynamic effects that -in many areas- produce a continuous water-table lowering, which in some cases may exceed 300 m, even when abstraction is less than recharge since outflow into the sea continues to be relatively important due to hydrodynamic conditions. Further problems are local seawater intrusion in some coastal aquifers, and quite important nitrate pollution derived mainly from agriculture. The Figure is a cartoon to show hydrogeological conditions in a volcanic-core island, as inspired in Gran Canaria (after Custodio & Cabrera, 2002). It shows how the progressive water-table lowering has a moderate effect in reducing outflow into the sea in part of the coastal areas, but may dramatically affect springs, wells and water galleries in mid and high altitudes.

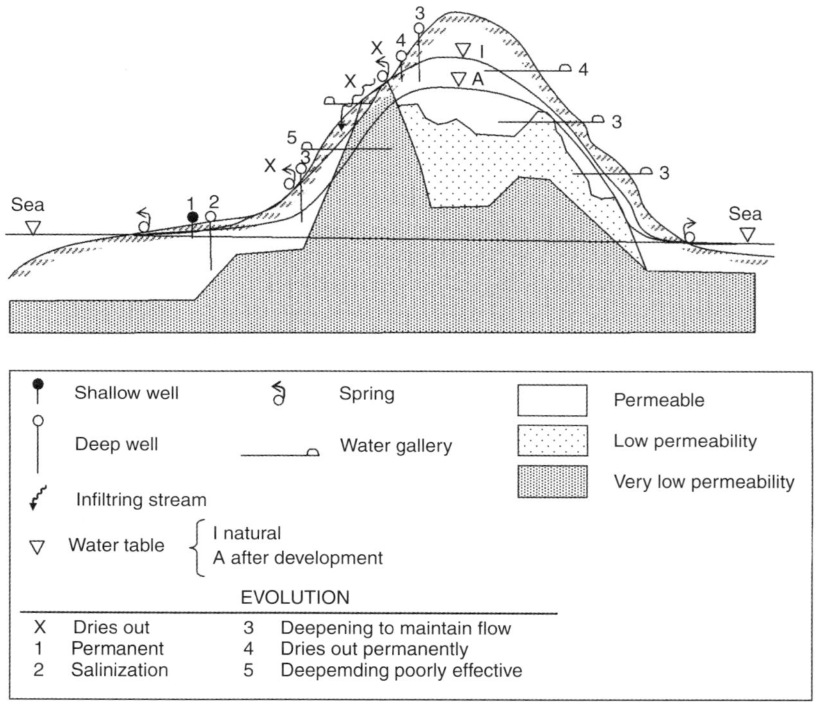

Some recent groundwater balances, derived from the island's Water Plans, are:

Island	Surface area km^2	Recharge hm^3/yr	Coastal outflow	Seawater inflow	Extraction	Irrigation return flows	Water storage decrease	Current average draw-dawn rate m/yr
Gran Canaria	1,500	100	40	5	80	15	30	0.5 – (0–2)
Tenerife	2,000	280	270	0	190	30	150	1.2 – (0–4)

Surface water resources are small, only 8% of total resources in Gran Canaria, the most favourable due to outcropping low permeability old volcanics, even if it has the world's largest density of large dams (higher than 15 m), but with rather small storage capacity due to the land stepness. Up to 25 hm^3/yr of desalinated water is made available in the eastern islands. There is an increasing treated waste water use for agricultural use in some areas, in some cases after reducing salinity through reverse osmosis. In Tenerife and La Palma, high altitude, long water galleries dominate. They yield continuously, so water produced in low demand seasons is wasted. To cope with this wastage of reserves, temporal storage reservoirs have been constructed, partly publicly sponsored. Besides, bulk-heads have been build-up recently in galleries to stop flow, but this needs favourable geological conditions to resist the very high differential pressures that may develop, up to tens of atmospheres.

Governmental and stakeholder actions are helping to reduce groundwater exploitation and decrease drawdown, and even redressing it to avoid current high water extraction costs. An important goal is preserving groundwater for high and medium altitude uses, where desalinated and reused water is not available or too costly. There is some management success in Gran Canaria where groundwater producers and users are being organized for self-control. Groundwater stakeholders have a quite numerous representation in the *Consejo Insular de Aguas* (Island Water Board) that allows them to influence decisions.

The policy is toward preserving aquifers for sustainable uses in critical areas for the economy, and for land and landscape conservation, while providing alternative water resources for urban supply at low altitude areas. But during the 20th century development was unsustainable, mining reserves.

Box 7. Groundwater reserve depletion: Small aquifers in southeastern Spain.

In the Alacant/Alicante, Murcia, and Almeria provinces, in southeastern Spain, climate is semi-arid but conditions for intensive cultivation of highly valued crops are excellent. Only small river basins and limestone aquifers exist since the high rangeland is close to the coast and tectonic disturbance of geological formations is high. Aquifers have been intensively exploited since the 1960s, and continue to be, in spite of the high cumulative groundwater level drawdown, as well as salinization in some cases. This is due to the high profitability of special cash crops that have a sure and lucrative market in Europe. In coastal areas, brackish groundwater is being desalted by reverse osmosis. Besides, seawater desalination plants have been made available, but the water price is currently higher than the direct cost of continuing to exhaust groundwater reserves, for which opportunity and environmental costs are not considered. Away from the coast, reserves are exhausted rapidly, waiting for importation of water resources from other areas, as in the Vinalopó area (Alacant/Alicante), where storage depletion since 1980 is currently about 120 to 150 hm^3/yr, and a water transfer is in an advanced stage of completion. Some figures by areas that include several of these small, quite isolated, thick aquifers, after data in diverse official sources (e.g. DGOH, ITGE,), are:

	Reserves in 1995, km^3			Exploitation, hm^3/yr		
Area	Used[(1)]	Remaining	Usable	Renewable	Reserves	Years to depletion
Almeria	0.8	1.1	0.7	50[(2)]	50[(2)]	15 (10–75)
Murcia Highlands	2	10	7 to 11	200	125 to 300	60 (10–800)
Campo de Cartagena	–	–	1–2	50	90	20 (10–40)
Alacant / Alicante	1	7	6 to 11	30	150	160 (10–200)

[(1)] Period 1980–1995

[(2)] Current figures of exploitation in the whole province are about 160 hm^3/yr, which correspond to the Campo de Dalías-Sierra de Gádor aquifer system.

For some of the Murcia Highlands small aquifers (Molina *et al.*, 2009):

Aquifer	R	B	ΔS	Δh	Rate
Cingla	13	30	17	25	1.3
Jumilla-Villena	13	46	31	115	3.5
Ascoy-Sopalmo	2	52	50	190	4.5
Serral-Salinas	5	18	13	130–290	4.9–10.5

R = estimated recharge rate, hm^3/yr; B = pumpage rate, hm^3/yr; ΔS = decrease rate of aquifer water storage, hm^3/yr; Δh = estimated total drawdown (m), from natural situation; Rate = current drawdown rate, m/yr

The groundwater level evolution is shown in the Figure, in which: a) shows the evolution of extractions in hm^3/yr; and b) the groundwater level drawdown referred to the situation in October 1980, when exploitation was already developed. Extraction is well above recharge. Even if pumpage is stopped, recovery will need many decades.

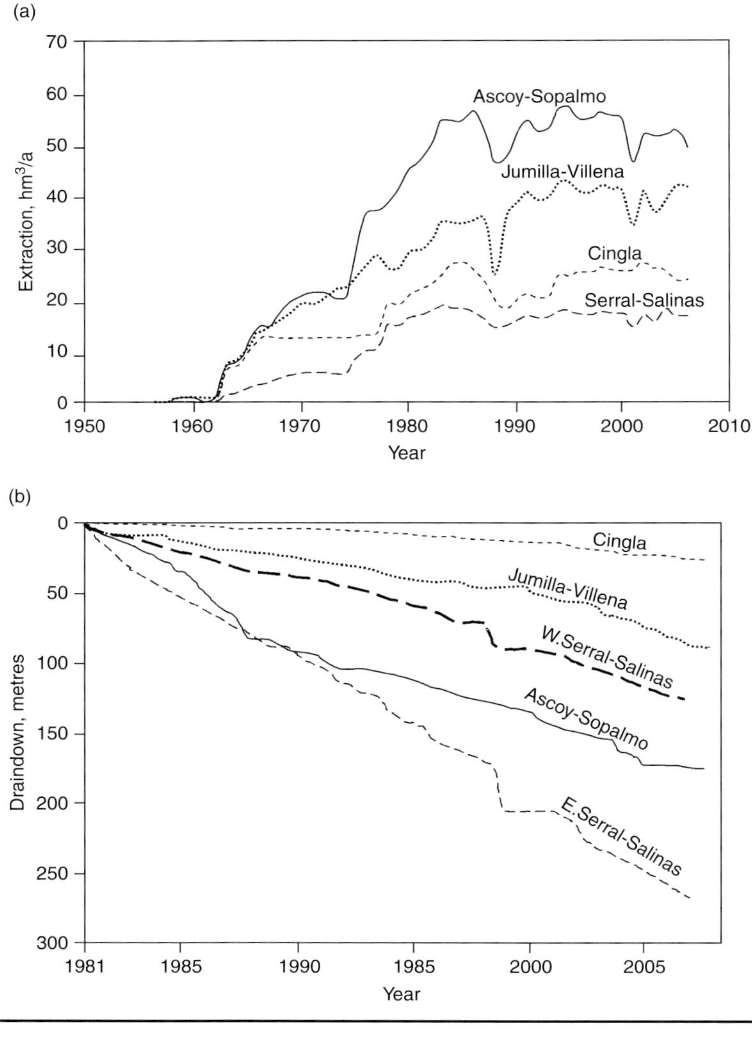

Box 8. Groundwater mining: The Great Man-Made River, Libya.

Libya is an arid country in which most of the population and economic activities are in the north, along the Mediterranean Sea coast. The southern area is a desert but holds one of the largest world's fresh groundwater reserves in several extensive aquifers. The most important are:

Aquifer	Emplacement	Surface area in Lybia (km^2)	Maximum thickness (m)	Water reserves (km^3)[(4)]
Kufra[(1)]	SE	350,000	2,000	20,000
Sirt[(2)]	Center-NE	300,000	1,000	10,000
Murzuk and Hammada[(3)]	SW	350,000	500	4,800

[(1)] It is part of the Nubian Sandstone Aquifer System
[(2)] They extend to the E and the SW
[(3)] They are in the western Sahara aquifer system (more than 1,000,000 km^2)
[(4)] Only a fraction is developable

A large project to extract deep groundwater (down to 1,000 m) from these aquifers was launched in 1984, to be conveyed to the north: 80% is to irrigate 150,000 ha, and 20% for urban and industrial areas. The project budget was 20×10^9 €, yielding water at a cost of about 0.10 €/m^3, which was assumed cheaper than alternative water sources, mostly seawater desalination (Zidan, 2007; Loucks, 2004; Salem, 2005a; 2005b; Gijsberg & Loucks, 1999), although some evaluations did not consider investments already carried out. The project has been developed over 25 years, in four stages, to convey 6.5 hm^3/day (2.3 km^3/yr) of water extracted by means of more than 1,300 wells, deeper than 500 m. There are 4,000 km of pipes in four large systems across the country, from south to north, and along the Mediterranean Sea coast, Benghazi and Tripoli being the main destinations, where local aquifers are overused (Salem, 2005b).

After numerical modelling was carried out, the environmental impact in the exploited areas was assumed small (Abdelrham *et al.*, 2008), but the effect on other areas, especially in confined sectors, seems poorly known. In the receiving areas important social changes (increased prices, more food availability, but increased wasteful use of water) and environmental changes (climate improvement, less local pressure on local aquifers, but increased pollution due to fertizers and pesticides, . . .) have been or will be produced.

This is an example of planned massive use of groundwater reserves, having both supporters and detractors. The project may produce large social benefits if correctly carried out with a long-term perspective, provided means to compensate current and future negative impacts are applied and then put aside.

Box 9. Groundwater mining: The Nubian Sandstone Aquifer System.

The Nubian Sandstone aquifer extends over more than 2,000,000 km^2, in eastern Libya, Egypt, northeastern Chad and northern Sudan. It is probably the largest of the world's fossil water aquifers. The area is arid, except in the SE, where recharge is significant. The aquifer system consists of a series of vertically and/or laterally related aquifers. In the southern part it crops or subcrops out, the rest being mostly a confined aquifer. Water is generally fresh but in some deep areas may contain saline water. Groundwater flows south to north where outflows from deep layers through discontinuities have created large evaporation areas with salt deposits. Results derived from a CEDARE/IFAD study of the area carried out early in the 2000s (Salem & Pallas, 2004), show:

	Area (km^2)	Freshwater storage (km^3)	Recoverable freshwater (km^3)	Extraction (hm^3/yr)
Nubian (1)	2,176,000	373,000		600
Pre-Nubian (2)	920,000	169,000		1,600
Total	3,096,000	542,000	15,000	2,200

(1) Nubian system: Paleozoic to Mesozoic sandstones
(2) Post-Nubian system: Miocene

Present annually extracted volume is a tiny fraction of recoverable freshwater reserves. Most of extracted water is for large irrigated agricultural areas around traditional oasis in Egypt and in Libya (Kufra aquifer), about 70 hm^3/yr to the Big Man-Made River (Box 8) to the branch heading toward the Benghazi area.

This development has actual and potential important impacts on lakes and oases. In some cases natural outflow may be substituted by pumped wells, with a progressively increasing cost. Dried up areas are sources on moving sands that increase dune fields, as in Chad, where people are forced to migrate. Soil salinization and contamination by fertilizers is an added problem. In confined areas, the groundwater head depressed areas may extend over distances of several tens of km. In the Egyptian oasis of Kharga, deep wells between 1960 and 1998 have produced a groundwater drawdown up to 60 m. Although in some cases development seems relatively safe in the short- and medium-term, long-term consequences need to be further assessed, taking into account the sustainability of current developments due to increasing extraction costs and the capacity to transfer part of the economic benefits to future generations.

The policy of resettlement of poor farmers from the Nile Valley – to relieve pressure for land – to the Nubian aquifer-sustained oases has to be compatible with the falling groundwater levels and increasing energy costs. Also, the use of valuable fossil groundwater resources, subsidized as regards infrastructure and energy costs, for not very competitive agriculture near Mediterranean areas is an issue, blurred by policies heavily influenced by politics.

Water-table drawdown desaturates formerly water saturated ground zones, thus allowing oxygen penetration. This favours mineral and organic matter oxidation. The increase in CO_2 enhances mineral hydrolysis. Reduced metals may be oxidized and dissolved, although metals are often re-precipitated afterwards as insoluble oxides and oxy-hydroxides, but not always, as may happen for arsenic (As) (Hering *et al.*, 2009), a current serious health concern. Also sorbed ammonia in the ground is often oxidized to nitrate and dissolved. Fluctuating water-table produces an alternatively saturated and desaturated ground zone in which chemical reactions are enhanced, especially redox ones.

In some agricultural areas, irrigation with imported water has raised the water table up to shallow depths from which evapoconcentration takes place, thus precipitating salts and producing soil salinization, and even alkalinization when sodium (Na) is abundant. This may reduce crop yield dramatically or may lead to barren areas. Groundwater exploitation may help to alleviate or prevent this situation by lowering the water table (Box 10).

Box 10. Groundwater extraction to control soil salinization: The Indus Plain, Pakistan.

The Indus Plain is an arid, large area, with about 13 million ha under irrigation, supplied by 43 large surface water canals from the Indus river basin, 15 of them with a flow of up to 280 to 600 m^3/s. The large, mostly unconfined aquifer below, contains about 350 to 500 km^3 of fresh to brackish water. The unsaturated thickness was initially up to 70 m in the interfluves. Irrigation started before 1910 and excess irrigation water has been recharging the aquifer and increasing groundwater levels, up to close to the land surface in many areas. This has damaged crops and salinized the soil through evaporation. In order to correct this by artificially lowering the water table, a project to drill and pump 32,000 wells was prepared by the Government, with the goal to pump up to 70 km^3 of water per year, which is about half the contributed surface water. The Government of Pakistan halted the project when farmers begun to carry out the task by themselves, combining the use of surface with extracted groundwater. Part of pumped water was intended to export excess salinity through disposal into the canals in the lower area. In large areas, below the freshwater lens there is brackish to salty groundwater of natural origin. Avoiding salinity increase of pumped water is quite a difficult task; scavenger wells could be needed (Stoner & Bakiewicz, 1993), which is quite a complex technology.

This is a case in which groundwater exploitation is useful to alleviate problems created by surface water use, but at a cost, and with technical problems in some areas to deal with current and future salinity problems in flat areas with difficult surface drainage, if canal salinization affects downflow users.

4.3 *Ecological effects*

The value and ecological diversity of many wetlands depends on sustained water availability to define and sustain the hydroperiod and its fluctuations. Wetlands provide important services to humans that may dwindle and even disappear by intensive groundwater development.

Groundwater is important to sustain springs and river baseflow as well groundwater-dependant wetlands, both freewater ones or shallow water tables attainable by vegetation (crypto-wetlands). Groundwater development may reduce – even dry out – wetlands, and increase the depth to the water table. This means ecological changes that may be compounded by important chemical changes. Box 11 refers to the wetlands of the Doñana area, in southwestern Spain, an area at risk. Box 12 shows a more acute situation in central-southern Spain, in which groundwater flow to wetlands ceased some time ago.

Box 11. Impact on wetlands: The Doñana aquifer system, southwestern Spain.

The Doñana area, in southwestern Spain, over the Provinces of Huelva and Sevilla, holds the largest wetland area of western Europe, a Ramsar Convention wetland. The area currently contains about 1,700 km^2 of surface-water fed marshes, bounded by 1,200 km^2 of sandy lands, partly with a recent eolian sand cover. This area is quite well recharged (Custodio *et al.*, 2009b) and sustains groundwater discharges to some permanent streams, the coastal area, and the contact fringe (ecotone) between the sands and the marshes, besides several hundreds of small lagoons all over the territory, from recharge to discharge areas. All of them are crucial for local fauna, especially during the dry season, and sustain important vegetation areas.

The area, a largely uninhabited one due to poor conditions for human settlement, has been subject to several kinds of groundwater development. The first one, from late the 1940s and mainly in the next two decades, was the introduction of planted eucalyptus forests that increased groundwater evapotranspiration over large areas. In the early 1970s a large irrigation project based on local groundwater extracted from the sands was initiated in areas close to the ecotones. At the same time a Biological Reserve and a National Park were created as a result of the growing Nature conservation awareness. Thus, a confrontation between development and preservation took place, with the result of the halt in new developments, except in the westernmost area. Currently, nature preservation and existing groundwater-demanding developments try to live together. The Park has been enlarged with new protection areas, with some agricultural land purchases and pressure to reduce irrigated areas for rice cultivation, a highly water consuming crop. Agriculture in the area is economically important. In the 1990s a large part of the eucalyptus forest was eradicated.

Groundwater abstraction leveled after a fast growing period. The Figure shows groundwater exploitation, except for the westernmost sector, with an initial accelerated development and a later stabilization, even before the irrigation plan was completed due to pressure to not damage ecosystems further.

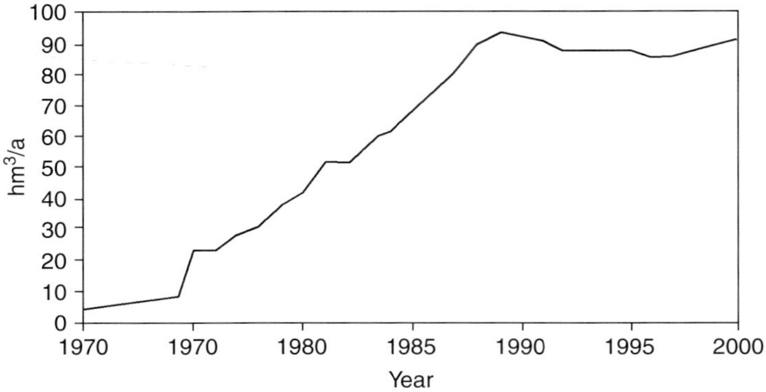

The consequence of groundwater development has been water-table drawdown, which is ecologically significant in many areas even if small, since they affect lagoons, mostly in recharge areas, and phreatophytes, and flowing wells used to water cattle and fauna along the ecotone. As numerical modeling shows, the time that it takes for the groundwater level and discharges to evolve from an initial value to half way towards the final steady state is about 20 years. This means that changes produced late in the 1970s are still evolving; depletion will continue for some time even if current extraction situation go on hold. An improvement was obtained after the eradication of the eucalyptus forest. Conspicuous seasonal and annual recharge changes can be observed, with periods in which it seems that Nature recovers the predevelopment aspect, but in reality it is not so, since the time between recovery events is increasingly longer. Agriculture and human activities are also a source of groundwater pollution which is poorly known, with current problems of nitrate build-up in some areas. This will affect ecological values in the future. Some effects seem to be already appearing (Manzano *et al.*, 2009a; 2009b). A part of the ecotone and a series of lagoons are seriously damaged but other parts of the ecotone and many lagoons survive, although with deteriorating health.

The current policy is towards management and water quantity and quality restoration, compatible with maintaining a reasonable groundwater development, reducing groundwater extraction and the use of agrochemicals, and purchasing critical agricultural areas.

Attempts to create governmental entities, further to the Park administration, have been largely unsuccessful since local people have been slow to understand the benefits of Nature protection and the importance of preserving their own groundwater resources. They have only been marginally involved in decisions and up to know have not been able to create representative associations, in part due to a lack of leadership. However this is partly compensated by quite a good level of monitoring of some variables, and good scientific hydrogeological studies.

The area has the potential to be correctly managed for a sustainable ecological and human situation, where groundwater is a key component. Efforts are under way.

Box 12. Impact on wetlands: The Upper Guadiana Basin, southern-central Spain.

The Upper Guadiana Basin, in southern-central Spain, about 16,000 km^2, is a rather flat platform with a poorly defined drainage network due to good soil permeability and high transmissivity of the water-table aquifer. Total usable water storage is about 3,000 hm^3. It is a semi-arid area with an average recharge less than 30 mm/yr, including streamflow infiltration (Martínez-Cortina, 2003; Bromley *et al.*, 2001). The area was well known for a large spring (*Ojos del Guadiana*) at the basin lower boundary, and for being an important Ramsar Convention wetland (*Las Tablas de Daimiel*). This and other minor wetland areas covered about 2,000 km^2. Wetlands depend partially on surface water but groundwater was the essential component for the hydroperiod.

This extensive rain-fed agricultural area started irrigation with local groundwater in the 1970s, with a fast expansion during the 1980s, with 130 km^2 of irrigated lands, using about 575 hm^3/yr of groundwater. This dramatically improved the local economy. However the rate of abstraction exceeded the estimated 415 hm^3/yr of recharge. The consequence has been the progressive lowering of the water table, down to 40 m in some areas, which means drying out the *Ojos del Guadiana* and many of the wetlands to the point that currently they depend only on surface water and occasionally on imported water to try to maintain artificially the hydroperiod in part of the area. Moreover, drained out peatlands started spontaneous ignition in dry periods. The Figures show: the water-table map evolution (a, b) with a cross-section (c); and the numerical model simulated evolution of water-table elevation (d), and the smoothened out wetland evapotranspiration rate (e). The actual (historical) evolution is shown together with what would be the undisturbed aquifer conditions. The slight decreasing trend of the undisturbed situation is due to a series of dry years in the 1980s and 1990s (modified from Martínez Cortina & Cruces, 2005).

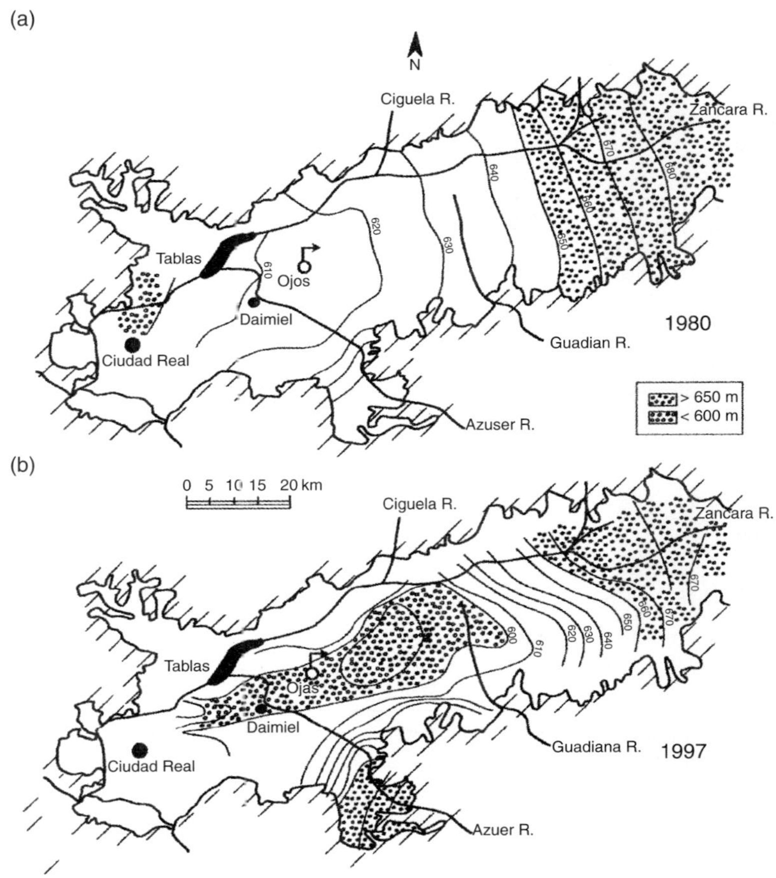

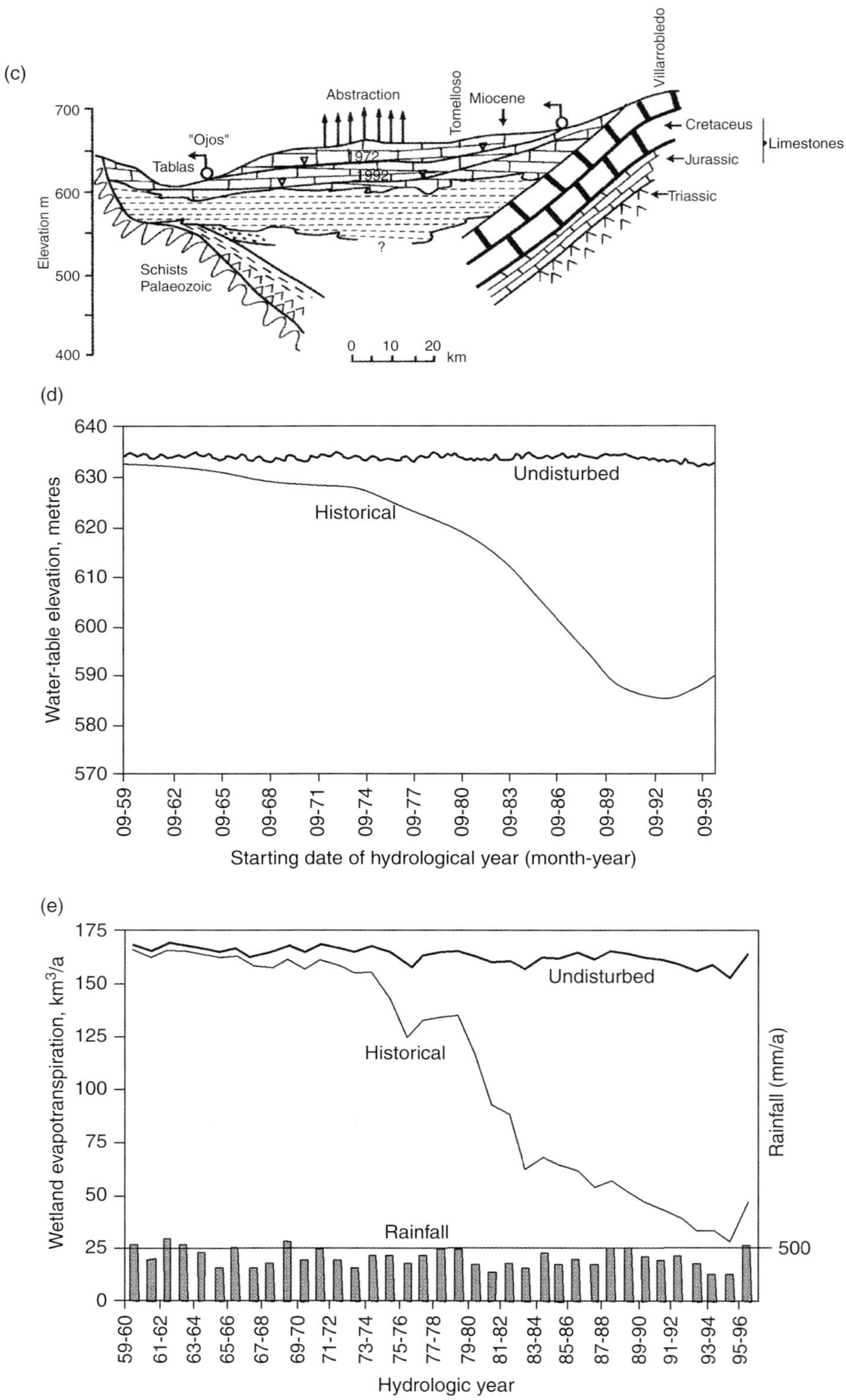
(c)
Villarrobledo
Tomelloso
Abstraction
Miocene
"Ojos"
Tablas
1972
1992
Cretaceus
Jurassic
Triassic
Limestones
Elevation m
700
600
500
400
Schists
Palaeozoic
?
0 10 20 km
(d)
Water-table elevation, metres
640
630
620
610
600
590
580
570
Undisturbed
Historical
09-59
09-62
09-65
09-68
09-71
09-74
09-77
09-80
09-83
09-86
09-89
09-92
09-95
Starting date of hydrological year (month-year)
(e)
Wetland evapotranspiration, km^3/a
175
150
125
100
75
50
25
0
Undisturbed
Historical
Rainfall
Rainfall (mm/a)
500
59-60
61-62
63-64
65-66
67-68
69-70
71-72
73-74
75-76
77-78
79-80
81-82
83-84
85-86
87-88
89-90
91-92
93-94
95-96
Hydrologic year

The water balance is (in rounded up figures):

	Inflow (Recharge), hm^3/yr			Outflow (Discharge), hm^3/yr				
	Rain	Streams	Total	Wetland evapotranspiration	Pumping	To rivers	Total	$\Delta S^{(1)}$
~1950	455	140	595	170	60	365	595	0
~1974	425	120	545	145	160	320	625	−80
~1990	330	115	445	45	540	85	670	−225

$^{(1)}\Delta S$ = water storage change (hm^3/yr)

The cumulated storage depletion due to extraction around 1990 is 3,500 hm^3, of which 1,000 hm^3 are due to natural causes related to the two long dry spells in the 1980s and 1990s (Martínez-Cortina & Cruces, 2005), and the remaining 2,500 hm^3 to the effect of exploitation.

Action undertaken to further try to control non-authorized new wells – a difficult task due to the large area, unstaffed water administration and the delayed creation of operative Groundwater Users Associations – was to give incentives for closing wells. There is some success but it will not solve the short- and mid-term problem. Even stopping all the extractions the system will not recover in more than a decade, and this is socially a highly difficult task. A politically and administratively difficult alternative is declassifying the area as a Ramsar Convention wetland and habilitating an equivalent surface of new protected wetlands in other not far-away areas, but at that moment this will have a high economic and social cost too. As in any other situations, a win-win solution has to be negotiated between social and ecological interests. A further constraint is to comply with the European Water Framework Directive (WFD, 2000), which is compulsory for all member States. It compels meeting good aquifer status by 2015, or at most obtain a negotiated extension to 2021 or 2027. Exceptions due to disproportionate economic and social burden is a highly debated issue (Görlach & Pielen, 2007), since this may imply economic advantages of some European Union countries with respect the others.

Ecological impacts have to be considered with the social component (Deb Roy & Shah, 2003) to put them in real perspective as the man as the final receiver of Nature benefits.

4.4 *Effects on land conditions*

A main result of groundwater head lowering, which means a pore water pressure decrease, is enhanced compaction of soft, recent, unconsolidated sediments. This results in land surface subsidence. Subsidence may be more or less homogeneous over large areas or present conspicuous spatial variations in heterogeneous sedimentary formations. This may produce deep cracks and faults that may disturb roads and railways, and break down pipes and sewers. Subsidence may be a worrying situation that favours flooding in flat areas and coastal plains, mostly recent sedimentary basins and deltas, as in Tokyo, Bangkok and Venice, or the coastline may retreat, as in the Llobregat delta, Catalonia, Spain. Subsidence of several meters is well known in the Central Valley of California, in Arizona, and in areas of Mexico such as the Capital and in Guanajuato State.

In karstified areas, underground cavities may form or be enlarged due to rock dissolution and internal erosion, especially where groundwater flow has been increased. In relatively shallow formations, land collapses may be a landscape feature, such as sinkholes and dolines. The frequency of these events may increase when the water table is lowered. This is well known in Florida, in shallow carbonates (Galloway *et al.*, 2001), and around Zaragoza, Spain, in gypsum-rich sediments.

Insert 3. Economic and social issues of groundwater development.

- Groundwater development involves benefits and costs: direct to the exploiter and indirect to other users and the Society.
- The cost of groundwater in the terrain is not nil; consider opportunity and environmental costs.
- The effects of development may be long delayed; decide how to value the future.
- Groundwater manifestations have social intangible values by themselves.
- Evaluations involve ethical aspects, informed by moral principles.
- Management is needed: governmental and by Society.
- Benefits are in most cases greater than costs.
- Sustainable use of aquifers is possible, generally at a cost less than no action.
- Groundwater is often the cheapest alternative when full cost is considered.
- Groundwater is a key piece for integrated water resources management.
- Groundwater management has to be linked with land use and energy policies.

4.5 *Effects in urban areas*

Groundwater is an important asset for many urban areas. It is often the main or the only source of freshwater. When feasible, wells are inside or around the urban area. Their exploitation may produce important water-table drawdown. Natural recharge may be drastically reduced, but it is usually replaced by water distribution and sewage network losses, and return flows from unsewered areas and green areas and orchards. Often the consequence is groundwater quality degradation, seawater intrusion in coastal areas, and in some cases, important land subsidence. Besides, soil dewatering in formerly shallow water-table areas allows to construct underground facilities (tunnels, building basements, parkings, . . .) in drained soils. Sooner or later these wells are abandoned and moved to peripheral areas or substituted by imported water. Then, the water-table recovers in the cities, even up to a position higher than the initial one. Then serious inundation problems may appear in underground structures, and instability and corrosion in some building basements may occur. Examples can be found in many places, like London, Mar del Plata (Argentina), Milan and Barcelona (Chilton, 1997; 1999). In Barcelona the problem is partly dealt with by pumping groundwater again, for municipal uses and also in some cases for drinking water after costly, advanced treatment.

5 ECONOMIC AND SOCIAL ISSUES

Economic and social issues are important aspects of groundwater development and use, as briefly shown in Insert 3. Problems related to groundwater development are common to any other natural resource development, although with different shades and timing. They have to be evaluated considering costs and benefits, not only from the point of view of an individual developer or a group of developers, but also from a social and environmental point of view. This last aspect of indirect costs has been, and often is neglected, while it is an essential part of the socio-economic analysis.

Often direct benefits and costs are the only ones considered in economic analyses, including taxation and subsidies. Under this point of view, groundwater development problems appear mostly as increased energy costs for pumping due to incremental water level lowering, and also due to early replacement or refitting of pumping machinery, energy transport lines, and wells. This may affect extractors if water is a significant part of production costs, but not so much when water is a small fraction, as for highly valued crops (Figure 3) or water supply to relatively rich urban and tourist areas. There are also direct costs related to water quality changes, mostly salinity increase, and the

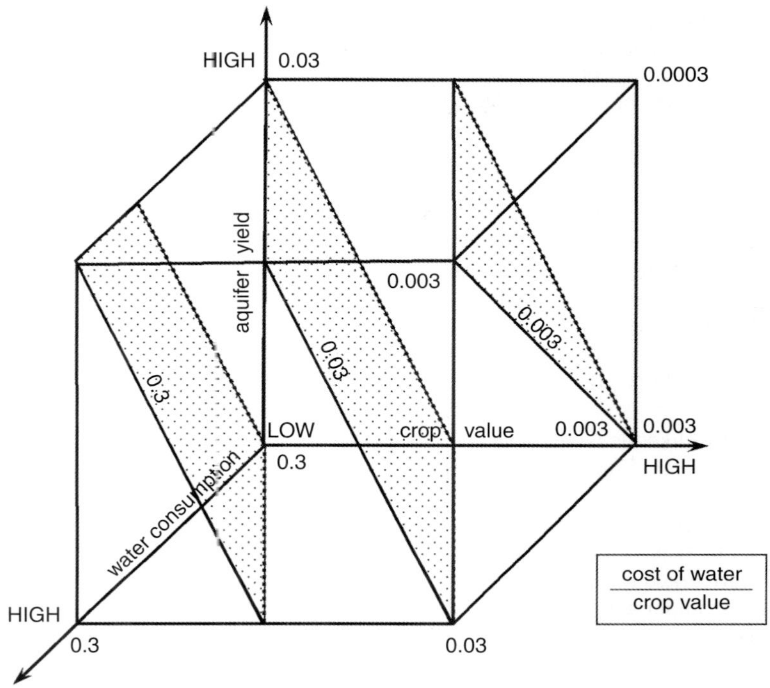

Figure 3. Simplified representation of indicative direct cost of irrigation with groundwater relative to crop value, depending on aquifer yield (low-high), crop water consumption (low-high) and crop value (low-high). The shaded planes are the surface of equal ratio, as indicated.

need for further water treatment to deal with increasing hardness, appearance of soluble iron (Fe) and/or manganese (Mn), arsenic (As), or fluoride (F). They are taken into account as an external burden for producing drinking or industrial water, but may not be taken into account for irrigation uses when crop yields need more water in order to avoid soil salinization, or early problems in irrigation devices appear. Some of these costs show up quite delayed, and this explains why these costs are often non considered in water cost computations. A correct economic evaluation should take them into account.

Also indirect costs and benefits have to be considered from a social point of view. These indirect costs are due to environmental impairment, to the effect of groundwater use on neighbours and downstream surface water users through decreased spring flow, river baseflow, and wetland area, and to land subsidence. If these costs are not internalized, some other must pay for these, or Society as a whole, now or in the future. There are also indirect benefits from exploitation-induced drawdown, such as increased aquifer recharge (if this is not damaging others), decrease of drainage efforts in underground infrastructures (tunnels, cellars, underground parking lots, . . .) and for keeping the water table low in order to avoid soil salinization due to direct evaporation.

As a general rule, water cost increases when groundwater development progresses, with magnitude and relevance depending on the case, although these costs appear gradually. In theory, this should not be a serious drawback since the effects can be anticipated and compensated if aquifer behaviour is taken into account from the beginning through management. Negative aspects are mostly the result of poor understanding and lack of developer awareness, and also poverty. Groundwater is cheaper at the early stages and becomes more expensive when benefits from development increase. Part of these benefits should be reserved to compensate for increased direct costs, and also for indirect costs, currently and in the future. However, this is not an easy task.

It needs institutions and awareness with a mid- to long-term perspective, but it can be done, and this has to be duly and realistically explained and made known; otherwise groundwater detractors have arguments to discourage groundwater use and send doomsday forecasts on aquifer use. Aquifers need protection, conservation, and restoration measures when damaged in their quantity and/or quality. This has a cost, which has to be accepted and supported by users although it is difficult in poor areas, and may need technical, economic and social support from richer areas.

In cost evaluation, the value of water in the aquifer is often considered nil. This is not true since the consequences of groundwater development, including environmental protection and restoration, if not paid by direct users, has to be paid by others or by Society in general. The evaluation of indirect costs – including intangible costs if there is some way to express them in monetary values – is neither an easy task, nor there is a well-established way to compensate Society and Nature. Taxation is a possibility, but how to put aside and apply the collected funds is also a complex issue. This highly depends on local legislation and political orientation, and need popular support.

Some countries like the Netherlands, Denmark, France and parts of Germany have introduced taxes for groundwater abstraction. Abstraction taxes or charges appear as more relevant for quantitative management of groundwater protection, e.g. leakage reduction, and only indirectly for the protection of groundwater quality (Görlach *et al.*, 2007). In the Netherlands levies may include costs involved in bank filtration and artificial recharge. In Germany, revenues from groundwater taxation are earmarked for improvements in municipal sewage treatment.

Timing is an important issue since effects may be long delayed. They may exceed normal political behaviour and even pass from one human generation to the next, and even beyond. This makes economic, social and environmental compensations and transfers a tricky affair. Really, the current generation is supporting the diseconomies from one or two human generations ago, when groundwater intensive extraction began. In the same way, the current generation is affecting the next ones, besides itself.

In the same way that current technology allows to deal with inherited diseconomies and afford increased costs, since the current generation is richer than the preceding ones – although this is not always true worldwide – it may be assumed that the coming generations will be under similar circumstances. But there are limits, both in technology and in the resource itself when water use is intensive. This is a poorly known and inexperienced field. Then, some caution is needed, although not leading to a paralysis that may over-increase the burden to current human generations and consequently affect future ones.

Economists try to consider delayed effects through the discount rate, but there are very different points of view on the long-term discount rate value that should be applied. This discount rate does not coincide with the financial interest and discount rates applied at a given moment by banks. It reflects what people and economists think in a given moment about how the future should be valued (Broome, 2008). A small discount rate favours future generations and gives preference to long-lasting infrastructures and resource conservation. A large discount rate favours the present generation and gives preference to short-term, non-lasting solutions, thus giving more weight to future generations capacity to develop improved technology, in a richer society. But science and technology development needs time and resources (López-Gunn & Llamas, 2008; Quevauviller, 2008), and may have limits.

Existing experience on socio-economic issues and sustainability of aquifer use is short and often poorly documented. Then, extrapolation towards the future is highly uncertain. This future has to be analyzed through different scenarios in which social perceptions of welfare, environment, solidarity in space and time, and ethics play an important role (Custodio, 2000b; 2009; Llamas, 2004; 2009; Llamas & Martínez-Cortina, 2009; Llamas *et al.*, 2009). This perception is also evolving and is different according to social groups and political systems, and also with background religious beliefs that consider humans have the duty of using natural resources properly, to improve living conditions of other persons, and to preserve the environment, but this is not the supreme value since humanity will transcend present life (Benedictus XVI, 2009).

6 A SILENT SOCIAL REVOLUTION FROM THE BOTTOM

Important surface water development generally need large economic investments that can only be carried out by governments, big enterprises or conglomerates of numerous, often rich persons and entities. This is not the most frequent case for groundwater development since it can be afforded at the local level, by individuals or small groups, often farmers, using 20th century technology. Economic resources often come from their savings or through relatively small loans (Llamas *et al.*, 2006).

This makes an enormous difference between surface and ground-water. Surface water development, which often involves large water works, is mostly from the top-down. Surface water is often managed by specialized teams, and users have to be organized into participative bodies in order to have access to water. In many areas of the world surface water is often in the public domain.

Instead, most groundwater developments, especially for irrigation and farming uses, are initiatives from the bottom, by a large number of mostly unrelated developers, often lacking a sufficient aquifer knowledge about the effects they produce and the effects in turn that they receive from other developers. This has to be combined with the circumstance that groundwater often has been, or still is, administratively a private affair, compounded with the difficulty of having to deal with a large number of unorganized owners and right holders. The result has been, and still often is, that public water administration is not interested in these affairs. Circumstances are changing fast in many areas due to already felt effects of intensive groundwater development, mostly on surface water, the environment, and recognized existing water rights (Llamas *et al.*, 2006).

Groundwater development, especially by farmers, but also for small town and rural water supply, has been growing almost exponentially, as already commented, at least in the initial stages. This may be viewed as a social *silent revolution* from the bottom (Fornés *et al.*, 2005; Llamas, 2007; Llamas & Martínez-Santos, 2005; 2006), mostly outside governmental plans and without using public funds. *Silent* refers to the almost imperceptible mode under which groundwater development is produced. *Revolution* reflects a conspicuous change in water availability, and the consequences in population health, and mostly in food production and farmers' income. This implies that developers and users pay the abstracted full direct water cost – with only some exceptions, as the subsidised energy cost in some states of India – and also common taxes. Instead surface water developments often pay only for operation and maintenance costs, sometimes with subsidies – wicked subsidies in many cases (Myers & Kent, 1998) – and tax exemptions, where the amortization costs of investments are not fully considered or even not accounted for at all.

As in most revolutions, some inconvenient issues crop up – perhaps some serious concerns in some areas – that have to be tamed and controlled in order to converge to a sustainable situation. But this sustainable situation is not easy to define, is the subject of different opinions and has to be established for a large area, even at world scale, in a wide social framework (Rogers *et al.*, 2006; Llamas *et al.*, 2007).

7 MANAGEMENT ISSUES AND CHALLENGES

Groundwater is one of the available water resources in a given region, varying from the most important in arid areas, mostly for irrigation, to only a convenient, safe and cheap commodity in wet areas, where it is used mostly for drinking water. They vary from fully renewable to non-renewable fossil reserve, depending on aquifer circumstances, recharge rate, and exploitation intensity. Management is needed to get a sustainable use and to preserve the benefits of its intensive development at the time the negative effects are bounded and compensated. This is a challenge (Insert 4). However, groundwater management is a sectorial aspect inside a wider framework, in which the conspicuous differential characteristics of groundwater have to be considered.

Insert 4. Management issues and challenges for groundwater.

- Groundwater is a kind of *common pool good.*
- Unrestricted use may lead to a *tragedy of the commons* situation.
- Management is needed for quantity and quality sustainability.
- Groundwater should be evaluated in the framework of whole water resources and land use.
- Management should be carried out by the joint action of government authorities, groundwater users and stakeholders, as well as civil society.
- Management means knowledge, monitoring, institutions from the top and from the bottom, legislation, co-responsibility and co-operation.
- Collective bodies appear as efficient tools for aquifer management and governance.

Water management is the set of activities, decisions, rules and means used by institutions that have capacity to carry out their job. It needs enough human and economic means, good knowledge and data, both in space and time, and a framework of accepted and respected rules. The institutions and organizations have to accomplish goals, such as providing water of adequate quantity and quality for human needs, preserving the environment and making sustainable and environmentally admissible the extraction of water resources. They may be from the government and public water administration, or from the stakeholders, and preferably by a well equilibrated and effective combination of them. All this should be placed inside the larger-scope framework of other goals such as land use and energy management, within the restrictions set by social constraints and ethics. Local management in a relatively small territory has to consider the links with other territories, even across political borders. This is provided by more general institutions, up to those contributing a worldwide, wide-scope perspective, such as water and food security, and virtual water trade as a means to compensate for water availability and cost heterogeneities.

In the case of groundwater, the often large number of developers – and also stakeholders – is a management challenge, since the starting point should be to raise a collective conscience that the aquifer is a common asset that needs to be managed. This is often a difficult task that needs many years, perhaps the passing of a full human generation. This cannot be simply imposed from the top since then they may be often rejected at the bottom, thus raising confrontation. Policies and undertakings should be adequately tailored to solve local problems.

Bottom-up pressure for management may be induced by careful action, involving some groundwater developers from the beginning, to act as seeds to drag other developers. This needs a well organized information and education effort at the local level, often with the initial help of government funds. Developers accept self control when they see that their water source – their income – is at risk of becoming a *tragedy of the commons* (Hardin, 1968; Nordhaus, 1994) – the situation in which access to the good is unbounded, thus leading to depletion–, and also when they trust experts and data. Identifying common interests and common threads is a key step to start action, as well as considering local administrative circumstances (see Box 13).

Box 13. Aquifer management: California and Arizona, USA.

Large urban and agricultural areas of California and Arizona (USA) depend on groundwater. They also receive surface water, if available, and imported water. Then, there is integrated water management in which the aquifers play a key role both as a source of water and by providing regulation through their large storage. In some areas the aquifer also provides water to large urban areas and to local supply wells. In some areas of California, drawdown is compensated through artificial recharge, and seawater intrusion is controlled through barriers of injection wells, in some cases with the help of saline water pumping barriers. This is the case of Los Angeles County, Santa Clara Valley and Orange County, through Water Districts (Box 14). This started late in the 1950s. In the 1970s, pilot tests and projects with treated sewage water were initiated. In Orange County injection of highly treated sewage water into the aquifer started in 1977 with the construction of a facility that included reverse osmosis desalinization. The reclaimed water was mixed with groundwater, to be injected into the 23 wells of the Coastal Barrier Project. This facility was the precursor of a recently completed much larger plant which uses only municipal effluent water treated through reverse osmosis to be injected in the seawater intrusion control hydraulic barrier. Also, both Orange County and Los Angeles County use reclaimed and imported water to replenish their aquifers by using water-spreading facilities located in Santa Ana, Rio Hondo and San Gabriel Rivers.

In Arizona, the Water Plan includes attaining sustainable use of aquifers by regulating excess exploitation. The 1980 Groundwater Management Act mandates the condition of safe yield to be attained by the year 2025. In the Phoenix metropolitan area the decline of groundwater of its underlying Salt River Basin aquifer is controlled by limiting the annual pumping volume of each well. Management is carried out by the Salt River Project and the Central Arizona Project through integrated use of local surface water from the reservoirs in the Salt and Verde Rivers and imported Colorado River water. The aquifer provides storage for artificially recharged water in basins by means of deep wells and unsaturated zone wells. Besides, highly treated urban waste water is recharged into the aquifer for later use in agriculture and recreation areas.

Management goals may include actions to preserve aquifers from a quantitative and qualitative point of view, including the damage derived from land use changes, new large constructions, and expanding urban areas. Insert 5 shows some main threads. It is not rare that management means reducing abstraction and integrating other water resources, and even carrying out artificial recharge (Harou & Lund, 2008). Aquifer system management in an integrated water resources framework has been and is being carried out in the Llobregat's Lower Valley and delta, at the southwestern boundary of Barcelona (Catalonia, northeastern Spain) by the Water Authority, with the cooperation of the Groundwater Users Association, as commented in Box 2.

Insert 5. Threads to groundwater sustainable use.

- Contamination from agriculture, livestock, urban and industrial sources.
- Unawareness of contamination risk due to its slow pace, which may translate into highly delayed recognition and policies.
- Contamination may affect large volumes.
- Quality degradation and related surface water degradation. It is often independent of groundwater development, but depends on land use and other human activities.
- Decreased recharge by land use changes and civil and mining dewatering.
- Winning cost becoming too high to sustain uses.
- Costly, very difficult, long-lasting restoration, if feasible at all.
- Poor knowledge and management.
- Poverty.
- Corrupt political behaviour.

The European Water Framework Directive (WFD, 2000) and the daughter Groundwater Directive (GWD, 2006) are key legal pieces for the European Union. Although these Directives are environmentally oriented, they impinge on groundwater quantity and quality management. Members have incorporated them into the country's water legislation, following the subsidiarity principle (Sahuquillo *et al.*, 2009; Molinero *et al.*, 2008). The subsidiarity principle means that public administration has to be carried out at the closest level to people and entities, and exceptionally at higher levels if the lower ones cannot carry out the job.

8 COLLECTIVE BODIES AS AN EFFECTIVE TOOL FOR AQUIFER MANAGEMENT AND GOVERNANCE

Groundwater management deals with numerous developers and stakeholders distributed all over the territory. This poses quite a large difficulty for management organizations whose officials are often not able to master the area, have no access to many sites and need an enormous monitoring effort. This may be carried out more easily and effectively by local people as corresponsible aquifer developers. These developers should have a voice in water management decisions in the Water Authority, share duties and receive funds, or at least a part of what they are paying as general and specific taxes.

Since the numerous stakeholders cannot participate individually, stakeholders need to be effectively represented in the water management body through agreed representatives. This means that users should be associated to be adequate represented. They should carry out their duties with their own economic resources, but also with funds transferred from the government to do specific tasks. Furthermore they should have a technical staff, a financial system, legal support, and a jury to punish deviations. Membership should be compulsory, or at least decisions adopted by a majority should bind other users.

Experience of these collective management bodies of stakeholders – or strictly the groundwater developers – is still small and recent, but there are encouraging examples in some countries such as USA, Spain, Mexico and India, with different contents and structure according to local circumstances. Collective groundwater management (Hernández Mora & Llamas, 2001; Schlager & López-Gunn, 2006) seems a necessary step toward groundwater management and governance, at least until worldwide experience develops. How to carry them out depends highly on local legal frameworks and habits. See Boxes 14 through 16 for experiences under very different circumstances. The existence of groundwater markets, even if imperfect and gray, may be first steps toward management, as in India (Shah, 1993), the Canary Islands and Eastern Spain, once the strict monitoring goals are overcome.

Box 14. Groundwater management institutions: California and Arizona, USA.

In California, landowners have the right to extract as much groundwater as can be put to beneficial use. The State cannot directly manage groundwater according to the California State Water Code. Thus, groundwater management programs have been developed including public (municipalities and counties) and private entity initiatives, in order to solve existing local problems. Agencies, adjudications and districts under special legislation have been created to allow users to manage groundwater inside their boundaries. Currently there are 12 districts. Some counties have a long tradition (Orange since 1933; Monterrey Peninsula since 1947; and Santa Clara Valley since 1951), and others are recent (Lassen since 1993) and still developing. Orange and Santa Clara Valley Water Districts rely on surface water and imported water, and can levy pump taxes to regulate groundwater extraction, but they do not have authority to regulate groundwater extraction by ordinance (WRD, 1996). Tasks vary from agency to agency, from monitoring to limiting extractions, including water imports to alleviate nitrate pollution and seawater intrusion problems, or to agree in exporting groundwater to other areas. After WRD (1996), an average 5 year time is currently needed from the start of the district steering group until the operative district board officially meets.

Orange and Santa Clara Valley Water Districts general policy is to transfer agricultural water to urban areas. They have carried out and operated aquifer recharge facilities in which imported water, and in some areas highly-treated municipal waste waters, are up-graded to drinking water standards, and then stored underground. This is carried out to control and mitigate seawater intrusion problems, and for water storage. The Orange County Groundwater Replenishment System Project, with a cost of US$ 485 million, went on line in January 2008, after almost 30 years of studies, tests and demonstration facilities.

The Los Angeles County Water Replenishment District has a long experience in operating artificial recharge facilities and seawater intrusion barriers since the 1960s, in order to improve conditions in the Central and West Coast basins.

Similar entities exist in other areas, including artificial recharge for temporal storage (the ASR, Aquifer Storage Recovery. An example is the Salt River Project (SRP), Phoenix area, Arizona (Lluria, 2005). Groundwater management by this Water District has been effective in reducing groundwater mining, partly by reducing irrigated area, but also through a better management of local and imported water. This is part of the Arizona Department of Water Resources Plan, which will last until 2025.

The goal of water management activities is to sustain the essential role of aquifers for local water supply and water distribution to users, and for temporal storage, as part of an integrated water resources management plan that, in some cases, may receive external water, as is the case in southern California and Arizona. The cost is rather high, but water demand can generally afford this, given the good economic productivity of water. In the mid- to long-term, management is cheaper than no action.

Box 15. Collective groundwater management: The COTAS of Mexico.

The central region of Mexico has a large population, well established irrigated agriculture, and important industrial developments. Groundwater is crucial for supply in the 192,000 km^2, high altitude, partially closed basin on the Lerma-Chapala river basin, in a semi-arid environment. It extends over five states that include the Federal District of Mexico and the State of Guanajuato to the north. There are about 100 aquifers that are considered overdrafted. The result at the local level is groundwater level drawdown, increasing water exploitation costs that are threatening some important agricultural developments, and land subsidence, as well as decreasing surface area of the large, relatively shallow Chapala Lake.

The *Comisión Nacional del Agua* (CNA, National Water Commission) is the federal government agency responsible for water management. It has been unable to deal effectively with problems. In order to foster better groundwater management at the local level, with more involvement of stakeholders, the CNA promoted and supported civil society organizations called COTAS (*Comités Técnicos de Aguas Subterráneas*; Technical Groundwater Committees) in 1992, with the goal of helping to address local groundwater resource management.

In the case of the State of Guanajuato, groundwater extraction exceeds local and seasonal recharge in 26 out of the 38 significant aquifers. Around 26,000 wells extract about 14 km^3/yr, while recharge is only 13 km^3/yr. Groundwater level depletion rate can be up to 2 to 3 m/yr. Groundwater supplies almost the whole urban and industrial needs, and about 60% of irrigation water demand. In Guanajuato, the COTAS have been developed more in depth. The State Government launched a complementary programme to confront groundwater resources problems. Each individual COTAS was given an office, three staff members, a vehicle, groundwater monitoring equipment and computer facilities, plus technical and juridical support. Total investment during the period 1998–2003 was about US$ 4 million; a second stage is now ongoing. The COTAS governing board is formed exclusively by groundwater users. The operational staff has to implement a yearly agreed work programme with the *Comité Estatal de Aguas de Guanajuato* (Guanajuato State Water Committee). Fund allocation could be retained if a COTAS does not comply with performance indicators. Groundwater resource management must rely on local social agreements as much as possible, to implement adaptive measures based on best-available scientific understanding. The potential activities envisaged by Guanajuato State, to be undertaken by the COTAS are (Foster *et al.*, 2004):

- Capacity building in support of groundwater management plans implementation.
- Promotion of management-related projects to solve specific water problems.

- Support to the federal government in groundwater rights administration.
- Improve public awareness of groundwater management needs.
- Assist well users in administrative requirements and improving efficiency.
- Achieve financial sustainability from members and public and private partners.
- Enhance recharge.
- Indemnify well owners that agree to close down their wells.
- Improving aquifer knowledge.

Some preliminary economic studies have shown that the cost associated to reducing extractions, improving groundwater use efficiency, and enhancing recharge is less that no action. After Foster *et al.* (2004), experience shows that up to the present, the smaller municipalities have shown more willingness in supporting the COTAS than the bigger ones. It is still too early to evaluate results and see if COTAS may survive without basic public funding. A few can be considered successful and many others do not since there is not enough support from stakeholders for what they consider a suspicious public administration initiative. However, the COTAS are a notable leap forward to attain aquifer sustainable use after the problems derived from the early development stages. Their full success needs to also consider groundwater quality, since nitrates are increasing in many areas. However, this has to be developed in the future.

The COTAS are an interesting local solution to address groundwater problems in any case, even when extraction is less than recharge. However, COTAS have to carefully avoid promoting further development or selling technology to users.

Box 16. Groundwater users associations: Experience in Spain.

Eastern and Central Spain is semi-arid but with excellent conditions for human settlements and highly productive agriculture. Communal water regulating infrastructures and rules to share and distribute water flows exist from the Middle Ages, so there is a long tradition for community action. In Valencia, a farmer's jury solves irrigators complaints on the spot; it meets once a week, at least since 700 years ago.

The development of groundwater for irrigation by means of pumped wells started late in the 19th century and mostly in mid the 20th century. In some cases this was to contribute water to irrigators' communities, although the development was often carried out by individuals, or small groups sharing expenses. This led to a rather uncontrolled aquifer development, with scarce public regulation since after the 1879 Water Act this was considered a private affair (Garrido & Llamas, 2009). Groundwater development was, and is, an important boost to water supply, industry, and especially agriculture, with clear private and social benefits, although with the drawbacks associated to intensive development. Something similar happened in the volcanic Canary Islands (Box 6), where the rather expensive groundwater mining prompted the creation of societies by shares to confront costs of construction and for the periodical refitting needed to keep yields.

To try to cope with groundwater intensive development, preserve the benefits and correct drawbacks, groundwater was declared a public domain in the 1985 Water Act. This means that new groundwater developers need a concession from the corresponding Water Authority. Already existing groundwater users were given the option to exchange their rights for a concession, or to remain just as they were. However, for *overexploited* aquifers (after the Water Act terminology) there is the possibility of a formal declaration by the Water Authority, which means that all groundwater users must adapt to norms, reduce extractions, and constitute an association.

Groundwater users associations are publicly recognized entities for collective management of aquifers, with regulations inspired in the tradition of surface water communities. The Water Act encouraged their formation. Currently there are more than 1,400 groundwater users associations registered, and hundreds of others organized as private corporations. But these are mostly for sharing water and manage irrigation networks (Hernández-Mora & Llamas, 2001), and so they cannot be considered as true institutions for the collective management of aquifers, except a few ones, as commented below.

The top-down creation of groundwater users associations (CUAS), even the compulsory ones in the areas declared as *overexploited*, have being largely a failure (Aragonés, 1995). But the good example of the Llobregat's association commented below, and the need to solve problems, has favoured bottom-up initiatives after making known to users the aquifer functioning, the current situation, and the benefits from collective action to try to solve existing problems, as well as the possibility to get public funding to complement their own financial resources. Promotion and assistance have come mostly from the Spanish Association of Groundwater Users, a private entity supported by the already formed CUAS. Some CUAS (Vall d'Uixó, Mancha Oriental, Lomas de Úbeda, Poniente Almeriense), besides those of the Llobregat, have been successful in reducing groundwater abstraction and improving water use efficiency in irrigated fields. Currently 12 groundwater users associations have been formed and are active, or are close to start. A fast expansion is foreseen. Each association is tailored to the specific situation of their area. They are dominated by irrigation interests.

An important success are the CUAS that exist in the small but important Llobregat's Lower Valley and delta aquifer system (Box 2). The detailed hydrogeological studies carried out from the 1960s by the Public Water Administration were make known to groundwater users, who decided to cope with existing and foreseeable problems. As a consequence, in 1975 a CUAS was created (CUADLL), well before groundwater was declared a public domain (Galofré, 2000; Codina, 2004), in an area in which aquifer management was already being carried out. Currently the CUADLL is a public entity supported by their members, with a technical staff capable of carrying out studies and monitoring. Currently it receives funds from the Water Authority by means of contracts and assignments to carry out specific jobs. It has been effective in abstraction control, including water drainage from large underground infrastructures, in controlling sand quarrying, in the virtual cease of waste disposal, and in promoting corrective action, as well as in reducing total groundwater abstraction and protecting associates' groundwater rights.

The success of the CUADLL prompted the creation in 1982 of a CUAS in the Cubeta of Sant Andreu de la Barca, and another in 2008 in the Cubeta of Abrera (Box 2).

A distinctive characteristic of these three CUASs is the dominance of urban supply and industrial groundwater users, while in the others dominate farmers, except in Vall d'Uixó, where there is a mix.

After Hernández-Mora & Llamas (2001), some common keys for the CUAS success are the understanding of the aquifer and the problems, an observation network and easy access to data, the ability to articulate common goals, and the capacity to establish mutually accepted rules regarding resource access and use. A good complement is long-term view of water resources and the environmental implications. The CUAS's area of influence should be large enough to be able to articulate effective solutions to existing problems, although the ability to agree on common goals is increasingly difficult the larger the area is and when more than one administrative or political jurisdiction is involved. The participation of users or stakeholders with economic means and technical know-how (large water suppliers, industrial complexes, important farmers groups) facilitates the creation and effective operation of CUAS, besides the collaboration of the Water Administration technical staff. Trusted leaders that understand the problems and are able to communicate, to organize and motivate others to cooperate are important. Existing social capital and tradition for creating representative civil associations are important too. An important issue is the attitude of the Water Authority towards the users, and the mutual relationships, which are not always easy due to personal attitudes or conflicting goals. A frequent complaint of existing groundwater users is their small weight in the Water Authority governing boards, generally dominated by surface water representatives and officials, except in the Canary Islands, where they dominate.

9 COMMENTS ON CLIMATE AND GLOBAL CHANGE ON GROUNDWATER RESOURCES

Groundwater resources depend essentially on rainfall recharge processes in the upper layers of the land, besides surface water infiltration, which may be increasingly important the drier the area is. Climate, and consequently rainfall and other recharge variables, have been changing along the Earth's history, and more significantly for groundwater, in the last thousands of years, including the recent 16th century to early 18th century small ice age in Europe, although less noticeable in other areas (Mann *et al.*, 2009). Climate will change in the future as a combination of the poorly known natural trend and the anthropic atmospheric changes, which are also uncertain. Besides, there are important land changes such as big forest fires, forest destruction and efforts for afforestation, expansion and reduction of agriculture, expanding urban areas, . . . All of them contribute to global change, whose effects on precipitation recharge mechanisms and stream flow is still not well known. Results of modeling by using established climatic change scenarios predict that recharge will increase in some areas and aquifers, mostly in mid and high latitudes, and will decrease in other areas, mostly at low latitude and around the Mediterranean Sea. Some significant changes seem to show out already in some basins, but the causes are not well defined and probably these are mostly due to land cover and use changes.

Actually, large aquifers contain a large proportion of groundwater recharged in the distant past, some -but not always- in pluvial epochs during the Pleistocene (thousands of years ago). Anthropic effects are important and may have both decreased and enhanced recharge. In the Murray River Basin, in southern Australia, and in other areas of central USA (Scanlon *et al.*, 2009), the transformation of native forest into grassland in mid the 20th century has notably increased local recharge. This is not a blessing because it has accelerated the downward movement of large bodies of climatically-generated saline water in the unsaturated zone toward the underlaying aquifer, thus currently enhancing salinity problems downstream. Something similar happened probably in semiarid areas of Spain in old times (about the 17th century in the Monegros, in the northeastern area) with an impact not recorded, but probably explaining some rather saline springs and streams. The effect last until the saline body is depleted, although this may last decades to centuries.

10 CONCLUSIONS

Groundwater is an important part of the hydrological cycle, with quite different behaviour with respect to other components. In most cases this has clear advantages to obtain fresh groundwater in adequate quantities and with small temporal changes, and also to improve integrated water resources availability and management. But intensive groundwater development has also drawbacks, most of which can be known and anticipated, provided there are enough studies and monitoring, and these can be compensated. One of the most serious future drawbacks is groundwater quality deterioration, and these are only partly due to groundwater development, and depend on land use. Impacts on surface water are also important.

Social and private benefits from groundwater development are in most cases quite large and thus they may compensate drawbacks, except in a few, local situations, in spite of alarming news on groundwater level drawdown, reserves depletion and quality impairment. Many of them are the result of the hydrodynamic evolution in the mid- and long-term, and the associated water cost increase can be progressively dealt with the economic and social benefits obtained from the development, even in cases in which there is groundwater mining. But caution is needed in order to arrive at a new, hydraulical and social sustainable situation. Often groundwater is the cheapest local water resource, and in cases where it is not, it is often due to hidden subsidies or unaccounted costs for other sources.

However, almost unrestricted access to groundwater resources and lack of development rules, carried out by persons and entities not aware of aquifer properties, characteristics, quality constraints, and mutual interrelationships, has often led, and is leading, to problems, sometimes serious ones, but still with large reserves left and environmental damage that can be reversed, at least partly. To deal with this, and save what is a common asset, groundwater management, jointly with the whole water resources, land use and energy management is needed. This takes time, perhaps more than one or two human generations, but is achievable with a combination of public administration efforts and groundwater stakeholder involvement and co-responsibility. This last aspect can be developed through representative water users associations. Experience is still scarce but will probably grow fast, if knowledge is made available, there is political will, civil organizations are involved, and users recognize there is a common asset and heritage to be managed, defended, and make available to future generations, under ethical and moral principles. A mid- and long-term view is always needed, within a time framework which depends on aquifer size, relevance to Society, and environmental values at stake, considering both water quantity and quality issues.

ACKNOWLEDGEMENTS

The author acknowledges the kind invitation of the Marcelino Botin Foundation to prepare this chapter, which derives from a presentation at the Seminar organized in Santander, September 2009, on "Re-Thinking Water and Food Security Paradigms".

Special thanks to M. Ramón Llamas for including this presentation. The author is highly indebted also to M.C. Cabrera, J. Carrillo, J. Fernández Bethencourt, S.S. Foster, L. Konikow, M. Lluria, M. Manzano, L. Martínez-Cortina, E. Queralt, A. Pulido Bosch, M. Rangel Medina, M. Sophocleous, for their useful comments and suggestions on the boxes content, to W. Struckmeier and J. van der Gun for helping in improving the tables, and to A. Sahuquillo for his comments on the full text. Also thanks to E. López-Gunn and L. Martínez-Cortina for language improvement and text upgrading.

REFERENCES

Abdelrham, I.M.; Rashid, K. & Ismail, A. (2008). Integrated management for Great River Project in Libya. *European Journal of Scientific Research*, 22(4): 562–569.

Aragonés, J.M. (1995). Las comunidades de aguas subterráneas y su participación en la gestión. *Las Aguas Subterráneas en la Ley de Aguas Española: Un Decenio de Experiencia*. Asociación Internacional de Hidrogeólogos-Grupo Español. ITGE, Madrid, Spain: 143–160.

Benedictus XVI (2009). *Encyclica Caritas in veritate*. [http://www.vatican.va].

Bocanegra, E.; Hernández, M. & Usunoff, E. (eds.) (2005). *Groundwater and human development*. International Association of Hydrogeologists, Selected Papers on Hydrogeology 6. Heise, Hannover, Germany. 278 pp.

Bocanegra, E.; Cardoso da Silva Jr., G.; Custodio, E.; Manzano, M. & Montenegro, S. (2010). State of knowledge of coastal aquifer management in South America. *Hydrogeology Journal*, 18(1): 261–267.

Bromley, J.; Cruces, J.; Acreman, M.; Martínez-Cortina, L. & Llamas, M.R. (2001). Problems of sustainable groundwater management in an area of over-exploitation: the Upper Guadiana catchment, Central Spain. *Water Resources Development*, 17(3): 379–396.

Broome, J. (2008). Ética del calentamiento climático. *Investigación y Ciencia*, August: 12–17.

Brown, L.R. (2002). *Water deficits growing in many countries. Lakes Directory*. [http://www.greatlakesdirectory.org/080902_water_shortages.htm].

Burke, J.J. & Moench, M. (2000). *Groundwater and society, resources, tensions and opportunities*. Themes in Groundwater Management for the 21th Century. United Nations, New York, USA. 170 pp.

Chilton, J. (ed.) (1997). *Groundwater in the urban environment: Problems, processes and management*. Balkema. Rotterdam, the Netherlands. 682 pp.

Chilton, J. (ed.) (1999). *Groundwater in the urban environment: Selected city profiles*. Balkema. Rotterdam, the Netherlands. 356 pp.

Codina, J. (2004). Las aguas subterráneas: una visión social: el caso de la Comunidad del Delta del Llobregat. *Revista Real Academia Ciencias Exactas, Físicas y Naturales*, 98(2): 323–329. Madrid, Spain.

Collin, J.J. & Margat, J. (1993). Overexploitation of water resources: overreaction or an economic reality? *Hydroplus*, 36: 26–37.

Custodio, E. (2000a). Groundwater-dependent wetlands. *Acta Geologica Hungarica*, 43(2): 173–202.

Custodio, E. (2000b). Some relevant ethical issues in relation to freshwater resources and groundwater. *Boletín Geológico y Minero*, 111(6): 121–130. Madrid, Spain.
Custodio, E. (2001a). Effects of groundwater development on the environment. *Boletín Geológico y Minero*, 111(6): 107–120. Madrid, Spain.
Custodio, E. (2001b). Aguas subterráneas y humedales. *Hidrogeología y Recursos Hidráulicos*, XXIV: 3–30. Madrid, Spain.
Custodio, E. (2002). Aquifer overexploitation, what does it mean? *Hydrogeology Journal*, 10(2): 254–277.
Custodio, E. (2003). Realities and myths of the intensive aquifer development. Aquifer Vulnerability and Risk. Salamanca, Guanajuato, Mexico. Eds: R. Ramírez, M. Civita & M. De Maio. Vol. 2: 207–222.
Custodio, E. (2005a). Coastal aquifers as important natural hydrogeological structures. In: E. Bocanegra, M. Hernández & E. Usunoff (eds.), *Groundwater and Human Development*. Selected Papers on Hydrogeology, 6: 15–38. International Association of Hydrogeologists. Balkema, Lisse, the Netherlands.
Custodio, E. (2005b). Groundwater as a key water resource. *Water Mining and Environment: Book Homage to Rafael Fernández Rubio*. Instituto Geológico y Minero de España, Madrid, Spain: 63–78.
Custodio, E. (2007). Groundwater protection and contamination. In: S. Ragone, A. de la Hera & N. Hernández-Mora (eds.), *The Global Importance of Groundwater in the 21*st *Century*. National Ground Water Association Press, Westerville, Ohio, USA: 219–231.
Custodio, E. (2008). Acuíferos detríticos costeros del litoral mediterráneo peninsular: Valle bajo y delta del Llobregat. *Las Aguas Subterráneas*. Revista Asociación Española Enseñanza de las Ciencias de la Tierra, 15(3): 295–304. Madrid, Spain.
Custodio, E. (2009). Aspectos éticos de la denominada crisis del agua. Implicaciones Éticas de Algunos Debates Científicos (Ed.: M.R. Llamas). Instituto de España: 91–114. Madrid, Spain.
Custodio, E. (2010). Coastal aquifers of Europe: an overview. *Hydrogeology Journal*, 18(1): 269–280.
Custodio, E. & Cabrera, M.C. (2002). ¿Cómo convivir con la escasez de agua? El caso de Canarias. *Boletín Geológico y Minero*, 113(3): 243–258. Madrid, Spain.
Custodio, E. & Cabrera, M.C. (2008). Síntesis de la hidrogeología de las islas Canarias. *Geotemas*, 10: 785–788.
Custodio, E. & Cardoso da Silva Jr., G. (2008). Conceptos básicos sobre o papel ambiental das águas subterrâneas e os efeitos da sua explotação. *Boletín Geológico y Minero*, 119(1): 93–106. Madrid, Spain.
Custodio, E. & Llamas, M.R. (1997). Consideraciones sobre la génesis y evolución de ciertos hidromitos en España. *En defensa de la Libertad, Homenaje a Víctor Mendoza*. Instituto de Estudios Económicos, Madrid, Spain: 167–179.
Custodio, E.; Kretsinger, V. & Llamas, M.R. (2005). Intensive development of groundwater: concept, facts and suggestions. *Water Policy*, 7: 151–162.
Custodio, E.; Llamas, M.R.; Hernández-Mora, N.; Martínez-Cortina, L. & Martínez-Santos, P. (2009a). Issues related to intensive groundwater use. In: A. Garrido & M.R. Llamas, *Water Policy in Spain*. CRC Press: 145–162.
Custodio, E.; Manzano, M. & Montes, C. (2009b). *El agua subterránea en Doñana: aspectos ecológicos y sociales*. Agencia Andaluza del Agua. Junta de Andalucía. Sevilla, Spain. 330 pp.
de Vries, J.J. (2007). History of groundwater hydrology. In: J.W. Delleur (ed.), *The Handbook of Groundwater Engineering*, 2nd edition. CRC Press. 39 pp.
Deb Roy, A. & Shah, T. (2003). Socio-ecology of groundwater irrigation in India. In: M.R. Llamas & E. Custodio (eds.), *Intensive Use of Groundwater. Challenges and Opportunities*. Balkema, Swets & Zeitlinger, Lisse, the Netherlands: 307–336.
Dennehy, K.F. (2000). *High Plains regional ground-water study*. U.S. Geological Survey, Fact Sheet FS-091-00. [http://co.water.usgs.gov/nawqa/hpgw/factsheets/DENNEHYFS1.html].
Eaton, L.; Melady, J. & Tolan, T. (2009). Successful implementation of ASR in basalt-hosted aquifers in the Pacific Northwest of the United States. *Boletín Geológico y Minero*, 120(2): 131–156. Madrid, Spain.
Fornés, J.M.; de la Hera, A. & Llamas, M.R. (2005). The silent revolution of groundwater intensive use and its influence in Spain. *Water Policy*, 7(3): 253–268.
Foster, S.S. & Loucks, P. (eds.) (2006). *Non-renewable groundwater resources: a guidebook on socially-sustainable management for water policy-makers*. UNESCO-IHP. Paris, France.
Foster, S.S.; Garduño, H. & Kemper, K. (2004). *Mexico – The "Cotas": progress with stakeholder participation in groundwater management in Guanajuato*. Sustainable Groundwater Management: Lessons from Practice, Case Profile Collection no. 10. GW–MATE, the World Bank, Washington D.C., USA. 16 pp.
Foster, S.S.; Garduño, H.; Kemper, K.; Tuinhof, A. & Nanni, M. (2005). Making intensive use of groundwater more sustainable – Key lessons from field experience. In: A. Sahuquillo, J. Capilla, L. Martínez-Cortina &

X. Sánchez-Vila (eds.), *Groundwater Intensive Use*. International Association of Hydrogeologists, Selected Papers on Hydrogeology, 7. Balkema, Leiden, the Netherlands: 105–126.

Galloway, D.; Jones, O.R. & Ingebritsen, S.E. (2001). *Land subsidence in the United States*. U.S. Geological Survey Circular, 1182. [http://water.usgs.gov/pubs/circ/circ1182].

Galofré, A. (2000). Las comunidades de usuarios de aguas subterráneas en la Cataluña del 2000: desarrollo histórico y realidad actual. In: N. Hernández-Mora & M.R. Llamas (eds.), *La Economía del Agua Subterránea y su Gestión Colectiva*. Ediciones Mundi Prensa – Fundación Marcelino Botín, Madrid, Spain: 323–366.

Garrido, A. & Llamas, M.R. (2009). *Water policy in Spain*. CRC Press/Balkema, Leiden, the Netherlands. 254 pp.

Gijsber, P.J.A. & Loucks, D.P. (1999). Libya's choices: Desalination or the Great Man-Made River Project. *Physics and Chemistry of the Earth, B, Hydrology, Oceans and Atmosphere*, 24(4): 285–389.

Görlach, B. & Pielen, B. (2007). *Disproportionate costs in the EC Water Framework Directive: The concept and its practical implementation*. Applied Economics Conference, London, UK.

Görlach, B.; Interwies, E.; Newcombe, J. & Johns, H. (2007). Assessing the cost-effectiveness of environmental policies in Europe. In: C. George & C. Kirkpatrick (eds.), *Impact Assessment and Sustainable Development: European Practice and Experience*. Edward Elgar Publishing, Chapter 11: 185–214.

GWD (2006). *Directive 2000/118/EC of the European Parliament and of the Council: Groundwater Directive*. Brussels, Belgium.

Hardin, G. (1968). The tragedy of the commons. *Science*, 162: 1243–1248.

Harou, J.J. & Lund, J.R. (2008). Ending groundwater overdraft in hydrologic-economic systems. *Hydrogeology Journal*, 16(6): 1039–1055.

Hering, J.G.; O'Day, P.A.; Ford, R.G.; He, Y.T.; Bilgin, A.; Reiniger, H.J. & Burns, O.H. (2009). MNA as a remedy for arsenic mobilized by anthropogenic inputs of organic carbon. *Ground Water Monitoring and Remediation*, 20(3): 84–92.

Hernández-Mora, N. & Llamas, M.R. (eds.) (2001). *La economía del agua subterránea y su gestión colectiva*. Fundación Marcelino Botin – Ediciones Mundi Prensa. Madrid, Spain. 550 pp.

Hernández-Mora, N.; Llamas, M.R. & Martínez–Cortina, L. (2001). Misconceptions in aquifer over-exploitation: implication for water policy in Southern Europe. In: C. Dosi (ed.), *Agricultural Use of Groundwater*. Kluwer, Dordrecht, the Netherlands: 107–125.

IGRAC (2009). Transboundary aquifers of the World. Map 1:50 M. For the 5th World Water Forum, Istambul, Turkey. International Groundwater Resources Assessment Centre (IGRAC), Utrecht, the Netherlands.

Kemper, K.; Mestre, E. & Amore, L. (2003). Management of the Guaraní aquifer system: moving towards the future. *Water International*, 28(2): 185–200.

Llamas, M.R. (1983). Las aguas subterráneas de Madrid y la política hidráulica española. *Estudios Territoriales*, 10: 113–130. Madrid, Spain.

Llamas, M.R. (2004). *Water and ethics: use of groundwater*. UNESCO-IHP/CONIST. Series on Water and Ethics, Essay 7. UNESCO Press, Paris, France. 33 pp.

Llamas, M.R. (2007). Aguas subterráneas: de la revolución silenciosa a los conflictos clamorosos. *Revista Real Academia Ciencias Exactas, Físicas y Naturales*, 101(1): 175–181. Madrid, Spain.

Llamas, M.R. (2009). Aspectos éticos de la gestión del agua en España. Revista Real Academia Ciencias Exactas, Físicas y Naturales, 102(1): 161–184. Madrid, Spain.

Llamas, M.R. & Custodio, E. (eds.) (2003). *Intensive use of groundwater: challenges and opportunities*. Balkema, Lisse, the Netherlands. 478 pp.

Llamas, M.R. & Garrido, A. (2007). Lessons from intensive groundwater use in Spain: economic and social benefits and conflicts. In: M. Giordano & K.G. Villholth (eds.), *The Agricultural Groundwater Revolution: Opportunities and Threats to Development*. CAB International, Wallingford, UK: 266–295.

Llamas, M.R. & Martínez-Cortina, L. (2009). Specific aspects of groundwater use in Water Ethics. In: M.R. Llamas, L. Martínez-Cortina & A. Mukherji (eds.), *Water Ethics*. Marcelino Botin Water Forum 2007. Taylor & Francis Group, London, UK: 187–203.

Llamas, M.R. & Martínez-Santos, P. (2005). Intensive groundwater use: silent revolution and potential sources of social conflicts. American Society of Civil Engineers (ASCE). *Journal of Water Resources Planning and Management*, 13(5): 337–341.

Llamas, M.R. & Martínez-Santos, P. (2006). Significance of the silent revolution of intensive groundwater use in world water policy. In: P.P. Rogers, M.R. Llamas & L. Martínez-Cortina (eds.), *Water Crisis: Myth or Reality?* Chapter 10: 163–180. Taylor & Francis, London, UK.

Llamas, M.R.; Mukherji, A. & Shah, T. (eds.) (2006). Social and economic aspects of groundwater governance. *Hydrogeology Journal*, 14(3): 269–432.

Llamas, M.R.; Martínez-Santos, P. & de la Hera, A. (2007). The manifold dimensions of groundwater sustainability. In: S. Ragone, A. de la Hera & N. Hernández-Mora (eds.), *The Global Importance of Groundwater in the 21st Century*. National Ground Water Association Press, Westerville, Ohio, USA: 105–116.

Llamas, M.R.; Martínez-Cortina, L. & Mukherji, A. (2009). *Water Ethics*. Marcelino Botin Water Forum 2007. Taylor & Francis Group, London, UK. 353 pp.

Lluria, M.R. (2005). Effective management of water resources in a region of intensive groundwater use: a lesson from Arizona's Valley of the Sun. In: A. Sahuquillo, J. Capilla, L. Martínez-Cortina & X. Sánchez-Vila (eds.), *Groundwater Intensive Use*. International Association of Hydrogeologists, Selected Papers on Hydrogeology, 7: 353–360. Balkema, Leiden, the Netherlands.

López-Gunn, E. & Llamas, M.R. (2008). Re-thinking water scarcity: can science and technology solve the global water crisis? *Natural Resources Forum*, 32: 228–238.

Loucks, D.P. (2004). *The Great-Man Made River in Lybia: does it make sense?* Department of Civil and Environment Engineering, Cornell University, UK.

Mann, M.E.; Zhang, Z.; Rutherford, S. *et al.* (2009). Global signatures and dynamical origins of the Little Ice Age and Medieval climate anomaly. *Science*, 326: 1256–1260.

Manzano, M.; Custodio, E.; Higueras, H. & Soler, A. (2009a). Influencia de la gestión del acuífero sobre los humedales del manto eólico de Doñana. *Boletín Geológico y Minero*, 120(3): 377–392. Madrid, Spain.

Manzano, M.; Custodio, E.; Montes, C. & Mediavilla, C. (2009b). Groundwater quality and quantity assessment through a dedicated monitoring network: the Doñana aquifer experience (SW Spain). In: Ph. Quevauviller, A-M. Fouillac, J. Grath & R.Ward (eds.), *Groundwater Monitoring*. Wiley: 273–287.

Margat, J. (2008). *Les eaux souterraines dans le monde*. BRGM Éditions/UNESCO-PHI, Paris, France. 187 pp.

Martínez-Cortina, L. (2003). Marco hidrológico de la cuenca alta del Guadiana. In: C. Coleto, L. Martínez-Cortina & M.R. Llamas (eds.), *Conflictos entre el Desarrollo de las Aguas Subterráneas y la Conservación de los Humedales: la Cuenca Alta del Guadiana*. Fundación Marcelino Botin – Ediciones Mundi-Prensa, Madrid, Spain: 3–68.

Martínez-Cortina, L. & Cruces, J. (2005). The analysis of the intensive use of groundwater in the Upper Guadiana Basin (Spain) using a numerical model. In: A. Sahuquillo, J. Capilla, L. Martínez-Cortina & X. Sánchez-Vila (eds.), *Groundwater Intensive Use*. International Association of Hydrogeologists, Selected Papers on Hydrogeology, 7. Balkema, Leiden, the Netherlands: 285–294.

McGuire, V.L. (2007). *Changes in water levels and storage in the High Plains aquifer: predevelopment to 2005*. U.S. Geological Survey, Washington D.C., USA. [http://pubs.usgs.gov/fs/2007/3029].

MIMAM (2000). *Libro Blanco del Agua en España* [White Book of Water in Spain]. Ministerio de Medio Ambiente. Madrid, Spain. 638 pp.

Molina, J.L.; García-Aróstegui, J.L.; Bromley, J.; Molina, M. & Benavente, J. (2009). Apoyo a la gestión de los recursos hídricos subterráneos mediante sistemas soporte de decisión basados en métodos probabilísticos: aplicación al Altiplano Murciano (España). In: L. Vives, E. Custodio & L. Rodríguez (eds.), *Planificación y Gestión de Aguas Subterráneas*. Asociación Internacional de Hidrogeólogos-Grupo Argentino. Facultad Ciencias Exactas y Naturales, Santa Rosa, Argentina: 133–142.

Molinero, J.; Custodio, E.; Sahuquillo, A. & Llamas, M.R. (2008). *DMA y la gestión del agua subterránea en España*. VI Congreso Ibérico sobre Gestión y Planificación del Agua: Los Nuevos Planes de Gestión de Cuencas, Una Oportunidad para la Recuperación de los Ciclos del Agua. Fundación Nueva Cultura del Agua, Madrid-Zaragoza, Spain. Sesión Paralela V: 9 pp.

Moreno Vázquez, J.L. (2006). *Por bajo del agua: sobreexplotación y agotamiento del acuífero de la Costa de Hermosillo, 1945-2005*. El Colegio de México, Mexico. 507 pp.

Mukherji, A. (2006). Is intensive use of groundwater a solution to world's water crisis? In: P.P. Rogers, M.R. Llamas & L. Martínez-Cortina (eds.), *Water Crisis: Myth or Reality?* Taylor & Francis, London, UK. Chapter 11: 181–193.

Myers, N. & Kent, J. (1998). *Perverse subsidies: their nature, scale and impacts*. International Institute for Sustainable Development, Winnipeg, Canada. 210 pp.

Niñerola, J.M.; Queralt, E. & Custodio, E. (2009). Llobregat delta aquifer. In: Ph. Quevauviller, A-M. Fouillac, J. Grath & R.Ward (eds.), *Groundwater Monitoring*. Wiley: 289– 301.

Nordhaus, W.D. (1994). *Managing the global commons: the economics of climate change*. MIT Press.

NWRA (2009). *Sana'a groundwater resources*. The Chairman of the National Water Resources Authority, Yemen. [http://www.un.org/esa/dsd/dsd_aofw_wat_pdfs/meetings/ws0109/z_yemen_salem.pdf].

Quevauviller, P. (ed.) (2008). *Groundwater science and policy: an international overview*. RSC Publ., Cambridge, UK. 754 pp.

Ragone, S.; de la Hera, A. & Hernández-Mora, N. (2007). *The global importance of groundwater in the 21st Century*. Proceedings of the International Symposium on Groundwater Sustainability. Alicante, Spain, 2006. National Ground Water Association Press, Westerville, Ohio, USA. 382 pp.

Rangel-Medina, M.; Monreal, R.; Morales, M.; Castillo, J. & Valenzuela, H. (2005). Importance of hydrogeological modeling in the management of groundwater, a case study: the coast of Hermosillo aquifer, Sonora. Mexico. In: A. Sahuquillo, J. Capilla, L. Martínez-Cortina & X. Sánchez-Vila (eds.), *Groundwater Intensive Use*. International Association of Hydrogeologists, Selected Papers on Hydrogeology, 7. Balkema, Leiden, the Netherlands: 277–283.

Rebouças, A. da C. & Amore, L. (2002). O sistema aqüífero Guaraní-SAG. *Revista Águas Subtérrâneas*, 16: 103–110. Curitiba, Brazil.

Rogers, P.P.; Llamas, M.R. & Martínez-Cortina, L. (2006). *Water crisis: myth or reality?* Taylor & Francis, London, UK. 331 pp.

Sahuquillo, A.; Capilla, J.; Martínez-Cortina, L. & Sánchez-Vila, X. (eds.) (2004). *Groundwater intensive use*. International Association of Hydrogeologists, Selected Papers on Hydrogeology, 7. Balkema, Leiden, the Netherlands. 393 pp.

Sahuquillo, A.; Custodio, E. & Llamas, M.R. (2009). La gestión de las aguas subterráneas. *Tecnología del Agua*, 305: 60–67; 306: 54–67.

Salem, O.M. (2005a). Aquifer overexploitation in Libya: the Gefara plain case. In: A. Sahuquillo, J. Capilla, L. Martínez-Cortina & X. Sánchez-Vila (eds.), *Groundwater Intensive Use*. International Association of Hydrogeologists, Selected Papers on Hydrogeology, 7. Balkema, Leiden, the Netherlands: 127–136.

Salem, O.M. (2005b). *Lybia after the Great Man-Made River Project*. College of Engineering, Al-Fatah University of Tripoli, Lybia.

Salem, O.M. & Pallas, P. (2004). The Nubian Sandstone Aquifer System (NSAS). In: B. Appelgren (ed.), *Managing shared aquifer resources in Africa*. IHP-VI, Series on Groundwater No. 8. Proceedings International Workshop. 2–4 June 2002, Tripoli, Lybia: 19–21.

Scanlon, B.R.; Stonestrom, D.A.; Reedy, R.C.; Leaney, F.W.; Gates, J. & Gresswelt, R.G. (2009). Inventories and mobilization of unsaturated zone sulphate, fluoride, and chloride related to land use change in semiarid regions, Southwestern United States and Australia. *Water Resources Research*, 45. W00A18, DOI: 10-1029/2008WR006963.

Schlager, E. & López-Gunn, E. (2006). Collective systems for water management: is the Tragedy of the Commons a myth? In: P.P. Rogers, M.R. Llamas & L. Martínez-Cortina (eds.), *Water Crisis: Myth or Reality?* Chapter 3: 43–58. Taylor & Francis, London, UK.

Shah, T. (1993). *Groundwater markets and irrigation development: political economy and practical policy*. Oxford University Press, Bombay, India.

Shah, T. (2005). Groundwater and human development: challenges and opportunities in livelihoods and environment. *Water, Science and Technology*, 51(8): 27–37.

Shah, T. *et al.* (2000). *The global groundwater situation: overview of opportunities and challenges*. International Water Management Institute, Colombo, Sri Lanka.

Shah, T.; Deb Roy, A.; Qureshi, A.S. & Wang, J. (2003). Sustaining Asia's groundwater boom: an overview of issues and evidence. *Natural Resources Forum*, 27: 130–141.

Shah, T.; Burke, J. & Villholth, K.G. (2007). Groundwater: a global assessment of scale and significance. In: D. Molden (ed.), *Water for Food, Water for Life*. Chapter 10: 395–423. International Water Management Institute (IWMI). Earthscan, London, UK.

Sophocleous, M. (2002). Interactions between groundwater and surface water: the state of the science. *Hydrogeology Journal*, 10: 52–67.

Sophocleous, M. (2010). Review: Groundwater management practices, challenges, and innovations in the High Plains aquifer, USA-lessons and recommended actions. *Hydrogeology Journal*, 18(3): 559–575.

Steinich, B.; Escolero, O. & Marin, L.E. (1998). Salt-water intrusion and nitrate contamination in the Valley of Hermosillo and El Sahuaral. *Hydrogeology Journal*, 6(4): 518–526.

Stoner, R.F. & Bakiewicz, E. (1993). Scavenger wells: 1.– Historic development. *Study and modelling of saltwater intrusion into aquifers* (Eds.: E. Custodio & A. Galofré). CIMNE, Barcelona. 545–556.

Szynkiewicz, A.; Rangel Medina, M.; Modelska, M.; Monreal, R. & Pratt, L.M. (2008). Sulfur isotopic study of sulfate in the aquifer of Costa de Hermosillo (Sonora, Mexico) in relation to upward intrusion of saline groundwater, irrigation dumping and land cultivation. *Applied Geochemistry*, 23(9): 2539–2588.

Tujchneider, O.; Pérez, M.; Paris, M.; D'Elia, M. (2007). The Guaraní aquifer system: state-of-the-art in Argentina. *Aquifer Systems Management: Darcy's Legacy on Impending Water Shortage*. International Association of Hydrogeologists, Selected Papers on Hydrogeology, 10. Chapter 10: 239–252. Balkema, Taylor & Francis.

UNESCO (2007). *Sistemas acuíferos transfronterizos en las Américas: evaluación preliminar*. ISARM–Americas. UNESCO-PHI/OEA. Paris, France.
WFD (2000). Directive 2000/60/EC of the European Parliament and of the Council establishing a framework for Community action in the field of water policy. Water Framework Directive, Brussels, Belgium.
WHYMAP (2004). Groundwater resources of the world 1:50,000 (Eds.: W.F. Struckmaier, W.H. Gilbrich, A. Richts & M. Zaepke). UNESCO–BGR–CGMW–IAH–IAEA. BGR, Hannover–UNESCO, Paris. Special edition for the 32nd International Geological Congress, Florence, Italy.
WHYMAP (2006). Groundwater resources of the world 1:50,000 (Eds.: W.F. Struckmaier, W.H. Gilbrich, J.S.D. Gun, T. Maurer, S. Puri, A. Richts, P. Winter & M. Zaepke). UNESCO–BGR–CGMW–IAH–IAEA–IGRAC–GRDC. BGR, Hannover–UNESCO, Paris. Special edition for the 4th World Water Forum, Mexico.
WRD (1996). *Water Facts*. Department of Water Resources, Resources Agency, State of California, report 4, January 1996: 1–5.
Zidan, A.M. (2007). The impact of the Great Man-Made river project on Lybia's agriculture activities and the environment. M.S. Master Thesis, Univ. Teknology Malaysia. [http://prints.utm.my/6377/1/AdelMohamedZidanMFAB2007TTT.pdf].

CHAPTER 15

Electricity reforms and its impact on groundwater use: Evidence from India

Aditi Mukherji
International Water Management Institute (IWMI), Colombo, Sri Lanka

Tushaar Shah
International Water Management Institute (IWMI), Anand, India

Shilp Verma
UNESCO-IHE, Delft, the Netherlands

ABSTRACT: Minimizing the negative impacts of groundwater over-exploitation, while preserving the benefits from such intensive use has emerged as the key natural resources management challenge in South Asia. Direct regulation of groundwater is not a feasible option in the region given over 20 million pumps and the huge transactions costs involved. In this context, indirect mechanism, such as regulation of electricity supply and changes in electricity pricing and subsidies can provide an effective tool for governing groundwater. This chapter documents two such cases of electricity reforms that have had profound impact on groundwater use in Indian states of Gujarat and West Bengal.

Keywords: groundwater, electricity, reforms, subsidies, Gujarat, West Bengal

1 INTRODUCTION

The electricity and groundwater sector in India is intricately linked by the fact that over 50% of pumps and tubewells in the country are electrically operated and electricity pricing affects farmers pumping behavior. Until the early 1970s, all state electricity boards (SEBs) charged their tubewell owners based on metered consumption and this was later changed to flat tariff in the 1980s, mostly on account of ease of administering flat tariff. The original intention was to keep increasing this flat tariff to reflect increase in cost of electricity generation. However, over time flat tariff rates became tool of political appeasement and remained perpetually low. The SEBs started making huge losses, both on account of low recovery from agriculture, but also due to their own inefficiencies. Low flat tariffs also led to over-exploitation of groundwater in arid and semi-arid states of India. Therefore, recently, there is a renewed interest in reforming the electricity sector and this has profound implications on the groundwater sector.

Given both water resources and electricity are state subjects in India, individual states have chosen to go about differently *vis-à-vis* power sector reforms, keeping in mind the political exigencies faced by these states. The state of West Bengal in the east has embarked upon a path of universal metering of agricultural electricity consumers, mainly because there are no strong farm lobbies in the state to oppose such a move. This shift from flat rate tariff (signifying zero marginal cost of pumping) to pro-rata tariff, altered the cost and incentive structure of the pump owners and hence affected their pumping behavior. On the other hand, the Government of Gujarat, in face of strident farmer opposition, decided not to meter tubewells, but instead separated agricultural feeders from non agricultural ones. They also improved the quality of power supply and rationed the number of hours of electricity to agriculture for only 8 hours in a day, thereby influencing farmer pumping behavior. This initiative of the Government of Gujarat is called the *Jyotirgram Yojana*.

The purpose of this chapter is to document the reforms and present a first cut analysis of the impact of these reforms on the pumping behavior of pump owners, on the informal groundwater markets through which water buyers would be impacted. The chapter is divided into five sections. After the first introductory section, the second section describes the groundwater and electricity situation in the two states. The third section discusses the process and implementation of the power sector reforms in each of these states, while the fourth section analyses the impact of these on groundwater use and on groundwater markets. The final section presents the conclusions and policy implications of this study.

2 GROUNDWATER AND ELECTRICITY SITUATION IN WEST BENGAL AND GUJARAT

West Bengal, an eastern state of India receives an annual rainfall of around 2,000 mm and has a groundwater potential of 31,000 Mm^3, most of which is available at shallow depths. Only 42% of the total available groundwater resources in the state has been utilized so far (WIDD, 2004). While West Bengal has plentiful groundwater resources that can be furthered developed, the state has for various political reasons (Mukherji, 2006) adopted one of the most stringent groundwater regulations in India. For instance, procuring electricity connection for tubewells needs permission from multiples sources, such as the State Water Investigation Directorate (SWID), village level bodies (*panchayats*), and the process is fraught with red-tape and corruption. The result is that West Bengal has the lowest proportion of electric tubewells to total tubewells in India (GOI, 2003). The farmers in West Bengal, till 2007, also paid the highest flat tariff (Rs. 2,160/HP/yr) [US$ 48/HP/yr][1] for electricity among all Indian states. Agricultural consumption of electricity accounted for only 6.1% of total electricity consumption (WBSEB, 2006), and unlike other states where electricity subsidy forms a major share of state fiscal deficits, in West Bengal this was negligible (Briscoe, 2005). Existence of very high flat tariff, coupled with small sized land holdings and abundant groundwater resources had led to the emergence of competitive informal groundwater markets, and small and marginal water buying farmers benefitted substantially through these markets. The main irrigated crop is the summer paddy, called *boro paddy*. Average annual pumping hour varies from 1,500 hours to 2,100 hours for centrifugal and submersible pumps respectively.

Gujarat, a western state of India, receives an average annual rainfall of 1,243 mm, though with wide regional variations. South Gujarat receives the bulk of the rainfall, while western parts of the states (Saurashtra and Kutch) are distinctly arid. The state has an annual replenishable groundwater potential of 15,810 Mm^3 of which 76% (11,490 Mm^3) is withdrawn every year. This is a state where groundwater is used intensively and 61% of the administrative blocks are over-exploited, critical or semi-critical as per the norms of the Central Groundwater Board (CGWB) [http://cgwb.gov.in/gw_profiles/st_Gujarat.htm]. North Gujarat, which on an average receives 500–700 mm/yr of rainfall, has deep alluvial aquifers and is a basket case of unsustainable use of groundwater. In many ways, the state of Gujarat epitomizes the groundwater crisis in India. Yet, the state has been registering an agricultural growth rate of 10% for the last 7–8 years and this surpasses that of other states better endowed with water resources (Gulati *et al.*, 2009). Here, farmers have also increasingly moved away from cereal crops to high value crops, such as Bt cotton, tobacco, dairy, orchard and commercial crops, so as to maximize value per drop of water. Gujarat also has strong farmer lobbies that have time and again successfully thwarted any attempt to curtail their access to groundwater (Mukherji, 2006). Gujarat, till the recent reforms, had one of the highest electricity subsidies in India. Given the heavy losses sustained by the state electricity board, there was a rapid deterioration of the quality of power supply in the state thereby negatively affecting the quality of life in rural areas. Gujarat, like West Bengal, also supports a vibrant groundwater market. Indeed, groundwater markets in Gujarat predate that of other regions in India (Shah, 1993). Figure 1 shows the location of these two states.

[1] 1 Rs (Indian Rupees) = 0.0221813 US$ (2010, May, 11th).

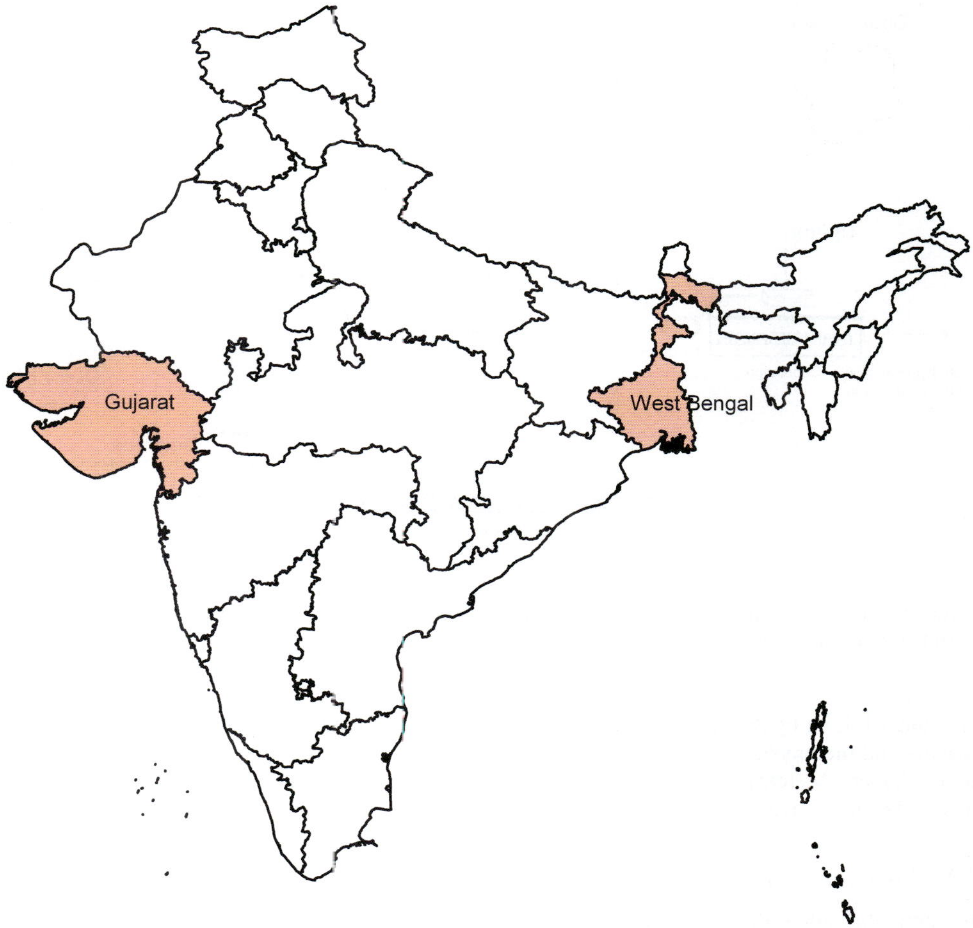

Figure 1. Location of study states.

3 THE PROCESS OF ELECTRICITY SECTOR REFORMS IN WEST BENGAL AND GUJARAT

3.1 *Metering in West Bengal*

The Government of West Bengal (GoWB) has adopted a hi-tech approach to metering through the installation of remotely sensed tamper-proof meters which operate on the *Time of the Day* (TOD) principle. TOD is a demand management tool whereby, by differentiating the cost of electricity during different times of the day, consumers are discouraged from using pumps during peak evening hours while they are encouraged to do the same during slack night hours. There are three metered tariff rates, namely, normal rates from 6 a.m. to 5 p.m. (Rs. 1.37/kW·h) [US$ 0.03/kW·h][1], peak rates from 5 p.m. to 11 p.m. (Rs. 4.75/kW·h) [US$ 0.11/kW·h][1] and off-peak rates from 11 p.m. to 6 a.m. (Rs. 0.75/kW·h) [less than US$ 0.02/kW·h][1]. On average these unit rates translates to around Rs. 6/hour [US$ 0.13/hour][1] inclusive of Rs. 22/month [US$ 0.49/month][1] as meter rent. The new meters use GIS and GSM technologies and are remotely read (Figure 2). These new meters solve many of the traditional problems of metering, namely, tampering, under-reporting

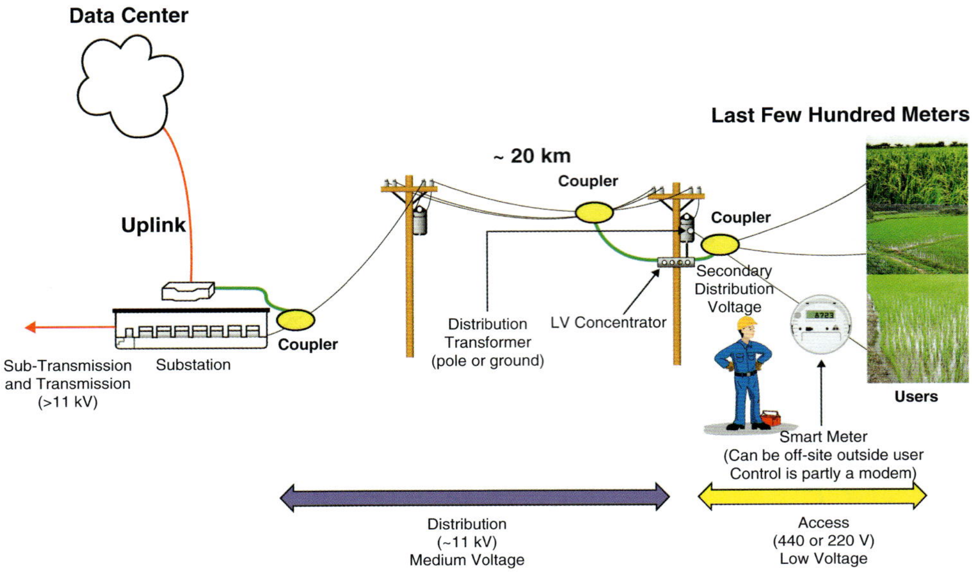

Figure 2. A schematic diagram of a generic IT Power Distribution System that is being used in West Bengal (adapted from Tongia, 2004).

and under-billing by the meter readers in collusion with the villagers, arbitrary power of the meter readers and the physical abuse that the meter readers were subject to at times at the hands of the irate villagers. Meters are now remotely read and reading is transmitted directly to the commercial office. The meter reader neither knows, nor can tamper with the meter reading.

3.2 *Jyotirgram Yojana in Gujarat*

In September 2003, the Government of Gujarat (GoG) pioneered a bold scheme –the *Jyotirgram Scheme* (JGS)– to separate agricultural feeders from non-agricultural ones. JGS was launched initially in 8 districts of Gujarat on pilot basis. The early results were so encouraging that by 2004, the scheme was extended to the entire state. By 2006, over 90% of Gujarat's 18,000 villages were covered under JGS. This involved total rewiring of rural Gujarat. 48,852 km of high tension lines and 7,119 km of low tension wires were added. 12,621 new transformer centres were installed. 1.2 million new electricity poles were used. 1,470 specially designed transformers were installed. 182,000 km of electricity conductors and 610,000 km of low tension PVC cables were used. 30,000 tonnes [1 tonne = 1,000 kg] of steel products were used. In short, under the JGS, the Gujarat Electricity Board (GEB) laid a parallel rural transmission network across the state at an investment of Rs. 11,700 million [US$ 260 million][1]. Feeders supplying agricultural connection were bifurcated from the supply to commercial and residential connection at sub-station level. Meters on distribution transformer centres were also installed on feeders to improve the accuracy for energy accounting.

Pre-JGS, at the lowest level, 11 kV feeders served a group of 2–5 villages wherein all connections in these villages (domestic, agricultural, as well as commercial) were through this feeder (see Figure 3a). Post-JGS however, the feeders were bifurcated into agricultural and non-agricultural feeders (Figure 3b). This meant that certain feeders only served farm consumers and connections, while the rest served the domestic and commercial customers. Meters were installed on each feeder, especially the agri-feeders to identify the source of any *significantly-greater-than-expected* demand at any particular feeder. Rural Gujarat thus rewired, the government put into place a new rural electricity regime that provided high quality, predictable, reliable, but *rationed* power supply

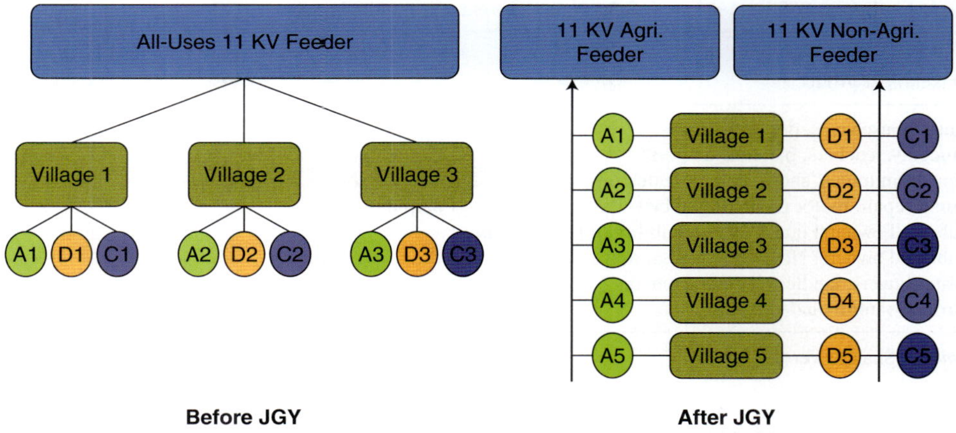

Figure 3. a) Electricity Network before JGS; b) Electricity Network after JGS.

to agriculture. Under the JGS, then: a) the villages began to be provided 24 hour power supply for domestic uses, in schools, hospitals, village industries; b) farmers began getting 8 hours of daily power supply but of full voltage and on a pre-announced schedule. Every village is to get agricultural power during the day and night in alternate weeks that are pre-announced.

4 IMPACT OF ELECTRICITY REFORMS ON GROUNDWATER USE

4.1 *Impact of metering in West Bengal*

In West Bengal, groundwater markets emerged in response to high flat rate tariff, whereby, the tubewell owners were under pressure to sell water just to recover the electricity bill given their own land holding was not sufficiently large to justify the high electricity cost. This compulsion on the part of tubewell owners also meant that water buyers, who happen to be mostly small and marginal farmers, had sufficient bargaining power over the water seller. That this reasoning is correct is shown by the fact while flat tariff rates increased around 10 fold from 1991 to 2006 (from Rs. 1,100/yr [US$ 24/yr][1] to Rs. 10,800/yr [US$ 240/yr][1]), water price only rose by 3 times from Rs. 300/acre [US$ 6.65/acre][1] [US$ 16.45/ha] in 1991 to Rs. 1,800/acre [US$ 40/acre][1] [US$ 99/ha] in 2006 for summer *boro paddy* (Mukherji, 2007). However, metering of electricity supply has changed the very incentive structure and now the water sellers are no longer under a compulsion to sell water, because they will pay only for as much as they pump. So, soon after metering, the pump owners have increased the rates at which they sell water by 30–50%, even though, assuming same hours of usage as under earlier flat tariff, we found that they would have to pay a lower electricity bill under metered tariff than before. The pump owners have therefore benefitted under the current meter tariff regime in two ways: a) by having to pay a lower electricity bill than before, for same hours of use; b) by being able to charge a higher water price than before and therefore increasing their profit margins for selling water. It has to be noted that there are only 100,000 or so electric pump owners in the state and they constitute less than 2% of the total farming households. It is this small group of relatively wealthier farmers who have benefitted directly from metering. On the other hand, the water buyers have lost out in two ways too: a) by having to pay a higher water charge than before; b) by having to face adverse terms of conditions for buying water (e.g. advance payments, not been able to get water at desired times, etc.). At the current tariff rates, and assuming same usage pattern, the SEB too will lose out in terms of revenues, but it may gain through decrease in transmission and distribution (T&D) losses. The actual impact of metering

Table 1. Impacts of the *Jyotirgram Scheme* on different stakeholder groups.

Stakeholder group	Positive (+) Negative (–)
Rural housewives, domestic users	+ + + + +
Students, teachers, patients, doctors	+ + + + +
Non-farm trades, shops, cottage industries, rice mills, dairy co-ops, banks, cooperatives	+ + + + +
Pump repair, motor rewinding, tubewell deepening, etc. (Pump mechanics)	– – – – –
Tubewell owners: quality and reliability of power supply	+ + +
Tubewell owners: No. of hours of power supply	– – –
Water buyers, landless laborers, tenants	– – – – –
Groundwater irrigated area	– – –

Source: Shah & Verma (2008).

on the size of groundwater markets (i.e. whether they will expand, contract, or remain the same), and volume of groundwater extracted cannot be predicted *a priori* and has to be answered only empirically (Mukherji *et al.*, 2009).

4.2 *Impact of JGS in Gujarat*

Jyotirgram has radically improved the quality of village life, spurred non-farm economic enterprises, halved power subsidy to agriculture and reduced groundwater draft. It has also offered a mixed bag to medium and large farmers but hit marginal farmers and the landless. These depend for their access to irrigation on water markets which have shrunk *post-Jyotirgram*; and water prices charged by tubewell owners have soared 30–50%. Table 1 summarizes the impact of the scheme on different groups of rural residents, including pump owners and water buyers.

Since over 90% of groundwater withdrawal in Gujarat occurs through electrified tubewells, electricity consumption is an accurate surrogate of aggregate groundwater withdrawal. Government figures suggest that farm power use on tubewells has fallen from 15,700 million kW·h/yr in 2001 to 9,900 million kW·h in 2006 – nearly a 37% decline. Unfortunately, pre-JGS figures on agricultural power use are residual figures, containing a portion of the T&D losses in other sectors, and therefore significantly inflate the extent of pre-JGS farm power use. However, even if we discount the 2001–02 figures, there is still a very substantial decline in agricultural power use; and halving of aggregate farm power subsidy, from US$ 788 million in 2001–02 to US$ 388 million in 2006–07. From this, we can infer that annual groundwater use in Gujarat agriculture has declined significantly during the same period. True, some of the decline may be caused by two successive good monsoons in 2005 and 2006; but there is unmistakable evidence of tubewell irrigation shrinking. Finally, it is evident that JGS has brought about unprecedented improvement in the quality of life of rural people, by creating a rural power supply environment qualitatively identical to urban one. The new era of round the clock high quality power supply in the countryside, without doubt, will unleash a myriad of impulses for socio-economic development and growth in the non-farm livelihoods in rural areas. Thus as a broad rural development intervention, JGS will prove instrumental in the future.

However, JGS's impact on the farming community has been generally negative and the intensity of this negative impact depends on the size of the land holding and the nature of the aquifer. In depleted alluvial aquifers of North Gujarat, where farmers can pump their deep tubewells continuously, feel adversely affected because the power ration restricts their irrigated area. But farmers in hard-rock areas are less affected because water available in their well during a day is a more binding constraint on their pumping than the hours of daily power supply. Small farmers owning tubewells are happy with improved power quality, although they miss their water selling business. Landless share croppers and water buyers are adversely affected everywhere because water markets have shrunk and water prices have soared 40–60%, driving many of them out of irrigated agriculture. The full import of rationed power supply has yet not been felt by the farmers because 2005 and 2006 were both good monsoon years when wells were full and water levels close to the ground. Come a drought year, and farmers will find the JGS ration of power too meager to meet their irrigation needs.

5 CONCLUSION AND POLICY IMPLICATIONS

India is in the midst of power sector reforms and states have chosen different pathways to reforms based on their political constituency. While most Indian states have resisted metering of agricultural tubewells, the state of West Bengal has embarked upon metering of all tubewells. In West Bengal, however, metering has benefitted a small section of wealthier pump owners at the cost of majority of small and marginal water buying farmers by changing the very incentive structure inherent in earlier flat tariff system, which encouraged pump owners to pro-actively sell water.

In view of this, our recommendations are the following:

a) In West Bengal, to safeguard the interest of the water buying farmers, the government should ease the process of electrification of tubewells and provide one time capital subsidy for constructing tubewells, especially for the small and marginal farmers. This will lead to increase in number of electric tubewells and enhanced competition in water markets through which water prices may come down in the future.
b) Village level governments (*panchayats*) can play an important role in West Bengal by regulating the price at which water is sold to the buyers.

The *Jyotirgram Scheme* in Gujarat has pioneered real-time co-management of electricity and groundwater irrigation. Its highly beneficial and liberating impacts on rural women, school children, village institutions and quality of rural life are all too evident; its impact on spurring the non-farm rural economy are incipient but all indicators suggest that this will be significant and deepen over time. But above all else, *Jyotirgram Scheme* has created a switch-on/off groundwater economy that is amenable to vigorous regulation at different levels. It can be used to reduce groundwater draft in resource-stressed areas and to stimulate it in water-abundant or water-logged areas; it can be used to stimulate conjunctive use of ground and surface water; it can be used to reward *feeder communities* that invest in groundwater recharge and penalize villages that overdraw groundwater.

Elsewhere in India and the rest of the world, groundwater managements have experimented with a diverse set of resource governance regimes – using water laws, tradable groundwater rights, economic incentives and disincentives – to achieve improved groundwater demand management for productivity, equity and sustainability. In their effectiveness, these regimes have proved ineffective, costly and time-consuming. In comparison, Gujarat under JGS has shown that effective rationing of power supply can indeed act as a powerful tool for groundwater demand management. And in so far as metering over 600,000 electric tubewells scattered over a large countryside may entail a very substantial transaction cost, which JGS saves, it may well be the *best* and not a second-best solution to the farm power imbroglio that all western and southern Indian states are confronted with.

As it is managed now, JGS has a big downside: its brunt is borne largely by marginal farmers, and landless because of the shrinking of water markets and of irrigated agriculture itself. There is no way of eliminating this completely, except by increasing hours of power supply – and subsidy, both of which will defeat the purpose of the entire initiative. However, JGS can significantly reduce the misery of the agrarian poor by replacing the present rationing schedule by an intelligent, demand-adjusted power rationing. The equity impact on the poor can be further enhanced by providing the daily power supply in two or more installments to respond to the behaviour of wells in hard rock areas. The equity impact can also be enhanced a great deal by charging a common flat tariff to all tubewells, regardless of whether metered or not. This would turn a large number of metered tariff paying tubewell owners from reticent sellers to aggressive water marketers to their poor neighbours.

ACKNOWLEDGMENTS

The authors gratefully acknowledge the financial support for this study from the Challenge Program for Water and Food (CPWF) project on "Strategic Analyses of India's National River Linking Project" and "Groundwater Governance in Asia" and "Indus-Ganges Basin Focal Project". This is a synthesis document based on Mukherji *et al.* (2009) published in *Energy Policy* and Shah & Verma (2008) published in the *Economic and Political Weekly*.

REFERENCES

Briscoe, J. (2005). *India's water economy: Bracing for a turbulent Future*. India Water Review, The World Bank, Washington, D.C., USA.

GOI (2003). All India report on minor irrigation census, 2001–02. Ministry of Agriculture, Government of India, New Delhi.

Gulati, A.; Shah, T. & Shreedhar, G. (2009). *Agriculture performance in Gujarat since 2000: Can it be a* divadandi *(Lighthouse) for other states?* International Water Management Institute (IWMI); International Food Policy Research Institute (IFPRI). Monograph.

Mukherji, A. (2006). Political ecology of groundwater: The contrasting case of water abundant West Bengal and water scarce Gujarat, India. *Hydrogeology Journal*, 14(3): 392–406.

Mukherji, A. (2007). The energy-irrigation nexus and its implications for groundwater markets in eastern Indo-Gangetic basin: Evidence from West Bengal, India. *Energy Policy*, 35(12): 6413–6430.

Mukherji, A.; Das, B.; Majumdar, N.; Nayak, N.C.; Sethi, R.R.; Sharma, B.R. & Banerjee, P.S. (2009). Metering of agricultural power supply in West Bengal: Who gains and who loses? *Energy Policy*, 37(12): 5530–5539.

Shah, T. (1993). Water markets and irrigation development: Political economy and practical policy. Oxford University Press, Bombay, India.

Shah, T. & Verma, S. (2008). Co-management of electricity and groundwater: An assessment of Gujarat's Jyotirgram Scheme. *Economic and Political Weekly*, 43(7): 59–66.

Tongia, R. (2004). What IT can and cannot do for the power sector and distribution in India: Link to reforms, incentives and management. Program on Energy and Sustainable Development, Stanford, University, Working Paper 19.

WBSEB (2006). *West Bengal State Electricity Board Yearbook*. Kolkata, India.

WIDD (2004). *Groundwater Resources of West Bengal: An Estimation by GEC-1997 Methodology*. Water Investigation and Development Department (WIDD), Government of West Bengal, Kolkata, India.

VI
Water, energy and technology for food security

CHAPTER 16

Water and energy nexus – Role of technology

Joppe F. Cramwinckel
Shell Exploration International and Production B.V., Rijswijk, the Netherlands

ABSTRACT: The aim of this chapter is to provide an overview of the inter-linkages between water, energy and the environment, and the role technology can play to reduce demand and impact and increase supply. Demand, supply, and effects on the environment are the main drivers of our energy and water systems, with all three of them set to undergo significant change. Stress on water and energy resources is exacerbated by low efficiencies, for example in the agricultural sector when it concerns water use and losses in water distribution systems. Boosting water and energy use efficiency through investments in relevant technologies and infrastructure is therefore critical.

To improve water efficiencies it is important to understand the complexity of water treatment. The appropriate water treatment technology will be determined by characteristics of the water sources available and quality characteristics for the discharge/re-use. It is concluded that technology alone cannot address the water and energy efficiency challenge. While water is a local issue, ensuring the allocation of sufficient supplies at the right time, in the right place, and of the right quality, sustainable water management requires increasingly consideration of the interconnectivity of larger contexts and many diverse stakeholders.

Keywords: energy, water, development, linkages, technology

1 THE WATER AND ENERGY SYSTEMS TODAY SET THE CONTEXT FOR THE FUTURE

Energy and water are both essential for most aspects of life: domestic use, food, industry, transportation, ecological services, leisure, cultural identity, etc. There are some hard truths about energy and water availability which taken together mean we are entering an era of revolutionary transitions. A step change in the rate of growth of energy demand is foreseen and increasingly it is recognized that this also will be true for water.

Shell Energy scenarios (Shell, 2008) estimated that in 2007 the primary energy the world needs is about 480 exajoule (EJ $= 10^{18}$ J) or 235 million barrel of oil equivalent (BOE $= 5.63 \times 10^9$ J; conversion factor used as a typical average) per day. In 2007 total final consumption is estimated to be around 335 EJ/yr or 160 million BOE/day. This means we lose approx. 30% of the primary energy we put into the system. And in the last 20 years it has been getting gradually worse (1987 = 28%). As people grow richer they use more energy (Figure 1) and it is expected that population and GDP will grow strongly in non-OECD countries, since China, India and Brazil are just starting on the energy ladder.

Easy access oil and gas (relative low cost and high availability) will not be able to match the pace of energy demand. In fact all energy sources together (renewable and non-renewable) will struggle to match demand, which will have to be met partly by new energy efficiency technology.

We are going to need all the energy we can get. This is due to a combination of constraints on the availability of resources, the huge infrastructure investments required to develop them and political constraints. For example, bio fuels need arable land, and to meet 10% of the USA's transport demand by ethanol from maize in 2020 requires 30% of USA's arable land. Coal will be constrained by logistical limits on the ability to construct new railways and ports to move massive amounts of coal from producing to consuming areas. Growth in nuclear power will become

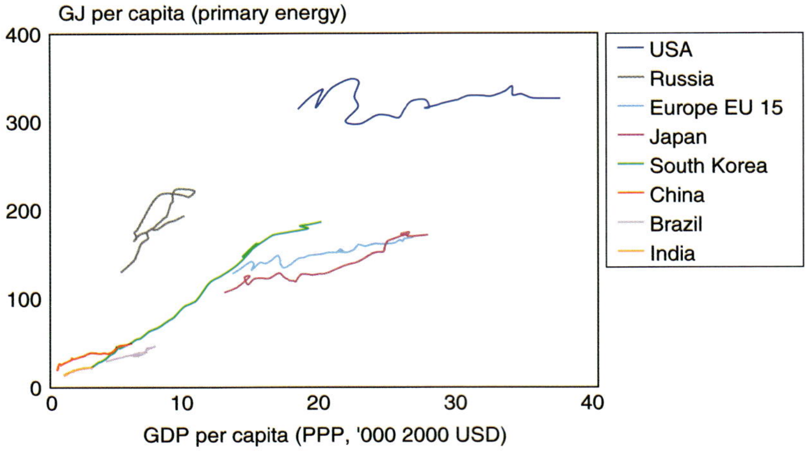

Figure 1. Per capita energy consumption (Shell, 2008).
Source: Shell International BV, Oxford Economics and Energy Balances OECD and Non-OECD Countries © OECD/IEA 2006.

constrained by political controversy, the high rate of decommissioning of old plants and the recreation or large scale expansion of three industries: uranium mining, plant construction and waste management

Less than 3% of the world water is fresh and most of it is locked up in Antarctica, the Arctic and glaziers and not available for man. Thus man and the environment must rely on only 5% of all freshwater. Where is this fresh water? 10×10^6 km^3 is stored in underground aquifers, 119,000 km^3 of rain falling on land after accounting of evaporation, 91,000 km^3 in natural lakes, over 5,000 km^3 in man made storage facilities, and 2,120 km^3 in rivers, constantly replaced from rainfall and melting of snow and ice (UNESCO, 2003).

A challenge in managing water resources is the scattered knowledge of the patterns in water use. The Third Edition of the United Nations World Water Development Report (UNESCO, 2009) states that there is a consensus that the estimate for total global freshwater use in 2007 is about 4,000 km^3/yr. Another 64,000 km^3 is directly used in agriculture.

Nature is the most important user of water. An estimated 70,000 km^3/yr of water is evaporated from forest, natural vegetation and wetlands.

There are also significant losses in the systems, however they are difficult to quantify, e.g. evaporation losses from human made reservoirs in arid areas.

While there is a strong relationship between water investment and growth, the relationship between quantity of water and growth is less conclusive. Many water poor economies have developed, while the ratio of water use to GDP in many developed countries is declining.

Taken into consideration uncertainties in population and GDP growth, freshwater withdrawals are predicted to increase by 50% by 2025 in developing countries, and 18% in developed countries (WBCSD, 2006). As traditional fresh water resources are under stress because of over exploitation, water supply for human, industrial and agriculture purposes will need to transition to greater use of non-traditional water resources like saline surface water and waste water. To balance supply and demand there will be a significant role for new water efficient technology.

The increase in environmental stresses, including loss of habitat, pollution and changes in biological processes (such as fish spawning) is another factor that intensifies this water and energy story. Maintaining environmental flows are critical to ensure river systems can supply water to human use and ecosystems. Impacts from climate change on both regional and global hydrological

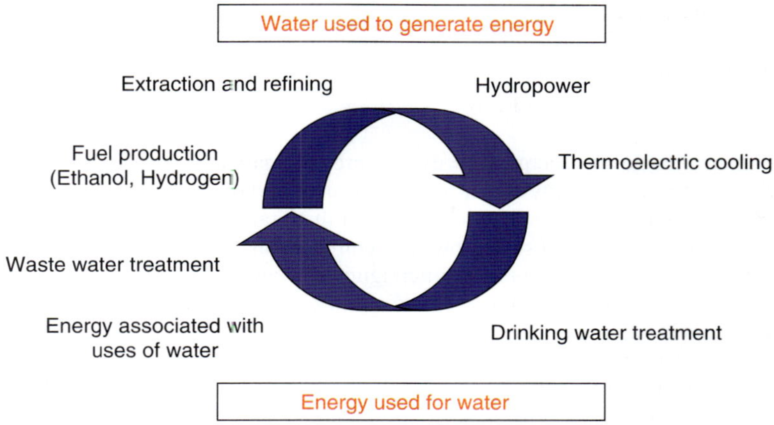

Figure 2. Water is used for energy generation and energy is used in water management.

systems will increase, bringing higher levels of uncertainty and risk about water availability, with some regions more impacted than others.

Increasing temperatures in northern climates impact melting of snow packs; globally rising sea levels and changes in precipitation patterns are already being seen. Small islands will be critical hotspots as well as coastal mega cities. It also means an increasing need to prepare for extreme weather events (droughts and floods) and the uncertainty in their occurrence.

2 WATER AND ENERGY ARE INCREASINGLY INTRINSICALLY LINKED

Water is used in a variety of ways in the energy sector. Water is used as an input into the production of both non-renewable energy sources such as coal, natural gas, nuclear fuels, and oil, and renewable energy sources such as biomass, geothermal, solar, wind, and hydropower. It is used for irrigation (energy crops), drilling, mining, transportation, fuel conversion, cooling, cleaning, and other technology-specific applications. On the other hand, energy is consumed for various water activities such as water extraction, conveyance, distribution, treatment, and desalination. Historically water and energy supply have been examined independently. As the world faces water stress in the face of growing energy demands, considering both sides of the water-energy nexus is essential to any major planning decision (Figure 2).

3 WATER DEMAND FOR PRIMARY ENERGY PRODUCTION

Data is increasingly published on the water footprint of the various energy pathways. Water foot printing is an emerging science. The water footprint is an indicator of water use that looks at both direct and indirect water use of a consumer or a producer. The water footprint of an individual, community or business is defined as the total volume of freshwater that is used to produce the goods and services consumed by the individual or community or produced by the business. Following is an overview of the current understanding of the water footprint of some of the key primary energy carriers and power generation. Significant more work is needed to have a good understanding in terms of their lifecycle, e.g. *from well to wheel* or *from cradle to grave*. In addition, the water footprint must be seen in the context of local impacts, for example whether a crop is rain-fed (green water) or irrigated (blue water).

This element of the water footprint is perhaps more important than the water use component itself. Knowing your water use without knowing where you are using it, is less valuable than knowing where you are using water, even if you do not know your water use.

3.1 *Biomass*

Driven by climate change and energy security concerns biomass is expected to grow as a primary energy carrier, bio fuels potentially becoming a significant part of it. The Shell energy scenarios to 2050 (Shell, 2008) predict that this may reach 15% of the total energy demand in 2050. While the current debate on the feasibility of such growth is centred about the land-use competition with food and ecosystem services, water linkages are often ignored. Water may be in some cases the limiting factor since the water footprint of bio-crops is significant. However, large differences in crop water requirements exist among countries due to different climates and production efficiencies. An illustrative range of the average water footprints for biomass production ranges from 24 m^3/GJ (24,000 m^3 per 1,000 GJ) (GJ = 10^9 J) in the Netherlands to 143 m^3/GJ (143,000 m^3 per 1,000 GJ) in Zimbabwe (Gerbens-Leenes *et al.*, 2008). These numbers are conservative as they do not take into account efficiency losses to secondary energy carriers.

3.2 *Enhanced oil recovery*

As easy oil is used up, it is common to use water for improved oil recovery. Water is injected by pumps to the required formation to recover the pressure in the reservoir and/or sweep the oil out of the reservoir. The water quality (and energy) requirements depend on the injected depth, hydrogeological parameters of the reservoir, and process efficiency. This injected water is also trapped in the reservoirs and as a consequence removed from the water cycle. To extract oil, including water for drilling, flooding and treating (Khatib, 2007), between 2 and 8 m^3 of water per 1,000 GJ have historically been required. However, when thermal steam injection or enhanced oil recovery, like water flooding is included in the process, this number can increase, on average, to 1,058 m^3 per 1,000 GJ.

3.3 *Hydropower*

Hydropower produced 89% of the world's renewable electricity in 2006, and 16.6% of total electricity generation worldwide. Two-thirds of worldwide economic potential remains unexploited – this resource is concentrated in the developing world (WBCSD, 2006).

The construction and operation of dams and hydropower plants can contribute to reducing greenhouse gas emissions, smooth operating peaks and store energy and water as well as provide improved ship traffic safety. Although hydropower does not directly consume water (consumption is limited to the loss by evaporation in the dams which vary as a function of the climatic conditions), its generation may conflict with other uses, especially irrigation, since its release schedule does not always correspond to the timing of other water needs. 25% of dams worldwide are used for hydropower and only 10% have hydropower as their main use. Most of them are used for flood control or irrigation, or for multiple purposes (Gerbens-Leenes *et al.*, 2008).

3.4 *Solar energy*

Concentrating solar power reflects and concentrates sunlight to create high-temperature heat, usually to create steam to drive traditional turbines. This power source is best harnessed through large, centralised power plants located regions with high direct normal radiation. There are plans currently to significantly increase the solar thermal power generation in such places as California, Arizona, Spain, Australia and Northern Africa. However, a solar thermal power plant water consumes about 1 m^3 of water per 1,000 kW·h (electric) or 277 m^3 of water per 1,000 GJ (Gleick, 1994). Concerns about water availability and environmental impacts have driven the Solar Millennium group to select dry cooling at their proposed project locations in Nevada and California, saving water consumption by up to 90%. In this technology the steam cycle is cooled with air, mainly from large ventilators.

Photovoltaic cells that produce electricity directly from sunlight are considered to have negligible water use.

3.5 *Canadian oil sands*

While 173,000 million barrels (BOE) of the Canadian oil sands resource rated economically recoverable with today's technologies, and therefore deemed proven, the ultimate potential of the Canadian oil sands is even greater given the estimated 1,700,000 million BOE in place[1]

Oil sands extraction is a water-based process. Large volumes of water are needed to separate bitumen molecules that are trapped in sand and clay. On average, the oil sands industry uses two to three barrels of fresh water to extract one barrel of bitumen. This fluctuates depending on recycling rates from tailings. About half to three quarters of a barrel of water is needed to convert the bitumen into synthetic crude oil.

The primary source of fresh water is the Athabasca River. Industry plans to increase oil sands mining production from 1.2 million barrels/day (current production) to 3.5 million barrels/day. This production level, on average, will use less than 3% of the mean annual flow in the Athabasca River. The Athabasca River is a natural river with no dams to moderate or interrupt natural seasonal flow variations. During winter low-flow periods poorly managed water withdrawal could impact the river in the future. However, regulatory limits are already in place to restrict water withdrawals by the oil sands industry during winter low-flow conditions. In addition, research programs and a multi-stakeholder committee have been established to recommend future withdrawal limits to ensure the Athabasca River remains protected.

4 WATER AND POWER GENERATION

There is a critical relationship between power generation, carbon capture and storage and water. Water consumption can vary by primary energy source (nuclear, coal, gas) and the type of carbon capture and cooling system employed.

The National Energy Technology Laboratory, in a report prepared for the Department of Energy of the USA (Khatib, 2007), calculated that there is almost a fourfold increase in water consumption per net kW·h between the lowest water consuming platform (Natural Gas Combined Cycle) and the highest (Nuclear). The addition of CO_2 capture and compression increases water consumption by 50% to 90%. Many of the advanced power platforms use less water and have a lower increase in water demand associated with incorporation of CO_2 capture equipment than do current technologies.

The water consumption factors are based on a cooling system in which the effluent cooling water is cooled in an evaporative cooling tower and re-circulated. *Consumption* represents water that must be made up to account for both evaporation in the cooling tower and a relatively small amount that is consumed in unit operations within the generation process. There is also the open loop water cooling, where water is withdrawn from a river, lake or the sea, and then returned to it after cooling. The average amount of water consumed is approximately zero. However a significant amount of water is required, which is then returned calculated at approximately 160 m^3/MW·h (equivalent to 44,444 m^3 per 1,000 GJ)[2].

5 WATER PRODUCED IN PRIMARY ENERGY PRODUCTION

Due to ageing reservoirs and increased oil recovery operations, pumping oil from reservoirs is now associated with more water production per amount of oil produced than ever before. The volume

[1] Government of Alberta, Energy, Oil Sands Statistics [http://www.energy.gov.ab.ca/OilSands/791.asp].

[2] Figures based on average values from EDF (*Électricité de France*) from nuclear power plants along rivers in France.

Table 1. Energy required to deliver 1 m^3 of clean water from different sources. Based on Scientific American (2008).

Source	Energy required ($kW \cdot h/m^3$)
Lake or river	0.37
Groundwater	0.48
Wastewater treatment	1.0–2.5
Waste water reuse	0.62–0.87
Seawater	2.58–8.5

of water produced worldwide from the oil and gas industry is still increasing at a rate of about 10% per year. Water to oil ratios ranged from <1 to up to 40 depending on maturity of the field, with the lowest ratios generally observed in the Middle East (UNESCO, 2009). The water quality varies depending on location, formation where it originates from, and the type of hydrocarbon being produced, e.g. oil or gas. Apart from hydrocarbons, salt content is another major issue when considering the beneficial reuse of this water.

Coalbed methane (CBM) or coalbed gas is a form of natural gas extracted from coal beds. In recent decades it has become an important source of energy in the USA, Canada, and other countries. Australia has rich deposits where it is known as coal seam gas.

Because of its large internal surface area, coal stores 6 to 7 times more gas than the equivalent rock volume of a conventional gas reservoir. In order for gas to be released from the coal, its partial pressure must be reduced and this is accomplished by removing water from the coal bed. Large amounts of water, sometimes saline, are produced from coal bed methane wells, especially in the early stages of production. Increasingly this water is seen as a resource rather than a waste, in particular in water scarce areas in Australia and the USA.

6 ENERGY FOR WATER MANAGEMENT

Energy is used to extract, transfer and treat water (Table 1). The numbers are illustrative and do not incorporate critical elements such as the distance the water is transported or the level of efficiency, which vary greatly from site to site. For example, water transfer over 350 km (horizontally) uses 3.6 $kW \cdot h/m^3$, or the same amount of energy needed to desalinate 1 m^3 of seawater. It is clear from this data that seawater desalination is the most energy intensive process to produce clean water, even though significant efficiency improvements have been achieved over the past decades.

Some studies have estimated that approximately 56,000 million kW·h are used for supply and treatment of drinking water in the USA, the equivalent of approximately 44.8 million tonnes (1 tonne = 1,000 kg) of greenhouse gas to the atmosphere at a cost of approximately US$ 4,000 million (EPRI, 2002).

7 POLICY CHOICES

Water policy therefore influences energy choices and vice versa creating a need for policy integration. Having the right price, policy and regulatory frameworks are critical to encourage behavioural changes, trigger innovation, ensure sustainable use of water and energy resources and to simultaneously adapt to and mitigate climate change. Within such frameworks, different solutions may be applied to local circumstances.

Research and knowledge have expanded and discussion progressed within technical circles. Some places in the world have successfully integrated both water and energy into planning, from investment to institutional decision-making. For example, in December 2008, the US

Environmental Protection Agency announced an inter-agency agreement between the offices of Air and Water to collaborate on energy and climate efforts at water utilities. Nevertheless, there is still a significant gap in communications addressing the linkages at a global scale. Only a limited number of publications, scenarios and perspectives about energy and climate change currently also address water issues.

Understanding both water and energy footprints for the different energy carriers and water users, and optimising both footprints by integrating them in the project design, should be undertaken. The different energy and water footprints have several economic, social, and environmental constraints (e.g. constraints with large water footprint associated with biofuels, or the large energy footprint associated with deep water injection, etc.) and so by considering the tradeoffs between the various footprints, this will allow a better utilization of the appropriate technologies to meet these constraints.

8 ROLE OF TECHNOLOGY

Demand, supply, and the effects on the environment are the three main drivers of our energy and water systems, with all three of them set to undergo significant change. Therefore we know that the next 50 years will see a revolution in the system and considerable turbulence along the way.

There are no ideal answers but some outcomes are clearly better than others, and while prices and technology will drive some of these transitions, political and social choices will be critical. Thus profound change is inevitable. Will national governments simply seek to secure their own energy and water supplies? Or will new coalitions emerge between various levels of societies and government, ranging from the local to the international that begin to add up to a new resource management framework?

Stress on water and energy resources is exacerbated by low efficiencies, for example in the agricultural sector, when it concerns water use and losses in the water distribution systems. The agriculture sector is facing outmoded water systems, poor regulatory enforcement, ineffective price signals, and the lack of incentives for change in behavior, particularly by those who claim historical rights to water access.

Boosting water and energy use efficiency through investments in relevant technologies and infrastructure is therefore critical. This efficiency challenge leads to the business challenge of innovation – not only in producing new products and services, but also in avoiding or addressing legacy constraints – for example, established infrastructure and technology standards, social habits and attitudes and standard business practices.

These behaviors and norms were appropriate for a bygone era (for example, a context of abundant cheap energy) and within a certain socio-economic and political context (for example, food security and priority for agricultural water uses), but not for current or future conditions (for example, increasing urbanization and post-industrial economies).

9 TECHNOLOGY TRENDS

In order to improve efficiencies, it is important to understand the complexity of water treatment. First one must have a good understanding of the characteristics of the water sources available such as: surface, groundwater or waste water. Then it is important to understand relevant quality characteristics for the discharge/re-use quality such as: salinity, solids, fines, dissolved and dispersed hydrocarbons, and bacterial activity. The water treatment technology will be determined by these two characteristics, namely source characteristics and discharge quality.

Reduction in the cost of membranes has been the principal driver for membrane technology becoming one of the technologies which is now extensively applied in water treatment. The pore size of the membrane, expressed in nanometers (nm) will determine what components can be removed as illustrated in Table 2.

Table 2. Membrane characteristics.

Filtration Spectrum	Pore size	Removal
Filtration	100–1,000 nm	giarda, crypto, bacteria
Ultra Filtration	10–100 nm	colloids, viruses
Nano Filtration	1–10 nm	colour, hardness, pesticide
Reverse Osmosis	<1 nm	salts

State-of-the-art sweet water purification units are now based on ultra filtration, however it needs large amounts of relatively hazardous chemicals to avoid bio-fouling and scaling. These chemicals might involve a health hazard, especially when a mistake is made during the cleaning process. It is therefore important not only to consider water treatment efficiency, but also the impact it may have on the environment in general, and on energy efficiency in particular. Some examples of game changing technologies that address both water and energy efficiency are:

- Biogas generated by the application of anaerobic treatment may be used for power generation, fuel for transport or heating.
- Heat transfer from groundwater and wastewater resources, respectively for both heating and cooling systems which provide generic solutions when the costs of fossil fuel based energy are on the increase.
- Microbial Fuels Cells (MFC) for green production of chemicals based on electricity produced directly during biodegradation of waste organic matter.
- New cooling systems can be designed in power plants to have an optimal trade-off between water and energy requirements and impacts (e.g. parallel condensing systems that combine wet and dry cooling systems).
- Reuse of output heat in the industrial processes as part of an ad-hoc recycling of waste streams (e.g. low grade steam in cooling systems, produced heat streams in desalination or waste water treatment processes).

10 AN EXAMPLE FROM THE OIL AND GAS SECTOR

Natural wetlands and mangroves mitigate flooding, and can be used for water storage, thus reducing the impacts of climate change, whilst also effective in improving water quality. The use of constructed wetlands, which mimic this natural phenomenon for treatment of surplus water associated with the production of oil and gas, has been pioneered by the Petroleum Development of Oman (PDO) (Society of Petroleum Engineers, 2004). In one of their oilfields more than 20 barrels of water are produced for every barrel of oil. This *produced* water is then disposed of via Deep Well Disposal (DWD). The dilemma faced by PDO is that from 2009 onwards the water disposal rate is expected to reach capacity constraints, which may result in oil deferment. Some 250,000 m^3 of water are produced each day; to dispose of this volume requires six pumping stations, each having two or three wells. The continued use of DWD is unsustainable because of the prohibitive and increasing cost, the high gas usage and the growing inaccessibility of low-pressure aquifers – in the face of ever increasing water production. Replacing DWD could reduce operating and capital expenditures and over 30 years, liberate gas for other uses. PDO has opted for constructed wetlands to treat this surplus water. Production water flows into the treatment beds through buried perforated pipes, then through the root zone of reed plants, and out through perforated drainage pipes in the base of the beds. Naturally occurring soil bacteria already present in the soil will degrade the hydrocarbons in the water in the time it takes for the water to flow through the beds.

Treated water is applied to the production beds via surface irrigation. Here a crop of reed plants will be grown. Initially, the crop will be mulched and left in the field. Later, the crop may be harvested for a commercial – non-food chain – use. Remaining water is evaporated in evaporation ponds, leaving salt. Salt may be buried at the end of facility life, but may also be harvested and sold if its composition is suitable. The application of constructed wetlands in Oman will be one the largest scale applications ever and has the potential to transform water management in Oman.

11 CONCLUSIONS

It is illustrated that the increased demand for water and the scarcity in supply leads to innovation – not only in producing new products and services, but also in avoiding or addressing legacy constraints – for example, established infrastructure and technology standards, social habits and attitudes, and standard business practices. These behaviors' and norms were appropriate for a bygone era of abundant cheap, low water intensity energy world. However technology alone is not the answer to the water and energy challenges. While water is a local issue, ensuring the allocation of sufficient supplies at the right time, in the right place, and of the right quality, increasingly requires consideration of the interconnectivity of larger contexts and many diverse stakeholders. Human security and development cannot be isolated from the health and viability of the earth's underlying life support systems. The interconnectivity challenge requires us to be able to think and act in terms of multiple geographies of connection, from nation states and city limits to watersheds and river basins, and in terms of multiple timeframes, in order to ensure that short-term interests do not foreclose longer-term possibilities. The interconnectivity challenge also requires us to take into account not only *blue water* issues of the water we see, such as the water in lakes and rivers, but also the so-called *green water* contained in healthy soils, and the *virtual* or embedded water contained in traded products and services. Human security and development also increasingly depend on the ability to consider links in actions and policies relating to food security, energy security, and water security.

REFERENCES

EPRI (2002). Water and Sustainability (Volume 4): U.S. Water Consumption for Water Supply and Treatment. The Next Half Century. Report 1006787, March. Electric Power Research Institute, Palo Alto, California, USA.

Gerbens-Leenes, P.W.; Hoekstra, A.Y & Van der Meer, T.H. (2008). Water Footprint of Bio-energy and other primary energy carriers. UNESCO-IHE Research Report Series No. 29, March.

Gleick, P.H. (1994). Water and Energy. Annual Review of Energy and the Environment, 19: 267–299.

Khatib, Z. (2007). Produced Water Management: Is it a Future Legacy or a Business Opportunity for Field Development? International Petroleum Technology Conference.

Scientific American (2008). Energy *vs*. Water. Why both crises must be solved together? Earth 3.0, Special Issue, October 2008, with amended data from GHD.

Shell (2008). Shell Energy Scenarios to 2050. Shell International B.V. [http://www.shell.com/scenarios].

Society of Petroleum Engineers (2004). Opportunities for re-use of produced water around desert oil fields A.C. Sluijterman, Y. Al-Lawati, S Al-Asmi, Petroleum Development Oman, and P.H.J. Verbeek, M.A.S. Schaapveld, J. Cramwinckel, Shell International Exploration and Production

UNESCO (2003). Water for People, Water for Life. United Nations World Water Development Report. Part II: A look at the world's fresh water resources. [http://www.unesco.org].

UNESCO (2009). Water in a Changing World. United Nations World Water Development Report, Third Edition.

WBCSD (2006). Powering a Sustainable Future: An agenda for concerted action, World Business Council for Sustainable Development. 26 pp.

CHAPTER 17

Energy supply for the coming decades and the consequent water demand

Ignasi Nieto
Economy and Engineering of Transport (INECO), Spain

ABSTRACT: Electricity production and water consumption are key factors for a country to achieve economic growth and greater welfare. However, there is an important need of water consumption for electricity production through most current technologies and there are a lot of regions suffering from water scarcity. In the future, technology innovation will be a key issue to overcome this problem.

Within the energy sectors, electricity is the one with significant water consumption, although from a global perspective (compared to other sectors like agriculture and industry) it is not a problem. The faster electricity production mix will shift towards technologies using less water (most renewable technologies and others such as combined-cycle) the lesser water will be consumed per energy unit produced or demanded.

Some renewable technologies like concentrating solar thermoelectric or biofuels that could be a good solution for poor countries must improve their efficiency in order to reduce their water demand and become a real alternative in areas where they suffer from water scarcity.

Keywords: freshwater use, renewable technologies, thermodynamic cycle, cooling processes

1 INTRODUCTION

Energy, as well as population growth, agricultural efficiency rates, water use[1] and resource use, is among the most important factors that could jeopardize the sustainability of our presence on the earth in the long run. Additionally, most of these factors are related, and the nexus energy-water is a clear example. A complete analysis of the nexus energy-water, should take into account both sides of the problem: *water for energy* and *energy for water*. Without an accurate analysis though, it could be concluded, in terms of sustainability, that the issue *water for energy* could be considered more important because it ends up affecting water consumption more than the issue *energy for water* does. Then, it could be thought that the more water we need in order to produce a unit of energy, the more water through energy we will need to produce/deliver a unit of water. This is a true statement, but it is not an important issue in terms of resource use. It is important to improve the efficiency of water use, both from the supply side (technology) and from demand side (responsible and efficient consumption) and so to reduce the use of energy to obtain water. While it is important to develop water efficient processes to obtain energy, water use for energy is not a critical point in terms of global sustainability as we will see in this chapter. Besides, the critical problems of long term sustainability of energy use has to do with other things but not to water use if we take a global perspective. However, from a local perspective it could be a problem to place

[1] During this chapter I will include the concepts of *use*, *consumption* and *withdrawal* of water. The last will be used to refer to that water taken from a source (river or lake) but given back to same source afterwards, and then, not consumed. The term *consumed water* will be used to refer to the real consumption of water, that is, the water that is finally lost. The term *water used* will be mentioned in a general way, that is, it could be referred to anyone of the other two terms or the sum of both.

electricity power plants in areas that are suffering from water scarcity. Most poor countries suffer from water scarcity or lack water infrastructure. As they need to grow, they will need energy which could be a problem that could jeopardize their potential economic growth or damage their, in some cases, already poor environmental equilibrium.

The relation between water and energy is very old in human history. From the first mills used to produce mechanical work to the steam machine, mankind has taken advantage of this relation. But it is due to the invention of electricity when the use of water to produce energy increased dramatically. The old physical concept of moving a wheel using water is applied now to move an alternator-generator and producing electricity. Then, in electricity, most of current technologies need water to complete their thermodynamic processes, while in the hydrocarbons sector we also need water for pumping and other processes. However the specific water consumption[2] in the hydrocarbons sector is minimal compared to that in the electricity sector.

While water use for energy is not a problem from a global perspective, it could be a problem locally, in areas where water resources are being used up, depleted or consumed over the natural rate of recovery. In order to demonstrate these conclusions, in this chapter, I will include a forecast of the energy demand over the coming decades and the energy supply and water needs to cover it, pointing out the main conclusions.

2 THE ENERGY ISSUE

Since the end of the past decade, we have been seeing what could be called *an energy revolution*. Technology improvements will, in the medium or longer term, ensure that economic growth can be based on energy sources that will allow us to reduce our dependence on oil and gas. The key issue that triggered this revolution was, without any doubt, the challenge of climate change. Up to this moment, in the energy political agendas, the most important issues were oil and gas prices as well as dependence. A happy consequence of these challenges (price, dependence and climate change) is that all of them take us to the same or similar strategies and actions to be taken. That is, to invest in new technologies and to foster technology change and substitution. This fact will make reaching political consensus and the common effort and policies that have to be launched easier. Then, the energy problem could be summarized as in the following statement. Firstly, we should achieve a decrease in our dependence on fuels like oil and gas, without risking the security of supply and in doing so avoiding a solution with too high energy prices that could jeopardize economic growth. We cannot forget that economic growth is not only needed to create wealth but also it is needed to carry out the huge investments in alternative technologies we need for the future. Lastly (last but not least at all), this revolution should allow us to evolve to a low emissions economy, up to the level of 450–550 ppm (IPCC, UN Intergovernmental Panel on Climate Change). According to IPCC, with this level of CO_2 emissions, it is expected that the average temperature increase will be at most between 2°C and 3°C.

3 THE FORESEEABLE ENERGY DEMAND FOR THE COMING DECADES AND HOW WE ARE GOING TO COVER IT

The International Energy Agency (IEA) considers that the total primary energy demand will grow at an average of 1.6% annually until the year 2013, while the final energy demand will grow at a rate of 1.4% annually. That means that in 2030 we will be consuming 17,014 Mtoe (million tonnes of oil equivalent) while in 2006 we consumed 11,730 Mtoe. This projection will make jump the CO_{2e} emissions[3] from $27,889 \times 10^6$ t (t = tonne = 1,000 kg) to $40,553 \times 10^6$ t in 2030,

[2] Consumed water to produce a unit of energy.

[3] CO_{2e} means CO_2 equivalent, in order to take into account the rest of the greenhouse gases, being CH_4(methane) the most known besides CO_2.

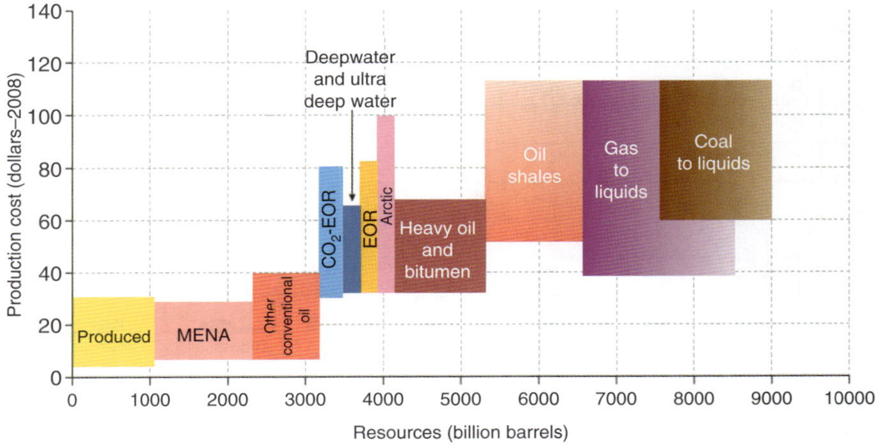

Figure 1. Long-term oil supply cost curve. Ultimately recoverable oil resources.[4]
Source: International Energy Agency (IEA, 2008).

with an annual growth rate of 1.6%. Electricity consumption accounts for 16.6% of our world current energy consumption. The IEA expects this figure to jump to 21.2% in 2030, which is a logical consequence of the higher average living standards of the world in 2030 compared to now. However, the higher weight of electricity on the total energy consumption means, keeping the current technology, more water consumption.

3.1 *Oil and gas sector*

From an strategic point of view, it is possible to maintain that the world is not running short of oil and gas and will not be at least for this century. The challenge facing both oil and gas sectors is related mainly with a lack of investment in production to match the demand growth, coming mainly from Asian countries like China and India. This fact will take us to a so-called peak-oil. That means that in the next decades we will live in a world with high oil and gas prices due to the pressure of demand over the supply. However, if we take a look on Figure 1, what is called *the ultimately recoverable conventional oil resources* (which include initial proven and probable reserves from discovered fields, reserves growth and oil that has yet to be found), amounts to a figure of 3.5 trillion (3.5×10^{12}) barrels. Only a third of this total has been produced up to now. If we take into account non conventional oil like extra-heavy oil, oil sands and oil shales, the figure amounts to 6.5 trillion (6.5×10^{12}), and finally, if we add coal-to-liquid and gas-to-liquid, the potential increases up to about 9 trillion (9×10^{12}) barrels. On the other hand, the production of these non conventional resources will bring the price significantly high. To sum up, the problem of oil (and it could be applied the same rational to gas, taking into account non conventional sources like methane hydrates), has not to do with reserves, but with the production rate and investment, the price and geostrategic issue due to reserves concentration in some areas of the world and, of course, the problem of climate change due to the CO_2 emissions. During the next decades, the world will keep on basing its growth on the conventional technologies or fuels like oil, trying to introduce alternative technologies in order to diminish its dependence. Yet, the speed of this progressive substitution will depend on governmental policies, prices and alternative technology improvements. The higher the price of conventional fuels the bigger the chances for alternative technologies. That is why, in spite of oil being the fuel that

[4] This curve draws the price without royalties and taxes. So, we have already produced around 1 trillion (10^{12}) barrels at an average cost of US$ 30/barrel (prices 2008).

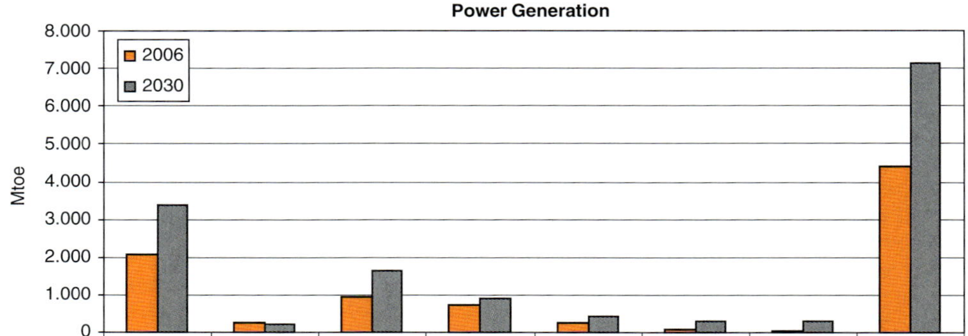

	Coal	Oil	Gas	Nuclear	Hydro	Biomass and Waste	Other renewable	TOTAL
2006	2.075	277	947	728	261	81	55	4.424
2030	3.375	204	1.664	901	414	286	287	7.130

Figure 2. Expected current and future (2030) power generation (electricity) in the world.
Source: International Energy Agency (IEA, 2008).

grows less, in the forecast of IEA for 2030, still it accounts for the biggest part of world total energy demand.

The Enhanced Oil Recovery (EOR) technologies (and other more advanced technologies like ERC and MRC: Extreme Reservoir Contact and Maximum Reservoir Contact), consist on the recovering oil from the fields above the actual average rate (35%). In the future, it will be used CO_2 to recover oil at this rate and above it, so that it will bring about significant water savings. However, in this chapter, in order to figure out the *water for energy* used in the oil sector in a worse or bad case scenario, we will consider that it is being used the current technology (that is, using water).

3.2 *Electricity sector*

Electricity will contribute during the next years with the main part of the solution to the energy challenges. This is because the main innovations on alternative technologies have taken place in this sector. There are technologies like wind, for example, that have achieved a good degree of economic viability over the last decades, something that has not happened in other subsectors within the energy sector. Besides, hydro power technology has been working all over the world for the last century. These facts, as well as others like the nuclear option, makes possible in the electricity sector, to invest in clean technologies and to achieve a good amount of electricity CO_2-free. Besides a whole bunch of renewable technologies which are close to achieve economic viability (taking into account technology improvement, high prices of conventional fuels and other issues like CO_2 prices), some not renewable but CO_2-free/minimizing technologies like CCS are advancing quickly in its way to economic competitiveness. This, together with the fact that societies increase the weight of electricity in their energy portfolio as they achieve higher levels of welfare, makes electricity technologies a key issue for the future. Besides, in this chapter we are focusing on water use for energy and it is the electricity sector, compared to other, the one that uses more water for a produced unit. Figure 2 shows that coal will be responsible for the main part of the electricity demand increase in the coming years.

3.3 *Energy key elements: EU, China, India and the USA*

Tackling successfully the challenge of climate change, will depend on the success achieved in three countries: China, India and USA. If we take a look at Figures 2 and 3, we realize that it is in these countries where the energy issue game is being played. Even if Europe were able to reduce

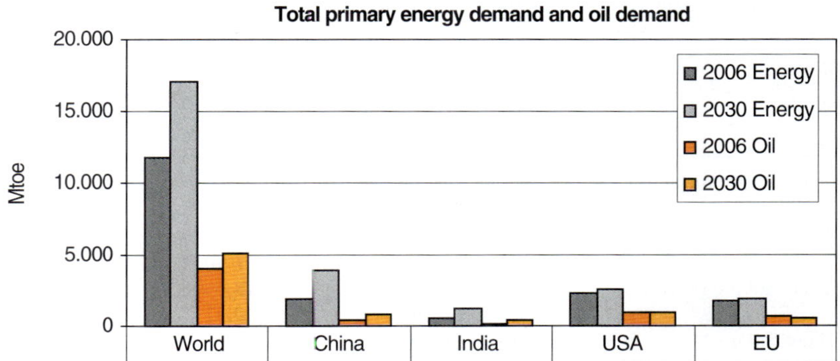

	World	China	India	USA	EU
2006 Energy	11.730	1.898	566	2.319	1.821
2030 Energy	17.014	3.885	1.280	2.566	1.903
2006 Oil	4.029	352	136	937	668
2030 Oil	5.109	814	346	881	602

Figure 3. Foreseeable total energy primary demand and oil demand.
Source: International Energy Agency (IEA, 2008).

a good amount of its emissions, China will write off this savings with some years growing at rates around 8%.

In 2030, the world will have doubled (in the IEA reference scenario) the CO_2 emissions from $20,000 \times 10^6$ t up to $40,500 \times 10^6$ t. Out of this increase, coal will contribute with an increase from $8,300 \times 10^6$ t to $18,600 \times 10^6$ t, that is, from a 42% to a 46% of the total emissions. China will increase its share on the total emissions from a 10.7% in 1990 to a 29% in 2030. USA will decrease its share from the 23.1% to a 14.3%, India from the 2.8% to an 8.1% and the EU from the 20% to a 9.2%. Besides, coal will be responsible of 77% of the China's emissions increase, 67% in India and 54% in USA.

Europe is making a huge effort to reduce its emissions. It is expected that in 2030, the EU emissions will be even below the 1990 level; however within the same period, in a business as usual scenario, China will have increased its emissions 32 times the decrease achieved by EU in the same period. To make the problem more difficult to solve, China will keep on relying on coal due to sovereignty as well as political and social issues. This is why, technologies like CCS are so important to try to curve the emissions of countries like China, India and USA, all of them producing a good amount of electricity from coal, and with a big amount of reserves on their land. This is the main reason why the production of electricity from coal will not decrease as it could be logical to think if we take into account that this is may be the worse conventional electricity technology from a CO_2 emissions point of view.

4 WHAT WILL BE THE WATER NEEDS FOR THE ENERGY PRODUCTION IN THE FUTURE?

As we said before, power plants use water to complete its thermodynamic cycle, using water as a heat carrier in most technologies. The shift towards combined-cycle gas turbine (CCGT) technologies over the last decade and the increase of this technology in the electric generation mix, entail an important amount of water savings, because CCGT avoids the use of water as a heat carrier in the first part of the cycle, so that, this kind of technology is more efficient in water use per MW·h produced than other conventional technologies.

Table 1. Water Withdrawal and Consumption for Electric Power Generation.

		Water Use Intensity (m^3/MW·h) Steam Condensing	
Plant-type	Cooling Process	Withdrawal	Consumption
Fossil/biomass steam turbine	Open-loop	75–190	0.75–1.14
	Closed-loop	1.14–2.28	1.14–1.82
Nuclear steam turbine	Open-loop	95–227	1.51
	Closed-loop	1.90–4.16	1.51–2.72
Natural Gas Combined-Cycle	Open-loop	28.4–76	0.38
	Closed-loop	0.87	0.68
Concentrating Solar	Closed-loop	2.83	2.80

Sources: EPRI (2002), MMA (2004) and other self-made calculations.

Besides the thermodynamic cycle, power plants use water for cooling purposes. For both reasons, power plants need to use water and to take it from a source like the sea, a river, etc. Most of this water that a plant needs to take or withdraw is used in the process but not consumed, because it is reused in the process or returned to the river where it had been taken. *Consumed water* would be the lost water after the whole process. There are plants that base its cooling process on water withdrawn from the sea, thus, our concern has to deal with those plants taking water from fresh sources.

Power plants also consume water for cycle make-up, washing equipment, coal ash transportation. However, with the exception of water used for fuel processing in gasification combined-cycle plant, among the mentioned process using water in a power plant, only cooling processes have a consumption of water that should be taken into account for the purpose of this chapter. The rest of processes have a very small consumption of water compared to it.

Then, water withdrawal will be strongly determined by the type of cooling process used. Power plants need to condense large amounts of low pressure steam water for return to plant's heat source for re-boiling. This step is achieved through an exchange of heat with a large quantity of cooling water. The warmed cooling water is either returned to the source where it was taken or cooled itself for re-use via evaporative heat transfer in a cooling tower or pond. In either case, a portion of water is lost or, what we have called *consumed water*, via evaporation to the atmosphere.

There are two main types of conventional cooling systems. In the once-through cooling or open loop process, a large quantity of water is withdrawn from its source and after returned to it at a quantity similar to that removed. The loss is due to the evaporation and only a small part (around 1%) could be considered as consumed. Usually, in these systems, the cooling process is design on the basis of a maximum allowable temperature increase above the natural water temperature of the source set by the environmental regulations.

In the closed loop systems, plants withdraw a much smaller quantity of water, because it is recirculated, but most of water is lost or consumed via evaporation in a cooling tower or pond. There are though other systems called *dry* using air-cooled condensers.

Then, the primary goal of cooling processes (used also in refineries, petrochemical plants and other industrial processes to remove heat) is to remove the heat absorbed in the circulating cooling water systems. Table 1 shows the water use and consumption in the power generation sector for each technology. The circulation rate of cooling water or water withdrawal in a typical 700 MW coal-fired power plant (working 6,000 equivalent hours) amounts to 75–190 m^3/MW·h if the plant uses an open-loop and 1.14–2.27 m^3/MW·h if it uses a closed-loop. The consumed water will be 0.75–1.14 m^3/MW·h and 1.14–1.82 m^3/MW·h respectively. The above mentioned plant will end up using through the year 315–800 Mm^3 with a water consumption of 3.1–4.8 Mm^3 if it uses an open-loop and 4.8–9.5 Mm^3 with a consumption of 4.8–7.6 Mm^3 if it uses a closed-loop. In the

Table 2. Water consumption for hydrocarbons and their substitutes.

Fuel type and process	Relationship to water quantity	Water consumption per-unit-energy (L/MW·h)[a]
Conventional Oil & Gas		
– Oil Refining	Water needed to extract and refine;	90.4–258.4
– NG Extraction Processing	water produced from extraction	25.8–28.7
Biofuels		
– Grain Ethanol Processing	Water needed for growing feedstock	155.04–2,067.2
– Corn Irrigation for EtOH	and for fuel processing	32,300–408,272
– Biodiesel Processing		51.7–64.6
– Soy Irrigation for Biodiesel		178,296–775,220
– Lignocellulosic Ethanol and other synthesized Biomass to Liquid (BTL) fuels	Water for procession; energy crop impacts on hydrologic flows	310–1,938 [b,c] (ethanol) 180.9–1,162.8 [b,c] (diesel)
Oil Shale	Water need to Extract/Refine	
– In situ retort		12.9–116.3[b]
– Ex situ retort		193.8–516.8[b]
Oil Sands	Water need to Extract/Refine	258.4–646

[a]Ranges of water use per unit energy largely based on data taken from the Energy-Water Report to Congress (DOE, 2007).
[b]Estimates based on unvalidated projections for commercial processing.
[c]Assuming rain-fed biomass feedstock production.

case a dry system were used, there would be no water consumption, but the efficiency of the plant would be lower than in the other cases, so that it would use more primary fuel to produce the same electricity and then the CO_2/MW·h ratio would be worse too.

Water consumption of petroleum and gas industry is difficult to quantify in the production or upstream activities. In the case of oil production some fields used to inject natural gas or are currently injecting water in order to extract oil and make the pressure up inside the field, but eventually, the quantities of fluid injected will depend on the geological structure of the field. For example, in fields that are using water to extract oil and need to make the field pressure up, like those on the North Africa, could be used a ratio of 8 L/barrel[5]. Besides, the oil extraction activity produces large amounts of water coming out together with the oil. The new technologies use CO_2 to do so, and mainly the EOR-CO_2 is one of the future ways to increase the efficiency of this activity. The variety of cases is so wide that to set a figure or a ratio for this activity could be misleading. In any case, the use of water in conventional oil production could be considered as not significant compared with other energy activities.

Petroleum refineries also have very large cooling systems. A typical refinery has a water consumption of 0.5 m^3/t of processed oil, although in some cases this figure could jump up to 1.5 m^3/t of processed oil.

When talking about other non conventional oil, like oil sands, the use of water becomes significant as shown in Table 2. Alternative fuels to oil have a significant consumption of water compared to it. For example, in the case of biofuels or synthetic fuels like CtL (Coal to Liquid), water is used for agricultural irrigation and fuel processing and to complete the synthesis process respectively. Biofuels need a good amount of water, but in the future, second generation biofuels can overcome this challenge and reducing significantly the water needs.

Tables 1 and 2 show a summary of water consumption by technologies and activities in the energy sector.

[5] The figure belongs to a real field although the reference is omitted for reasons of confidentiality.

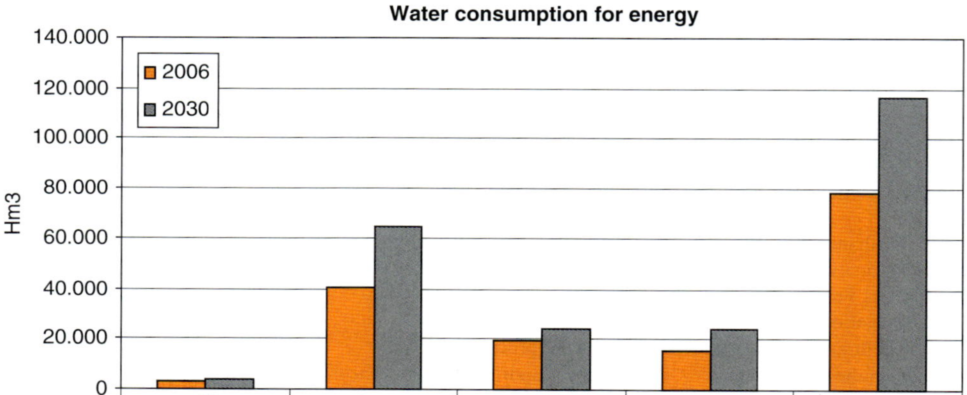

	Oil	Coal	Nuclear	Nat. Gas (CCGT)	TOTAL
2006	2.982	40.322	19.230	15.894	78.428
2030	3.781	64.822	23.800	24.240	116.642

Figure 4. Water consumption for energy in the main electricity production technologies.

Using the projections and forecast of energy demand for the next decades we can approximately draw the water consumption of water for energy in the next decades and draw some conclusions from it.

In Figure 4, it has been drawn some of the energy activities along with their water consumption. Oil has a not significant consumption and most of it takes place in the refining activity. As it has been pointed out through this chapter, the main water consumption takes place in the electric generation activity. In absolute figures (Mm^3 of water consumption), coal is the first consumer, followed by nuclear and natural gas power plants. In order to simplify the analysis and draw the Figure 4, it has been taken average specific data of water consumption per energy unit produced for every technology. This assumption implies a worse case scenario because technology will improve in the next years and environmental requirements will push companies to choose those systems that minimize water consumption. In any case, although the water consumption, in absolute figures, is huge (in the case of coal, for example, its consumption is even bigger than the overall consumption of fresh water in a country like Spain), they are very small if we compare them with the world water consumption in other activities like agriculture or industry.

For example, we can compare these data with the consumption of water in Spain which is around 35,000–40,000 Mm^3 (MMA, 2004), or with the world consumption of blue and green water to produce food which is 5,300–6,000 km^3/yr, or with the water consumed in irrigated land which is 2,000–2,500 km^3/yr (according with different figures obtained in several chapters of this book). We can see that the water consumption in electricity production, now and in the future, is not enough significant as to be considered as a problem.

However, the amount of water a plant needs does not represent a problem in global terms, but it can be a real problem in some areas of the world where they suffer from water scarcity. Even in the area of renewable energy, one of technologies with a very brilliant future as it is the solar concentrating thermoelectric power plant needs a good amount of water to work. A typical 50 MW plant needs 1 Mm^3 of water withdrawal with a 25% of consumption. Taking into account that this technology works with direct radiation and in some areas where there is high sun radiation (like in the South of Europe or North of Africa), there is also water scarcity, a high target of installed capacity in those areas will be a challenge that we have to overcome. The only way to do it is through the improvement of technology in order to make it more efficient in its water consumption needs.

5 CONCLUSIONS

The fast growth of fresh water demand due to increase of population, economic development and environmental requirements have made many areas of the world vulnerable to water shortages. This fact affects agriculture, industry but also energy industry and mainly electricity production which needs to be produced close to demand and uses water to complete its processes. The dependence on electricity production and water availability can jeopardize some economies to grow and to achieve a good level of wealth or standard of living. In the future, technology innovation will be a key issue to overcome these challenges. While water consumption from a global perspective is not a problem, it will in some areas suffering from water scarcity, adversely affecting these areas in their potential economic growth.

Within the energy sectors, electricity is the one with significant water consumption, although from a global perspective (compared to other sectors like agriculture and industry) it is not a problem. The faster electricity production mix will shift towards technologies using less water (most of renewable technologies and others like combined-cycle) the less water will be consumed per energy unit produced or demanded.

Some of renewable electricity technologies like concentrating solar thermoelectric must improve its efficiency in order to reduce its water demand and become a real alternative in areas where they suffer from water scarcity. Biofuels, aside from other challenges like deforestation and feedstock prices interference, will have to look for processes reducing water consumption too. In this case, second generation biofuels can solve this problem.

REFERENCES

DOE (2007). *Energy Demands on Water Resources. Report to Congress on the Interdependency of Energy and Water*. Department of Energy, USA. Available at: [http://www.sandia.gov/energy-water/congress_report.htm].

EPRI (2002). *Water & Sustainability (Volume 3): U.S. Water Consumption for Power Production – The Next Half Century*. Electric Power Research Institute (EPRI), Palo Alto, California, USA.

IEA (2008). *World Energy Outlook 2008*. Organisation for Economic Co-operation and Development (OECD)/International Energy Agency (IEA), Paris, France. 578 pp. Available at: [http://www.worldenergyoutlook.org/docs/weo2008/WEO2008.pdf].

MMA (2004). *Guía de mejores técnicas disponibles en España del sector refino de petroleo* [Guide of the best available technologies in the oil refining industry]. Spanish Ministry of the Environment (MMA). 258 pp.

CHAPTER 18

The economics of desalination for various uses

Xavier Bernat
CETaqua, Water Technology Center. Barcelona, Spain

Oriol Gibert
CETaqua, Water Technology Center. Barcelona, Spain
UPC, Universitat Politècnica de Catalunya. Barcelona, Spain

Roger Guiu & Joana Tobella
CETaqua, Water Technology Center. Barcelona, Spain

Carlos Campos
CETaqua, Water Technology Center. Barcelona, Spain
AGBAR Group. Barcelona, Spain

ABSTRACT: Water scarcity is a major problem that needs to be efficiently solved to ensure water availability for future generations. Desalination is an alternative technology for water production based on salt separation from water. Reverse osmosis (RO) and multi stage flash (MSF) are by far the most prevalent desalination processes. Desalination operating costs depend on several factors such as pre-/post-treatment requirements, selected technology, feed water quality, plant location, energy availability and brines disposal. These costs are usually lower for RO than MSF, moreover, MSF has been stated to be at its technological limit whilst RO can be developed and improved in the future. Despite its higher cost compared to conventional water sources, desalination can play an important role when developing a full-cost assessment of the integrated water cycle as it avoids some of the resource costs incurred when demand cannot be met. The aim of this study is to give a general overview on available desalination technologies and market, desalination for water reuse applications, desalination costs (capital, operation, resource and environmental) and economic instruments available to integrate desalination into water cycle economics.

Keywords: desalination costs, desalination technologies, desalination in water treatment, full-cost assessment, economic instruments

1 INTRODUCTION

Water supply for domestic, agricultural and industrial uses comes from many different conventional sources such as rivers, lakes and aquifers. This water often needs to be treated and purified at a treatment plant in order to meet quality requirements depending on its final use.

When conventional water resources become chronically scarce, as happens in many parts of the world, governments usually respond by 1) restricting water demand; 2) using existing resources more effectively; and/or 3) exploiting non-conventional water supplies such as desalination and water reuse. Non-conventional resources are well recognized as being able to potentially contribute to mitigate quantitative and qualitative stress on conventional water resources, especially in arid regions. This can be achieved by supplying water that can be used after proper treatment for different purposes such as agricultural irrigation, industrial uses, recreational activities, environment enhancement and aquifer recharge. An illustration of the water supply cycle is shown in Figure 1.

Among non-conventional water supplies, desalination has become an efficient alternative resource to mitigate water shortage in many countries throughout the world. The world production

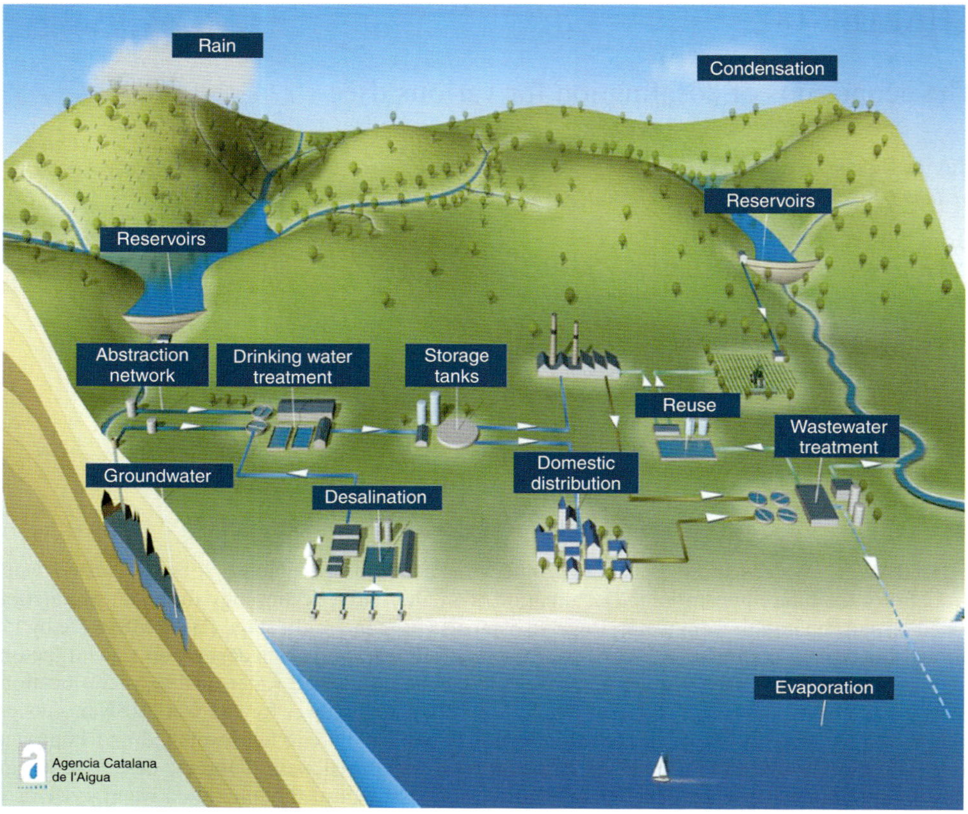

Figure 1. Diagram of the global water cycle (ACA, 2009).

of desalted water, which includes desalted sea and brackish water, was 48 Mm^3/day (enough for supplying potable water to 240 million people) in 2006 and about 65 Mm^3/day in 2008 (AEDyR, 2009). Desalted water production is expected to keep growing in the near future as water demand increases. A production of desalted water of 95 Mm^3/day for 2011 and 140–160 Mm^3/day for 2025 has been predicted (AEDyR, 2009).

The aim of this chapter is to present both the current situation and the future challenges of water desalination. For this purpose, aspects such as desalination technologies, the desalination market and uses of desalted water (including water reuse) are analyzed. The chapter includes a full-cost assessment of desalination considering capital and operating costs of the selected technology as well as the effects of the introduction of desalted water on the water cycle, in terms of resource and environmental costs. Furthermore, the document outlines how the economic instruments associated should be readdressed when introducing non-conventional water supplies.

2 AVAILABLE DESALINATION TECHNOLOGIES

There are several desalination technologies able to technically fulfill the water quality standards required for its final use. Table 1 shows the most common desalination processes organized by the source of energy used (electrical and/or thermal) and the separation mechanism involved in (according to whether water is separated from salts or viceversa).

Table 1. Classification of desalination processes (AEDyR, 2009).

Separation mechanism	Energy	Process	Name
Water separation	Thermal + Electrical	Evaporation	Multi Stage Flash (MSF) Multi Effect Distillation (MED) Thermal Vapor Compression (TVC) Solar Desalination (SD)
		Crystallization	Freezing Formation of hydrates
		Evaporation and filtration	Membrane Distillation (MD)
	Electrical	Evaporation	Mechanical Vapor Compression (MVC)
		Ionic filtration	Reverse Osmosis (RO)
Salt removal	Electrical	Ionic migration	Electrodialysis (ED)
	Chemical	Others	Ion Exchange (IX) Solvent Extraction (SE)

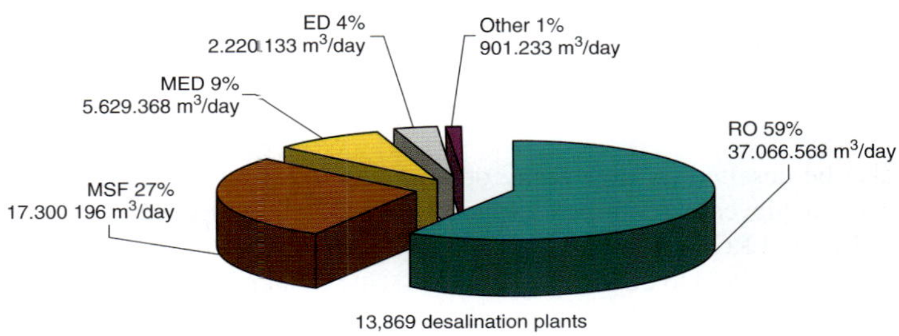

Figure 2. Distribution of desalination technologies by total installed capacity (GWI DesalData/IDA, 2008).

Selection of the appropriate technology has to be done considering several factors. Site-specific conditions such as energy and labor costs, available area, electric power cost and availability, impact of brines disposal, feed water salinity and desalted water quality will influence the desalination technology choice (Fritzmann *et al.*, 2007).

Reverse osmosis (RO) and multi stage flash (MSF) processes together account nowadays for more than 85% of the total installed desalination capacity, as shown in Figure 2. It has been predicted that most large-scale desalination plants over 40,000 m^3/day built during the next quarter of the century will be based on these technologies (Blank *et al.*, 2007).

MSF is by far the most robust desalination technology. Like any thermal evaporation process, it essentially consists of heating a saline solution to generate water vapor, which is then condensed to liquid water with lower salt concentration. Though a small percentage of the feed water is converted into vapor and condensed, MSF is capable of producing significant amounts of high-purity desalted water. The main disadvantage of MSF is that it is very energy intensive. For this reason, MSF is the preferred choice in the Middle East, where local energy costs are low. In other parts of the world, where oil and other fossil fuels cost is much higher and thermal processes are expensive, other desalination technologies such as RO are preferred.

RO, like any membrane process, consists in selectively restricting the passage of certain ions (including salts) through a membrane, resulting in a desalted water stream (permeate) and a waste stream containing the salts left behind (brine or concentrate). RO technology can be used for the removal not only of salts but also of non-saline contaminants (e.g. organic matter, bacteria, viruses, etc.).

Table 2. General features of MSF and RO (adapted from Blank *et al.*, 2007; Reddy & Ghaffour, 2007; Nisan & Benzarti, 2008).

	MSF	RO
Physico-chemical principle	Flash evaporation	Solution-diffusion
Energy consumption (including auxiliaries)	Electrical: 2.5–5.0 kW·h/m^3 Thermal: 40–120 kW·h/m^3	Electrical: 3.5–4.5 kW·h/m^3 Thermal: None
Total capital costs	High	Low
Operation costs	High	Low at 250 mg/L TDS product High at 10 mg/L TDS product
Replacement parts requirement	Low-Medium (large special pumps)	High (large pumps, membrane replacement)
Maintenance requirements	Medium	High
Chemical consumption	High	High
Conversion rate*	0.1–0.2	0.3–0.5
Potential for further requirements	Low (at technological limit)	High
Most needed R&D areas	Low-cost materials Low-cost and more efficient heat transfer materials	Pre-treatment, high-durable membranes, low-fouling/-energy materials

* Conversion rate: ratio between product to total feed water flow.

Water can also be desalted through many other processes, as shown in Table 1 and Figure 2. None, however, has achieved the commercial success of MSF and RO. Given that other technologies different from MSF and RO account all together for less than 15% of the total installed capacity, and that they are not expected to be installed in large-scale desalination plants (Blank *et al.*, 2007), these technologies will not be further discussed in this study.

The differences between MSF and RO are grouped in Table 2. As shown in the table, operating principles, process limitations, costs and perspectives are significantly different between MSF and RO.

The main advantage of MSF is the available knowledge of the technology. However, it is probably at its technological development limit. Due to this limitation, together with the expected development for RO, it can be estimated that MSF will probably not continue growing and being improved like RO will apparently do. In fact, it is foreseen that research into new pre-treatments, high-resistant elements manufacture, low-fouling/-energy materials formulation, together with the increase of the RO knowledge and transfer to plant operation levels, will enable the RO technology to be further adapted to specific water sources and operation constraints. Nevertheless, vast efforts still need to be invested in developing the RO process further in order to achieve a more efficient and cost-effective desalination technology.

3 DESALTED WATER AND RECLAIMED WATER USES

3.1 *Desalted water sources and uses*

The quality of the water to be desalted is a crucial parameter which determines its final use. Figure 3 shows the typical water resources being desalted and their final use. As the figure shows, seawater represents 62% of total desalted water, followed by brackish water (19%). With regards to the final use, more than 65% of desalted water is used for municipal purposes followed by industrial applications (around 20%), while irrigation only accounts for 2% of the total.

Desalted water use is conditioned by the quality standards specified in water quality regulations. Hence, for instance, the high percentage of municipal uses of desalted water is justified by the fact that RO technology is best suitable for the production of drinking water within the chemical

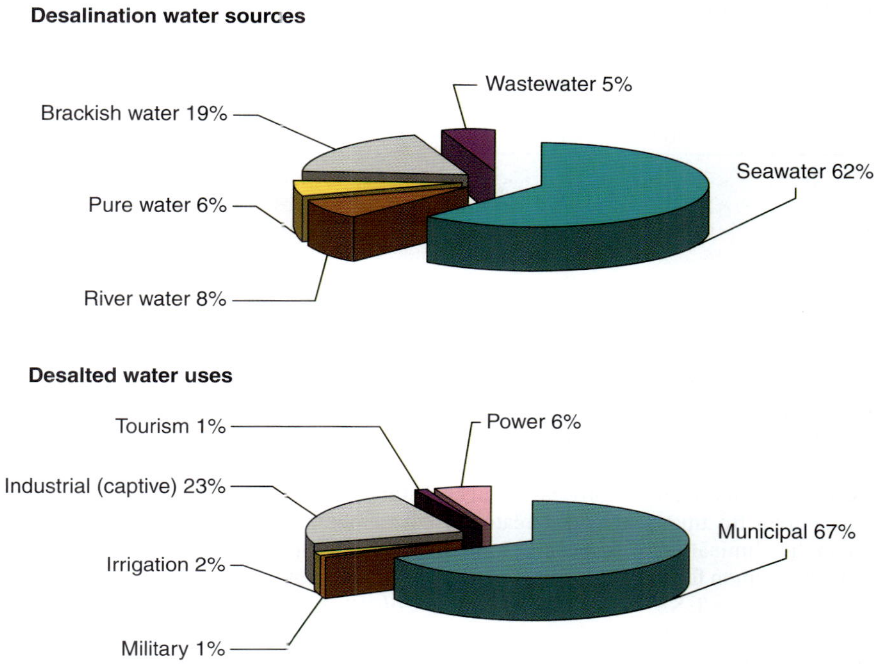

Figure 3. Sources and uses of desalted water in the world (GWI Desaldata/IDA, 2008).

and microbiological quality standards according to the European Directive 2000/60/CE (Water Framework Directive). However, two operational aspects have to be taken into account. The first is related to the pH of drinking water, which needs to be between 6.5 and 9.5. As RO permeate has an acidic pH, it has to be accordingly adjusted to satisfy the standards. The second limitation is related to boron presence in drinking water, whose concentration cannot be beyond 1 mg/L. Seawater typically contains an average of 4.6 mg/L boron. When desalting seawater with RO (one-step operation), boron concentration in permeate ranges between 0.8 and 1.5 mg/L. Hence, the system works close to the limits of the drinking water quality standards. Two possible solutions can help avoid this situation. One is based on increasing the pH up to 9.5–10 so that boron is in the form of borate and accordingly almost completely retained by RO in one-step operation. The other option consists in the installation of two RO steps. This process configuration allows fulfilling the drinking water quality standards in terms of boron concentration because the first-step permeate is purified in a subsequent RO step. Two-step operation is more adequate than pH adjustment because the latter could precipitate some of the ions present in the feed water causing membrane fouling (AEDyR, 2009).

3.2 *Water reclamation technologies and uses*

The possible uses of reclaimed water are related to their qualities and depend on the user country. For instance, in Spain, the Royal Decree 1620/2007 dictates the quality levels required for reclaimed water depending on its final use. The possible uses have been classified as industrial, agricultural, urban, recreational and environmental. Each group has specific subcategories and the legislation defines the quality standards of each subcategory.

The minimal quality standards can be organized in three main groups: microbiological parameters (*Ancylostoma*, *Escherichia Coli*, etc.), physico-chemical parameters (turbidity, suspended solids,

Table 3. Quality of tertiary effluents in function of the membrane technology installed (Prats Rico, 2009).

	Tertiary effluent quality			
	SS (mg/L)	Turbidity (NTU)	*E. Coli* (UFC/100 mL)	Nematodes (eggs/10 L)
MF	<7	<1	<1,000	Absence
UF	<5	<1	Absence	Absence
UF+RO	0	<0.3	Absence	Absence
EDR	<5	<2	10	Absence

etc.) and specific quality parameters for certain uses (*Legionella spp.*, total phosphorous, etc.). There are several technologies capable of treating wastewater and rendering it able to be used for a certain application. However, primary (generally flocculation/coagulation) and secondary treatments (aerobic oxidation) are installed in wastewater treatment plants. Nevertheless, for reuse applications, a tertiary treatment is required.

Three main tertiary treatment processes can be distinguished to achieve efficient water regeneration: conventional treatment, advanced treatment and membrane treatment. Conventional treatment is based on the elimination of suspended solids and turbidity followed by disinfection. This treatment is most common for treating water from the secondary treatment where the effluents present a good quality (<10 mg/L SS, <5 NTU, <10 UFC *E. Coli*/100 mL, absence of nematodes eggs/10 L). Advanced treatments are efficient for the elimination of biorefractory organic compounds contained in the secondary treatment effluent. They include adsorption and advanced oxidation processes (Fenton, ozonation, etc.). Finally, membrane technologies can also be satisfactorily used as tertiary treatment. For instance, microfiltration (MF) and ultrafiltration (UF) can be installed after secondary treatment to obtain higher-quality product water. However, when residual waters have a high salt content, RO is also needed to reduce it. Thus, desalination processes are frequently present in water reclamation processes and will gain relevance to satisfy water sustainability criteria in terms of water regeneration and reusability. In order to decrease membrane fouling, which is a common drawback in membrane processes, RO is generally preceded by UF pre-treatment (e.g. Wastewater Treatment Plants of Rincón de León and Benidorm in Spain).

Table 3 shows that the final quality of reclaimed water depends upon the membrane technology. Thus, the final water use dictates the right membrane filtration range (or a combination of them) to be installed as tertiary treatment (Prats Rico, 2009).

Apart from the previous membrane technologies, membrane bioreactors (MBR) can also be employed as simultaneous secondary and tertiary treatment. MBR are based on the use of a biological treatment coupled to membrane filtration (MF or UF). With the intensified technology, the decanter after the secondary treatment is not required. Hence, it can be stated that MBR are not only able to treat wastewater achieving the required discharge limits but also to simultaneously reclaim treated wastewater achieving the quality standards required for reusing it. It has to be pointed out that, as MBR are based on MF or UF membranes, the systems are unable to decrease water salinity. In order to reduce salts content, an additional RO unit would be required.

4 WORLD DESALINATION INDUSTRY

The evolution of the installed and contracted global desalination capacity is shown in Figure 4. As can be observed, overall desalination capacity has been continuously growing during the last twenty years and, as stated above, it is not expected to stabilize in the near future. Nowadays, the contracted capacity of desalted water in the world is around 65 Mm^3/day (installed capacity: 52 Mm^3/day). The geographical distribution of the desalination capacity on a worldwide basis is

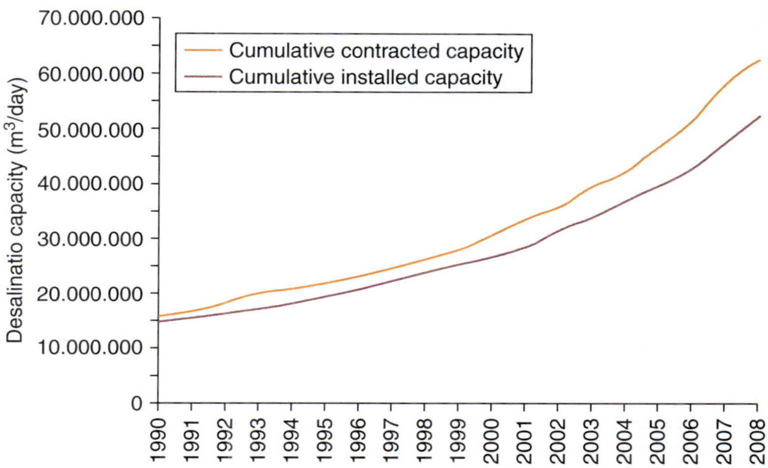

Figure 4. Installed and contracted desalination capacities evolution (GWI Desaldata/IDA 2009).

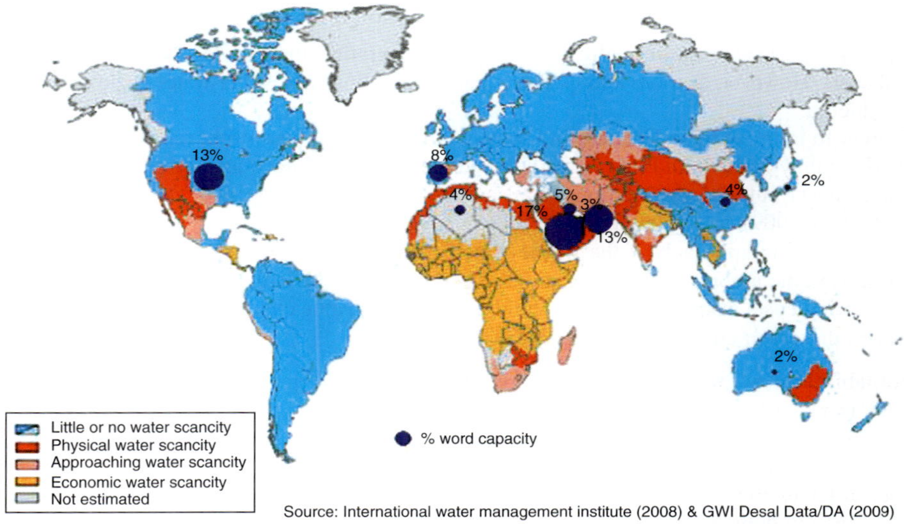

Figure 5. Major desalted water producing countries (IWMI, 2006; GWI DesalData/IDA, 2009).

presented in Figure 5. As can be observed, Saudi Arabia is the largest desalted water producer (17% of the global water production), followed by United Arab Emirates (UAE) and USA (13% of the desalted water production each) and Spain (8%).

Spain's pioneering involvement with desalination spans several decades. The installation of the first desalination plant on the island of Lanzarote in the Canary Islands dates back to 1964. Since then, a growing number of other plants have been installed to such an extent that nowadays the desalted water capacity in Spain is 8% of the global desalination capacity while its population represents only 0.6% of the world population (AEDyR, 2009). The Spanish Mediterranean coast currently hosts the largest desalination plants in Europe, for both drinking water (in Barcelona with a capacity of 200,000 m^3/day) and water for irrigation (in Torrevieja with a capacity of

Table 4. Largest desalination plants for potable water production from seawater (GWI DesalData/IDA, 2009).

RO		MSF	
Plant	Capacity (m^3/day)	Plant	Capacity (m^3/day)
Ashkelon (Israel)	326,144	Jebel Ali (UAE)	529,000
El Hamma (Algeria)	200,000	Shoaiba (Saudi Arabia)	524,000
Barcelona(Spain)	200,000	Fujairah 1 (UAE)	454,000

220,000 m^3/day). In some coastal regions in Spain, desalted seawater accounts for between 30 and 50% of water consumed (GWI Desaldata/IDA, 2009).

The desalination plants with largest capacities in the world are summarized in Table 4. As discussed above, desalination plants using MSF operate at higher capacities than those using RO and are geographically more concentrated around the Persian Gulf. Desalination plants using RO usually have lower production capacities and are geographically more distributed throughout the world.

5 ECONOMIC ELEMENTS OF DESALINATION FULL COST ASSESSMENT

Water supply costs have traditionally been defined as operating and maintenance costs (O&M) and capital costs. However, as more knowledge is gathered in this field, other water cycle costs need to be considered.

In order to improve the efficiency and sustainability of integrated water resources management (IWRM), policy-makers and actors in the water sector should account not only for capital and operation costs but also for environmental and resource costs (Rogers *et al.*, 2002; WATECO, 2003; OECD, 2009), as schematized in Table 5.

From a holistic approach of the whole water cycle, beyond the production plant, total costs are quite variable among different locations. Therefore, it is difficult and not reliable to compare unit operating costs from locations which do not share the same characteristics. Overall, the most important drivers of cost heterogeneity are:

- Availability of water resources, and more specifically, their quantity, quality and proximity to final users.
- Desalination technology and plant capacity.
- Environmental and energy cost.
- Special levies from water sector regulation.
- Efficiency in delivery service.
- Investments in network maintenance. It must be noted that recent networks do not require significant expense in maintenance, but on the other hand, higher amortization allowances must be faced as the value of the asset is significant.

In the following sections, desalination costs, introduced below in Table 5, are presented according to the whole water cycle approach.

6 CAPITAL AND OPERATING COSTS

6.1 *Desalination cost evolution*

The operating costs of producing desalinated water have steadily decreased during the last few decades thanks to continuous technological advances. Though still a costly water supply option compared to natural water resources such as ground- or surface water, desalination may soon be

Table 5. General principles for the cost of public water supply service. (adapted from Rogers *et al.*, 2002)

Cost component			
Capital charges	Full supply cost	Full use cost (= Economic cost)	Full cost
Operating and maintenance costs			
Resource costs			
Environmental cost			

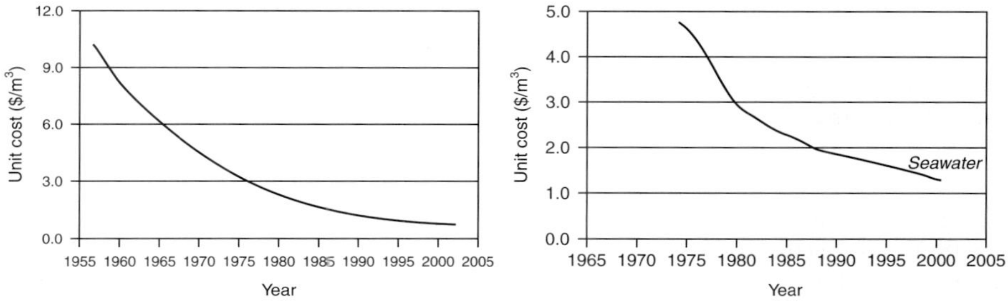

Figure 6. Evolution of the unit desalted water cost by: a) MSF; and b) RO, over the last decades (adapted from Reddy & Ghaffour, 2007).

a competitive alternative even in non-water stressed regions. It is important to note that costs for conventional water sources are expected to increase, while costs for desalting are expected to decrease as technology improves (Reddy & Ghaffour, 2007).

Figure 6 shows the evolution of desalination costs over the past few decades for both MSF and RO. As can be seen, costs for MSF in the early 1960s were between 7 and 9 US$/m^3, while they dropped down to below 1 US$/m^3 by 2000. A similar trend is observed for RO-desalted water, whose cost was between 3 and 5 US$/m^3 in the 1970s and is currently less than 1 US$/m^3. Recent examples for MSF and RO installations are Abu Dhabi's Taweelah (United Arab Emirates) desalination plant and Israel's Ashkelon desalination plant, which produce water at 0.7 US$/m^3 and 0.53 US$/m^3, respectively.

This sharp cost reduction has occurred in the three main cost areas: capital, energy and operation and maintenance. For desalination technology as a whole, the reduction of unit water cost is due mainly to:

- Technological developments.
- Increasing size of plants.
- Lower interest rate and energy costs.
- Changes in managing enterprise performance.
- Intense competition between equipment suppliers worldwide.

Specific improvements that have contributed to the cost reduction, among many others, have been optimization of process design (with regards to equipment and configurations) and thermodynamic efficiency, use of newer materials with better heat transfer properties, and development of new construction and transportation techniques for MSF; technological improvements of membranes (with higher surface per unit volume, higher salt rejection factors and fluxes, and extended life-span), optimization of pre-treatment options and use of energy recovery devices for RO (Fritzmann *et al.*, 2007; Reddy & Ghaffour, 2007).

Figure 7 illustrates desalination technological improvements by representing the evolution of the energy consumption necessary per m^3 of desalted seawater in Spain during the last 40 years. The figure clearly shows how the energy needed has been reduced from 22 kW·h/m^3 to less than 3 kW·h/m^3,

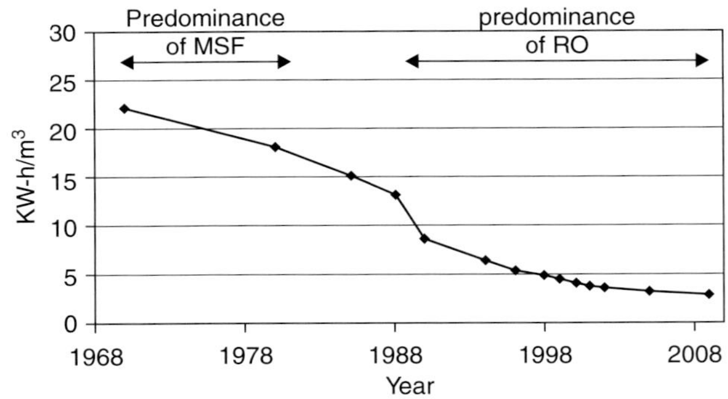

Figure 7. Evolution of the energy consumption necessary per m^3 of desalted seawater in Spain over the last decades (adapted from AEDyR, 2009).

i.e. the energy required per m^3 of desalted water has been reduced by almost 90%. It can also be seen in the figure that the emergence of the RO technology in the market notably helped in reducing the energy needs for desalination. These lower costs associated with RO promoted the expansion of RO facilities around the world (Reddy & Ghaffour, 2007; Karagiannis & Soldatos, 2008).

It must be pointed out that water costs of existing desalination plants vary widely and are not completely comparable because of differences in the cost-determining factors. This explains why values in Figure 7 are scattered, providing a general trend instead of giving definitive cost values. Site-specific cost-determining factors depend on a number of issues, such as feed water characteristics, product water quality, plant capacity, plant reliability, concentrate disposal, space requirements, operation and maintenance aspects, geographic location, energy availability, etc. (Blank *et al.*, 2007; Reddy & Ghaffour, 2007). Among them, the most critical factor is clearly energy consumption and availability, and this dictates not only the final cost but often also the desalination method. Hence, most MSF plants operate in oil-producing countries (basically in the Persian Gulf), whereas RO plants are more common worldwide in those places where there is a lack of adequate freshwater supplies and a good source of available seawater (e.g. the Mediterranean basin or Asia-Pacific region) (Lattemann & Höpner, 2008). Despite the variability of factors affecting the final cost of desalination, some general remarks on costs can be made according to the type of feed water, plant capacity, energy source and desalination technology.

6.2 *Water desalination cost with regards to feed water nature and plant capacity*

The amount of salt to be removed greatly affects desalination cost. In general, the more salts to be removed, the more expensive the desalination process. The cost of desalting seawater is about 3 to 5 times the comparable cost of desalting brackish water. Besides the salinity influence, the plant capacity also impacts desalination cost, with larger plants generally being more economical.

Table 6 summarizes the costs of desalination according to the nature of feed water and plant capacity. As can be seen, brackish water can be desalted more economically (at cost of about 0.21–0.43 €/m^3 at large scale) than seawater (at cost of about 0.40–0.80 €/m^3 also at large scale). At smaller capacity plants (<1,000 m^3/day), desalination costs increase to about 0.63–1.06 €/m^3 and 1.78–9.00 €/m^3 for brackish and seawater, respectively, or even more if equipment is not operated under optimal conditions.

6.3 *Water desalination cost with regards to energy source*

Desalination processes require significant quantities of energy to achieve separation of salts from seawater. Existing MSF and RO plants are powered mainly by conventional energy sources (gas,

Table 6. Cost of water produced according to the type of water desalted (Karagiannis & Soldatos, 2008).

Type of feed water	Plant capacity (m^3/day)	Cost (€/m^3)
Brackish	<1,000	0.63–1.06
	5,000–60,000	0.21–0.43
Seawater	<1,000	1.78 – 9.00
	1,000–5,000	0.56–3.15
	12,000–60,000	0.35–1.30
	> 60,000	0.40–0.80

Table 7. Cost of water produced according to the type of energy supply system (Karagiannis & Soldatos, 2008).

Type of feed water	Type of energy used	Cost (€/m^3)
Brackish	Conventional	0.21–1.06
	Photovoltaic	4.50–10.32
	Geothermal	2.00
Seawater	Conventional	0.35–2.70
	Wind	1.00–5.00
	Photovoltaic	3.14–9.00
	Solar collectors	3.50–8.00

oil electricity) because they still represent the most economic way to satisfy the energetic needs of desalination facilities. However, the coupling of renewable energy sources (RES) and desalination systems holds great promise as a feasible solution to water scarcity in remote areas where drinking water and conventional energy infrastructure is currently lacking. Several renewable energy powered desalination plants that operate only on solar, wind or geothermal energy to produce freshwater have been installed throughout the world and the majority have been successfully in operation for a number of years.

Table 7 compares desalination costs according to the source of energy used. It is clear that RES are still much more expensive than the conventional sources, although the higher cost of RES is counterbalanced by its environmental benefits. Hence, while the cost of desalination using a conventional source of energy ranges between 0.21 and 2.70 €/m^3 depending on whether the feed water is brackish or seawater, the cost of desalination when RES are used rises up to between 1.00 and 10.32 €/m^3, depending on the type of feed water and also on the type of renewable energy used.

Renewable energy powered desalination systems may not compete nowadays with conventional systems in terms of the direct cost of water produced. Nevertheless, their application is steadily expanding in certain areas and it seems clear that they will become a competitive alternative to conventional energy powered plants in the future as fuel prices keep rising and fuel supplies decreasing (Reddy & Ghaffour, 2007).

6.4 *Water desalination cost with regards to type of desalination technology*

Generally, MSF systems are more cost intensive but usually have larger production capacities than RO (Fritzmann *et al.*, 2007; Karagiannis & Soldatos, 2008; Lattemann & Höpner, 2008). As stated, they are mostly used in oil-producing countries where fuel availability is not a limiting factor, such as the Persian Gulf countries, where thermal desalination processes account for 90% of overall desalted water production (Lattemann & Höpner, 2008). RO emerged as a cheaper and more

Table 8. Cost of water produced according to the desalination method (Karagiannis & Soldatos, 2008).

Desalination method	Type of feed water	Plant capacity (m^3/day)	Cost (€/m^3)
MSF	Seawater	23,000–528,000	0.42–1.40
RO	Brackish	<20	4.50–10.32
		20–1,200	0.62–1.06
		40,000–46,000	0.21–0.43
	Seawater	<100	1.20–15.00
		250–1,000	1.00–3.14
		1,000–4,800	0.56–1.38
		15,000–60.000	0.38–1.30
		100,000–320,000	0.36–0.53

flexible desalination technology and has been gaining ground worldwide in the last years as the cost of membranes and their operation decreases. During the last few years, and thanks to these reduced costs, RO has become increasingly installed at larger plants and applied for the desalination of both brackish and seawater, as opposed to MSF, which is rarely used for the desalination of brackish water due to its much higher energy consumption at low salinity in comparison with RO (Fritzmann *et al.*, 2007; Reddy & Ghaffour, 2007).

Table 8 shows desalted water cost in relation to the desalination method used and the size of the plant. The desalination cost of seawater ranges between 0.42 and 1.40 €/m^3 for MSF. For comparative purposes, it must be taken into account that MSF plants have usually larger production capacities than RO and that this low cost is favored by economies of scale. The desalination cost through RO is between 0.38 and 1.38 €/m^3 for a mid-sized RO plant producing 1,000–60,000 m^3/day, while it is between 0.36 and 0.53 €/m^3 for larger plants producing between 100,000 and 320,000 m^3/day. For brackish water desalination using RO, the cost is even lower (0.21–0.43 €/m^3).

By inspecting data in Table 8, one can conclude that there are no significant differences in the cost per m^3 of product water when using RO or MSF for seawater desalination. However, as summarized in Table 2, the energy demand per m^3 of product water is significantly higher when operating with MSF technology than with RO. This is primarily due to the absence of thermal energy demand in RO plants. Thus, desalination costs should be evaluated in terms of energy demand per m^3 of product water. The cost similarities between RO and MSF shown in Table 8 are caused by the variability of energy costs depending on the country where the desalination plant is installed.

6.5 *Desalination cost breakdown*

The economic analysis of a desalination plant is based on a number of technical and economic determining factors such as capital, energy, labor, chemicals, materials, and consumables that are specific to the location, the process type, the plant capacity, the feed water components, the energy availability, the pre-treatment required, the product salinity desired and other site-related costs for land, plant, and brine disposal. Figure 8 shows the cost composition for standard MSF and RO desalination plants.

While amortization clearly represents a significant portion of the total cost (approximately one third of the total for both MSF and RO), the greatest cost driver actually arises from the energy consumption, which amounts up to 55.2% for MSF (including both fuel and electricity consumption) and to 42.6% for RO. Other components that contribute to the overall cost are labor and the use of chemicals (which amount to 9.3% and 6.5% for RO, much higher than 3.5% for MSF). For RO, the contribution to the total cost of the membrane replacement is 4.7%, a percentage which has decreased over the years owing to the improvements in membranes. Figure 8 compares well with those published by Reddy & Ghaffou (2007) and Fritzmann *et al.* (2007).

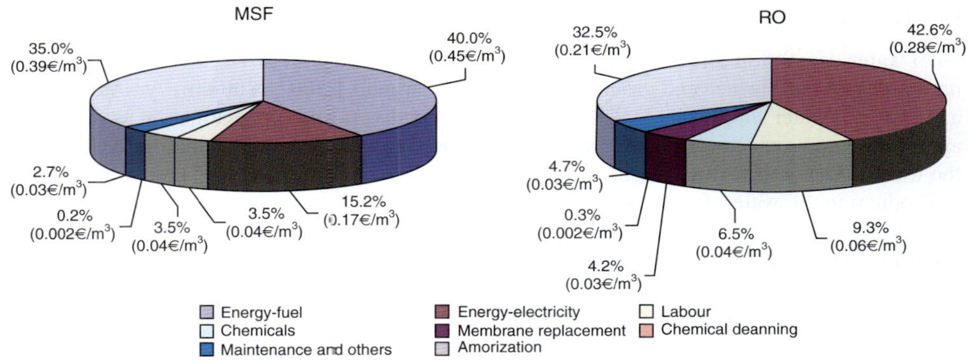

Figure 8. Comparison of freshwater production costs from seawater for MSF and RO (AEDyR, 2009).

7 RESOURCE COSTS

Supplying enough water to fulfill current demand in a certain region is the essential element of water policy. Water deficit in domestic, industrial, agricultural, recreational or environmental uses should be measured or accounted as a resource cost, not just as a probability (Gaya, 2008).

Policy-makers must properly manage water demand and supply. For this reason, water demand policies are acquiring more and more relevance (Dinar & Albiac, 2009). As water supply guarantee is not free of cost, a compromise between benefits gained and investment associated is required.

In this field, desalination can play an important role in most water scarce regions worldwide. In spite of its relatively high costs, desalination seems to appear as one of the most suitable alternative resources to satisfy water demand.

Several benefits derived from integrated water resources management have been identified (Mujeriego, 2007):

- Increase of own resources availability, which can decrease the dependence on water transfers from distant locations or from other uses, such as irrigation.
- Higher direct control on the available volume of water resources by the regulator or operators.
- Increased protection against delivery service failure due to circumstances beyond control, such as natural disasters or other threats on current water resources.
- Diversification of water sources.

Boxes 1 and 2 examine water management actions in Israel and Singapore, showing how resource costs play an essential role when policy-makers readdress water policy and choose whether to foster desalination or not.

8 ENVIRONMENTAL IMPACT AND COSTS

Adverse impacts on ecosystems caused by water related issues have usually been associated only with water uses. However, it is well known that water treatment technologies themselves also origin environmental costs, which are entailed in the final water use.

The environmental costs of desalination are not negligible and there is much debate about its extent. Basically, there are two main impacts on ecosystems that need to be addressed:

- Brine discharges and their impact on the receiving water bodies.
- Greenhouse gas emissions arising from high energy requirements.

Box 1. Israel: Cost-effective supply through desalination

Total water demand in Israel is expected to reach 2,500 Mm^3/yr in 2010, while the available natural resources (surface and ground water) are only around 1,500–1,800 Mm^3/yr. As there is a big gap for water supply, the Israeli government recently prioritized desalination as the most appropriate non-conventional water source as opposed to wastewater reuse and reductions in the water allocations for irrigation.

Production of water from surface and ground water resources represents about 0.15–0.45 US\$/m^3 and 0.37–0.45 US\$/m^3 for desalted water produced from the Sea of Galilee. Environmental and resource costs, which are not included in the costs given above, are estimated to be high because water withdrawal and distribution (carrier of 130 km along one of the tensest regions in the world) consume considerable amounts of energy. On the other hand, supplying reclaimed wastewater and brackish water is still a more expensive system (0.50–0.60 US\$/m^3).

Ashkelon desalination plant is the first step of a wider plan in order to supply enough water in Israel and, therefore, reduce the resource costs that eventually could challenge the country. Desalination presents additional benefits such as the improvement of water quality (softer and less mineralized, with a reduction on costs for customers' domestic facilities and subsequent wastewater treatment for reuse) and savings in pumping energy from the North provision. For these reasons, desalination is expected to provide 580 Mm^3/yr of freshwater in 2015, which will mean 22% of total supply.

Source: Dreizin (2006)

Box 2. Singapore: Self-sufficiency from water reuse and desalination

Water supply in Singapore has traditionally relied on different small reservoirs and on water imports from Malaysia, which represented the main source. The government has developed two new action plans to become self-sufficient in water by 2060: water reuse and desalination. However, these alternatives may imply higher costs than current water transfers (the price of transferred water from Malaysia is only 0.20 US\$/m^3). Therefore, the determining factor in this scenario is the resource cost, which depends on a third country, along with constraints from other alternatives to meet the increasing water demand.

The seawater RO plant, with a capacity to produce 136,000 m^3/day, in operation since 2005, under a public-private scheme is producing desalted water at 0.78 US\$/m^3. Reclaimed water, which is produced in three plants with similar technology, is reused for indirect potable use (discharge in urban reservoirs) and direct non-potable use (urban landscape irrigation) at 1.00 US\$/m^3, including both production and distribution costs.

Source: GWI (2009); Tortajada (2006)

8.1 *Brine discharges*

Brine discharges can damage the environment due to: 1) their high salinity and density, which can impact on benthic sea life, being the case of the *Posidonia Oceanica* (this sea grass plays a key role in marine sustainability); and 2) the possible oxygen depletion when sodium bisulphite is used as chlorine control. Pre-treatment residual turbidity/TSS can also have an impact on photosynthesis activity by lowering the transmissivity of the water body at the spilling point. In order to reduce the environmental impact of desalination, in the USA and Australia, there are regulations limiting the content of TDS and chlorides in brine discharges. The United States Environmental Protection Agency (USEPA) sets a limit of 10% for TDS while Australian desalination projects are requested to guarantee less than 1–2% increase of TDS at the discharge zone limit. Temperature variations can also have an adverse impact on marine ecosystems. Thus, when dealing with thermal desalination processes (RO brine temperature increase is expected to be only between 1 and 2°C), brine temperature impact on the receptor environment has to be considered and efficiently reduced.

Table 9. Estimated quantities of GHG emissions on a world basis by diverse fossil fuelled plants and desalination processes (Nisan & Benzarti, 2008).

	MSF (Mt/yr)				RO (Mt/yr)			
Power plant	CO_2	SO_x	NO_x	Particles	CO_2	SO_x	NO_x	Particles
Coal fired	264.45	0.33	0.54	0.04	32.25	0.04	0.07	0.005
Oil fired	216.22	1.31	0.3	0.03	25.74	0.16	0.04	0.003
Gas turbine, CC	141.58	0.01	0.23	0.01	12.87	0.001	0.02	0.001

Mt: million tonnes [1 tonne = 1,000 kg]

With the aim of minimizing the impact of brines disposal on receiving water bodies, transport direction and the potential impact range of a discharge plume is controlled by site-specific oceanographic conditions such as currents, tides, water depth and the bottom and shoreline topography. As for the characterization of environmental and operational conditions, they can be investigated by hydrodynamic computer models.

Quantifying this environmental impact and the associated cost has caught the interest of many researchers (Lattemann & Höpner, 2008). In order to properly manage this environmental cost and consider it in the full-cost approach, further estimations on the effects of brines discharge need to be carried out via monetization of the impact.

8.2 *Greenhouse gas emissions*

The significant quantities of energy required by conventional desalination plants and the consequent high greenhouse gas (GHG) emissions are major issues with desalination that need to be addressed. Table 9 gives the estimated quantities of GHG by diverse fossil fuelled plants and desalination processes for a representative plant with a total desalting capacity of 20.1 Mm^3/day (Nisan & Benzarti, 2008). In view of the values in Table 9, it seems clear that energy consumption in desalination contributes to environmental pollution, including global warming.

It is possible to roughly approximate the value of the environmental costs through the international carbon market. In this market, a value for an emitted tonne of CO_2 is agreed and considered as a cost for polluting industries, which need to buy extra credit in order to continue with their economic activity. However, the carbon market is not applicable for calculating the environmental costs generated by SO_x, NO_x and particles. The data obtained by the carbon market can be used as a proxy of the cost of pollution. For rapid calculations, the average value for 2008 of a tonne of CO_2 is 16.78 US$ (World Bank, 2009). By using the data for oil-powered plants, it can be found that, as Table 9 shows, the environmental cost from GHG emissions is 0.49 US$/$m^3$ for MSF and 0.06 US$/$m^3$ for RO. Therefore, the consideration of these costs in the production stage of the water cycle implies an important impact on overall costs, especially for MSF.

8.3 *Actions to mitigate the environmental impact*

Once the external costs of desalination over ecosystems have been identified, a number of actions need to be taken in the near future to reduce its impact. These actions are listed below:

- Brine discharges: waste streams generated in desalination processes can be efficiently treated and its impact reduced if they are diluted with the effluent from near wastewater treatment plants prior being discharged into the sea. This mitigation method is used in the desalination plant of Barcelona (Spain).
- GHG emissions: increasing the energy efficiency of desalted water production and integrating non-pollution renewable energy sources is crucial for decreasing environmental impact of desalination. Nisan & Benzarti (2008) compare RO processes using nuclear power and other fuel fossils and conclude that the desalination costs of nuclear options, which include environmental costs, are 10–80% lower than the cheapest fuel fossil options. However, it has to be pointed out

that nuclear wastes, which have an adverse impact onto the environment, are generated during nuclear energy production.

9 ECONOMIC INSTRUMENTS IN AN INTEGRATED WATER CYCLE

The above sections identify the costs imposed by the final use of water in the whole cycle and assess how desalination can be incorporated in this framework. At this point, it is time to raise the most controversial question: who should be charged for this, and how? The background behind this economic analysis is the need to fulfill the principle of cost recovery: all the agents involved in the water public service would have to participate in their appropriate part of full cost.

Legislation worldwide is incorporating these relatively new concepts into water policy. Policy-makers are gradually modifying the economic instruments intended to recover all the costs, the most noteworthy being the so-called 3Ts: tariffs, taxes and transfers (OECD, 2009). The right combination of these instruments contributes to a sustainable cost recovery, which is a more realistic and practical policy principle than the traditional financial basis.

a) *Tariffs* are a basic element of pricing policy and are paid directly by consumers. Tariffs can take on various forms but they are usually focused on and designed according to the final users characteristics (industry, agricultural, etc.). Mathys (2008) gives an exhaustive and complete assessment of the different tariff structures currently applied, their main policy objectives and their consequences on the water service and users. Due to the increased need of using alternative water resources, conventional tariff structures need to be adapted considering the integrated water resources pricing approach (Rogers *et al.*, 2002). Moreover, as alternative water sources full production costs impact differently on the average supply cost, a redefinition of the tariffs is needed.

   The price of each m^3 delivered to the user should reflect the water source in that particular moment (surface water, groundwater, seawater…) which in turn depends on the resource availability. The sense of this tariff variability is shown by the following equation, where the tariff (P) is a function of both variables (volume and specific cost) of each source available.

$$P = f[(r_{surface}, a_{surface}); (r_{groundwater}, a_{groundwater}); (r_{desalination}, a_{desalination}); (r_{reuse}, a_{reuse}); (r_{others}, a_{others})]$$

   where *r* refers to the volume supplied by a particular source and *a* refers to its specific costs.

b) *Taxes* and other instruments related to water taxation can have several objectives: environmental, economic, social and/or territorial sustainability (Sangrà, 2008). Nevertheless, sustainability, which is mostly provided and regulated by the public sector, needs to be financed as well.

   These instruments can take different forms and be applied along the cycle usually following the user-pays and polluter-pays principle. For instance, the water agency of the State of São Paulo (Brazil) collects flat charges from all users that extract and discharge water (farmers, industries and water utilities), whose financial resources are used to restore the quality and availability within the basin (GWP, 2009). On the other hand, in France there are different taxes applied by the basin agencies and the State. For instance, charges applied to consumers polluting and withdrawing water are used for network renovations. The State also collects a levy for financing the navigable river routes.

   Finally, another diversified example is Catalonia (Spain) (Sangrà, 2008), where a water canon is levied by the regulator and to be paid by domestic and industrial users. The levy has an ecological aim: pollution prevention and maintenance of environmental flows. At the same time, it finances the water management and wastewater treatment (investment and operation) and permits a sense of equity due to increasing block-rates. The agency also applies a canon for water abstractions and a charge for financing water related infrastructures, which is paid by the direct users of these investments. It has to be noted that the construction and amortization of a desalination plant in this region is partly financed by this charge.

c) *Transfers* consist in the grants given by other public sector levels, supranational levels or institutions outside the basin or State. In particular, official development assistance from multilateral sources, financing programs provided by international finance institutions or regional development banks and the funds given by the European Union to its member States, are some examples of this kind of economic instrument.

The role played by transfers is not trivial since it represents sometimes the major or unique source for the States or regulators to finance the required infrastructures for guaranteeing water supply and the essential sanitation facilities. Recently, most of the desalination plants and urgent waterworks in the European Union have been implemented thanks to the cohesion and structural funds allocated by the European Commission. Indirectly, those investments are being paid by a wider range of taxpayers. This makes economic sense because all the steps taken for stronger water supply security will benefit all citizens and potential consumers and not only the direct users of the new waterworks.

10 CONCLUSIONS

Desalination represents a potential alternative technology for the efficient production of freshwater from many sources. MSF and RO are the most common desalination technologies and their geographical distribution strongly depends, among other factors, on energy availability. MSF is preferred in arid regions with fuel availability at low cost whilst RO is most installed in regions with lack of good-quality freshwater and with good-quality seawater.

The costs associated with desalination depend on many factors such as capital, energy, labor, chemicals, materials, and consumables that are specific to the location, process configuration, plant capacity, feed water components, energy availability, pre-treatment necessities, product salinity and other site-related costs for land, plant and brine disposal. A detailed analysis of each situation is thus required to estimate desalination costs. It could be stated that RO cost is lower than MSF one in energy and environmental terms but the available knowledge on the latter is higher than on the former. Both technologies are by far predominant in the desalination market, accounting for more than 85% of the total installed desalination capacity. Nevertheless, these technologies need to be further improved to satisfy future water demands at sustainable cost. As MSF seems to be at its technological limit, current R&D efforts are being invested mostly in RO. New-generation membranes having higher permeabilities, rejection factors, fouling resistance, etc. have to be developed and installed to decrease desalination costs. In general, knowledge on RO has to be gained and transferred from labs-to-plants so that operators can efficiently operate membrane plants and understand mechanisms involved in the desalination process.

From an economic point of view, when integrating desalination technology in water cycle economics, policy-makers should consider capital, operating and maintenance, resource and environmental costs in the decision process. Although desalination emerges as a moderate resource cost technology, environmental impact of desalination is still high and thus needs to be further minimized.

The integration of alternative water resources implies rethinking and restructuring the traditional financing framework, focusing on tariff structures. An integrated water resource pricing, where the final price varies in function of the source and where the different production costs are considered, would better adapt to the current water policies and legislation. Public transfers and environmental taxes can also be defined to foster the development of desalination technologies.

In general, it can be concluded that R&D efforts need to be invested in minimizing capital, operating and maintenance, resource and environmental costs of desalination. Furthermore, integrated water resource pricing needs to be implemented in order to properly quantify desalted water cost.

ACKNOWLEDGEMENTS

The authors are indebted to Prof. E. Custodio and Prof. M.R. Llamas for their helpful discussions and valuable cooperation. The *Fundación Marcelino Botín* is also acknowledged for inviting the authors to participate in the 4th Water Workshop *Re-thinking paradigms: water and food security*.

REFERENCES

ACA (2009). Agència Catalana de l'Aigua, [http://aca-web.gencat.cat/aca]. Accessed on September 2009.

AEDyR (2009). Official notes of desalination course. Asociación Española de Desalación y Reutilización (AEDyR). Madrid, Spain.

Blank, J.E.; Tusel, G.F. & Nisan, S. (2007). The real cost of desalted water and how to reduce it further. *Desalination*, 205: 298–311.

Dinar, A. & Albiac, J. (2009). *Policy and strategic behaviour in water resource management*. 1st ed., United Kingdom.

Dreizin, Y. (2006). Ashkelon seawater desalination: off-taker's self costs, supplied water costs, total costs and benefits. *Desalination*, 190: 104–116.

Fritzmann, C.; Löwenberg, J.; Wintgens, T. & Melin, T. (2007). State-of-the-art of reverse osmosis desalination. *Desalination*, 216: 1–76.

Gaya, J. (2008). La sequera i el proveïment d'aigua. In: *Aigua: font de vida, font de risc*. Report from Observatory of Risk, Institut d'Estudis de la Seguretat, Barcelona, Spain.

GWI (2009). *Municipal water reuse markets 2010*. Global Water Intelligence (GWI) & Media Analytics, United Kingdom.

GWI DesalData/IDA (2008). Global Water Intelligence DesalData/International Desalination Association, *Desalination in 2008 - Global Market Snapshot. USA*.

GWI DesalData/IDA (2009). Global Water Intelligence DesalData/International Desalination Association, [http://www.desaldata.com]. Accessed on August 2009.

GWP (2009). *A handbook for Integrated Water Resources Management in basins*. Global Water Partnership (GWP) & International Network of Basin Organization. Sweden.

IWMI (2006). International Water Management Institute, [http://www.iwmi.cgiar.org/]. Accessed on July 2009.

Karagiannis, I.C. & Soldatos, P.G. (2008). Water desalination cost literature: review and assessment. *Desalination*, 223: 448–456.

Lattemann, S. & Höpner, T. (2008). Environmental impact and impact assessment of seawater desalination. *Desalination*, 220: 1–15.

Mathys, A. (2008). Appropriateness of tariff structure: lessons from experience. Working paper for: Managing and planning water services, IWA World Water Conference, Vienna, Austria.

Mujeriego, R. (2007). La reutilización, la regulación y la desalación en la gestión integrada del agua. In: *Scarcity in Spain: Directives to minimize its impact* (original title: *La Sequía en España: Directrices para Minimizar su Impacto*). Ministry of Environment. Madrid, Spain: 155–202.

Nisan, S. & Benzarti, N. (2008). A comprehensive economic evaluation of integrated desalination systems using fossil fuelled and nuclear energies and including their environmental costs. *Desalination*, 229: 125–146.

OECD (2009). *Managing water for all: an OECD perspective on pricing and financing*. OECD Publishing.

Prats Rico, D. (2009). Water treatment technologies for reuse - quality conditioning (original title: Las tecnologías de tratamiento del agua para la reutilización - acondicionamiento de la calidad). Conferencia Nacional sobre Reutilización del Agua.

Reddy, K.V. & Ghaffour, N. (2007). Overview of the cost of desalinated water and costing methodologies. *Desalination*, 205: 340–353.

Rogers, P.; de Silva, R. & Bhatia, R. (2002). Water is an economic good: How to use prices to promote equity, efficiency and sustainability. *Water Policy*, 4: 1–17.

Sangrà, J.M. (2008). Qui paga l'aigua? In: *Aigua: font de vida, font de risc*. Report from Observatory of Risk, Institut d'Estudis de la Seguretat, Barcelona, Spain.

Tortajada, C. (2006). Water management in Singapore. *International Journal of Water Resources Development*, 22: 227–240.

WATECO (2003). Economics and the environment: the implementation challenge of the Water Framework Directive. Common implementation strategy for the WFD, Guidance document No 1. WATECO (Working group for the economic studies of the WFD). Office for Official Publications of the European Communities, Luxembourg.

World Bank (2009). State and trends of the carbon market 2009. World Bank Institute, Washington, D.C., USA.

POSTSCRIPT

The impossibilities of water in agriculture: An increasingly unreal world. What we have to do about it?[1]

Margaret Catley-Carlson
Global Water Partnership

> *Alice laughed: "There's no use trying", she said; "one can't believe impossible things".*
>
> *"I daresay you haven't had much practice", said the Queen. "When I was younger, I always did it for half an hour a day. Why, sometimes I've believed as many as six impossible things before breakfast".*[2]

1 THE VIEW THROUGH OUR LOOKING GLASS: HERE ARE SOME OF THE INESCAPABLE REALITIES OF THE WATER/AGRICULTURE RELATIONSHIPS FACING OUR WORLD

- Population is growing – world population has now past 6,800 million; and in water stressed areas, the population grows fastest.[3]
- Climate change forecasts are firming up to show real regional shortfalls.
- Meanwhile agricultural water demand is sharply increasing.
- Municipal and industrial water use has shown a sharp increase.
- Recent food price doubling, even tripling showed a widespread atavistic response.[4]
- Groundwater depletion – we are headed for a brick wall.
- There is no new investment in low yield areas.
- Globally 70 River Basins are closing where 1,400 million people live there is no water left for more development – Yellow River, Colorado, Amu/Syr Darya, Murray-Darling, Egypt's Nile, Lerma-Chapala, Jordan, Gediz, Zayanda Rud, Indus, Cauvery, Krishna, Chao Phraya, …[5]

And here of some of the equally inescapable *impossibilities* – impossible in the sense that if food and water are essential to life itself, it seems impossible that we as an intelligent innovative species, given the above changes, countenance these developments.

- The investments in agricultural research are falling.
- There are few or negative changes in incentives for agricultural change.
- Population growth is a taboo subject; which is also attracting little or no investment.

[1] This Appendix corresponds to the notes prepared and distributed by Margaret Catley-Carlson to all the participants of the Workshop of Santander (September 2009) as possible *hot issues* related to *Water and Food Security*. These notes have been here included due to its interest in order to stimulate the debate.
[2] Lewis Carroll – Alice through the Looking Glass.
[3] Digital Journal, July 11, 2009.
United Nations chief addresses one solution to the challenge of overpopulation on World Population Day, as the world's human population now reaches 6.8 billion people.
[4] IFPRI – *Coping with current high food prices: what, who and how of Proposed Policy Actions*. 2008.
[5] The Comprehensive Assessment of Water in Agriculture – Water for food, Water for life – D. Molden (ed.).

- The Doha trade round is stuck, and is being replaced by bilateral agreements.
- Yet the subsidies remain, even in high price climate and these are a major factor in agriculture production decisions, including water use.
- There is however, a closed door, closed minds to drought/temperature resistant GMOs.
- Meanwhile the maps of potential conflict areas grow.
- In the case of fisheries – ocean and freshwater are at a limit.

The essential Food/Agriculture/Water Dilemmas …

- In the future there will be 2,000 million more to feed – with a protein diet it is important to consider that 1 calorie takes 1 liter of water.
- Water scarcity is a threat to food security but we are NOT running out of drinking water – we are running out of economic water – agricultural, energy, industry, tourism uses are in competition. The poor and the environment tend to lose in this competition for economic water.
- We are not appreciably moving towards a solution.

So,

- What needs to be done?
- Why it is so difficult to do what needs to be done?
- What might be the responsibility of those in this room – different responsibilities than have existed in the past?
- Are there lessons from those who have created real change from policy prescriptions?

II WHAT NEEDS TO BE DONE? THIS IS THE PART WE ALL LIKE BEST – SETTING OUT THE SOLUTIONS

- More crop per drop (or more benefits to nature per drop).
- Multifunctional Integrated Agriculture-Aquaculture Systems. Livestock/fish.
- Expand policies and take key actions to upgrade rainfed systems – this might offer the highest potential for poverty reduction and water productivity gains.
- Some new technology:
  - Water harvesting.
  - Supplemental irrigation.
  - Field water conservation to reduce evaporation.
- Build human capacity to understand, participate, and manage systems.
- Ensure secure access (including water rights).
- Targeted investments in pro-poor technologies.

What are the enabling conditions for these things to happen? This is a zone of much less comfort.

- Cost & affordability.
- Price and profitability.
- Risk – market, climate, water availability.
- Market reform.
- Individual behaviour – eating, consuming.
- Reform incentives and institutions.
- Education.
- Whole sectoral reforms needed.
  - Craft solutions suited to local needs.
- No blueprints.

Glossary

Blue virtual-water content (V_b)
Volume of surface and groundwater consumed as a result of the production of a good or service (m^3/t). Consumption refers to the volume of freshwater used and then evaporated or incorporated into a product. It also includes water abstracted from surface- or ground-water in a catchment and returned to another catchment or the sea. It is the amount of water abstracted from ground- or surface water that does not return to the catchment from which it was withdrawn (Hoekstra *et al.*, 2009).

Blue water
Fresh surface- and ground-water (Hoekstra & Chapagain, 2008).

Crop water requirement (CWR)
It is defined as the total water needed for evapotranspiration, from planting to harvest for a given crop in a specific climate regime, when adequate soil water is maintained by rainfall and/or irrigation, so that it does not limit plant growth and crop yield (mm/time period) (Allen *et al.*, 1998).

Effective rainfall (P_{eff})
The portion of the total precipitation that is retained by the soil so that it is available for crop production (mm/time period) (FAO, 2009).

Environmental flow (EF)
It is the water regime provided within a river, wetland or coastal zone to maintain ecosystems and their benefits (Dyson *et al.*, 2003).

Environmental flow (or water) requirement (EFR or EWR)
It refers to the quantity, quality and timing of water flows required to sustain freshwater and estuarine ecosystems, and the human livelihoods and well-being that depend on these ecosystems (The Brisbane Declaration, 2007).

External water footprint (WF_e)
It is defined as the annual volume of water resources used in other countries or regions to produce goods and services consumed by the inhabitants of the country or region concerned (km^3/yr, m^3/yr per capita) (Hoekstra & Chapagain, 2008).

Food security
Food security exists when all people, at all times, have physical and economic access to sufficient, safe and nutritious food to meet their dietary needs and food preferences for an active and healthy life (WFS, 1996; FAO, 2001; FAO, 2009).

Green virtual-water content (V_g)
Green virtual-water content of a product is the volume of rainwater that evaporated during the production process. This is particularly relevant for agricultural or forestry products, where it refers to the total rainwater evapotranspiration (from fields and plantations) plus the water incorporated into the harvested crop or wood (Hoekstra *et al.*, 2009).

Green water
It is the rainwater stored in the soil as soil moisture, also called soil water. Eventually, this part of precipitation evaporates or transpires through plants. Green water can be made productive for crop growth (but not all green water can be taken up by crops, because there will always be evaporation from the soil and because not all periods of the year or areas are suitable for crop growth) (Falkenmark, 1995; Hoekstra *et al.*, 2009).

Gross value added (GVA)
It is the value of goods and services produced in an economy at different stages of the productive process (million €). The gross value added at market prices is equal to the gross output (value of production) minus the intermediate consumption.

Internal water footprint (WF_i)
It is defined as the use of domestic water resources to produce goods and services consumed by inhabitants of a country or region (km^3/yr, m^3/yr per capita) (Hoekstra & Chapagain, 2008).

Net Income
It is equal to the income that a firm or a nation has after subtracting costs and expenses from the total revenue. Net income is an accounting term. It refers to the GVA plus subsidies and taxes, minus the consumption of fixed capital and salary payments, rentals and interests.

Value of production
It is defined as the total economic value received for the commodities sold in the market (total €).

Virtual-water content (V)
The virtual-water content of a product (a commodity, good or service) is the volume of freshwater used to produce the product, measured at the place where the product was actually produced (production-site definition) (m^3/t) (Allan, 1997; Hoekstra & Chapagain, 2008). The use of the term *embedded water* as a *virtual water* synonym could be misleading since the word *embedded* seems to refer to the actual water content of the finished product.

Virtual-water flow or trade
The virtual-water flow between two geographically delineated areas (e.g. two nations) is the volume of virtual water that is being transferred from one to another area as a result of product trade (Hoekstra *et al.*, 2009).

Water footprint (WF)
It is an indicator of water consumptive use that looks at both direct and indirect water use of a consumer or producer (Hoekstra, 2003). The water footprint of an individual, community or business is defined as the total volume of freshwater that is used to produce the goods and services consumed by the individual or community or produced by the business (km^3/yr, m^3/yr per capita). Water use is measured in terms of water volumes consumed (evaporated) and/or polluted per unit of time. A water footprint can be calculated for a particular product, for any well-defined group of consumers (e.g. an individual, family, village, city, province, state or nation) or producers (e.g. a public organization, private enterprise or economic sector). The water footprint is a geographically explicit indicator, not only showing volumes of water use and pollution, but also the locations (Hoekstra *et al.*, 2009).

Water scarcity taking EWR into account
It refers to the proportion of water consumption with respect to water available to human use. Water available to human use is equal to the total amount of water available in the basin minus the estimated environmental water demand (the water needed by the ecosystem to sustain its integrity)

(Smakhtin *et al*., 2004). Water scarcity occurs where there are insufficient water resources to satisfy long-term average requirements (EEA, 2009).

REFERENCES

Allan, J.A. (1997). *Virtual water: A long term solution for water short Middle Eastern economies?* Paper presented at the 1997 British Association Festival of Science, 9 September. University of Leeds, UK.

Allen, R.G., Pereira, L.S., Raes, D. & Smith, M. (1998). *Crop Evapotranspiration – Guidelines for Computing Crop Water Requirements*. FAO Irrigation and Drainage, Paper 56. Food and Agriculture Organization (FAO), Rome, Italy.

Dyson, M., Bergkamp, G. & Scanlon, J. (2003). *Flow: The Essentials of Environmental Flows*. International Union for Conservation of Nature (IUCN), Gland, Switzerland; and Cambridge, UK.

EEA (2009). *Water scarcity*. European Environmental Agency [online] Available from: [http://www.eea.europa.eu/themes/water/featured-articles/water-scarcity].

Falkenmark, M. (1995). *Coping with Water Scarcity under rapid Population growth*. Conference of SADC Ministers, November, 23–24. Pretoria, South Africa.

FAO (2001). *Food Insecurity: when people live with hunger and fear starvation. The State of Food Insecurity in the World 2001*. SOFI Third Edition. Food and Agriculture Organization (FAO), Rome, Italy. 58 pp.

FAO (2009). *AQUASTAT*. Food and Agriculture Organization [online]. Rome, Italy. Available from: [http://www.fao.org/nr/water/aquastat/countries/index.stm].

Hoekstra, A.Y. (ed.) (2003). *Virtual water trade*. Proceedings of the International Expert Meeting on Virtual Water Trade. Delft, the Netherlands, December, 12–13, 2002. Value of Water Research Report Series, No.12. UNESCO-IHE, Delft, the Netherlands.

Hoekstra, A.Y. & Chapagain, A.K. (2008). *Globalization of water: Sharing the planet's freshwater resources*. Blackwell Publishing, Oxford, UK

Hoekstra, A.Y., Chapagain, A.K., Aldaya, M.M. & Mekonnen, M.M. (2009). *Water footprint manual: State of the art 2009*. Water Footprint Network [online], Enschede, the Netherlands. Available from: [http://www.waterfootprint.org/downloads/WaterFootprintManual2009.pdf].

Smakhtin, V., Revenga, C. & Döll, P. (2004). *Taking into account environmental water requirements in global-scale water resources assessments*. Comprehensive Assessment Research Report 2. International Water Management Institute (IWMI), Colombo, Sri Lanka.

The Brisbane Declaration (2007). *Environmental Flows are Essential for Freshwater Ecosystem Health and Human Well-Being*. Declaration proclaimed at the 10th International Riversymposium and International Environmental Flows Conference, Brisbane, Australia, 3–6 September.

WFS (1996). *World Food Summit Plan of Action*. World Food Summit, 13–17 November, FAO headquarters, Rome, Italy.

Author Index

(Bold numbers indicate complete chapters)

Subject Index

(Bold numbers indicate complete chapters)

Geographic Index

(Bold numbers indicate complete chapters)